MATH-
STAT.
LIBRARY

Institute of Mathematical Statistics

LECTURE NOTES–MONOGRAPH SERIES

Volume 22

Stochastic Inequalities

Moshe Shaked and Y. L. Tong, Editors

Institute of Mathematical Statistics
Hayward, California

Seplae
Math

The production of the *IMS Lecture Notes–Monograph Series* is managed by the IMS Business Office: Jessica Utts, IMS Treasurer, and Jose L. Gonzalez, IMS Business Manager.

Library of Congress Catalog Card Number: 92-56246

International Standard Book Number 0-940600-29-3

Printed in the United States of America

Preface

As noted by Pólya (1967), "Inequalities play a role in most branches of mathematics and have different applications." This is certainly true in statistics and probability. Applications of stochastic inequalities can be found in statistical inference, multivariate analysis, reliability theory and queueing theory, constituting an integral part of statistical research.

On the other hand, the theory of stochastic inequalities has intrinsic interest and importance and need not rely solely upon specific applications. The general study of stochastic inequalities is, of course, closely related to the developments of inequalities in mathematics. As Mitronović (1970) pointed out, although "the theory of inequalities (in mathematics) began its development from (the days of) C. F. Gauss, A. L. Cauchy and P. L. Cebysev," it is "the classical work *Inequalities* by G. H. Hardy, J. E. Littlewood and G. Pólya (1934, 1952) ... which transformed the field of inequalities from a collection of isolated formulas into a systematic discipline." During the past two decades, research activities on stochastic inequalities have been growing at an accelerated rate. In addition to the publication of many important research papers, several research monographs and books in this area have become available (e.g., Karlin (1968), Barlow and Proschan (1975), Marshall and Olkin (1979), Tong (1980, 1984), Eaton (1987), Dharmadhikari and Joag-Dev (1988), Block, Sampson and Savits (1990) and Mosler and Scarsini (1991)).

The conference "Stochastic Inequalities" (held in July 1991 in Seattle, Washington as one of the 1991 AMS–IMS–SIAM Joint Summer Research Conferences) focused on the recent developments in the theory and applications of stochastic inequalities with special emphasis on the following topics:

(a) Convexity-related, majorization-related inequalities and stochastic convexity,

(b) Dependence-related probability and moment inequalities,

(c) Optimal stopping-related and prophet inequalities,

(d) Inequalities in multivariate distributions and multivariate analysis,

(e) Inequalities in reliability theory and queueing theory,

(f) Applications in business and economics, operations research, and other related areas.

This volume is a collection of papers which are based on the lectures given at that conference.

All the papers submitted were subjected to intensive refereeing, in most cases each paper was carefully read by two referees. Based on the referees' input, and space limitations, not all the papers submitted were accepted for publication. We thank all authors and conference participants for their notable contributions. We also owe a debt of gratitude to the conscientious referees, and a list of the referees can be found on p. vii of this volume.

We are particularly thankful to Morris L. Eaton who handled the refereeing process for the paper that we submitted, and to the anonymous referees who commented on it.

The Organizing Committee comprised of J. H. B. Kemperman (Rutgers University), Albert W. Marshall (University of British Columbia) and Frank Proschan (Florida State University), to whom we are grateful. Special thanks go to Ingram Olkin (Stanford University) who strongly encouraged us to go on with the organization and was very helpful throughout the various stages of this venture. The financial support for this meeting was provided by the National Science Foundation through the American Mathematical Society, and we thank the AMS Conferences Department for hosting this conference and for the fine logistic arrangements.

We also thank Annette Rohrs who did an extraordinary TEXing job preparing this volume.

References

Barlow, R. E. and Proschan, F. (1975). *Statistical Theory of Probability and Life Testing.* Holt, Rinehart and Winston, New York.

Block, H. W., Sampson, A. R. and Savits, T. H., eds. (1990). *Topics in Statistial Dependence.* Institute of Mathematical Statistics, Hayward, CA.

Dharmadhikari, S. and Joag-Dev, K. (1988). *Unimodality, Convexity and Applications.* Academic Press, New York.

Eaton, M. L. (1987). *Lectures on Topics in Probability Inequalities.* Centrum voor Wiskunde en Informatica, Amsterdam.

Hardy, G. H., Littlewood, J. E. and Pólya, G. (1934, 1952). *Inequalities*, 1st ed., 2nd ed. Cambridge University Press, Cambridge.

Karlin, S. (1968). *Total Positivity*, Vol. 1. Stanford University Press, Stanford, CA.

Marshall, A. W. and Olkin, I. (1979). *Inequalities: Theory of Majorization and Its Applications.* Academic Press, New York.

Mitronović, D. S. (1970). *Analytic Inequalities.* Springer-Verlag, Berlin and New York.

Mosler, K. and Scarsini, M. eds. (1991). *Stochastic Orders and Decision Under Risk.* Institute of Mathematical Statistics, Hayward, CA.

Pólya, G. (1967). Inequalities and the principle of nonsufficient reason. In *Inequalities*, O. Shisha, ed. Academic Press, New York. 1–15.

Tong, Y. L. (1980). *Probability Inequalities in Multivariate Distributions.* Academic Press, New York.

Tong, Y. L. ed. (1984) (with the cooperation of I. Olkin, M. D. Perlman, F. Proschan and C. R. Rao). *Inequalities in Probability and Statistics.* Institute of Mathematical Statistics, Hayward, CA.

Moshe Shaked and Y. L. Tong
November, 1992

Conference Participants

Elja Arjas, University of Oulu
Barry C. Arnold, University of California at Riverside
Asit P. Basu, University of Missouri
Z. W. Birnbaum. University of Washington
Henry W. Block, University of Pittsburgh
Philip J. Boland, University College, Dublin
J. Bondar, Carleton University
Frans A. Boshuizen, Vrije Universeteit Amsterdam
Mark Brown, City College, City University of New York
Arthur Cohen, Rutgers University
Herbert A. David, Iowa State University
Crisanto A. Dorado, Florida State University
Ramon Durazo, University of Arizona
Richard L. Dykstra, University of Iowa
Morris L. Eaton, University of Minnesota
James D. Esary, Naval Postgraduate School
Stergios B. Fotopoulos, Washington State University
David Gilat, Tel Aviv University
Joseph Glaz, University of Connecticut
Pushpa L. Gupta, University of Maine
Ramesh C. Gupta, University of Maine
Theodore P. Hill, Georgia Institute of Technology
Donald R. Jensen, Virginia Politechnic Institute and State University
Harry Joe, University of British Columbia
Martin Jones, College of Charleston
J. H. B. Kemperman, Rutgers University
Robert P. Kertz, Georgia Institute of Technology
Samuel Kotz, University of Maryland
Henry A. Kreiger, Harvey Mudd College
Mei-Ling Ting Lee, Harvard University
Claude Lefèvre, Université Libre de Bruxelles
Haijun Li, University of Arizona
Jiann-Hua Lou, National University of Singapore
Robert S. Maier, University of Arizona
Albert W. Marshall, University of British Columbia
Lutz Mattner, Universität Hamburg
Laurie M. Meaux, University of Arkansas
Ludolf E. Meester, Technical University of Delft
Roger B. Nelsen, Lewis and Clark College
Ingram Olkin, Stanford University

Michael D. Perlman, University of Washington
Arthur O. Pittenger, University of Maryland Baltimore County
András Prékopa, Rutgers University
Donald Richards, University of Virginia
Yosef Rinott, University of California at San Diego
Sheldon M. Ross, University of California at Berkeley
Harold B. Sackrowitz, Rutgers University
Allan R. Sampson, University of Pittsburgh
Ester Samuel-Cahn, Hebrew University
Stephen M. Samuels, Purdue University
Thomas H. Savits, University of Pittsburgh
Marco Scarsini, Università D'Annunzio Pescara
Joshua P. Seeger, BBN Communications
Jayaram Sethuraman, Florida State University
Moshe Shaked, University of Arizona
J. George Shanthikumar, University of California at Berkeley
Ton Steerneman, University of Groningen
Y. L. Tong, Georgia Institute of Technology
Erik N. Torgersen, University of Oslo
Joseph S. Verducci, Ohio State University
Richard A. Vitale, University of Connecticut
John C. Wierman, John Hopkins University
Haolong Zhu, University of Arizona

Referees[1]

Elja Arjas, University of Oulu
Barry C. Arnold*, University of California at Riverside
Asit P. Basu, University of Missouri
Henry W. Block*, University of Pittsburgh
Philip J. Boland, University College, Dublin
Frans A. Boshuizen, Vrije Universeteit Amsterdam
Herbert A. David, Iowa State University
Richard L. Dykstra, University of Iowa
Morris L. Eaton, University of Minnesota
James D. Esary, Naval Postgraduate School
David Gilat, Tel Aviv University
Joseph Glaz*, University of Connecticut
Theodore P. Hill, Georgia Institute of Technology
Donald R. Jensen*, Virginia Politechnic Institute and State University
Harry Joe, University of British Columbia
Martin Jones, College of Charleston
Robert P. Kertz, Georgia Institute of Technology
J. H. B. Kemperman, Rutgers University
Samuel Kotz, University of Maryland
Mei-Ling Ting Lee, Harvard University
Claude Lefèvre, Université Libre de Bruxelles
Haijun Li, University of Arizona
Robert S. Maier, University of Arizona
Albert W. Marshall*, University of British Columbia
Ludolf E. Meester, Technical University of Delft
Ingram Olkin*, Stanford University
Michael D. Perlman, University of Washington
András Prékopa, Rutgers University
Yosef Rinott*, University of California at San Diego
Allan R. Sampson*, University of Pittsburgh
Ester Samuel-Cahn, Hebrew University
Stephen M. Samuels, Purdue University
Thomas H. Savits*, University of Pittsburgh
Marco Scarsini*, Università D'Annunzio Pescara
Jayaram Sethuraman*, Florida State University
J. George Shanthikumar*, University of California at Berkeley
Erik N. Torgersen, University of Oslo
Joseph S. Verducci, Ohio State University
Haolong Zhu, University of Arizona

[1] a * indicates refereeing of more than one paper.

Conference Program

Sunday, July 7

(A) Multivariate Moment-Related Inequalities
Chair: Ingram Olkin, Stanford University
8:30 - 9:10 *On the Multi-Dimensional Moment Problem*
J. H. B. Kemperman, Rutgers University
9:10 - 9:50 *Inequalities on Probabilities and Expectations Based on the Knowledge of Multivariate Moments*
András Prékopa, Rutgers University
9:50 - 10:00 Discussion

10:00 -10:20 Coffee

(B) Inequalities in Probability
Chair: Sheldon M. Ross, University of California at Berkeley
10:20 - 11:00 *Inequalities for Rare Events in Time Reversible Markov Chains*
David J. Aldous, University of California at Berkeley and Mark Brown*, City College, CUNY
11:00 - 11:40 *Partitioning Inequalities in Probability and Statistics*
Theodore P. Hill, Georgia Institute of Technology
11:40 - 11:50 Discussion

(C) Applications of Inequalities, I
Chair: J. George Shanthikumar, University of California at Berkeley
1:30 - 2:10 *Comparing Two Groups of Ranked Objects by Matching Pairs*
Herbert A. David* and Jingyu Liu, Iowa State University
2:10 - 2:50 *Applications of $\mu(F), \lambda(F)$ Inequalities*
Chris Dorado* and Myles Hollander, Florida State University
2:50 - 3:00 Discussion

3:00 - 3:20 Coffee

(D) Inequalities in Multivariate Analysis
Chair: Morris L. Eaton, University of Minnesota
3:20 - 4:00 *The Multivariate Correlation Inequality and Unbiased Tests*
Arthur Cohen* and Harold B. Sackrowitz, Rutgers University
4:00 - 4:40 *Concentration Inequalities for Multivariate Distributions*
Morris L. Eaton, University of Minnesota and Michael D. Perlman*, University of Washington
4:40 - 4:50 Discussion

8:00 - 10:00 **Wine and Cheese Reception**
Patio - McMahon Hall

Monday, July 8

(E) Inequalities in Reliability Theory, I
Chair: Mark Brown, City College, CUNY
8:30 - 9:10 *Optimal Allocations in Additive Exponential Systems*
Sheldon Ross, University of California at Berkeley
9:10 - 9:50 *Stochastic Comparisons for Maintenance Policies*
Henry W. Block*, University of Pittsburgh,
Naftali A. Langberg, University of Haifa
and Thomas H. Savits, University of Pittsburgh
9:50 - 10:00 Discussion

10:00 -10:20 Coffee

(F) Univariate Moment-Related Inequalities
Chair: Yosef Rinott, University of California at San Diego
10:20 - 11:00 *Orderings Arising from Expected Extremes, with an Application*
Peter J. Downey and Robert S. Maier*,
University of Arizona
11:00 - 11:40 *Skewness and Kurtosis Orderings*
Barry C. Arnold, University of California at Riverside
11:40 - 11:50 Discussion

(G) Extremal Problems and Covariance Inequalities
Chair: Michael D. Perlman, University of Washington
1:30 - 2:10 *Extremal Problems for Probability Distributions*
Lutz Mattner, University of Hamburg
2:10 - 2:50 *Covariances of Symmetric Statistics*
Richard A. Vitale, University of Connecticut
2:50 - 3:00 Discussion

3:00 - 3:20 Coffee

(H) Stochastic Orderings and Multivariate Majorization

Chair: Elja Arjas, University of Oulu

3:20 - 4:00 *Towards a Theory of Stochastic Functions with Applications to Stochastic Optimization*
J. George Shanthikumar, University of California at Berkeley

4:00 - 4:40 *Multivariate Majorization by Positive Combinations*
Harry Joe, University of British Columbia
and Joseph S. Verducci*, Ohio State University

4:40 - 4:50 Discussion

Tuesday, July 9

(I) Optimal Stopping–Related Inequalities

Chair: David Gilat, Tel Aviv University

8:30 - 9:10 *Optimal Stopping Values and Prophet Inequalities for Negatively Dependent Random Variables*
Yosef Rinott, University of California at San Diego
and Ester Samuel-Cahn*, Hebrew University

9:10 - 9:50 *Secretary Problems As a Source of Benchmark Bounds*
Stephen M. Samuels, Purdue University

9:50 - 10:00 Discussion

10:00 -10:30 Group Picture

10:30 -10:50 Coffee

(J) New Research Problems and Informal Presentations

Chair: Henry W. Block, University of Pittsburgh

10:50 - 11:50 Presenters:
Frans A. Boshuizen, Vrije Universeteit
Jiann–Hua Lou, National University of Singapore
Donald Richards, University of Virigina
Halong Zhu, University of Arizona

Afternoon: Open

Wednesday, July 10

(K) Convexity-Related and Majorization Inequalities

Chair: Albert W. Marshall, University of British Columbia

8:30 - 9:10 *Hyperbolic-Concave Functions and the Hardy-Littlewood Maximal Function*
Robert P. Kertz*, Georgia Institute of Technology and Uwe Rösler, University of Göttingen

9:10 - 9:50 *Generalized Majorization Orderings and Applications*
Harry Joe, University of British Columbia

9:50 - 10:00 Discussion

10:00 -10:20 Coffee

(L) Matrix-Related Inequalities

Chair: J. H. B. Kemperman, Rutgers University

10:20 - 11:00 *Correlation and Permanental Inequalities and Matrix Majorization*
Yosef Rinott* and Michael Saks
University of California at San Diego

11:00 - 11:40 *Matrix Extremes and Related Stochastic Bounds*
Donald R. Jensen, Virginia Polytechnic Institute and State University

11:40 - 11:50 Discussion

(M) Dependence-Related Inequalities, I

Chair: Philip J. Boland, University College Dublin

1:30 - 2:10 *Applications of Dependence to Burn-In*
Henry W. Block, J. Mi, and Thomas H. Savits*, University of Pittsburgh

2:10 - 2:50 *Extreme Order Statistics for a Sequence of Dependent Random Variables*
Joseph Glaz, University of Connecticut

2:50 - 3:00 Discussion

3:00 - 3:20 Coffee

(N) Applications of Inequalities, II

Chair: Samuel Kotz, University of Maryland

3:20 - 4:00 *On Dynamic Forms of Probabilistic Causality*
Elja Arjas, University of Oulu

4:00 - 4:40 *A Multivariate Stochastic Ordering by the Mixed Descending Factorial Moments with Applications*
Claude Lefèvre, University of Brussels
and Philippe Picard, University of Lyon

4:40 - 4:50 Discussion

7:00 Salmon Barbecue
Terrace - McMahon Hall

Thursday, July 11

(O) Comparison of Experiments and Information Inequalities

Chair: Erik N. Torgensen, University of Oslo

8:30 - 9:10 *A Group Action on Covariances with Applications to the Comparison of Linear Normal Experiments*
Morris L. Eaton, University of Minnesota

9:10 - 9:50 *Comparison of Experiments Based on Dependent Random Variables*
Moshe Shaked*, University of Arizona
and Y. L. Tong, Georgia Institute of Technology

9:50 - 10:00 Discussion

10:00 -10:20 Coffee

(P) Dependence-Related Inequalities, II

Chair: Thomas H. Savits, University of Pittsburgh

10:20 - 11:00 *Lower Bounds on Multivariate Distributions with Preassigned Marginals*
Samuel Kotz*, University of Maryland
and Joshua Seeger, BBN Communications

11:00 - 11:40 *Dependence of Stable Random Variables*
Mei-Ling Lee*, Harvard University,
Svetlozar T. Rachev, University of California at Santa Barbara
and Gennady Samorodnitsky, Cornell University

11:40 - 11:50 Discussion

(Q) Inequalities in Reliability Theory, II

Chair: Y. L. Tong, Georgia Institute of Technology

1:30 - 2:10 *A Study of the Role of Modules in the Failure of Systems*
E. El-Neweihi, University of Illinois at Chicago Circle
and J. Sethuraman*, Florida State University

2:10 - 2:50 *Stochastic Inequalities for a Redundancy Enhancement to a Series or Parallel System*
Philip J. Boland, University College Dublin

2:50 - 3:00 Discussion

3:00 - 3:20 Coffee

(R) Applications of Inequalities, III

Chair: Moshe Shaked, University of Arizona

3:20 - 4:00 *Some Statistical Applications of Stochastic Inequalities*
Allan R. Sampson, University of Pittsburgh

4:00 - 4:40 *Orderings for Capacities and Their Applications*
Marco Scarsini, University of Rome

4:40 - 4:50 Discussion

Table of Contents

PREFACE iii
CONFERENCE PARTICIPANTS v
REFEREES vii
CONFERENCE PROGRAM viii

Inequalities for Rare Events in Time–Reversible Markov Chains I
David J. Aldous and Mark Brown 1

Skewness and Kurtosis Orderings: An Introduction
Barry C. Arnold and Richard A. Groeneveld 17

Stochastic Inequalities for a Redundancy Enchancement to a Series or Parallel System
Philip J. Boland 25

Some Remarks on a Notion of Positive Dependence, Association, and Unbiased Testing
Arthur Cohen and H. B. Sackrowitz 33

Further Aspects of the Comparison of Two Groups of Ranked Objects by Matching in Pairs
H. A. David and Jingyu Liu 38

Inequalities for the Parameters $\lambda(F), \mu(F)$ *with Applications in Nonparametric Statistics*
Crisanto Dorado and Myles Hollander 50

Orderings Arising from Expected Extremes, with an Application
Peter J. Downey and Robert S. Maier 66

A Group Action on Covariances with Applications to the Comparison of Linear Normal Experiments
Morris L. Eaton 76

The Role of a Module in the Failure of a System
Emad El–Neweihi and Jayaram Sethuraman 91

Extreme Order Statistics for a Sequence of Dependent Random Variables
Joseph Glaz 100

Partitioning Inequalities in Probability and Statistics
Theodore P. Hill 116

Matrix Extremes and Related Stochastic Bounds
D. R. Jensen 133

Generalized Majorization Orderings and Applications
Harry Joe 145

Multivariate Majorization by Positive Combinations
Harry Joe and Joseph Verducci 159

Covariance Spaces for Measures on Polyhedral Sets
J. H. B. Kemperman and Morris Skibinsky 182

Hyperbolic–Concave Functions and Hardy–Littlewood Maximal Functions
Robert P. Kertz and Uwe Rösler 196

Lower Bounds on Multivariate Distributions with Preassigned Marginals
S. Kotz and J. P. Seeger 211

Dependence of Stable Random Variables
Mei–Ling Ting Lee, Svetlozar T. Rachev and Gennady Samorodnitsky 219

A Multivariate Stochastic Ordering by the Mixed Descending Factorial Moments with Applications
Claude Lefèvre and Philippe Picard 235

Allocation Through Stochastic Schur Convexity and Stochastic Transposition Increasingness
Liwan Liyanage and J. George Shanthikumar 253

Extremal Problems for Probability Distributions: A General Method and Some Examples
L. Mattner 274

Concentration Inequalities for Multivariate Distributions: II. Elliptically Contoured Distributions
Michael D. Perlman 284

Inequalities on Expectations Based on the Knowledge of Multivariate Moments
András Prékopa 309

On FKG–Type and Permanental Inequalities
Yosef Rinott and Michael Saks 332

Optimal Stopping Values and Prophet Inequalities for Some Dependent Random Variables
Yosef Rinott and Ester Samuel–Cahn 343

Some Applications of Monotone Transformations in Statistics
Allan R. Sampson 359

Secretary Problems as a Source of Benchmark Bounds
Stephen M. Samuels 371

Comparison of Experiments of Some Multivariate Distributions with a Common Marginal
Moshe Shaked and Y. L. Tong 388

On the Bias of the Jackknife Estimate of Variance
Richard A. Vitale 399

Author Citation Index 404

Key Words and Phrases Index 410

Stochastic Inequalities
IMS Lecture Notes – Monograph Series
Volume 22 (1993)

INEQUALITIES FOR RARE EVENTS IN TIME–REVERSIBLE MARKOV CHAINS I

By DAVID J. ALDOUS[1] and MARK BROWN[2]

University of California, Berkeley and City College, CUNY

The distribution of waiting time until a rare event is often approximated by the exponential distribution. In the context of first hitting times for stationary reversible chains, the error has a simple explicit bound involving only the mean waiting time ET and the relaxation time τ of the chain. We recall general upper and lower bounds on ET and then discuss improvements available in the case $ET \gg \tau$ where the exponential approximation holds. In a sequel, Stein's method will be used to get explicit bounds on the Poisson approximation for the number of non-adjacent visits to a rare subset.

1. Introduction

The Poisson approximation for numbers of rare events which actually occur, and the exponential approximation for the waiting time until first occurrence of a rare event, are useful throughout many areas on probability – one view of this big picture is presented in Aldous (1989). Here we study explicit bounds in these approximations, in the special setting of hitting times of stationary reversible Markov chains. This paper deals with the exponential approximation and bounds on the mean waiting time; a sequel (Aldous and Brown (1991)) studies Poisson approximations using an implementation of the Chen-Stein method.

The following set-up and notation will be used throughout. $(X_t; t \geq 0)$ is an irreducible finite-state reversible Markov chain in continuous time. The state space is I and the transition rate matrix is $Q = (q(i,j); i,j \in I)$ where $q(i,i) = -\sum_{j \neq i} q(i,j)$. Let π be the stationary distribution. The symmetrizable matrix $-Q$ has real eigenvalues $0 = \lambda_0 < \lambda_1 \leq \lambda_2 \leq \ldots$. Call $\tau = 1/\lambda_1$ the *relaxation time* of the chain. Let A be a fixed (proper, non-empty) subset of I, and let T_A be the first hitting time on A. So $0 < E_\pi T_A < \infty$.

[1]Research supported by National Science Foundation Grant MCS90–01710.
[2]Research supported by U.S. Air Force Office of Scientific Research Grant 89–0083.
AMS 1991 *subject classifications.* 60J27.
Key words and phrases. Markov chain, waiting time, hitting time, exponential approximation, completely monotone.

THEOREM 1

$$|P_\pi(T_A/E_\pi T_A > t) - e^{-t}| \le \frac{\tau/E_\pi T_A}{1+\tau/E_\pi T_A} \le \tau/E_\pi T_A \quad \textit{for all } t > 0.$$

In the special case where A is a singleton this was proved by Brown (1990). That proof exploits the completely monotone property of the distribution of T_A. In section 3 we show how the general case can be reduced to the special case. In section 7 we give bounds on the density function $f(t)$ of T_A. Lemma 13 gives explicit bounds in the natural approximation

$$f(t) \approx \frac{1}{E_\pi T_A} \text{ on } \tau \ll t \ll E_\pi T_A.$$

Use of Theorem 1 and Lemma 13 requires an upper bound on τ, and bounds on $E_\pi T_A$. We have nothing new to say about the much-studied issue of upper bounding τ, and refer the reader to Diaconis and Stroock (1991) for interesting recent work. The problem of bounding $E_\pi T_A$ in terms of readily computable quantities has apparently not been studied much, aside from the simple and well-known general bounds stated in Lemma 2 below. The left two inequalities, and (1), follow easily from complete monotonicity (Keilson (1979) Theorem 6.9C). The rightmost inequality is an easy consequence of extremal characterizations: see the proof of Lemma 10 below.

LEMMA 2

$$\frac{1-\pi(A)}{q(A,A^c)} \le \frac{E_\pi T_A}{1-\pi(A)} \le E_\alpha T_A \le \frac{\tau}{\pi(A)}$$

where

$$q(A,A^c) = \sum_{i\in A}\sum_{j\notin A} \pi(i)q(i,j)$$

and where α is the quasistationary distribution defined at (12).

The extreme bounds in Lemma 2 are often crude. In our setting where $\tau \ll E_\pi T_A$ the intermediate inequality, and its distributional version

$$P_\pi(T_A > t) \le (1-\pi(A))\exp(-t/E_\alpha T_A) \tag{1}$$

is often fairly sharp, but in practice is hard to use. The point is that, because of the extremal characterization (13) of $E_\alpha T_A$ as a *sup*, it is often easy to get a good *lower* bound on $E_\alpha T_A$, but much harder to get an upper bound. Our main new result is the following usable inequality.

THEOREM 3

$$P_\pi(T_A > t) \ge \left(1 - \frac{\tau}{E_\alpha T_A}\right)\exp\left(\frac{-t}{E_\alpha T_A}\right), \; t > 0.$$

Integrating over t gives

COROLLARY 4 $E_\pi T_A \geq E_\alpha T_A - \tau$.

The point is that these results may be applied without any prior estimate of $E_\pi T_A$, merely a lower bound on $E_\alpha T_A$. Note also that we may use Corollary 4 to rewrite Theorem 1 as

COROLLARY 5

$$|P_\pi(T_A/E_\pi T_A > t) - e^{-t}| \leq \frac{\tau}{E_\alpha T_A}, \; t \geq 0.$$

Note also that Theorem 3 improves on the simple complete monotonicity result underlying the left inequality in Lemma 2:

$$P_\pi(T_A > t) \geq (1 - \pi(A)) \exp(-t \frac{q(A, A^c)}{1 - \pi(A)}).$$

In section 4 we record results bounding $E_{\pi_0} T_A$ in terms of $E_\pi T_A$ for non-stationary initial distributions π_0. Let us emphasize that our results are "absolute" inequalities, i.e. do not involve any unspecified constants depending on the Markov chain under consideration. Without going into details, it follows that our results extend unchanged to *continuous*-space stationary reversible processes (under weak regularity conditions – say strong Markov and cadlag paths). For such processes we may have $\tau = \infty$, but the results are only interesting when $\tau < \infty$. One explanation is that the proofs extend, unchanged except for terminology. Another explanation is that, given a continuous-space process, one may express it as a weak limit of finite-state processes in such a way that relaxation times, first hitting times and the other parameters of interest converge.

Finally, we repeat that the existence of some exponential approximation has nothing to do with reversibility or even Markovianness: it is merely that in the reversible Markov context one can hope to get sharper general bounds. The non-reversible case, and applications to queueing networks, has recently been discussed by Iscoe and McDonald (1991), who give general explicit exponential approximations in terms of the spectral gap. Charles Stein, lecturing at Stanford in June 1991, outlined how to use the Chen-Stein method in the non-reversible case, giving bounds in terms of non-explicit "coupling times" which need ad hoc arguments to estimate. In older work, Flannery (1986) gave complicated bounds in terms of the maximal correlation function. Aldous (1982) gave bounds in terms of a uniform mixing coefficient. Exponential limits *without* explicit bounds can be proved in great generality. For instance, Korolyuk and Silvestrov (1984) and Cogburn (1985) prove exponential limits for hitting times to receding subsets of a fixed Harris-recurrent chain.

2. Background

It has recently become fashionable to treat reversible chains in terms of the associated *Dirichlet form* $\mathcal{E}$. That is, for functions $g : I \to R$ define

$$\begin{aligned}\mathcal{E}(g,g) &= \frac{1}{2}\sum_i \sum_{j\neq i} \pi(i)q(i,j)(g(j)-g(i))^2 \\ &= \frac{1}{2}\lim_{t\to 0} t^{-1}E(g(X_t)-g(X_0))^2 \\ &= \lim_{t\to 0} t^{-1}Eg(X_0)(g(X_0)-g(X_t)).\end{aligned}$$

(X denotes the *stationary* chain). Write also

$$[g,g] = \sum_i \pi(i)g^2(i) = Eg^2(X_0).$$

Various parameters of the chain have extremal characterizations, for example (Diaconis and Stroock (1991))

$$\tau = \sup\{[g,g]/\mathcal{E}(g,g) : \sum_i \pi(i)g(i) = 0\} \tag{2}$$

A more probabilistic interpretation is via the following *maximal correlation* property. For the stationary chain,

$$\begin{aligned}\rho(t) &= \max\{\text{cor}(Z_1,Z_2) : Z_1 \in \mathcal{F}(X_s, s\leq 0), Z_2 \in \mathcal{F}(X_s, s \geq t)\} \\ &= \max_{h,g} \text{cor}(h(X_0), g(X_t)) \\ &= \exp(-t/\tau). \end{aligned} \tag{3}$$

As a standard consequence, if $Eg(X_0) = 0$ and $||g|| \equiv \sqrt{Eg^2(X_0)} < \infty$ then

$$||\mathbf{P}_t g|| \leq e^{-t/\tau}||g||, \quad \text{where } (\mathbf{P}_t g)(i) = E(g(X_t)|X_0 = i). \tag{4}$$

Finally, we often use the fact that the tail distribution function $P_\pi(T_A > t)$ and the corresponding density function f are *completely monotone* (CM) functions, i.e. are of the form $\sum_i c_i \exp(-\gamma_i t)$ for nonnegative c_i, γ_i. Elementary qualitative properties of CM functions will be used without comment.

3. Proof of Theorem 1

Write

$$\rho_A = \frac{E_\pi T_A^2}{2(E_\pi T_A)^2} - 1.$$

Brown (1990) section 7 shows that for singletons $A = \{a\}$

(5) $$|P_\pi(T_a/E_\pi T_a > t) - e^{-t}| \le \frac{\rho_a}{\rho_a + 1}.$$

(6) $$\rho_a \le \tau/E_\pi T_a.$$

To prove Theorem 1 it suffices to show these remain true for general subsets A:

(7) $$|P_\pi(T_A/E_\pi T_A > t) - e^{-t}| \le \frac{\rho_A}{\rho_A + 1}.$$

(8) $$\rho_A \le \tau/E_\pi T_A.$$

Write $A^c = I\backslash A$. Consider the chain $\hat{X}$ in which the set A has been collapsed to a singleton a. Precisely, the state space is $\hat{I} = A^c \cup \{a\}$ and the transition rates are

$$\begin{array}{lll} \hat{q}(i,j) = & q(i,j), & i,j \in I\backslash A \\ \hat{q}(i,a) = & \sum_{j\in A} q(i,j), & i \in I\backslash A \\ \hat{q}(a,i) = & \sum_{k\in A} \pi_k q(k,i)/\pi(A), & i \in I\backslash A \\ \hat{q}(a,a) = & -\sum_{i\in A^c} \hat{q}(a,i) & \end{array}$$

It is straightforward to verify (e.g. Keilson (1979, p. 41)) that $\hat{X}$ is reversible and has stationary distribution

$$\begin{array}{lll} \hat{\pi}_i = & \pi_i, & i \in A^c \\ \hat{\pi}_a = & \pi(A) & \end{array}$$

By construction the P_π-distribution of T_A is the same as the $\hat{P}_{\hat{\pi}}$-distribution of $\hat{T}_a$. So (7) is immediate from (5). To get (8) the key fact, proved below, is

(9) $$\hat{\tau} \le \tau.$$

Then

$$\rho_A = \hat{\rho}_A \le \frac{\hat{\tau}}{E_{\hat{\pi}}\hat{T}_a} \le \frac{\tau}{E_\pi T_a}$$

which is (8).

Inequality (9) is one of a group of fairly well-known consequences of the extremal characterization (2) of the relaxation time. Given a function $\hat{g}$ on $\hat{I}$, consider its natural extension to a function g on I:

$$\begin{array}{lll} g(i) & = \hat{g}(i), & i \in A^c \\ & = \hat{g}(a), & i \in A. \end{array}$$

It is straightforward to verify

$$\sum_i \pi_i g(i) = \sum_i \hat{\pi}_i \hat{g}(i)$$

$$[g,g] = [\hat{g},\hat{g}]$$
$$\mathcal{E}(g,g) = \hat{\mathcal{E}}(\hat{g},\hat{g}).$$

Then (9) follows from (2).

4. Means from Nonstationary Starts

If $E_\pi T_A \gg \tau$ then the approximation

$$E_{\pi_0} T_A \approx E_\pi T_A$$

holds for many non-stationary initial distributions π_0. In this section we study bounds for $E_{\pi_0} T_A$ obtainable by standard methods.

LEMMA 6 *Let* $h(i) = E_i T_A$. *Then* $\mathrm{var}(h(X_0)) \leq \tau E_\pi T_A$.

It is easier to interpret this if we consider the standardized mean hitting time

$$\bar{h}(i) = E_i T_A / E_\pi T_A$$

for which Lemma 6 implies

$$\mathrm{var}\ (\bar{h}(X_0)) \leq \tau / E_\pi T_A.$$

So Chebyshev's inequality says that for "most" i the mean hitting time started at i is about the same as starting with the stationary distribution.

PROOF OF LEMMA 6 Applying (2) to $g(i) = h(i) - E_\pi T_A$ gives

$$\mathrm{var}\ (h(X_0)) \leq \tau \mathcal{E}(h,h).$$

But the third formulation of $\mathcal{E}$, along with a routine dominated convergence argument, gives

$$\mathcal{E}(h,h) = Eh(X_0) \times 1 = E_\pi T_A.$$

Proposition 7 gives an upper bound on ET from an arbitrary initial distribution π_0. Define

$$\chi^2(\pi_0) = \sum_i \frac{(\pi_0(i) - \pi(i))^2}{\pi(i)} = ||\frac{\pi_0 - \pi}{\pi}||^2.$$

If X_0 has distribution π_0 then X_t has distribution π_t which satisfies (after a brief calculation)

$$\frac{\pi_t - \pi}{\pi} = \mathbf{P}_t \left(\frac{\pi_0 - \pi}{\pi} \right)$$

and then by (4)

(10) $$\chi(\pi_t) \leq \chi(\pi_0) e^{-t/\tau}.$$

PROPOSITION 7 *For any initial distribution* π_0

$$E_{\pi_0} T_A \leq E_\pi T_A + \tau + \frac{\tau}{2} \log^+ \left(\chi^2(\pi_0) \frac{E_\pi T_A}{\tau} \right).$$

Taking $\pi_0 = \delta_i$ gives

COROLLARY 8

$$E_i T_A \leq E_\pi T_A + \tau + \frac{\tau}{2} \log^+ \left(\frac{1 - \pi(i)}{\pi(i)} \frac{E_\pi T_A}{\tau} \right).$$

Another Corollary is given below.

PROOF OF PROPOSITION 7 For any function g

$$\begin{aligned} E_{\pi_0} g(X_t) - E_\pi g(X_t) &= \sum_i \pi(i) \frac{\pi_t(i) - \pi(i)}{\pi(i)} (g(i) - E_\pi g(X_t)) \\ &\leq \sqrt{\chi^2(\pi_t) \mathrm{var}_\pi g(X_0)} \text{ by Cauchy-Schwarz} \\ &\leq e^{-t/\tau} \chi(\pi_0) \sqrt{\mathrm{var}_\pi g(X_0)} \text{ by (10).} \end{aligned}$$

Now put $g(i) = E_i T_A$, observe that

$$E_\pi g(X_t) = E_\pi T_A, \quad E_{\pi_0} T_A \leq t + E_{\pi_0} g(X_t)$$

and apply Lemma 6 to get

$$E_{\pi_0} T_A - E_\pi T_A \leq t + \chi(\pi_0) \sqrt{\tau E_\pi T_A}\, e^{-t/\tau}.$$

Minimizing the right side by putting

$$t = \frac{\tau}{2} \log^+ \left(\chi^2(\pi_0) \frac{E_\pi T_A}{\tau} \right)$$

gives the result.

The next Corollary shows that in our setting i's with large $E_i T_A$ are exponentially rare with respect to π. It is easy to see that no similar result can hold for *small* $E_i T_A$.

COROLLARY 9 *For all* $b \geq 0$

$$\pi\{i : E_i T_A \geq E_\pi T_A + \tau + b\} \leq e^{-2b/\tau} \frac{E_\pi T_A}{\tau}.$$

PROOF Fix a subset B and set $\pi_B(i) = \pi(i|B)$. Then

$$\chi^2(\pi_B) = \frac{1 - \pi(B)}{\pi(B)} \leq \frac{1}{\pi(B)}$$

and so Proposition 7 gives

$$E_{\pi_B} T_A \leq E_\pi T_A + \tau + \frac{\tau}{2} \log^+ \left(\frac{E_\pi T_A}{\tau \pi(B)} \right). \tag{11}$$

Now fix $b > 0$ and consider

$$B = \{i : E_i T_A \geq E_\pi T_A + \tau + b\}$$

which we may assume to be non-empty. Then

$$E_\pi T_A + \tau + b \leq E_{\pi_B} T_A.$$

Combining with (11) gives the inequality below:

$$\frac{2b}{\tau} \leq \log^+ \left(\frac{E_\pi T_A}{\tau \pi(B)} \right) = \log(\frac{E_\pi T_A}{\tau \pi(B)})$$

the equality because the left side is positive. Rearrange.

REMARK One can use Proposition 7 to give an upper bound on $E_\alpha T_A$, but the bound is weaker than the bound $E_\alpha T_A \leq E_\pi T_A + \tau$ given by Corollary 4. So the (novel ?) argument in section 6 seems more powerful than the standard arguments above.

5. Properties of the Quasistationary Distribution

Let Q_A be the transition rate matrix Q restricted to A^c. Let λ_A be the smallest eigenvalue of $-Q_A$ and let α be the corresponding eigenvector

$$\alpha Q_A = -\lambda_A \alpha \tag{12}$$

normalized to be a probability distribution on A^c. Then α is the *quasistationary distribution*, and

$$P_\alpha(T_A > t) = \exp(-\lambda_A t)$$

$$E_\alpha T_A = 1/\lambda_A.$$

And there is a variational interpretation:

$$E_\alpha T_A = \sup\{[g,g]/\mathcal{E}(g,g) : g \geq 0, g = 0 \text{ on } A\} \tag{13}$$

where the sup is attained by

$$g(i) = \alpha(i)/\pi(i). \tag{14}$$

It is often easy to find *lower* bounds on $E_\alpha T_A$ by evaluating (13) for convenient g.

Lemma 10 gives easy relations between $E_\pi T_A$ and $E_\alpha T_A$. Part (b) implies that $\alpha \approx \pi$ when $\tau / E_\pi T_A$ is small, using the general identity

$$\sum \frac{(\alpha_i - \pi_i)^2}{\pi_i} = \sum \frac{\alpha_i^2}{\pi_i} - 1.$$

LEMMA 10 *(a)* $\frac{E_\pi T_A}{1-\pi(A)} \leq E_\alpha T_A \leq \tau / \pi(A)$.
(b) Now suppose $\tau < E_\pi T_A$. *Then*

$$\sum_i \alpha^2(i)/\pi(i) \leq (1 - \tau / E_\pi T_A)^{-1}.$$

PROOF Assertion (a) repeats part of Lemma 2. We shall give the usual proof of the right inequality, because ingredients are needed for subsequent parts. As at (14) set $g(i) = \alpha(i)/\pi(i)$, so

$$E_\alpha T_A = [g, g] / \mathcal{E}(g, g). \tag{15}$$

Applying (2)

$$\tau \geq \frac{[g-1, g-1]}{\mathcal{E}(g-1, g-1)} = \frac{[g,g]-1}{\mathcal{E}(g,g)}.$$

Using the expression for $\mathcal{E}(g, g)$ given in (15), we get

$$\frac{1}{[g,g]} \geq 1 - \frac{\tau}{E_\alpha T_A}. \tag{16}$$

Since α is a probability distribution on A^c we have

$$1 = E 1_{A^c}(X_0) g(X_0)$$

and so by Cauchy-Schwarz

$$1^2 \leq (E 1_{A^c}(X_0)) \times [g, g] = (1 - \pi(A))[g, g].$$

Using (16),

$$1 - \frac{\tau}{E_\alpha T_A} \leq 1 - \pi(A)$$

which gives the right inequality in (a).

Now suppose $\tau / E_\pi T_A < 1$. Since $E_\alpha T_A \geq E_\pi T_A$ we can invert (16) to get

$$\begin{aligned} [g, g] &\leq (1 - \tau / E_\alpha T_A)^{-1} \\ &\leq (1 - \tau / E_\pi T_A)^{-1} \end{aligned} \tag{17}$$

which is (b).

6. Proof of Theorem 3

We now come to Theorem 3, the main result which we believe to be new. There is no loss of generality in supposing that A is a singleton $\{a\}$, since (as in the proof of Theorem 1) we could replace the original chain by the "collapsed" chain $\hat{X}$.

The proof involves study of a certain irreducible reversible Markov chain on states $\{0,1,2,\ldots,r\}$ where the only allowable transitions are $0 \leftrightarrow i$. In graph theory terminology, the graph of such allowable transitions is a "star", so we shall call the chain X^* and write $\pi^*, \tau^*, q^*(i,0)$, etc, for quantities associated with X^*. We shall construct X^* such that $\mathcal{L}_{\pi^*}T_0^* = \mathcal{L}_\pi T_a$, (i.e. the distribution of T_0^* under π^* coincides with the distribution of T_a under π) and $E_{\alpha^*}T_0^* = E_\alpha T_a$. From the explicit form of X^* it is easy to see that the analog of Theorem 3 holds for X^*, and to complete the proof we demonstrate in Lemma 12 that $\tau^* \leq \tau$.

We shall make use of standard facts from the spectral representation for finite state irreducible reversible Markov chains. Keilson (1979, Chapter 3 and section 6.9) covers the relevant material. We first recall that

$$P_\pi(T_a > t) = \sum_{i=1}^{m} p_i \exp(-\gamma_i t) \tag{18}$$

where $p_i \geq 0$ and $0 < \gamma_1 < \gamma_2 < \ldots < \gamma_m$ are the distinct eigenvalues of $-Q_a$, the restriction of $-Q$ to $\{a\}^c$. We first show that the term in (18) corresponding to failure rate γ_1 has strictly positive coefficient p_1, and this identifies γ_1 as $1/E_\alpha T_a$.

LEMMA 11 $p_1 > 0$.

PROOF For $x \neq a, y \neq a$ write $q_t(x,y) = P_x(X_t = y, T_a > t)$. By the spectral representation for Q_a

$$q_t(x,x) = \sum_{j=1}^{m} c_j(x,x)\exp(-\gamma_j t)$$

where $c_1(x,x) \geq 0$ and where $\sum_{x \neq a} c_1(x,x) =$ multiplicity of $\gamma_1 \geq 1$. Thus $c_1(x_0,x_0) > 0$ for some $x_0 \neq a$. Again by the spectral representation, for $y \neq x$

$$q_t(x,y) = \sum_{j=1}^{m} c_j(x,y)\exp(-\gamma_j t)$$

for certain constants $c_j(x,y)$. It follows that

$$0 \le \lim_{t\to\infty}(\exp(\gamma_1 t)q_t(x,y)) = c_1(x,y).$$

Since $P_\pi(T_a > t) = \sum_{x,y\neq a}\pi(x)q_t(x,y)$, it follows that

$$p_1 = \sum_{x,y\neq a}\pi(x)c_1(x,y) \ge \pi(x_0)c_1(x_0,x_0) > 0$$

establishing the Lemma.

Now eliminate the zero-coefficient terms from (18) and relabel, so (18) becomes

$$P_\pi(T_a > t) = \sum_{i=1}^{r} p_i \exp(-\gamma_i t) \tag{19}$$

where $p_i > 0$, $1 \le i \le r$, and $0 < 1/E_\alpha T_a = \gamma_1 < \gamma_2 < \ldots < \gamma_r$ and $\sum_{i=1}^{r} p_i = P_\pi(T_a > 0) = 1-\pi(a)$. Each γ_i is an eigenvalue of $-Q_a$.

Let X^* be the Markov chain on states $\{0,1,\ldots,r\}$ with transition rate matrix Q^* defined by

$$\begin{aligned}
q^*(i,0) &= \gamma_i,\ i\neq 0\\
q^*(0,i) &= \frac{\gamma_i p_i}{\pi(a)},\ i\neq 0\\
q^*(i,i) &= -\gamma_i,\ i\neq 0\\
q^*(0,0) &= -\sum_{i=1}^{r}\frac{\gamma_i p_i}{\pi(a)}\\
q^*(i,j) &= 0 \text{ elsewhere.}
\end{aligned}$$

Here the quantities γ_i, p_i are those appearing in (19). The chain X^* is irreducible and reversible, with stationary distribution $\pi^* = (p_0, p_1, \ldots, p_r)$, where $p_0 = \pi(a)$.

Clearly the eigenvalues of Q_0^* are the (γ_i), and so in particular

$$E_{\alpha^*}T_0^* = 1/\gamma_1 = E_\alpha T_a \tag{20}$$

using Lemma 11. We also have the distributional identity

$$\mathcal{L}_{\pi^*}T_0^* = \mathcal{L}_\pi T_a \tag{21}$$

because

$$\begin{aligned}
P_{\pi^*}(T_0^* > t) &= \sum_{i=1}^{r}\pi^*(i)P_i^*(T_0 > t)\\
&= \sum_{i=1}^{r} p_i \exp(-\gamma_i t)\\
&= P_\pi(T_a > T) \text{ by (19).}
\end{aligned}$$

Next, note that the star chain has Dirichlet form

$$\mathcal{E}(g,g) = \sum_{i=1}^{r} p_i \gamma_i (g(i) - g(0))^2. \tag{22}$$

The particular function $g(i) = 1_{(i=1)} - p_1$ satisfies $\sum_{i=0}^{r} \pi^*(i) g(i) = 0$, and so by (2) and (22)

$$\tau^* \geq \frac{[g,g]}{\mathcal{E}(g,g)} = \frac{\sum_{i=0}^{r} p_i g^2(i)}{\sum_{i=1}^{r} p_i \gamma_i (g(i) - g(0))^2} = \frac{p_1(1-p_1)}{p_1 \gamma_1} = \frac{1-p_1}{\gamma_1}.$$

In other words

$$p_1 \geq 1 - \gamma_1 \tau^*. \tag{23}$$

Granted Lemma 12 below, we have

$$\begin{aligned} P_\pi(T_a > t) &\geq p_1 \exp(-\gamma_1 t) \text{ by (19)} \\ &\geq (1 - \gamma_1 \tau^*) \exp(-\gamma_1 t) \text{ by (23)} \\ &\geq (1 - \gamma_1 \tau) \exp(-\gamma_1 t) \text{ by Lemma 12} \\ &= (1 - \frac{\tau}{E_\alpha T_a}) \exp(-\frac{t}{E_\alpha T_a}) \text{ by (20)} \end{aligned}$$

establishing Theorem 3.

LEMMA 12 *The eigenvalues of Q^* form a subset of those of Q. Consequently $\tau^* \leq \tau$.*

PROOF A classical Markov chain identity (Keilson (1979, p. 77)) relates Laplace transforms of hitting times to Laplace transforms of transition densities. Writing

$$\tilde{p}(s) = \int_0^\infty P_a(X_t = a) e^{-st} dt$$

the identity asserts that for arbitrary initial distribution μ

$$\tilde{p}(s) \int_0^\infty e^{-st} dP_\mu(T_a < t) = \int_0^\infty P_\mu(X_t = a) e^{-st} dt.$$

Taking $\mu = \pi$ the right side becomes $\pi(a)/s$ and the identity shows that the distribution $\mathcal{L}_\pi T_a$ determines $\tilde{p}(s)$. So the distributional identity (21) implies

$$\tilde{p}(s) = \tilde{p}^*(s) \equiv \int_0^\infty P_0(X_t^* = 0) e^{-st} ds.$$

Now in terms of the matrix Q

$$\begin{aligned}\tilde{p}(s) &= \int_0^\infty e^{-st}(e^{Qt})_{aa}\,dt \\ &= \left(\int_0^\infty e^{(Q-sI)t}\,dt\right)_{aa} \\ &= -(Q-sI)^{-1}_{aa} \\ &= -\frac{\det(Q_a - sI)}{\det(Q-sI)}.\end{aligned}$$

Equating this with the corresponding expression for the star chain gives

$$\frac{\det(Q-sI)}{\det(Q^*-sI)} = \frac{\det(Q_a-sI)}{\det(Q_0^*-sI)}. \tag{24}$$

The eigenvalues $(\gamma_1, \ldots, \gamma_r)$ of Q_0^* are also, by (21), eigenvalues of Q_a. So the characteristic polynomial of Q_0^* is a divisor of the characteristic polynomial of Q_a. But then by (24) we see that the characteristic polynomial of Q^* is a divisor of the characteristic polynomial of Q. Thus the eigenvalues of Q^* are also eigenvalues of Q, establishing the Lemma.

QUESTION. Does the relation between X and X^* fall into the general notion of "duality" described in Liggett (1985) page 84?

7. Bounds on the Density Function

For use in Aldous and Brown (1991) we need delicate estimates of the density function $f(t), t > 0$ of $T = T_A$ for the stationary chain, which imply that $f(t) \approx 1/E_\pi T_A$ on the range $\tau \ll t \ll E_\pi T_A$. By scaling we can reduce to the case $ET = 1$.

LEMMA 13 *If $ET = 1$ then*
(a) $f(t) \leq 1 + \tau/(2t)$, $t > 0$.
(b) $f(t) \geq 1 - 2\tau - t$, $t > 0$.
(c) $-f'(t) \leq 2$ for $t \geq \tau(5 + 2\log(1/\tau))$, provided $\tau \leq 1$.

PROOF f is CM and so is convex (Keilson (1979, p. 66)). Then

$$\begin{aligned}2tf(t) &\leq \int_0^{2t} f(s)ds \text{ by convexity} \\ &= P_\pi(0 < T < 2t) \\ &\leq P_\pi(0 \leq T < 2t) \\ &\leq 1 - e^{-2t} + \tau \text{ by Theorem 1} \\ &\leq 2t + \tau\end{aligned}$$

giving (a).

To prove (b), note that Corollary 4 and Lemma 2 imply

$$1 \leq E_\alpha T_A \leq 1 + \tau. \tag{25}$$

By CM, the "hazard rate" is decreasing (Keilson (1979, p. 75)):

$$\frac{f(t)}{P_\pi(T_A > t)} \downarrow \frac{1}{E_\alpha T_A} \text{ as } t \to \infty.$$

So

$$\begin{aligned} f(t) &\geq \frac{1}{E_\alpha T_A} P_\pi(T_A > t) \\ &\geq \frac{1}{E_\alpha T_A}(\exp(-\frac{t}{E_\alpha T_A}) - \tau) \text{ by Theorem 3} \\ &\geq \frac{1}{E_\alpha T_A}(1 - \frac{t}{E_\alpha T_A} - \tau) \\ &\geq \frac{1}{E_\alpha T_A}(1 - t - \tau) \text{ by (25)} \\ &\geq \frac{1}{1+\tau}(1 - t - \tau) \text{ by (25)} \\ &\geq 1 - t - 2\tau \end{aligned}$$

giving (b). In the inequalities above we implicitly assumed $1 - t - \tau \geq 0$, since otherwise the result is trivial.

The proof of (c) has several ingredients. Excursions of the stationary chain inside A alternate with excursions outside A. Let $r(l)dl$ be the rate of excursions outside A of length $\in (l, l + dl)$. Then an easy renewal theory argument (Aldous and Brown (1991, Lemma 3)) shows

$$f(l) = -f'(l). \tag{26}$$

Now consider the joint density $\theta(t_1, t_2)$ of (T_A^-, T_A), where $T_A^- = \min\{t \geq 0 : X_{-t} \in A\}$ and the stationary chain is extended to $-\infty < t < \infty$. By conditioning on X_0 we see

$$\theta(t_1, t_2) = E f_{t_1}(X_0) f_{t_2}(X_0).$$

But in the notation of (26), $\theta(t_1, t_2) = r(t_1 + t_2)$. So putting $t_1 = t_2 = t$,

$$E f_t^2(X_0) = \theta(t, t) = r(2t) = -f'(2t) \tag{27}$$

the final equality by (26).

For the second ingredient, let $f_t(i)$ be the density of T_A under P_i. We shall show

$$E_\pi f_{s+t}^2(X_0) \leq f^2(t) + e^{-2s/\tau} E_\pi f_t^2(X_0). \tag{28}$$

For $s, t > 0$ consider

$$g_{s,t}(i)dt = P_i(X_u \in A^c \text{ for } s \leq u \leq s+t, X_{s+t+dt} \in A).$$

Then $g_{s,t} = \mathbf{P}_s f_t$ in the notation of (4), and the assertion of (4) gives (for the stationary chain)

$$\text{var } g_{s,t}(X_0) \leq e^{-2s/\tau} \text{var } f_t(X_0).$$

Plainly $f_{s+t} \leq g_{s,t}$ and $Eg_{s,t}(X_0) = Ef_t(X_0) = f(t)$, so

$$\begin{aligned} Ef^2_{s+t}(X_0) &\leq Eg^2_{s,t}(X_0) \\ &= f^2(t) + \text{var } g_{s,t}(X_0) \\ &\leq f^2(t) + e^{-2s/\tau} \text{var } f_t(X_0) \end{aligned}$$

and (28) follows. Next we shall show

$$-f'(2t) \leq t^{-2}/2. \tag{29}$$

For the function $-f'(u)$ is decreasing, so

$$f(u) \geq (2t-u)(-f'(2t)) \text{ on } 0 < u < 2t$$

and hence

$$1 \geq \int_0^{2t} f(u)du \geq 2t^2(-f'(2t))$$

giving (29).

Combining (27-29) shows that, for any $s, t > 0$,

$$-f'(2s+2t) \leq f^2(t) + e^{-2s/\tau} t^{-2}/2.$$

Putting $t = 2\tau$ and appealing to part (a)

$$-f'(2s+4\tau) \leq (5/4)^2 + e^{-2s/\tau}/(8\tau^2).$$

Then putting $s = \tau(\log(1/\tau) + 1/2)$ establishes part (c).

References

Aldous, D. J. (1982). Markov chains with almost exponential hitting times. *Stochastic Proc. Appl.* **13** 305–310.

Aldous, D. J. (1989). *Probability Approximations via the Poisson Clumping Heuristic.* Springer–Verlag, New York.

Aldous D. J. and Brown, M. (1991). Inequalities for rare events in time–reversible Markov chains II. *Stochastic Proc. Appl.* to appear, 1992.

Brown, M. (1990). Consequences of Monotonicity for Markov Transition Functions. Tech. Report, City College, CUNY.

Cogburn, R. (1985). On the distribution of first passage and return times for small sets. *Ann. Probab.* **13** 1219–1223.

Diaconis, P. and Stroock, D. (1991). Geometric bounds for eigenvalues of Markov chains. *Ann. Appl. Probab.* **1** 36–61.

Flannery, B. R. (1986). *Quasi–stationary Distributions for Markov Chains.* Ph.D. thesis, University of California, Berkeley.

Iscoe, I. and McDonald, D. (1991). Asymptotics of Exit Times for Markov Jump Processes with Applications to Jackson Networks. Tech. Report, University of Ottawa.

Keilson, J. (1979). *Markov Chain Models — Rarity and Exponentiality.* Springer–Verlag, New York.

Korolyuk, D. V. and Sil'vestrov, D. S. (1984). Entry times into asymptotically receding regions for processes with semi–Markov switching. *Theory Prob. Appl.* **29** 558–563.

Liggett, T. M. (1985). *Interacting Particle Systems.* Springer–Verlag, New York.

Department of Statistics
University of California, Berkeley
Berkeley, CA 94720

Department of Mathematics
City College, CUNY
New York, NY 10031

Stochastic Inequalities
IMS Lecture Notes – Monograph Series
Volume 22 (1993)

SKEWNESS AND KURTOSIS ORDERINGS: AN INTRODUCTION

By BARRY C. ARNOLD and RICHARD A. GROENEVELD

University of California, Riverside and Iowa State University

Competing skewness orderings are surveyed. It is argued that those based on natural skewness functionals are preferable to those related to convex orderings. Analogous kurtosis orderings are also discussed. Here the role of convex and Lorenz orderings appears more natural.

1. Introduction

What is skewness? An analogous question regarding inequality led Dalton eventually down the majorization path via the enunciation of clearly agreed upon inequality reducing transformations. Can a similar analysis be performed with skewness? In a sense the answer is easy; skewness is asymmetry, plain and simple. It is of course easy to recognize symmetric distributions but not so easy to decide whether one non symmetric distribution is more unsymmetric than another. Robin Hood (i.e. rich to poor) transfers are at the heart of the accepted inequality orderings. It is natural to search for analogous basic operations which will increase skewness. The present paper surveys suggested skewness orderings (although not in the detail provided by MacGillivray (1986)) but puts its major focus on promoting a particular group of skewness orderings. Clear parallels may be discerned between some of these orderings and the Lorenz inequality ordering generated by Robin Hood operations.

What is Kurtosis? This is a bit harder. To quote our dictionary (Webster's of course) it is the state or quality of peakedness or flatness of the graphic representation of a statistical distribution. Again a plethora of competing orderings have been proposed (see Balanda and MacGillivray (1988) for a recent survey). Again we champion a particular ordering related to Lorenz ordering.

AMS 1991 *subject classifications.* Primary 62E10, 60E99.

Key words and phrases. Asymmetry, peakedness, Lorenz order, star order, convex order, skewness functionals.

2. A Budget of Skewness Orderings

To simplify discussion, we restrict attention to the class $\mathcal{L}$ of distributions with median 0 and finite first absolute moment. The moment condition is not crucial for some of the orderings but several orderings will only be well-defined when first moments exist and it is convenient to restrict our focus to such distributions.

For any distribution $F \epsilon \mathcal{L}$, we define its quantile (or inverse distribution) function F^{-1} by

$$F^{-1}(u) = sup\{x : F(x) \leq u\}, \quad 0 < u < 1. \tag{2.1}$$

The zero median condition is equivalent to

$$F^{-1}(\frac{1}{2}) = 0 \tag{2.2}$$

and the first absolute moment is expressible as

$$E(|X|) = \int_0^{\frac{1}{2}} [F^{-1}(1-u) - F^{-1}(u)]du. \tag{2.3}$$

The class of median zero distributions with finite first moment can be identified conveniently with the class of all non-decreasing functions defined on $(0,1)$ satisfying (2.2) and (2.3). A symmetric distribution can be characterized by the requirement that

$$F^{-1}(1-u) + F^{-1}(u) = 0, \quad \forall u \epsilon (0, \frac{1}{2}). \tag{2.4}$$

Skewness corresponds to the violation of condition (2.4). Positive values of (2.4) for some, most or all values of u will be associated with positive skewness or skewness to the right. Negative skewness is associated with negative values of (2.4). We will define skewness orderings denoted by a symbol $\leq$ with a variety of subscripts and will in a cavalier fashion write them in terms of random variables or distribution functions i.e. $X \leq Y \iff F_X \leq F_Y$.

It is generally conceded that measures of skewness and related skewness orderings should be scale invariant, i.e. for any *positive* constant c, X and cX exhibit the same degree of skewness. Accepting this viewpoint it is defensible to divide any random variable by its first absolute moment and effectively focus on skewness orderings defined on the class of random variables with

$$F_X^{-1}(\frac{1}{2}) = 0 \tag{2.5}$$

and

$$E|X| = 1. \tag{2.6}$$

Denote the class of distribution functions satisfying (2.5) and (2.6) by $\mathcal{L}_0$.

A strong or uniform skewness ordering on $\mathcal{L}_0$ is provided by

$$\begin{aligned} X \leq_s Y \text{ iff } F_X &\leq_s F_Y \text{ iff } F_X^{-1}(1-u) + F_X^{-1}(u) \\ &\leq F_Y^{-1}(1-u) + F_Y^{-1}(u) \ \forall u \epsilon (0, \tfrac{1}{2}). \end{aligned} \tag{2.7}$$

A natural skewness functional is implicit in (2.7), namely

$$s_X(u) = F_X^{-1}(1-u) + F_X^{-1}(u), \quad 0 < u < \frac{1}{2}. \tag{2.8}$$

Rather than insist on a uniform domination of $s_X(u)$ by $s_Y(u)$, we might be willing to be forgiving of minor local violations. Two alternative skewness functionals are proposed namely

$$\begin{aligned} \nu_X(u) &= \int_0^u s_X(v)dv \\ &= \int_0^u [F_X^{-1}(1-v) + F_X^{-1}(v)]dv, \quad 0 < u < \frac{1}{2} \end{aligned} \tag{2.9}$$

and

$$\lambda_X(u) = \int_0^u [F_X^{-1}(\frac{1}{2}+v) + F_X^{-1}(\frac{1}{2}-v)]dv, \quad 0 < u < \frac{1}{2}. \tag{2.10}$$

Using these functionals we define analogous skewness orderings

$$X \leq_\nu Y \text{ iff } \nu_X(u) \leq \nu_Y(u), \ \forall u \epsilon (0, \frac{1}{2}) \tag{2.11}$$

and

$$X \leq_\lambda Y \text{ iff } \lambda_X(u) \leq \lambda_Y(u), \ \forall u \epsilon (0, \frac{1}{2}). \tag{2.12}$$

None of the above three skewness orderings seems to have been given close attention in the literature. Though certainly the uniform ordering has been implicitly considered.

Much of the literature on skewness orderings focusses on a variety of weakenings of Van Zwet's (1964) convex ordering. Our presentation is a somewhat simplified version of MacGillivray's (1986) section 2. Simplification is possible because of our focus on $\mathcal{L}_0$, the class of distributions with median 0 and unit first absolute moment.

The strong convex ordering of Van Zwet is defined by

$$\begin{aligned} X \leq_1 Y \quad &\text{iff} \quad F_Y^{-1}(F_X(x)) \text{ is convex on the support of } F \\ &\text{iff} \quad F_X^{-1}(u) \text{ and } [\text{a } F_Y^{-1}(u) + b] \text{ cross at most twice on} \\ &\qquad (0,1) \text{ for any } a > 0, b \epsilon \mathbb{R} \text{ with sign sequence of} \\ &\qquad F_X^{-1}(u) - aF_Y^{-1}(u) - b \text{ being } -, +, -. \end{aligned} \tag{2.13}$$

A star ordering (related to Oja's (1981) ordering) is defined by

$$X \leq_2 Y \text{ iff } \frac{F_Y^{-1}(u)}{F_X^{-1}(u)} \uparrow \text{ on } (0,1) - \{\frac{1}{2}\} \tag{2.14}$$

where the symbol $\uparrow$ is to be read here and henceforth as non-decreasing. Progressively further weakening yields the following orders.

(2.15) $X \leq_3 Y$ iff

$$F_Y^{-1}(u) - \frac{f_X(0)}{f_Y(0)} F_X^{-1}(u) \text{ is } \downarrow \text{ on } (0, \frac{1}{2}) \text{ and } \uparrow \text{ on } (\frac{1}{2}, 1)$$

$$X \leq_4 Y \text{ iff } F_Y^{-1}(u) f_Y(0) \leq F_X^{-1}(u) f_X(0) \text{ on } (0,1) - \{\frac{1}{2}\} \tag{2.16}$$

and

$$X \leq_5 Y \text{ iff } \frac{F_X^{-1}(1-u)}{F_X^{-1}(u)} \geq \frac{F_Y^{-1}(1-u)}{F_Y^{-1}(u)}, \quad 0 < u < \frac{1}{2}. \tag{2.17}$$

In the above definitions f_X and f_Y are the densities corresponding to F_X and F_Y.

An ordering based on stochastic ordering of the positive and negative parts of X and Y has been proposed:

$$X \leq_6 Y \text{ iff } X^+ \leq_{st} Y^+ \text{ and } Y^- \leq_{st} X^- \tag{2.18}$$

Instead of stochastic ordering in (2.18) we might invoke Lorenz ordering. Thus we have

$$X \leq_7 Y \text{ iff } X^+ \leq_L Y^+ \text{ and } Y^- \leq_L X^-. \tag{2.19}$$

A convenient reference for the definition of $\leq_L$ is Arnold (1987). Second and higher order stochastic dominance could of course be used as a basis of a skewness order definition but we will not pursue that possibility.

Finally we mention the David and Johnson (1956) skewness functional related to but distinct from $s_X(u)$ defined in (2.8). It takes the form

$$\tilde{s}_X(u) = \frac{F_X^{-1}(1-u) + F_X^{-1}(u)}{F_X^{-1}(1-u) - F_X^{-1}(u)}, \quad 0 < u < \frac{1}{2} \tag{2.20}$$

Analogs to (2.9) and (2.10) are available

$$\tilde{\nu}_X(u) = \int_0^u \left[\frac{F_X^{-1}(1-v) + F_X^{-1}(v)}{F_X^{-1}(1-v) - F_X^{-1}(v)}\right] dv, \quad 0 < u < \frac{1}{2} \tag{2.21}$$

and

$$\tilde{\lambda}_X(u) = \int_0^u \frac{F_X^{-1}(\frac{1}{2}+v) + F_X^{-1}(\frac{1}{2}-v)}{F_X^{-1}(\frac{1}{2}+v) - F_X^{-1}(\frac{1}{2}-v)} dv, \quad 0 < u < \frac{1}{2}. \tag{2.22}$$

An advantage of the David-Johnson functionals is that they are well defined without any assumption regarding the existence of $E|X|$. All three functions are uniformly bounded in absolute value by 1. Skewness orderings $\leq_{\tilde{s}}, \leq_{\tilde{\nu}}$ and $\leq_{\tilde{\lambda}}$ are defined in the natural way using these functionals. Note that $\leq_{\tilde{s}}$ is equivalent to $\leq_5$.

We remark finally without further comment on the possibility of introducing a non-negative weight function $\psi(v)$ inside the integral in definitions (2.9), (2.10), (2.21) and (2.22).

3. Inter–Relationships Among the Orderings

Thirteen skewness orderings were introduced in Section 2. How are they related? As MacGillivray (1986) noted, convex ordering is very strong and implies many reasonable definitions of skewness. Specifically from her paper we have the relations

$$\leq_1 \Longrightarrow \leq_2 \Longrightarrow \leq_3 \Longrightarrow \leq_4 \Longrightarrow \leq_5 . \tag{3.1}$$

It is not difficult to verify that we also have

$$\leq_6 \Longrightarrow \leq_s \begin{cases} \Longrightarrow \leq_\nu \\ \Longrightarrow \leq_\lambda \end{cases} \tag{3.2}$$

$$\leq_5 \Longleftrightarrow \leq_{\tilde{s}} \begin{cases} \Longrightarrow \leq_{\tilde{\nu}} \\ \Longrightarrow \leq_{\tilde{\lambda}} \end{cases} \tag{3.3}$$

and

$$\leq_2 \Longrightarrow \leq_7 \tag{3.4}$$

The key observations justifying (3.4) are that $X \leq_2 Y$ implies $X^+ \leq^* Y^+$ and $Y^- \leq^* X^-$ and star-ordering implies Lorenz ordering (see for example Arnold (1987, p. 78)).

The convex ordering $\leq_1$ is much stronger than necessary. It is our contention that the candidate orderings most worthy of consideration are $\leq_5$ (equivalently $\leq_{\tilde{s}}$), $\leq_6$ and $\leq_s$. Additionally it is felt that the concept of skewness does not necessarily involve comparison of positive and negative parts of random variables. On this basis $\leq_6$ is not as appealing. We are left with $\leq_s$ and $\leq_{\tilde{s}}$ together with the more forgiving integrated versions provided by $\leq_\nu, \leq_{\tilde{\nu}}, \leq_\lambda$ and $\leq_{\tilde{\lambda}}$.

At this juncture, if we were forced to select a single skewness ordering to recommend, it might well be $\leq_\nu$. Note that $\nu_X(u) \leq \nu_Y(u) \ \forall \ u \epsilon (0, \frac{1}{2})$ is equivalent to

$$E(X|X \leq F_X^{-1}(u)) + E(X|X \geq F_X^{-1}(1-u))$$

$$\leq E(Y|Y \leq F_Y^{-1}(u)) + E(Y|Y \geq F_Y^{-1}(1-u))$$

i.e. the average of the right and left tails of Y exceeds the average of the right and left tails of X (at each percentile). This is not an unreasonable definition of Y being more skewed to the right than X.

Details regarding the relationships (3.2)–(3.4) together with counterexamples to any other inter-relationships among the orderings will appear in a more extensive report. The example in the following section highlights the potential clash between $\leq_1$ and $\leq_s$.

4. Examples of the Skewness Functionals

Consider a random variable X whose distribution is given by

$$F_X(x) = 1 - (x + 2^b)^{-(1/b)}, \quad x > 1 - 2^b \tag{4.1}$$

where $b \epsilon (0, 1)$. The resulting inverse distribution function is

$$F_X^{-1}(u) = (1-u)^{-b} - 2^b, \quad 0 < u < 1$$

and hence

$$\begin{aligned} E(|X|) &= \int_0^{\frac{1}{2}} [F_X^{-1}(1-u) - F_X^{-1}(u)]du \\ &= (2^b - 1)/(1-b) \end{aligned}$$

so that our skewness functionals assume the form

$$s_X(u) = \frac{(1-u)^{-b} + u^{-b} - 2^{b+1}}{(2^b-1)/(1-b)},$$

$$\nu_X(u) = \frac{1 + u^{1-b} - (1-u)^{1-b} - (1-b)2^{b+1}u}{2^b - 1},$$

and

$$\lambda_X(u) = \frac{(\frac{1}{2}+u)^{1-b} - (\frac{1}{2}-u)^{1-b} - (1-b)2^{b+1}u}{2^b - 1}.$$

The functional ν_X is monotonically increasing in b. The larger the values of b, the more positively skewed is the resulting distribution as measured by ν_X.

Note that if X_1 and X_2 have distribution (4.1) with corresponding parameters $b_1 < b_2$, then $X_{b_1} \leq_1 X_{b_2}$ and as observed above $X_{b_1} \leq_\nu X_{b_2}$. However numerical calculations indicate that, for a given b_1, there exists $b_2 > b_1$ such that $X_{b_1} \not\leq_s X_{b_2}$ and $X_{b_1} \not\leq_\lambda X_{b_2}$.

5. Skewness Accentuating Transformations

We may reasonably seek to characterize all transformations $g : \mathbb{R} \to \mathbb{R}$ which have the property that for any $X \epsilon \mathcal{L}_0$ we have X less skew than $g(X)$. It is evidently true that if we choose a function g such that $g(0) = 0$ and both $g(x)$ and $g(x)/x \uparrow$ on $\mathbb{R} - \{0\}$ then $X \leq_1 g(X)$ for any $X \epsilon \mathcal{L}_0$. Thus transformations of this kind accentuate skewness using the strongest (Van Zwet) skewness ordering. They thus accentuate skewness using any of the other orderings implied by the Van Zwet orderings. These transformations however do not necessarily accentuate skewness as measured by $\leq_6$.

6. Kurtosis

An analogous variety of kurtosis orderings exist. Most restrict attention to symmetric distributions. A reason for this is the difficulty of interpretation of the concept of kurtosis in the absence of symmetry. Setting aside such niceties for the moment, it is possible to provide kurtosis orderings analogous to several of the skewness orderings described in this paper. Some of these reduce to already known kurtosis orderings if symmetry is imposed. As in our skewness discussion we standardize all variables to have median 0 and first absolute moment equal to 1; i.e. we restrict attention to $\mathcal{L}_0$. To distinguish our kurtosis ordering from the corresponding parallel skewness orderings we place a superscript k above the inequality sign. Thus $\leq_2^k$ will be a kurtosis ordering analogous to the skewness ordering $\leq_2$. Here is the list (as usual $X = X^+ - X^-$ where $X^+ \geq 0$ and $X^- \geq 0$).

$$X \leq_1^k Y \text{ iff } F_{Y^+}^{-1}(F_{X^+}(x)) \text{ is convex on the support of } X^+ \text{ and } F_{Y^-}^{-1}(F_{X^-}(x)) \text{ is convex on the support of } X^-. \tag{6.1}$$

$$X \leq_2^k Y \quad \text{iff} \quad F_{Y^+}^{-1}(u)/F_{X^+}^{-1}(u) \uparrow \text{ on } (0,1) \quad \text{and} \quad F_{Y^-}^{-1}(u)/F_{X^-}^{-1}(u) \uparrow \text{ on } (0,1) \tag{6.2}$$

$$X \leq_7^k Y \quad \text{iff} \quad X^+ \leq_L Y^+ \text{ and } X^- \leq_L Y^- \tag{6.3}$$

$$X \leq_\lambda^k Y \quad \text{iff} \quad \int_0^u [F_X^{-1}(\frac{1}{2} + v) - F_X^{-1}(\frac{1}{2} - v)]dv \geq \int_0^u [F_Y^{-1}(\frac{1}{2} + v) - F_Y^{-1}(\frac{1}{2} - v)]dv \;\; \forall u \epsilon (0, \frac{1}{2}) \tag{6.4}$$

In the case in which X and Y are symmetric random variables, this last ordering is equivalent to $|X| \leq_L |Y|$. In the absence of symmetry, the Lorenz order of absolute values may be considered to be candidate variant kurtosis order. We may define

$$X \leq_8^k Y \text{ iff } |X| \leq_L |Y|. \tag{6.5}$$

This ordering has an attractive simplicity. It certainly captures some of the idea of kurtosis when the random variables are symmetric. Interpretation in the asymmetric case is potentially more problematic. It is not difficult to construct an asymmetric example in which $X \leq_\lambda^k Y$ but $X \not\leq_8^k Y$ and an example in which $X \leq_8^k$ but $X \not\leq_\lambda^k Y$. One advantage of the absolute Lorenz ordering ($\leq_8^k$) is its potential for straightforward extension to higher dimensions. For m dimensional random vectors $\underline{X}$ and $\underline{Y}$ centered to have medians $\underline{0}$, we can define $\underline{X} \leq_8^k \underline{Y}$ if and only if $d(\underline{X}, \underline{0}) \leq_L d(\underline{Y}, \underline{0})$ where d is a metric in $\mathbb{R}^m$.

More details on these kurtosis orderings and related summary measures of kurtosis will appear in a separate report.

References

Arnold, B. C. (1987) *Majorization and the Lorenz Order: A Brief Introduction.* Lecture Notes in Statistics **43**, Springer-Verlag, Berlin.

Balanda, K. P. and MacGillivray, H. L. (1988) Kurtosis: a critical review. *Amer. Statist.* **42** 111–119.

David, F. N. and Johnson, N. L. (1956) Some tests of significance with ordered variables. *J. Royal Stat. Soc.* **B18** 1–20.

MacGillivray, H. L. (1986) Skewness and asymmetry: measures and orderings. *Ann. Statist.* **14** 994–1011.

Oja, H. (1981) On location, scale, skewness and kurtosis of univariate distributions. *Scand. J. Statist.* **8** 154–168.

Van Zwet, W. R. (1964) *Convex Transformations of Random Variables.* Mathematisch Centrum, Amsterdam.

Department of Statistics
University of California, Riverside
Riverside, CA 92502

Department of Statistics
Iowa State University
Ames, IA 50010

Stochastic Inequalities
IMS Lecture Notes – Monograph Series
Volume 22 (1993)

STOCHASTIC INEQUALITIES FOR A REDUNDANCY ENHANCEMENT TO A SERIES OR PARALLEL SYSTEM

By PHILIP J. BOLAND

University College, Dublin

We consider the question of where to allocate a redundant spare in a series or parallel system of components in order to stochastically optimize the resulting performance of the system. Both active (or warm or parallel) and standby (or cold) redundancy are considered. We show that if the components are stochastically ordered in the usual sense, then an active redundancy allocation to the weakest (strongest) component is stochastically optimal for a series (parallel) system. The situation is more delicate for standby redundancy. If the components are ordered according to the likelihood ratio ordering, then it is stochastically optimal to make a standby redundancy to the strongest component in a parallel system. It is shown however that even for this stronger sense of component ordering, the stochastically optimal redundancy allocation in a series system is not necessarily to the weakest component.

1. Introduction

We let $T_1, \ldots, T_n$ be random variables representing the lifetimes of n components which make up a series or parallel system. We will assume the lifetimes are independent, and that they are stochastically ordered (usually increasing) in some sense. There are two types of redundancy enhancements to the system that we consider: (1) an **active** (also called a **warm** or **parallel**) redundancy, and (2) a **standby** (also called a **cold**) redundancy. An active redundant spare is one which is actively working in parallel with one of the components in the system, while a standby spare is one which only begins to operate when the component for which it is 'standing by' ceases to operate. In any case the system performance as a whole is enhanced by a redundancy, and we will be interested in placing the redundant spare in the system so as to stochastically maximize its resulting lifetime. Consideration of an active redundancy leads one to study the maximum of random variables while that of a standby redundancy leads to the study of convolutions.

AMS 1991 *subject classifications.* 62N05, 90B25.

Key words and phrases. Stochastic ordering, likelihood ratio ordering, series and parallel systems, active and standby redundancy.

For either type of redundancy enhancement we will consider the situation where the available spare component is (1) **common** or (2) **like.** By a **common** spare we will mean there exists a component with independent lifetime T which can be placed in redundancy with any of the components in the system. We will be in a position to consider using 'like' spares if there are spare components with respective lifetimes $T'_1, \ldots, T'_n$ where $T_i \stackrel{d}{=} T'_i$ (equal in distribution) for $i = 1, \ldots, n$. The redundant spare for the ith component will then have lifetime T'_i.

The lifetime of a parallel system with components $T_1, \ldots, T_n$ is given by

$$\tau_P(T_1, \ldots, T_n) = \max\{T_1, \ldots, T_n\}$$

while the lifetime of the series system with the same components would be

$$\tau_S(T_1, \ldots, T_n) = \min\{T_1, \ldots, T_n\}.$$

Stochastic results for active redundancy enhancement by either a 'common' or 'like' spare are reasonably straightforward when the lifetimes $T_1, \ldots, T_n$ are stochastically increasing in the usual sense ($T_1 \stackrel{st}{\leq} \cdots \stackrel{st}{\leq} T_n$). (Remember that $X \stackrel{st}{\leq} Y \Leftrightarrow F_X(t) \geq F_Y(t)$ for all t where F_X and F_Y are respectively the distribution functions of X and Y.)

In section 2 it is noted that for a common spare with independent lifetime T, $\tau_S(T_1, \ldots, \max(T_i, T), \ldots, T_n)$ is stochastically decreasing in i, while the distribution of $\tau_P(T_1, \ldots, \max(T_i, T), \ldots, T_n)$ is independent of i. In the situation where independent like spares $T'_1, \ldots, T'_n$ are available, one may show that

$$\tau_S(T_1, \ldots, \max(T_i, T'_i), \ldots, T_n)$$

is stochastically decreasing in i, while

$$\tau_P(T_1, \ldots, \max(T_i, T'_i), \ldots, T_n)$$

is stochastically increasing in i.

The topic of standby redundancy is treated in section 3. Here we make use of a result of Brown and Solomon (1973) (see also Ross (1983)) concerning the likelihood ratio ordering. Their result states that if X, Y are independent nonnegative random variables with respective densities f and g such that $X \stackrel{lr}{\leq} Y$ (i.e. $g(x)/f(x) \uparrow$ in x), then for any function $h(x, y)$ with the property that $h(x, y) \geq h(y, x)$ whenever $x < y$, it follows that

$$h(X, Y) \stackrel{st}{\geq} h(Y, X).$$

This result implies that if $X_1, \ldots, X_n$ are independent where $X_1 \stackrel{lr}{\leq} \cdots \stackrel{lr}{\leq} X_n$ and h is any arrangement increasing function of n variables, then

$$h(X_1, \ldots, X_n)$$

is 'stochastically' arrangement increasing. By an arrangement increasing function of n variables we shall mean a function which increases in value as the order of the components approaches the situation where the coordinates are increasing. (For more on arrangement increasing functions see Marshall and Olkin (1979).) Ross's result is useful in making stochastic statements with respect to standby redundancy. For the situation when a common spare T is available, one may show that if $T_1 \overset{lr}{\leq} \cdots \overset{lr}{\leq} T_n$, then

$$\tau_P(T_1, \ldots, T_i + T, \ldots, T_n)$$

is stochastically increasing in i, while

$$\tau_S(T_1, \ldots, T_i + T, \ldots, T_n)$$

is stochastically decreasing in i. When a 'like' spare situation is relevant, one may show that

$$\tau_P(T_1, \ldots, T_i + T_i', \ldots, T_n)$$

is stochastically increasing in i, but a similar result for the series system is not valid.

We give an example of Gamma distributed random variables $T_1, \ldots, T_n$ where $T_1 \overset{lr}{\leq} \cdots \overset{lr}{\leq} T_n$ does not imply that

$$\tau_S(T_1, \ldots, T_i + T_i', \ldots, T_n)$$

is stochastically decreasing in i. Although other examples of a positive nature are given, it remains to find sufficient conditions on the stochastic ordering of components in order to insure that it is stochastically optimal to allocate a like spare component to the weakest component in a series system.

The results in this paper summarize much of the recent work in the area of stochastic ordering and redundancy allocation to series and parallel systems.

2. Stochastic Order for an Active Redundancy

Initially we consider an active redundancy allocation of a 'common' component with life distribution T. A k out of n system is a system of n components which functions if k or more of the components function. A parallel system is a 1 out of n system while a series system is an n out of n system. Boland, El–Neweihi and Proschan (1992) show that when the components in a k out of n system are stochastically ordered so that $T_1 \overset{st}{\leq} \cdots \overset{st}{\leq} T_n$, then it is always stochastically optimal to improve the weakest component with a common active redundancy when given the choice. In particular they obtain:

Theorem 2.1 *Let $T_1, \ldots, T_n$, T be independent lifetimes where $T_1 \overset{st}{\leq} \cdots \overset{st}{\leq} T_n$. Then*

a) Series System: $\tau_S(T_1, \ldots, \max(T_i, T), \ldots, T_n)$ is stochastically decreasing in i, and

b) Parallel System: $\tau_P(T_1, \ldots, \max(T_i, T), \ldots, T_n)$ is (clearly) independent of i.

Now let us suppose a k out of n system is composed of independent components with lifetimes $T_1, \ldots, T_n$ where $T_1 \overset{st}{\leq} \cdots \overset{st}{\leq} T_n$. Moreover let us assume that a set of independent 'like' spares with lifetimes $T_1', \ldots, T_n'$ is available ($T_i \overset{d}{=} T_i'$ for $i = 1, \ldots, n$) for active redundancy with the original system. In the case where only one active redundancy allocation is permitted, a natural question is to determine where this might be done in order to give the greatest improvement to the system. Unfortunately there is no general stochastic result for arbitrary k out of n systems. Boland, El–Neweihi and Proschan (1992) give an example of a 2 out of 3 system where the answer might be with either component 1, 2 or 3 depending on the specific distributions of T_1, T_2 and T_3 and the point in time being considered. For series and parallel systems however we have the following result:

Theorem 2.2 *Let $T_1, \ldots, T_n, T_1', \ldots, T_n'$ be independent lifetimes where $T_i \overset{d}{=} T_i'$ for $i = 1, \ldots, n$ and $T_1 \overset{st}{\leq} \cdots \overset{st}{\leq} T_n$. Then*

a) Series System

$$\tau_S(T_1, \ldots, \max(T_i, T_i'), \ldots, T_n)$$

is stochastically decreasing in i, while

b) Parallel System

$$\tau_P(T_1, \ldots, \max(T_i, T_i'), \ldots, T_n)$$

is stochastically increasing in i.

Proof Let F_i be the distribution function for T_i and T_i'. For any $t \geq 0$ and $i = 1, \ldots, n-1$ we have that $F_i(t) \geq F_{i+1}(t)$. Hence

$$(1 - F_i^2(t))\overline{F}_{i+1}(t) = (1 - F_i^2(t))(1 - F_{i+1}(t)) \geq \overline{F}_i(t)(1 - F_{i+1}^2(t)),$$

and it follows that

$$\begin{aligned} &\text{Prob}[\tau_S(T_1, \ldots, \max(T_i, T_i'), \ldots, T_n) > t] \\ &\quad \geq \text{Prob}[\tau_S(T_1, \ldots, \max(T_{i+1}, T_{i+1}'), \ldots, T_n) > t]. \end{aligned}$$

This proves the result for a series system and similarly the parallel result follows.

Before passing to the next section on the mathematically more delicate problem of standby redundancy, we note that for active redundancy considerations (when the components are stochastically ordered) we can now say the weakest component is the most important **stochastically** in a series system while it is the strongest component in a parallel system. Further results in active redundancy may be found in Boland, El–Neweihi and Proschan (1988) and Xie and Shen (1989) and (1991).

3. Stochastic Order for a Standby Redundancy

In this section we consider the allocation of a **standby** redundant spare (where the available spare is either **common** or **like** in the sense previously described) in a series or parallel system. As in the previous section, one would expect in particular that when the components are stochastically ordered and independent, then an allocation to the weakest component in a series system and the strongest in a parallel system will be stochastically optimal. Some results of this type are in fact true, but only on the assumption that the components are stochastically ordered according to the likelihood ratio ordering (which is a stronger ordering than the usual stochastic ordering). We will also see however that examples exist where the components in a series system are increasing in the likelihood ratio sense, but it is not stochastically optimal to make a 'like' standby redundancy with the weakest component.

If X and Y are independent with respective densities f and g, we say X is less than Y in the likelihood ratio sense ($X \overset{lr}{\leq} Y$) if

$$g(x)/f(x)$$

is increasing over the common support of X and Y. It is well known that $X \overset{lr}{\leq} Y \Rightarrow X \overset{st}{\leq} Y$, but not conversely (see Ross (1983)). Brown and Solomon (1973) proved that if $X \overset{lr}{\leq} Y$ and h is a function with the property that

$$h(x,y) \geq h(y,x) \quad \text{whenever} \quad x \leq y,$$

then $h(X,Y) \overset{st}{\geq} h(Y,X)$.

We now use this result of Brown and Solomon to prove the following Theorem of Boland, El–Neweihi and Proschan (1992) concerning standby redundancy of a **common** spare in a series or parallel system.

THEOREM 3.1 *Let $T_1, \ldots, T_n$, T be independent lifetimes where $T \overset{lr}{\leq} \cdots \overset{lr}{\leq} T_n$. Then*

a) Series System: $\tau_s(T_1, \ldots, T_i + T, \ldots, T_n)$ is stochastically decreasing in i, and

b) Parallel System: $\tau_p(T_1, \ldots, T_i + T, \ldots, T_n)$ is stochastically increasing in i.

PROOF For any $t \geq 0$, we define the functions

$$\begin{aligned} h_1(t_1, t_2) &= \min(t_1 + t, t_2) \quad \text{and} \\ h_2(t_1, t_2) &= \max(t_1, t_2 + t). \end{aligned}$$

Then clearly for any $t_1 \leq t_2$, $h_i(t_1, t_2) \geq h_i(t_2, t_1)$ for $i = 1, 2$. It follows by conditioning on the values of T and using the result of Ross that

$$\begin{aligned} \tau_S(T_1 + T, T_2) &\overset{st}{\geq} \tau_S(T_1, T_2 + T) \qquad \text{and} \\ \tau_P(T_1 + T, T_2) &\overset{st}{\leq} \tau_P(T_1, T_2 + T). \end{aligned}$$

The result for general $n > 2$ follows by independence.

Examples may be easily constructed (see Boland, El–Neweihi, Proschan (1992)) to show the results of Theorem 3.1 are not valid when the likelihood ratio ordering of the components is relaxed to the ordinary stochastic ordering.

We now consider the question of making one standby redundancy allocation when 'like' spares are available. The parallel case is easy, as the following Theorem (Boland, El–Neweihi, and Proschan (1992) or Shaked and Shanthikumar (1990)) demonstrates.

THEOREM 3.2 *Let $T_1, \ldots, T_n, T_1', \ldots, T_n'$ be independent lifetimes where $T_i \overset{d}{=} T_i'$ for $i = 1, \ldots, n$, and $T_1 \overset{lr}{\leq} \cdots \overset{lr}{\leq} T_n$. Then for a parallel system*

$$\tau_P(T_1, \ldots, T_i + T_i', \ldots, T_n)$$

is stochastically increasing in i.

PROOF

$$\begin{aligned} &\tau_P(T_1, \ldots, T_i + T_i', T_{i+1}, \ldots, T_n) \\ &\overset{st}{\leq} \tau_P(T_1, \ldots, T_i, T_{i+1} + T_i', \ldots, T_n) \quad \text{(by Theorem 3.1b)} \\ &\overset{st}{\leq} \tau_P(T_1, \ldots, T_i, T_{i+1} + T_{i+1}', \ldots, T_n) \quad (\text{since } 0 \leq T_i' \overset{st}{\leq} T_{i+1}'). \end{aligned}$$

Shaked and Shanthikumar (1990) implicitly prove that if $T_1, \ldots, T_n$, $S_1, \ldots, S_n$ are independent lifetimes where $T_1 \overset{lr}{\leq} \cdots \overset{lr}{\leq} T_n$ and $S_1 \overset{lr}{\leq} \cdots \overset{lr}{\leq} S_n$, then

$$\tau_P(T_1, \ldots, T_i + S_i, \ldots, T_n)$$

is stochastically increasing in i [see their Lemma 3.2]. This in particular implies the results of Theorem 3.1b) and Theorem 3.2 above. This interesting paper of Shaked and Shanthikumar presents many other results about resource allocations to parallel and series systems.

What about the corresponding result of Theorem 3.2 for series systems. Natvig (1985) showed that if $T_1, \ldots, T_n, T_1', \ldots, T_n'$ are independent Gamma random variables with the same shape parameter m, and where $T_i \overset{d}{=} T_i'$, then $T_1 \overset{st}{\leq} \cdots \overset{st}{\leq} T_n$ implies that

$$\tau_S(T_1, \ldots, T_i + T_i', \ldots, T_n)$$

is stochastically decreasing in i. Boland, Proschan and Tong (1990) show that if each T_i is uniformly distributed on $[0, \theta_i)$ where $\theta_1 \leq \cdots \leq \theta_n$, then a similar result holds. Other examples of a parametric nature also exist. However Boland, Proschan and Tong (1990) also gave the following example:

EXAMPLE 3.1 Let T_1, T_2, T_1', T_2' be independent random variables where $T_1 \overset{d}{=} T_1' \equiv \Gamma(m, \theta)$, and $T_2 = T_2' = \Gamma(m+1, \theta)$, for some $m \geq 1$ and $\theta > 0$. Then $T_1 \overset{lr}{\leq} T_2$ but

$$\tau_S(T_1 + T_1', T_2) \overset{st}{\not\geq} \tau_S(T_1, T_2 + T_2').$$

This is demonstrated by showing that the difference between the reliability of the first and second systems is positive for small t and negative for large t.

We note in conclusion that the likelihood ratio ordering is not strong enough to imply that when independent components are ordered in this sense in a series system the stochastically optimum choice for a like redundant spare allocation is with the weakest component. One might ask then, is there a stronger stochastic ordering for which this intuitively plausible result is true?

REFERENCES

BOLAND, P.J., EL–NEWEIHI, E. AND PROSCHAN, F. (1988). Active redundancy allocation in coherent systems. *Prob. Eng. Inform. Sci.* **2** 343–353.

BOLAND, P. J., EL–NEWEIHI, E. AND PROSCHAN, F. (1991). Redundancy importance and allocation of spares in coherent systems. *J. Statist. Plann. Inf.* **29** 55–66.

BOLAND, P. J., EL–NEWEIHI, E. AND PROSCHAN, F. (1992). Stochastic order for redundancy allocation in series and parallel systems. *Adv. Appl. Prob.* **24** 161–171.

BOLAND, P. J., PROSCHAN, F. AND TONG, Y. L. (1990). Standby redundancy policies for series systems. Tech. Report, University College, Dublin.

BROWN, M. AND SOLOMON, H. (1973). Optimal issuing policies under stochastic field lives. *J. Appl. Prob.* **10** 761–768.

MARSHALL, A. W. AND OLKIN, I. (1979). *Inequalities: Theory of Majorization and its Applications.* Academic Press, New York.

NATVIG, B. (1985). New light on measures of importance of system components. *Scand. J. Statist.* **12** 43–54.

ROSS, S. M. (1983). *Stochastic Processes.* Wiley, New York.

SHAKED, M. AND SHANTHIKUMAR, J. G. (1990). Optimal allocation of resources to nodes of parallel and series systems. Tech. Report, University of Arizona.

SHEN, K. AND XIE, M. (1991). The effectiveness of adding standby redundancy at system and component levels. *IEEE Trans. Reliab.* **40** (1) 53–55.

XIE, M. AND SHEN, K. (1989). On ranking of system components with respect to different improvement actions. *Microelectron. Reliab.* **29** (2) 159–164.

DEPARTMENT OF STATISTICS
UNIVERSITY COLLEGE, DUBLIN
BELFIELD, DUBLIN 4, IRELAND

Stochastic Inequalities
IMS Lecture Notes – Monograph Series
Volume 22 (1993)

SOME REMARKS ON A NOTION OF POSITIVE DEPENDENCE, ASSOCIATION, AND UNBIASED TESTING[1]

By ARTHUR COHEN and H. B. SACKROWITZ

Rutgers University

A new notion of positive dependence is studied. The new notion implies association of k random variables but is weaker than the notion of conditionally increasing in sequence.

1. Introduction

Tong (1990) discusses various notions of positive dependence of a collection of k random variables $(X_1, X_2, \ldots, X_k)$. Among the notions are multivariate totally positive of order 2 (MTP_2), conditionally increasing in sequence (CIS), and (positively) associated (A). In Proposition 5.1.2 on page 95 of Tong (1990), it is noted that

$$MTP_2 \Rightarrow CIS \Rightarrow A. \tag{1.1}$$

For applications it can be important to know whether a set of k random variables is MTP_2 or CIS, sometimes because the property implies A and sometimes because of other probability statements or inequalities that can be achieved. In Cohen, Kemperman, and Sackrowitz (CKS) (1992) another notion of positive dependence was introduced. This new notion, which we will call weak conditionally increasing in sequence ($WCIS$) is implied by CIS but implies A. Thus $WCIS$ is weaker than CIS but yet still $WCIS$ implies A. CKS (1992) used $WCIS$ to establish a class of unbiased tests for testing whether the natural parameters of k exponential family PF_2 distributions lie on a line against the alternative that the parameters are convex, i.e., their weighted second order differences are nonnegative.

Association is frequently a method of establishing unbiasedness of classes of tests. See for example, Perlman and Olkin (1980), Cohen and Sackrowitz (1987), and Cohen, Perlman and Sackrowitz (1991). CKS (1992) prove unbiasedness by establishing A via $WCIS$. In that study both MTP_2 and CIS fail to hold for the relevant variables.

[1]Research supported by National Science Foundation Grant DMS-9112784.

AMS 1991 *Subject Classification.* Primary 60E99, Secondary 62F03.

Key words and phrases. Unbiased tests, association, positive dependence, stochastic ordering, ordered alternatives.

In this note we formally define $WCIS$ and show that

$$MTP_2 \Rightarrow CIS \Rightarrow WCIS \Rightarrow A. \tag{1.2}$$

An example is given where a random vector $\mathbf{X} = (X_1, \ldots, X_k)'$ is $WCIS$ but not CIS and another example is given in which $\mathbf{X}$ is A but not $WCIS$.

For a $k \times 1$ normal random vector with covariance matrix Σ, we indicate necessary and sufficient conditions for it to have the CIS property or $WCIS$ property. These conditions are useful when it is of importance to establish the CIS or $WCIS$ property for its own use as opposed to establishing it for the purpose of establishing A. If one were seeking to establish A for a normal random vector, the known necessary and sufficient condition that $\Sigma \geq 0$ would be used. This is a result of Pitt (1982).

In the next section, we give the results on $WCIS$.

2. Weak Conditionally Increasing in Sequence ($WCIS$)

Let $\mathbf{X}^{k \times 1}$ be a random vector with density function $f_{\mathbf{X}}(\mathbf{x})$.

DEFINITION 2.1 The random vector $\mathbf{X}$ and its density are said to be MTP_2 if

$$f_{\mathbf{X}}(\mathbf{x} \vee \mathbf{y}) f_{\mathbf{X}}(\mathbf{x} \wedge \mathbf{y}) \geq f_{\mathbf{X}}(\mathbf{x}) f_{\mathbf{X}}(\mathbf{y}), \quad \text{for all } \mathbf{x}, \mathbf{y}, \tag{2.1}$$

where $\mathbf{x} \vee \mathbf{y} = (\max(x_1, y_1), \ldots, \max(x_k, y_k))$ and $\mathbf{x} \wedge \mathbf{y} = (\min(x_1, y_1), \ldots, \min(x_k, y_k))$.

DEFINITION 2.2 The random vector $\mathbf{Y}$ is stochastically greater than or equal to the random vector $\mathbf{X}$ ($\mathbf{X} \leq^P \mathbf{Y}$) if

$$Eh(\mathbf{X}) \leq Eh(\mathbf{Y}), \tag{2.2}$$

for all nondecreasing functions h for which the expectations in (2.2) exist. (h is nondecreasing if it is nondecreasing in x_i while $x_1, \ldots, x_{i-1}, x_{i+1}, \ldots, x_k$ are fixed, $i = 1, 2, \ldots, k$.)

DEFINITION 2.3 The random variables in $\mathbf{X} = (X_1, \ldots, X_k)'$ are said to be conditionally increasing in sequence (CIS) if for $j = 1, \ldots, k-1$

$$\begin{aligned} &[X_{j+1} | (X_1 = x_1, X_2 = x_2, \ldots, X_j = x_j)] \\ &\leq^P [X_{j+1} | (X_1 = x_1^*, \ldots, X_j = x_j^*)], \end{aligned} \tag{2.3}$$

for $x_1 \leq x_1^*, x_2 \leq x_2^*, \ldots, x_j \leq x_j^*$. (See Tong (1990, p. 92).)

DEFINITION 2.4 The random variables in $\mathbf{X}$ are said to be weak conditionally increasing in sequence ($WCIS$) if for $j = 1, \ldots, k-1$,

$$\begin{aligned} &[(X_{j+1}, \ldots, X_k) | X_1 = x_1, \ldots, X_j = x_j] \\ &\leq^P [(X_{j+1}, \ldots, X_k) | X_1 = x_1, \ldots, X_j = x_j^*], \end{aligned} \tag{2.4}$$

for $x_j \leq x_j^*$.

In other words, $WCIS$ means that for each $j = 1, \ldots, k$, the random vector $(X_{j+1}, \ldots, X_k)$ given $X_1, \ldots, X_j$ need only be stochastically nondecreasing in $X_j = x_j$.

DEFINITION 2.5 The random variables in $\mathbf{X} = (X_1, \ldots, X_k)'$ are said to be (positively) associated (A) if

$$Eh_1(\mathbf{X})h_2(\mathbf{X}) \geq Eh_1(\mathbf{X})Eh_2(\mathbf{X}) \tag{2.5}$$

holds for all nondecreasing functions h_1, h_2 for which the expectations in (2.5) exist.

The relation (2.5) is sometimes called the multivariate correlation inequality.

Now we prove

PROPOSITION 2.6 *The following implications are true:*

$$MTP_2 \overset{(a)}{\Rightarrow} CIS \overset{(b)}{\Rightarrow} WCIS \overset{(c)}{\Rightarrow} A.$$

Furthermore, all implications are strict for $k \geq 3$. (Implication (a) requires $f_{\mathbf{X}}(\mathbf{x}) > 0$.)

PROOF The implication (a) is proven in Tong (1990, p. 95). The implication (c) is proven in CKS (1992), Theorem 2.5. To prove the implication (b), we need to show that (2.3) implies (2.4). Fix j and let $h(x_{j+1}, \ldots, x_k)$ be a nondecreasing function. For each $m = j, \ldots, k$, define

$$h_{(m)}(X_1, \ldots, X_m) = E[h(X_{j+1}, \ldots, X_k) | X_1, \ldots, X_m].$$

We make the following three observations:

$$h_{(j)}(X_1, \ldots, X_j) = E[h(X_{j+1}, \ldots, X_k) | X_1, \ldots, X_j], \tag{2.6}$$

$$\begin{aligned} h_{(m)}(X_1, \ldots, X_m) &= E[h(X_{j+1}, \ldots, X_k) | X_1, \ldots, X_m] \\ &= E\{E[h(X_{j+1}, \ldots, X_k) | X_1, \ldots, X_{m+1}] | X_1, \ldots, X_m\} \\ &= E[h_{(m+1)}(X_1, \ldots, X_{m+1}) | X_1, \ldots, X_m], \\ &\qquad \text{for } m = j, \ldots, k-1, \end{aligned} \tag{2.7}$$

$$h_{(k)}(X_1, \ldots, X_k) = h(X_{j+1}, \ldots, X_k). \tag{2.8}$$

We claim that, due to (2.7), if $h_{(m+1)}$ is nondecreasing as a function of $X_1, \ldots, X_{m+1}$, then $h_{(m)}$ is nondecreasing as a function of $X_1, \ldots, X_m$. To see this, let $x_i \leq x_i^*$, $i = 1, 2, \ldots, m$. Then by (2.7),

$$\begin{aligned} h_{(m)}(x_1, \ldots, x_m) &= E[h_{(m+1)}(X_1, \ldots, X_{m+1}) | X_1 = x_1, \ldots, X_m = x_m] \\ &= E[h_{(m+1)}(x_1, \ldots, x_m, X_{m+1}) | X_1 = x_1, \ldots, X_m = x_m] \\ &\leq E[h_{(m+1)}(x_1^*, \ldots, x_m^*, X_{m+1}) | X_1 = x_1, \ldots, X_m = x_m] \end{aligned} \tag{2.9}$$

whenever $h_{(m+1)}$ is nondecreasing in all its arguments. Furthermore, the *CIS* property implies that the last expression in (2.9) is less than or equal to

$$E[h_{(m+1)}(x_1^*, \ldots, x_m^*, X_{m+1}) | X_1 = x_1^*, \ldots, X_m = x_m^*] = h_{(m)}(x_1^*, \ldots, x_m^*).$$

Thus, $h_{(m+1)}$ nondecreasing implies $h_{(m)}$ nondecreasing for all $m = j, \ldots, k-1$. From (2.8) we see that $h_{(k)}$ is nondecreasing as it is equal to h. Therefore, $h_{(j)}$ is nondecreasing in $X_1, \ldots, X_j$. This completes the proof of implication (b) since (2.4) only requires that $h_{(j)}$ be nondecreasing in x_j.

To show that all implications are strict for $k \geq 3$, it suffices to give a counterexample for each case. Tong (1990, p. 96) gives an example for implication (a).

(b) Let $\mathbf{X}^{3 \times 1} \sim N(\mathbf{0}, \Sigma)$ where

$$\Sigma = \begin{pmatrix} 14 & 8 & 3 \\ 8 & 5 & 2 \\ 3 & 2 & 1 \end{pmatrix}.$$

We claim $\mathbf{X}$ is *WCIS* but not *CIS*. Note first that normal random vectors $\mathbf{U}^{m \times 1}$ and $\mathbf{V}^{m \times 1}$ with the same covariance matrix are such that $\mathbf{U} \leq^P \mathbf{V}$ if and only if $EU_i \leq EV_i$, $i = 1, 2, \ldots, m$. Therefore, to show $\mathbf{X}$ is *WCIS*, we must show $E(X_2, X_3 | X_1 = x_1)$ is nondecreasing in x_1 and $E(X_3 | X_1 = x_1, X_2 = x_2)$ is nondecreasing in x_2.

Now $E(X_2, X_3 | X_1 = x_1) = (\frac{8}{14}, \frac{3}{14})' x_1$ which is increasing in x_1 and

$$E(X_3 | X_1 = x_1, X_2 = x_2) = -(\frac{1}{6}) x_1 + (\frac{2}{3}) x_2, \tag{2.10}$$

which is increasing in x_2. Note also from (2.10) that $E(X_3 | X_1 = x_1, X_2 = x_2)$ is decreasing in x_1 which proves that $\mathbf{X}$ is not *CIS*.

(c) Let $\mathbf{X}^{3 \times 1} \sim N(0, \Sigma)$, where

$$\Sigma = \begin{pmatrix} 5 & 8 & 2 \\ 8 & 14 & 3 \\ 2 & 3 & 1 \end{pmatrix}.$$

Here $E(X_3|X_1 = x_1, X_2 = x_2) = (\frac{2}{3})x_1 - (\frac{1}{6})x_2$ which is decreasing in x_2. Thus this $\mathbf{X}$ is not $WCIS$. Yet clearly $\mathbf{X}$ is A by virtue of Pitt (1982). □

REMARK 2.7 The implication (c) should be interpreted as follows: If there exists a permutation $\{j_1, \dots, j_k\}$ of $\{1, 2, \dots, k\}$ such that $(X_{j_1}, \dots, X_{j_k})$ are $WCIS$, then $(X_1, \dots, X_k)$ are associated.

REMARK 2.8 The proof of implication (b) given above also demonstrates that CIS given in (2.3) for $j = 1, \dots, k-1$ is equivalent to

$$\begin{aligned} &[(X_{j+1}, \dots, X_k)|X_1 = x_1, \dots, X_j = x_j] \\ &\leq^P [(X_{j+1}, \dots, X_k)|X_1 = x_1^*, \dots, X_j = x_j^*]. \end{aligned} \tag{2.11}$$

REMARK 2.9 Let $\mathbf{X}^{k\times 1} \sim N(\underline{\mu}, \Sigma)$. Let $\Sigma = \begin{pmatrix} \Sigma_{11} & \Sigma_{12} \\ \Sigma_{21} & \Sigma_{22} \end{pmatrix}$, where Σ_{22} is of order $p \times p$. Then a necessary and sufficient condition for $\mathbf{X}$ to be CIS is that $\Sigma_{21}\Sigma_{11}^{-1} \geq 0$ for $p = 1, \dots, k-1$. A necessary and sufficient condition for $\mathbf{X}$ to be $WCIS$ is that the last column of $\Sigma_{21}\Sigma_{11}^{-1}$ have all nonnegative elements for $p = 1, \dots, k-1$. The statement follows by noting that

$$E\mathbf{X}^{(2)}|\mathbf{X}^{(1)} = \underline{\mu}^{(2)} + \Sigma_{21}\Sigma_{11}^{-1}(\mathbf{x}^{(1)} - \underline{\mu}^{(1)}),$$

where $\mathbf{X}^{(1)} = (X_1, \dots, X_p)'$, $\mathbf{X}^{(2)} = (X_{p+1}, \dots, X_k)'$.

REFERENCES

COHEN, A., KEMPERMAN, J.H.B., AND SACKROWITZ, H. B. (1992). Positive dependence, stochastic ordering, and unbiased testing in exponential family regression. Submitted.

COHEN, A., PERLMAN, M., AND SACKROWITZ, H. B. (1991). Unbiasedness of tests for homogeneity when alternatives are ordered. *Topics in Statistical Dependence*, H. W. Block, A. R. Sampson and T. H. Savits, eds. Institute of Mathematical Statistics, Hayward, CA. 135-145.

COHEN, A. AND SACKROWITZ, H. B. (1987). Unbiasedness of tests of homogeneity. *Ann. Statist.* **15** 805–816.

PERLMAN, M. AND OLKIN, I. (1980). Unbiasedness of invariant tests for MANOVA and other multivariate problems. *Ann. Statist.* **8** 1326-1341.

PITT, L. D. (1982). Positively correlated normal variables are associated. *Ann. Probab.* **10** 496-499.

TONG, Y. L. (1990). *The Multivariate Normal Distribution.* Springer–Verlag, New York.

DEPARTMENT OF STATISTICS
RUTGERS UNIVERSITY
NEW BRUNSWICK, NJ 08903

Stochastic Inequalities
IMS Lecture Notes – Monograph Series
Volume 22 (1993)

FURTHER ASPECTS OF THE COMPARISON OF TWO GROUPS OF RANKED OBJECTS BY MATCHING IN PAIRS[1]

By H. A. DAVID and JINGYU LIU

Iowa State University

Suppose $\Gamma_X = (X'_{(1)}, \ldots, X'_{(n)})$ and $\Gamma_Y = (Y'_{(1)}, \ldots, Y'_{(n)})$ are two groups of stochastically ordered rv's, representing, say, the increasing strengths of the members of two chess teams. Let $\pi = (\pi_1, \ldots, \pi_n)$ be a permutation of $(1, \ldots, n)$. Then the statistic $S(\pi) = \sum_{i=1}^n I(Y'_{(i)} > X'_{(\pi_i)})$ measures the superiority of Γ_Y over Γ_X in matchings under π, where $I(y > x)$ is an indicator function. The dependence of $ES(\pi) = \sum_{i=1}^n P(Y'_{(i)} > X'_{(\pi_i)})$ on π, especially when $\pi = (1, \ldots, n)$ and when π is randomly given, has been studied in Liu and David (1992) under two different models. After a review of the main results of that paper, some new optimality results are developed. In addition, a threshold model is used to treat tied comparisons.

1. Introduction

Tournaments in which n players or teams are compared by being matched up in pairs have been studied by mathematicians and statisticians at least since Zermelo (1929). The eminent author proposed and examined a method for evaluating the strengths of contestants in a round robin chess tournament that had to be broken off before each pair of players could meet. Independently, statisticians became interested through the connection between tournaments and the method of paired comparisons. In the latter, typically, n flavors are compared by being tasted in pairs, pairwise comparison providing maximal discrimination.

The method of paired comparisons goes back to the psychometrician Thurstone (1927), other notable early contributions being Kendall and Babington Smith (1940), Bradley and Terry (1952), and Kendall (1955). The last paper is perhaps the first to stress the connection between tournaments and paired comparisons, a point pursued in David (1959), where

[1]Work supported by the U. S. Army Research Office.

AMS 1991 *subject classifications.* Primary 62G30, 62J15, 62E15, 15A39; Secondary 62P99.

Key words and phrases. Stochastic inequalities, order statistics, linear preference model, majorization, unimodality, tied comparisons, threshold model.

knock–out tournaments are also studied. Since then many authors have dealt with questions of design and analysis for round robin, knock–out, and related tournament–type situations. A review of this literature is given in David (1988).

The present paper is in the same spirit although it does not, strictly speaking, involve a tournament. We examine the standard method for comparing two chess teams, namely to match pairs of players having the same rank within their teams. Questions that arise are: Is this *ordered matching* really an optimal procedure and what are its properties? How does it compare with random matching, or with other possible matchings of the two teams? What kind of matchings are fair? Some answers are provided in Liu and David (1992) under two different probability models. After reviewing the main features of that paper, referred to as LD from here on, we present some new optimality results and deal with the complication of tied games (draws in chess).

It should be noted that, in more general terms, we are concerned with the nonparametric comparison of two groups of n objects by pairwise matching, when there is good knowledge of the within–group ranking.

Further results are given in Liu (1991).

2. The Probability Models

Let $\Gamma_X = (X'_{(1)}, \ldots, X'_{(n)})$ and $\Gamma_Y = (Y'_{(1)}, \ldots, Y'_{(n)})$ be two groups of stochastically increasing random variables (not necessarily order statistics) which represent the increasing "strengths" of the ordered objects in the two groups. Let $F_i(x)$ and $G_i(x)$ represent the continuous cdf's of $X'_{(i)}$ and $Y'_{(i)}$, respectively, $i = 1, 2, \ldots, n$. Here we assume $F_i(x) \geq F_j(x)$ for all x and any $1 \leq i < j \leq n$, i.e., $X'_{(i)}$ is stochastically smaller than $X'_{(j)}$ or $X'_{(i)} \leq_{\text{st}} X'_{(j)}$. Usually, we also assume that Γ_X and Γ_Y are independent; however, we do not assume independence within Γ_X and Γ_Y.

Let $\pi = (\pi_1, \ldots, \pi_n)$ be a permutation of $\pi_0 = (1, \ldots, n)$. Then for each given π, we can define a matching by comparing $X'_{(\pi_i)}$ and $Y'_{(i)}$, $i = 1, 2, \ldots, n$. So we will simply speak of a matching π. Correspondingly, we suppose that in a particular matching hypothetical realizations $x'_{(\pi_i)}$ of $X'_{(\pi_i)}$, and $y'_{(i)}$ of $Y'_{(i)}$, $i = 1, \ldots, n$, are compared. While $x'_{(\pi_i)}$ or $y'_{(i)}$, cannot be observed, we can make the (usually subjective) judgment whether $y'_{(i)} > x'_{(\pi_i)}$. For the moment, we assume a clear decision; the possibility of a tie is considered in Section 5. Thus we prefer the Y–group, Γ_Y, to the X–group,

Γ_X, in this particular matching if

$$\sum_{i=1}^{n} I(y'_{(i)} > x'_{(\pi_i)}) > \frac{1}{2}\, n$$

where

$$I(y > x) = \begin{cases} 0 & \text{if } y \leq x \\ 1 & \text{if } y > x \end{cases}.$$

Write

$$S(\pi) = \sum_{i=1}^{n} I(Y'_{(i)} > X'_{(\pi_i)}).$$

Then $S(\pi)$ denotes the random number of preferences for objects in Γ_Y. We regard Γ_Y as *superior* to Γ_X under matching π if the expectation

$$E[S(\pi)] = \sum_{i=1}^{n} P(Y'_{(i)} > X'_{(\pi_i)}) > \frac{1}{2}\, n.$$

It is clear that some matchings may favor one of the groups. Both ordered and random matching are clearly *fair*, i.e., $E[S(\pi)] = \frac{1}{2}\, n$, when Γ_X and Γ_Y have the same distribution. Other matchings are not necessarily fair (see LD). Let V_1 and V_2 be the values of $E[S(\pi)]$ under ordered and random matching, respectively.

Our first model is the *order statistics model.* In this model we assume that $X'_{(i)}$ and $Y'_{(i)}$ have the same *marginal* distributions as $X_{(i)}$ and $Y_{(i)}$, the ith order statistics in random samples of size n from F and G, respectively. We use $X'_{(i)}$ rather than $X_{(i)}$ since we generally want to permit $P(X'_{(i)} > X'_{(j)}) > 0$ for $i < j$. The joint distribution of the $X'_{(i)}$ may, in fact, have any dependence structure, including independence.

However, our measure of superiority of Γ_Y over Γ_X, viz. $E[S(\pi)]$, depends only on the marginal distributions of $X'_{(i)}$ and $Y'_{(j)}$. We will therefore replace $X'_{(i)}$ and $Y'_{(j)}$ by the order statistics $X_{(i)}$ and $Y_{(j)}$ from here on in discussions of the order statistics model. For an ordered matching we have

$$V_1 = E[S(\pi^0)] = \sum_{i=1}^{n} P(Y_{(i)} > X_{(i)}).$$

Note that under random matching we simply have $V_2 = nP(Y > X)$, with $X \sim F$ and $Y \sim G$.

The question of whether ordered matching in the order statistics model is more effective than random matching may now be reduced to the question of whether $V_1 \geq V_2$ if $X \leq_{\text{st}} Y$. The answer is yes under certain conditions. We deal with this and related issues in Section 3.

The second model is the *linear preference model.* In this model, we assume that

$$(1) \qquad X'_{(i)} \sim F(x - \lambda_{(i)}) \quad \text{and} \quad Y'_{(i)} \sim F(x - \mu_{(i)}), \quad i = 1, \ldots, n$$

where $F(x)$ is a continuous distribution function and $\lambda_{(1)} \leq \cdots \leq \lambda_{(n)}$ and $\mu_{(1)} \leq \cdots \leq \mu_{(n)}$. The model is based on the linear model much used in the method of paired comparisons (e.g., David, 1988, p. 7). At times, we will assume that $F(x)$ is in the class of unimodal distribution functions, that contains almost all the common useful distribution functions.

It is easy to see that when $\mu_{(i)} = \lambda_{(i)}$, $i = 1, \ldots, n$, and both Γ_X and Γ_Y are groups of independent random variables, $S(\pi^0)$ has a Binomial $\left(n, \frac{1}{2}\right)$ distribution. In general, there is no closed form for the distribution of $S(\pi)$.

In this model, we are still interested in comparing V_1 and V_2, as well as V_1 and $E[S(\pi)]$. Under certain conditions, we get similar results to those in the order statistics model. However, there are significant differences between the two models.

3. Order Statistics Model

We consider the case $G(x) = F(x - \mu)$, where $\mu \geq 0$, and write $p_{ij} = P(Y'_{(i)} > X'_{(j)}) = P(Y_{(i)} > X_{(j)})$. The following basic result is obtained in LD.

THEOREM 3.1 *Let* $(X_{(1)}, \ldots, X_{(n)})$ *and* $(Y_{(1)}, \ldots, Y_{(n)})$ *be the order statistics in two independent random samples from populations with continuous cdf's* $F(x)$ *and* $F(x - \mu)$, *respectively, where* $\mu \geq 0$. *Then for any* $1 \leq i, j \leq n$, *we have*

$$p_{ii} + p_{jj} \geq p_{ij} + p_{ji}.$$

Under the conditions of the theorem it is then shown that $V_1 \geq V_2$, meaning that ordered matching has at least as much power to identify the stronger group Γ_Y as does random matching.

We now need two definitions

DEFINITION 3.1 π is said to be a *simple matching* (or permutation) if it can be obtained from π^0 by interchanging pairs of the components of π^0, with no component involved in more than one interchange.

DEFINITION 3.2 $\pi = (\pi_1, \ldots, \pi_n)$ is said to be a *symmetric matching* (or permutation) if $\pi_{n-i+1} = n - \pi_i + 1$ for $i = 1, \ldots, n$.

For example, (3,4,1,2) is a simple matching and (2,4,1,3) is a symmetric matching.

It is shown in LD that simple and symmetric matchings, as well as combinations thereof, are fair.

Moreover, the following result is obtained.

THEOREM 3.2 *Under the conditions of* Theorem 3.1

(a)
$$\sum_{i=1}^{n} p_{ii} \leq \sum_{i=1}^{n} p_{i,\pi_i}$$

for any simple permutation π.

(b) *If* $F(x)$ *is the cdf of a symmetric rv, then* (a) *holds for any symmetric permutation* π.

4. Linear Preference Model

The specific assumptions for this model have been given in (1). If X and Y are iid with cdf $F(x)$, then $X'_{(i)} \stackrel{d}{=} X + \lambda_{(i)}$ and $Y'_{(i)} \stackrel{d}{=} Y + \mu_{(i)}$. Also $X - Y$ is symmetrically distributed about zero, with cdf $U(x)$, say.

We have

$$P(Y'_{(i)} > X'_{(i)}) = U(\mu_{(i)} - \lambda_{(i)})$$

and

$$E[S(\pi)] = \sum_{i=1}^{n} P(Y'_{(i)} > X'_{(\pi_i)}) = \sum_{i=1}^{n} U(\mu_{(i)} - \lambda_{(\pi_i)}).$$

It is easy to see that ordered, random, and simple matching is still fair, i.e., $E[S(\pi)] = \frac{n}{2}$ if $\lambda_{(i)} = \mu_{(i)}$, $i = 1, \ldots, n$. Symmetric matching is also fair for any symmetric spacing, i.e., $\lambda_{(i)} = \mu_{(i)}$ and $\lambda_{(i)} + \lambda_{(n-i+1)} =$ constant, $i = 1, \ldots, n$.

For this model we now assume that $F(x)$ is a unimodal cdf, that is, there exists x_0 such that $F(x)$ is convex on $(-\infty, x_0)$ and concave on (x_0, ∞). With X, Y iid unimodal, $X - Y$ is also unimodal (e.g., Dharmadhikari and Joag-dev (1988, p. 15)). This is needed for the proof in LD of the following result.

THEOREM 4.1 *Let* $(X'_{(1)}, X'_{(2)})$ *and* $(Y'_{(1)}, Y'_{(2)})$ *be independent with* $X'_{(i)} \sim F(x - \lambda_{(i)})$ *and* $Y'_{(i)} \sim F(x - \mu_{(i)})$, $i = 1, 2$, *where* $\lambda_{(1)} \leq \lambda_{(2)}$, $\mu_{(1)} \leq \mu_{(2)}$, *and* $F(x)$ *is an absolutely continuous unimodal distribution. If* $\mu_{(1)} + \mu_{(2)} \geq \lambda_{(1)} + \lambda_{(2)}$, *then we have*

$$U(\mu_{(1)} - \lambda_{(1)}) + U(\mu_{(2)} - \lambda_{(2)}) \geq U(\mu_{(1)} - \lambda_{(2)}) + U(\mu_{(2)} - \lambda_{(1)}) \tag{2}$$

Note that if either $\mu_{(1)} = \mu_{(2)}$ or $\lambda_{(1)} = \lambda_{(2)}$, then equality holds in (2).

For any fixed $\lambda_{(1)} \le \cdots \le \lambda_{(n)}$ and $\mu_{(1)} \le \cdots \le \mu_{(n)}$, let $\tilde{\pi} = (\tilde{\pi}_1, \ldots, \tilde{\pi}_n)$ be a permutation for which

$$E[S(\tilde{\pi})] = \max_{\pi} \sum_{i=1}^{n} U(\mu_{(i)} - \lambda_{(\pi_i)}).$$

Theorem 4.1 tells us that if $\sum_{i=1}^{n} \mu_{(i)} \ge \sum_{i=1}^{n} \lambda_{(i)}$, then $\pi^0 = \tilde{\pi}$ for $n = 2$. However, in contrast to the order statistics model, $\pi^0 = \tilde{\pi}$ no longer necessarily holds for $n > 2$, even within the class of simple matchings.

In general, $\sum_{i=1}^{n} \mu_{(i)} \ge \sum_{i=1}^{n} \lambda_{(i)}$ does not imply

$$\sum_{i=1}^{n} U(\mu_{(i)} - \lambda_{(i)}) \ge \frac{n}{2}. \tag{3}$$

The following sufficient condition is established in LD.

THEOREM 4.2 *Let* $\{a_1, \ldots, a_m\} = \{\mu_{(i)} - \lambda_{(i)}, \text{ s.t. } \mu_{(i)} - \lambda_{(i)} \ge 0,\ i = 1, \ldots, n\}$ *and* $\{b_1, \ldots, b_{n-m}\} = \{\lambda_{(i)} - \mu_{(i)}, \text{ s.t. } \mu_{(i)} - \lambda_{(i)} < 0,\ i = 1, \ldots, n\}$. *If* $m \ge \left[\frac{n}{2}\right]$ *and*

$$\sum_{i=1}^{k} a_{(i)} \ge \sum_{i=1}^{k} b_{(i)} \qquad (k = 1, \ldots, n-m), \tag{4}$$

then (3) *holds.*

It is also shown in LD that if $\mu_{(i)} + \mu_{(j)} \ge \lambda_{(i)} + \lambda_{(j)}$ for all $1 \le i < j \le n$, then ordered matching is superior to random and simple matching.

4.1. *Some New Optimality Results*

The following preliminaries are needed (e.g., Marshall and Olkin (1979)).

DEFINITION 4.1 Write $x_{[i]} = x_{(n+1-i)}$, $i = 1, \ldots, n$. Then for any $x, y \in R^n$, x is majorized by y ($x \prec y$) if

(a) $\sum_{i=1}^{k} x_{[i]} \le \sum_{i=1}^{k} y_{[i]} \qquad k = 1, \ldots, n-1$

(b) $\sum_{i=1}^{n} x_{[i]} = \sum_{i=1}^{n} y_{[i]}$.

LEMMA 4.1 (Hardy, Littlewood, and Pólya (1952)). *The inequality* $\sum_{i=1}^{n} g(x_i) \le \sum_{i=1}^{n} g(y_i)$ *holds for all continuous convex functions* $g : \mathbb{R} \to \mathbb{R}$ *if and only if* $x \prec y$.

We have noted a variety of situations in which ordered matching is most effective in detecting the superior group. Now we obtain a sufficient condition for a matching $\tilde{\pi} = (\tilde{\pi}_1, \ldots, \tilde{\pi}_n)$ to be optimal, i.e., $\tilde{\pi}$ to be such that

$$\sum_{i=1}^{n} U(\mu_{(i)} - \lambda_{(\tilde{\pi}_i)}) \geq \sum_{i=1}^{n} U(\mu_{(i)} - \lambda_{(\pi_i)}) \tag{5}$$

for any permutation π.

Let

$$\{x_1, \ldots, x_p\} = \{\mu_{(i)} - \lambda_{(\tilde{\pi}_i)} \text{ s.t. } \mu_{(i)} - \lambda_{(\tilde{\pi}_i)} > 0 \quad i = 1, 2, \ldots, n\},$$

$$\{y_1, \ldots, y_q\} = \{\lambda_{(\tilde{\pi}_i)} - \mu_{(i)} \text{ s.t. } \lambda_{(\tilde{\pi}_i)} - \mu_{(i)} > 0 \quad i = 1, 2, \ldots, n\}.$$

Similarly, let

$$\{x'_1, \ldots, x'_{p'}\} = \{\mu_{(i)} - \lambda_{(\pi_i)} \text{ s.t. } \mu_{(i)} - \lambda_{(\pi_i)} > 0 \quad i = 1, 2, \ldots, n\},$$

$$\{y'_1, \ldots, y'_{q'}\} = \{\lambda_{(\pi_i)} - \mu_{(i)} \text{ s.t. } \lambda_{(\pi_i)} - \mu_{(i)} > 0 \quad i = 1, 2, \ldots, n\}.$$

Then, since $U(-x) = 1 - U(x)$, we have

$$\begin{aligned} \sum_{i=1}^{n} U(\mu_{(i)} - \lambda_{(\tilde{\pi}_i)}) &= \sum_{j=1}^{p} U(x_j) + q - \sum_{j=1}^{q} U(y_j) + (n - p - q)U(0) \\ &= \sum_{j=1}^{p} U(x_j) - \sum_{j=1}^{q} U(y_j) + (n - p + q)U(0) \end{aligned}$$

and

$$\sum_{i=1}^{n} U(\mu_{(i)} - \lambda_{(\pi_i)}) = \sum_{j=1}^{p'} U(x'_{(j)}) - \sum_{j=1}^{q'} U(y'_{(j)}) + (n - p' + q')U(0).$$

Therefore (5) is equivalent to

$$\begin{aligned} &\sum_{j=1}^{p} U(x_j) \quad + \quad \sum_{j=1}^{q'} U(y_{j'}) + (n - p + q)U(0) \\ \geq &\sum_{j=1}^{p'} U(x_{j'}) \quad + \quad \sum_{j=1}^{q} U(y_j) + (n - p' + q')U(0). \end{aligned}$$

Write

$$a_i = x_i \qquad (i = 1, \ldots, p), \qquad a_{p+i} = y'_i \qquad (i = 1, \ldots, q'),$$

$$a_{p+q'+i} = 0 \qquad (i = 1, \ldots, n - p + q),$$

$$b_i = x'_i \quad (i = 1, \ldots, p'), \qquad b_{p'+i} = y'_i \quad (i = 1, \ldots, q),$$
$$b_{p'+q+i} = 0 \quad (i = 1, \ldots, n - p' + q').$$

By

$$\sum_{i=1}^{n} (\mu_{(i)} - \lambda_{(\tilde{\pi}_i)}) = \sum_{i=1}^{n} (\mu_{(i)} - \lambda_{(\pi_i)}),$$

we have $\sum_{i=1}^{n+q+q'} a_i = \sum_{i=1}^{n+q+q'} b_i$. Also, we can write (5) as

$$\sum_{i=1}^{n+q+q'} U(a_i) \geq \sum_{i=1}^{n+q+q'} U(b_i). \tag{6}$$

Now $-U(x)$ is a convex function on $[0, +\infty)$. Therefore, by Lemma 4.1, we see that (6) holds if $\sum_{i=1}^{k} a_{[i]} \leq \sum_{i=1}^{k} b_{[i]}$ for $k = 1, 2, \ldots, n + q + q' - 1$. Summarizing the above argument, we have the following result.

THEOREM 4.3 *A sufficient condition for $\tilde{\pi}$ such that* (5) *holds is*

$$\sum_{i=1}^{k} a_{[i]} \leq \sum_{i=1}^{k} b_{[i]} \quad \textit{for } k = 1, \ldots, n + q + q' - 1.$$

If there exist $\mu_{(i)}$ and $\mu_{(j)}$ such that $\mu_{(i)} + \mu_{(j)} > \lambda_{(\pi_i)} + \lambda_{(\pi_j)}$, where $i < j$ and $\pi_i > \pi_j$, then by Theorem 4.1, we can increase $\sum_{i=1}^{n} U(\mu_{(i)} - \lambda_{(\pi_i)})$ by interchanging $\lambda_{(\pi_i)}$ and $\lambda_{(\pi_j)}$. Also, if there exist $\mu_{(i)}$ and $\mu_{(j)}$ such that $\mu_{(i)} + \mu_{(j)} < \lambda_{(\pi_i)} + \lambda_{(\pi_j)}$, where $i < j$ and $\pi_i < \pi_j$, we can increase $\sum_{i=1}^{n} U(\mu_{(i)} - \lambda_{(\pi_i)})$ by interchanging $\lambda_{(\pi_i)}$ and $\lambda_{(\pi_j)}$. Therefore, Theorem 3.1 gives us a way to increase $\sum_{i=1}^{n} U(\mu_{(i)} - \lambda_{(\pi_i)})$. We can also see that if $\sum_{i=1}^{n} U(\mu_{(i)} - \lambda_{(\tilde{\pi}_i)}) > \sum_{i=1}^{n} U(\mu_{(i)} - \lambda_{(\pi_i)})$ for any other permutation π, then for any $i < j$, either $\mu_{(i)} + \mu_{(j)} > \lambda_{(\tilde{\pi}_i)} + \lambda_{(\tilde{\pi}_j)}$ with $\tilde{\pi}_i < \tilde{\pi}_j$ or $\mu_{(i)} + \mu_{(j)} < \lambda_{(\tilde{\pi}_i)} + \lambda_{(\tilde{\pi}_j)}$ with $\tilde{\pi}_i > \tilde{\pi}_j$.

We therefore have the following Lemma.

LEMMA 4.2 *Suppose there exists a unique permutation $\tilde{\pi}$ such that for any $i < j$, if $\mu_{(i)} + \mu_{(j)} > \lambda_{(\tilde{\pi}_i)} + \lambda_{(\tilde{\pi}_j)}$, then $\tilde{\pi}_i < \tilde{\pi}_j$, and if $\mu_{(i)} + \mu_{(j)} < \lambda_{(\tilde{\pi}_i)} + \lambda_{(\tilde{\pi}_j)}$, then $\tilde{\pi}_i > \tilde{\pi}_j$. For such a $\tilde{\pi}$ it follows that* (5) *holds for any permutation π.*

5. The Treatment of Ties

In practice, some pairwise comparisons may result in ties. A tie occurs when the performances x and y of two objects being compared are too close to tell the difference. Accordingly, we introduce the indicator function

$$I(y > x; \tau) = \begin{cases} 0 & \text{if } y - x < -\tau \\ 1/2 & \text{if } |y - x| \leq \tau \\ 1 & \text{if } y - x > \tau \end{cases}$$

where $\tau (\geq 0)$ is called a *threshold parameter* (Glenn and David (1960)).

5.1. *Order Statistics Model*

For any permutation $\pi = (\pi_1, \ldots, \pi_n)$, we now define

$$S_\tau(\pi) = \sum_{i=1}^{n} I(Y_{(i)} > X_{(\pi_i)}; \tau)$$

as a measure of the performance of Γ_Y relative to Γ_X under the matching π, with ties permitted. Then

$$\begin{aligned} (7) \qquad ES_\tau(\pi) &= \sum_{i=1}^{n} [P(Y_{(i)} > X_{(\pi_i)} + \tau) + \frac{1}{2} P(|Y_{(i)} - X_{(\pi_i)}| \leq \tau)] \\ &= \frac{1}{2} \sum_{i=1}^{n} [P(Y_{(i)} > X_{(\pi_i)} + \tau) + P(Y_{(i)} > X_{(\pi_i)} - \tau)]. \end{aligned}$$

For $\pi^0 = (1, \ldots, n)$ corresponding to ordered matching, we write

$$(8) \qquad V_1^\tau = ES_\tau(\pi^0).$$

Let V_2^τ be the expectation of $S_\tau(\pi)$ under random matching. Then we have

$$\begin{aligned} V_2^\tau &= \frac{1}{n} \sum_{i=1}^{n} \sum_{j=1}^{n} EI(Y_{(i)} > X_{(j)}; \tau) \\ &= \frac{1}{2n} \left\{ \sum_{i=1}^{n} \sum_{j=1}^{n} [P(Y_{(i)} > X_{(\pi_i)} + \tau) + P(Y_{(i)} > X_{(\pi_i)} - \tau)] \right\} \end{aligned}$$

so that

$$(9) \qquad V_2^\tau = \frac{n}{2} [P(Y > X + \tau) + P(Y > X - \tau)],$$

where X and Y are independent with respective cdf's $F(x)$ and $F(x - \mu)$.

The questions that concern us here are how V_1^τ and V_2^τ compare, and also their relations to V_1 and V_2, respectively.

LEMMA 5.1 *Let $X \sim F(x)$, where $F(x)$ is an absolutely continuous unimodal cdf. If $Y \stackrel{d}{=} X + \mu$ with $\mu \geq 0$, and X and Y are independent, then for any $\tau \geq 0$,*

$$P(Y > X) \geq \frac{1}{2}\,[P(Y > X + \tau) + P(Y > X - \tau)]. \tag{10}$$

PROOF Let $U(x)$ be the cdf of $X_1 - X_2$, where X_1 and X_2 are iid with cdf $F(x)$. As noted earlier, $X_1 - X_2$ is symmetrically distributed about 0 and $U(x)$ is a unimodal cdf. Then (10) may be written

$$U(\mu) \geq \frac{1}{2}\,[U(\mu + \tau) + U(\mu - \tau)],$$

a result which is evident from the unimodality of U. □

COROLLARY *If $X_{(i)}$ has a unimodal distribution ($i = 1, \ldots, n$), then under the conditions of* Lemma 5.1

$$P(Y_{(i)} > X_{(i)}) \geq \frac{1}{2}\,[P(Y_{(i)} > X_{(i)} + \tau) + P(Y_{(i)} > X_{(i)} - \tau)].$$

We have thus shown that under the conditions stated $V_2 > V_2^{\tau}$ and $V_1 > V_1^{\tau}$. For the corollary we need $X_{(i)}$ to have a unimodal distribution. Alam (1972) shows that if the density function $f(x)$ of X satisfies the condition that $1/f(x)$ is convex, then the order statistics have unimodal distributions. This condition is satisfied by the following distributions (among others): normal, logistic, Cauchy, uniform, and the gamma and Weibull families for shape parameters ≥ 1.

THEOREM 5.1 *Let X and Y be independent absolutely continuous rv's with respective cdf's $F(x)$ and $F(x - \mu)$, where $\mu \geq 0$.*

(a) If $0 \leq \tau \leq \mu$, then $V_1^{\tau} \geq V_2^{\tau}$ and

$$V_1^{\tau} \geq ES_{\tau}(\pi) \tag{11}$$

for any simple permutation π. If X is a symmetric rv, then (11) *holds also for any symmetric permutation.*

(b) If $\mu = 0$ and $\tau \geq 0$, then

$$ES_{\tau}(\pi) = \frac{1}{2}\,n \tag{12}$$

for any simple permutation. If X is a symmetric rv, then (12) *holds also for any symmetric permutation.*

PROOF (a) From (7) we have

$$V_1^\tau = \frac{1}{2}\sum_{i=1}^{n}[P(Y_{(i)} - \tau > X_{(i)}) + P(Y_{(i)} + \tau > X_{(i)})].$$

Note that $Y_{(i)} \mp \tau$ $(i = 1, \ldots, n)$ is the ith order statistic from a population with cdf $F[x - (\mu \mp \tau)]$, where $\mu \mp \tau \geq 0$. Then since $V_1 \geq V_2$, we have

$$\sum_{i=1}^{n} P(Y_{(i)} > X_{(i)} \pm \tau) \geq nP(Y > X \pm \tau),$$

which establishes $V_1^\tau \geq V_2^\tau$.

The remaining results in (a) follow from Theorem 3.2 and (b) is easily proved. □

Numerical work for the standard normal shows that $V_1^\tau \geq V_2^\tau$ does not necessarily hold if $\tau > \mu$.

5.2. *Linear Preference Model*

Corresponding to (7) we now have

$$\begin{aligned} ES_\tau(\pi) &= \sum_{i=1}^{n}[P(Y'_{(i)} > X'_{(\pi_i)} + \tau) + \frac{1}{2}P(|Y'_{(i)} - X'_{(\pi_i)}| \leq \tau)] \\ &= \frac{1}{2}\sum_{i=1}^{n}[U(\mu_{(i)} - \lambda_{(\pi_i)} - \tau) + U(\mu_{(i)} - \lambda_{(\pi_i)} + \tau)] \end{aligned}$$

Then $V_1^\tau = ES_\tau(\pi^0)$ and

$$V_2^\tau = \frac{1}{2n}\sum_{i=1}^{n}\sum_{j=1}^{n}[U(\mu_{(i)} - \lambda_{(j)} - \tau) + U(\mu_{(i)} - \lambda_{(j)} + \tau)],$$

which does not simplify further, in contrast to (9).

The next lemma follows directly from Theorem 4.1.

LEMMA 5.2 *If* $\mu_{(i)} + \mu_{(j)} \geq \lambda_{(i)} + \lambda_{(j)} + 2\tau$ *for any* $1 \leq i, j \leq n$, *then*

(a) $V_1^\tau \geq V_2^\tau$;

(b) $V_1^\tau \geq ES_\tau(\pi)$ *for any simple matching* π.

We conclude with the following easily proved result.

LEMMA 5.3 *(a) If* $\lambda_{(i)} = \mu_{(i)}$ *for* $i = 1, 2, \ldots, n$, *then*

$$ES_T(\pi) = \frac{1}{2}\, n \tag{13}$$

for any simple matching π.

(b) If $\lambda_{(i)} = \mu_{(i)}$ *and* $\lambda_{(i+1)} - \lambda_{(i)} = \lambda_{(n-i+1)} - \lambda_{(n-i)}$ *for* $i = 1, \ldots, n$ *then* (13) *holds for any symmetric matching* π.

REFERENCES

ALAM, K. (1972). Unimodality of the distribution of an order statistic. *Ann. Math. Statist.* **43** 2041–4.

BRADLEY, R. A. AND TERRY, M. E. (1952). The rank analysis of incomplete block designs. I. The method of paired comparisons. *Biometrika* **39** 324–45.

DAVID, H. A. (1959). Tournaments and paired comparisons. *Biometrika* **46** 139–49.

DAVID, H. A. (1988). *The Method of Paired Comparisons.* 2nd ed., Charles Griffin, London; Oxford University Press, New York.

DHARMADHIKARI, S. AND JOAG–DEV, K. (1988). *Unimodality, Convexity, and Applications.* Academic Press, Boston, MA.

GLENN, W. A. AND DAVID, H. A. (1960). Ties in paired–comparison experiments using a modified Thurstone–Mosteller model. *Biometrics* **16** 86–109.

HARDY, G. H., LITTLEWOOD, J. E. AND PÓLYA, G. (1952). *Inequalities.* 2nd ed. Cambridge University Press, London and New York.

KENDALL, M. G. (1955). Further contributions to the theory of paired comparisons. *Biometrics* **11** 43–62.

KENDALL, M. G. AND SMITH, B. B. (1940). On the method of paired comparisons. *Biometrika* **31** 324–45.

LIU, J. (1991). Comparing Two Groups of Ranked Objects by Pairwise Matching. Ph.D. Thesis, Iowa State University.

LIU, J. AND DAVID, H. A. (1992). Comparing two groups of ranked objects by pairwise matching. *J. Statist. Plann. Inf.* **33** (to appear).

MARSHALL, A. W. AND OLKIN, I. (1979). *Inequalities: Theory of Majorization and its Applications.* Academic Press, New York.

THURSTONE, L. L. (1927). A law of comparative judgment. *Psychol. Rev.* **34** 273–86.

ZERMELO, E. (1929). Die Berechnung der Turnier–Ergebnisse als ein Maximumproblem der Wahrscheinlichkeitsrechnung. *Math. Zeit.* **29** 436–60.

DEPARTMENT OF STATISTICS
IOWA STATE UNIVERSITY
AMES, IA 50011

Stochastic Inequalities
IMS Lecture Notes – Monograph Series
Volume 22 (1993)

INEQUALITIES FOR THE PARAMETERS $\lambda(F)$,$\mu(F)$ WITH APPLICATIONS IN NONPARAMETRIC STATISTICS[1]

By CRISANTO DORADO and MYLES HOLLANDER

Florida State University

The parameter $\lambda(F) = P(X_1 < X_2+X_3-X_4; X_1 < X_5+X_6-X_7)$, where $X_1,\ldots,X_7$ are independent and identically distributed (iid) according to a continous distribution F, was first considered by Lehmann (1964) in the context of certain nonparametric methods for the two–way layout. The parameter $\mu(F) = P(X_1 < X_2; X_1 < X_3 + X_4 - X_5)$ was first studied by Hollander (1966), also in the context of nonparametric techniques for the two–way layout. The best known bounds on these probabilities are

$$.28254 \approx 89/315 \leq \lambda(F) \leq 7/24 \approx .29167,$$

and

$$3/10 \leq \mu(F) \leq (\sqrt{2}+6)/24 \approx .30893.$$

The upper bound on $\lambda(F)$ is due to Lehmann (1964), the lower bound on $\lambda(F)$ to Spurrier (1991), the upper bound on $\mu(F)$ to Hollander (1967), and the lower bound on $\mu(F)$ to Spurrier (1991). We briefly review the development of these bounds and then present some new applications motivated by the recent bounds due to Spurrier. The applications include studying the extent to which the new bounds can improve large sample approximations to certain nonparametric test statistics and providing tighter upper and lower bounds on certain correlation coefficients involving these parameters.

1. Introduction

Consider the two–way layout with one observation per cell. Let

$$(1.1)\quad X_{ij} = \mu + b_i + \theta_j + e_{ij}, i = 1,\ldots,n, j = 1,\ldots,k \ (\Sigma b_i = \Sigma\theta_j = 0)$$

where the θ's are the parameters of interest, the b's are the nuisance parameters, and the e's are iid according to a common continuous distribution F. Let $Y_{uv}^{(i)} = |X_{iu} - X_{iv}|$ and $R_{uv}^{(i)}$ = rank of $Y_{uv}^{(i)}$ in the ranking from

[1]Research supported by U.S. Air Force Office of Scientific Research Grant 91–0048.
AMS 1991 *subject classifications.* 60E15, 62G99.
Key words and phrases. $\lambda(F)$ inequalities, $\mu(F)$ inequalities, ordered alternatives test, selecting the best treatment.

least to greatest of $\{Y_{uv}^{(i)}\}_{i=1}^{n}$. The Wilcoxon signed–rank statistic between treatments u and v is

$$T_{uv} = \sum_{i=1}^{n} R_{uv}^{(i)} \Psi_{uv}^{(i)}$$

where

$$\Psi_{uv}^{(i)} = \begin{cases} 1 & \text{if } X_{iu} < X_{iv} \\ 0 & \text{otherwise} \end{cases}$$

Hollander (1966) showed that under $H_0 : \theta_1 = \theta_2 = \cdots = \theta_k$, the null correlation coefficient $\rho_0^n(F)$ between T_{uv} and T_{uw} ($u \neq v$, $u \neq w$) is given by

$$\begin{aligned} \rho_0^n(F) &= [(24\lambda(F) - 6)n^2 + (48\mu(F) - 72\lambda(F) + 7)n \\ &\quad + (48\lambda(F) - 48\mu(F) + 1)][(n+1)(2n+1)]^{-1} \end{aligned} \tag{1.2}$$

where

$$\mu(F) = P(X_1 < X_2; X_1 < X_3 + X_4 - X_5) \tag{1.3}$$

and

$$\lambda(F) = P(X_1 < X_2 + X_3 - X_4; X_1 < X_5 + X_6 - X_7) \tag{1.4}$$

where $X_1, \ldots, X_7$ are iid according to F.

To our knowledge this was the first appearance of the parameter $\mu(F)$. Lehmann (1964) had introduced $\lambda(F)$ earlier in a related context. Lehmann considered

$$Y_{st} = \text{med}[\frac{1}{2}(X_{is} - X_{it} + X_{js} - X_{jt})]$$

where the median is over all $i \leq j$, as an estimator of $\theta_s - \theta_t$. Let G denote the common distribution of the difference between two e's. Then if G has a density g satisfying the regularity conditions of Lemma 3(a) of Hodges and Lehmann (1961), Lehmann showed that the joint limiting distribution of the Y_{st}'s is the $\binom{k}{2}$–variate normal distribution with zero mean and covariance matrix $\Sigma^* = (\sigma_{st,uv}^*)$ where the variances are given by

$$\sigma_{st,st}^* = 1/12(\int g^2(x)dx)^2 \quad \text{all } s,t \tag{1.5}$$

and the covariances by

$$\begin{aligned} \sigma_{st,uv}^* &= 0 && \text{if } s,t,u,v \text{ are distinct} \\ &= \{\lambda(F) - 1/4\}/\{\int g^2(x)dx\}^2 && \text{if } s = u \text{ or } t = v \\ &= \{1/4 - \lambda(F)\}/\{\int g^2(x)dx\}^2 && \text{if } s = v \text{ or } t = u\,. \end{aligned} \tag{1.6}$$

The parameters $\mu(F)$, $\lambda(F)$ try very hard to be distribution–free but just don't make it. To get a feel for these parameters consider first

$$\alpha(F) = P(X_1 < X_2; X_1 < X_3)$$

where X_1, X_2, X_3 are iid according to the continuous distribution F. Of course $\alpha(F)$ is distribution–free being equal to 1/3, the chance that X_1 is the smallest of X_1, X_2, X_3.

Consider

$$\mu(F) = P(X_1 < X_2; X_1 < X_3 + X_4 - X_5)$$

and note that $X_3 + X_4 - X_5$ is a little like X_3 itself (because $X_4 - X_5$ is symmetric about 0) but the variance of $X_3 + X_4 - X_5$ is three times that of X_3. Thus it is slightly harder for X_1 to be simultaneously smaller than both X_2 and $X_3 + X_4 - X_5$ than it is to be simultaneously smaller than X_2 and X_3. Thus we should expect the value of $\mu(F)$ to be pulled slightly below the 1/3 value of $\alpha(F)$. A similar argument indicates the values of $\lambda(F)$ should be pulled down even further. That this is the case can be observed in Table 1.

F	Uniform	Normal	Logistic	Exponential	Cauchy
$\mu(F)$	0.3083	0.3075	0.3064	0.3056	0.3043
$\lambda(F)$	0.2909	0.2902	0.2898	0.2894	0.2879

Table 1. Values of $\mu(F)$ and $\lambda(F)$ for Various Distributions

In Section 2 we review the best known bounds for $\mu(F)$ and $\lambda(F)$, and indicate how they were derived. Section 3 is devoted to some applications. The bounds, motivated by the improved bounds due to Spurrier (1991), enable us to

(i) present improved upper and lower bounds for the correlation coefficient $\rho_0^n(F)$ defined by (1.2),

(ii) present improved upper and lower bounds for the correlation coefficient $r_0^n(F)$ defined by (2.3) of Section 2,

(iii) improve Hollander's (1967) estimator of the asymptotic null variance of his Y–test for ordered alternatives (see Section 3.3) and compare observed levels of this modified test with their asymptotic nominal values,

(iv) fine–tune Hsu's (1982) selection procedure for the best treatment (see Section 3.4) and compare observed coverage probabilities of this fine–tuned version of Hsu's procedure with their asymptotic nominal values.

2. Bounds for $\mu(F)$, $\lambda(F)$

The best bounds to date on $\mu(F)$,$\lambda(F)$ are

$$3/10 \leq \mu(F) \leq (\sqrt{2}+6)/24 \approx .30893, \tag{2.1}$$

and

$$.28254 \approx 89/315 \leq \lambda(F) \leq 7/24 \approx .29167. \tag{2.2}$$

The upper bound on $\mu(F)$ is due to Hollander (1967), the lower bound on $\mu(F)$ to Spurrier (1991), the upper bound on $\lambda(F)$ to Lehmann (1964) and the lower bound on λ(F) to Spurrier (1991). It remains an open question whether any of these bounds in (2.1) and (2.2) are best possible. We briefly sketch their derivations, referring the reader to the original articles for more details.

THEOREM 1 (Hollander (1967)) $\mu(F) \leq (\sqrt{2}+6)/24$.

PROOF Let $X_1, X_2, \ldots, X_n, Y_1, Y_2, \ldots, Y_n$ be iid according to the continuous distribution F. Let U_1 be the Mann–Whitney–Wilcoxon statistic, $U_1 = \sum_{i=1}^n \sum_{j=1}^n \phi(X_i, Y_j)$ where $\phi(a,b) = 1$ if $a < b$, 0 otherwise and let U_2 denote Wilcoxon's signed rank statistic applied to a random pairing of the X's with the Y's. Using a representation due to Tukey, U_2 can be represented as $U_2 = \sum_{i<j}^n \phi(X_i + X_j, Y_i + Y_j) + \sum_{i=1}^n \phi(X_i, Y_i)$. The correlation coefficient between U_1 and U_2 is directly obtained to be

$$r_0^n(F) = \frac{[n^2(24\mu(F)-6) + n(23-72\mu(F)) + (48\mu(F)-14)]}{(2n+1)[n(n+1)/2]^{1/2}} \tag{2.3}$$

and its limiting value is

$$r^*(F) \stackrel{\text{def}}{=} \lim_n r_0^n(F) = (24\mu(F)-6)/\sqrt{2}.$$

The result follows since $r^*(F) \leq 1$.

THEOREM 2 (Spurrier (1991)) $\mu(F) \geq 3/10$.

PROOF Spurrier's method, motivated by Mann and Pirie (1982), is to exhibit an unbiased estimator $\hat{\mu}(F)$ of $\mu(F)$ that assumes just two values. These values are 3/10 and 19/60 and since $E(\hat{\mu}(F)) = \mu(F)$ it follows that $\mu(F) \geq 3/10$. Spurrier's estimator is as follows. Let

$$\begin{aligned} \hat{\mu}(F) &= \sum I(X_{i_1}, \ldots, X_{i_5})/120 \\ &= \sum I(X_{(i_1)}, \ldots, X_{(i_5)})/120, \end{aligned}$$

where $I(a,b,c,d,e) = 1$ if $a < e$ and $a < b+c-d$; 0 otherwise, $X_{(1)} < \cdots < X_{(5)}$ are the order statistics of $X_1, \ldots, X_5$ and the Σ is over all permutations $(i_1, \ldots, i_5)$ of (1,2,3,4,5). The function I is invariant under interchanges of its second and third arguments, thus $\hat{\mu}(F)$ can be rewritten as

$$\hat{\mu}(F) = \sum I(X_{(i_1)}, \ldots, X_{(i_5)})/60, \tag{2.4}$$

where the summation is over all permutations $(i_1, \ldots, i_5)$ of (1,2,3,4,5) such that $i_2 < i_3$. Spurrier partitions the 60 summands of (2.4) into four groups. Group 1 consists of 13 summands which identically equal 1, group 2 consists of 37 summands which identically equal 0, and group 3 consists of 8 summands which can be organized into 4 pairs of summands, with the total of the two summands in each pair equal to 1. Thus the first three groups have 58 I functions summing to 17 w. p. 1. The last group consists of two summands and form the random part of $\hat{\mu}(F)$. At least 1 of these 2 must equal 1. Thus either 18 or 19 of the 60 summands equal 1 and $\hat{\mu}(F) = 3/10$ or $19/60$.

THEOREM 3 (Lehmann (1964)) $\lambda(F) \leq 7/24$.

PROOF Consider the covariance matrix defined by (1.5) and (1.6) for the random variables Y_{12}, Y_{13}, Y_{23}. The determinant of the matrix is proportional to $(1+\gamma)^2(1-2\gamma)$ where $\gamma = 3(4\lambda(F) - 1)$. The determinant can be nonnegative only if either $\gamma = -1$ in which case $\lambda = 1/6$ or if $\gamma \leq 1/2$ in which case $\lambda \leq 7/24$. The case $\lambda = 1/6$ is ruled out directly as Lehmann used Schwarz' inequality to show $\lambda \geq 1/4$.

THEOREM 4 (Spurrier (1991)) $\lambda(F) \geq 89/315$.

PROOF The proof is similar to Spurrier's proof of Theorem 2 but is more tedious and involves a computerized evaluation of numerous cases. Let $X_{(1)} < \cdots < X_{(7)}$ denote the order statistics of $X_1, \ldots, X_7$ and let

$$\hat{\lambda}(F) = \sum J(X_{(i_1)}, \ldots, X_{(i_7)})/5040,$$

where $J(a,b,c,d,e,f,g) = 1$ if $a < b+c-d$ and $a < e+f-g$; 0 otherwise, and the Σ is over all permutations $(i_1, \ldots, i_7)$ of (1,2,...,7). Using invariance and symmetry $\hat{\lambda}(F)$ can be rewritten as

$$\hat{\lambda}(F) = \sum J(X_{(i_1)}, \ldots, X_{(i_7)})/630, \tag{2.5}$$

where the summation is over all permutations $(i_1, \ldots, i_7)$ of (1,...7) such that $i_2 < i_3$, $i_4 < i_7$ and $i_5 < i_6$. Spurrier partitions the 630 summands in (2.5) into four groups. The first group contains 89 summands identically equal to 1, the second group contains 331 summands identically equal to 0, and the third group contains 132 summands which can be partitioned into pairs such that one pair is 0 and the other is 1. The 552 summands in the first three groups sum to 155. The remaining 78 summands comprise the fourth group and they are the random component of $\hat{\lambda}(F)$. Using a computer, Spurrier shows the minimum sum of those 78 is 23. Thus the minimum possible value of $\hat{\lambda}(F)$ is $(155+23)/630 = 89/315$. Since $\hat{\lambda}(F)$ is an unbiased estimator of $\lambda(F)$, it follows that $\lambda(F) \geq 89/315$.

3. Applications

3.1. *Upper and Lower Bounds on the Null Correlation Coefficient Between Overlapping Signed Rank Statistics*

The null correlation coefficient between two overlapping signed rank statistics $\rho_0^n(F)$, given by (1.2), depends on F except for $n = 1$ when its value is 1/3. Upper bounds ρ_U^n can be obtained by substituting the upper bounds for $\mu(F)$ and $\lambda(F)$ given respectively by the right–hand inequalities of (2.1) and (2.2), into (1.2). Lower bounds ρ_L^n can be obtained by substituting the lower bounds for $\mu(F)$ and $\lambda(F)$, given respectively by the left–hand–inequalities of (2.1) and (2.2), into (1.2). These bounds are displayed in Table 2.

n	1	2	3	4	5	6	7	8	9	10
ρ_U^n	.3333	.3886	.4163	.4330	.4441	.4521	.4581	.4627	.4665	.4695
ρ_L^n	.3333	.3600	.3701	.3752	.3784	.3804	.3819	.3830	.3839	.3845
n	11	12	13	14	15	20	25	40	50	∞
ρ_U^n	.4720	.4741	.4760	.4776	.4790	.4840	.4871	.4918	.4934	.5000
ρ_L^n	.3851	.3856	.3859	.3863	.3866	.3876	.3881	.3890	.3893	.3905

Table 2. Upper bounds ρ_U^n and lower bounds ρ_L^n for $\rho_0^n(F)$

3.2. *Upper and Lower Bounds on the Null Correlation Coefficient Between the Mann–Whitney–Wilcoxon Statistic and the Randomly Paired Signed Rank Statistic*

The null correlation between U_1 and U_2 (defined in the proof of Theorem 1) is given by (2.3). The null correlation $r_0^n(F)$ depends on F except for $n = 1$ and $n = 2$ where its values are 1 and .9238, respectively. Upper bounds r_U^n can be obtained by substituting the upper bound for $\mu(F)$ given by the right–hand inequality of (2.1) into (2.3). Lower bounds r_L^n can be obtained by substituting the lower bound for $\mu(F)$ given by the left–hand inequality of (2.1) into (2.3). These bounds are displayed in Table 3.

n	1	2	3	4	5	6	7	8	9	10
r_U^n	1.0000	.9238	.9231	.9306	.9382	.9448	.9503	.9549	.9587	.9620
r_L^n	1.0000	.9238	.8981	.8854	.8779	.8729	.8693	.8667	.8646	.8630
n	11	12	13	14	15	20	25	40	50	∞
r_U^n	.9648	.9673	.9694	.9712	.9729	.9790	.9828	.9890	.9911	1.0000
r_L^n	.8616	.8605	.8596	.8588	.8581	.8557	.8542	.8521	.8514	.8485

Table 3. Upper bounds r_U^n and lower bounds r_L^n for $r_0^n(F)$

3.3. *Observed Levels of Hollander's Test*

For testing, in the two–way layout, $H_0 : \theta_1 = \theta_2 = \cdots = \theta_k$ versus the ordered alternatives $H_a : \theta_1 \leq \theta_2 \leq \cdots \leq \theta_k$ (with at least one inequality strict), Hollander (1967) proposed tests based on

$$Y = \sum_{u<v}^{k} T_{uv} \tag{3.1}$$

where the T_{uv} are the Wilcoxon signed–rank statistics defined in Section 1. The statistic Y, suitably standardized, is asymptotically normal but it is not distribution–free under H_0 as the finite–dimensional joint distributions of the $\{ T_{uv} \}$ depend on F; in particular the null variance $\sigma_0^2(Y)$ depends on F as

$$\sigma_0^2(Y) = n(n+1)(2n+1)k(k-1)(3+2(k-2)\rho_0^n(F))/144 \tag{3.2}$$

where $\rho_0^n(F)$ is given by (1.2). Y is not asymptotically distribution–free as the asymptotic null variance of Y depends on F through $\lambda(F)$. Through a Monte Carlo study, we determined the levels of a modification of Hollander's (1967) test which rejects H_0 at the approximate α–level if $Y \geq k(k-1)n(n+1)/8 + z_\alpha\hat{\sigma}_0(Y)$, and accepts otherwise. Here z_α is the upper α percentile point of a $N(0,1)$ distribution, $\hat{\sigma}_0^2(Y)$ is obtained by replacing $\rho_0^n(F)$ by $\hat{\rho}_n = 12\hat{\lambda}(F) - 3$ in (3.2), and $\hat{\lambda}(F)$ is Lehmann's estimator given in (3.3) below provided $89/315 \leq \hat{\lambda}(F) \leq 7/24$, otherwise we take $\hat{\lambda}(F)$ to be 89/315 if (3.3) goes below the lower bound or 7/24 if (3.3) goes above the upper bound. This is a modification of Hollander's test because we are using Spurrier's new lower bound for $\lambda(F)$ to fine–tune $\hat{\lambda}(F)$. Results for the simulation are given in Table 4. For the range of distributions considered, the observed levels are reasonably close to the nominal α's and indicate the asymptotic test can be trusted in applications. (In Table 4, w is the number of times $\hat{\lambda}$ was within bounds.)

Lehmann's original estimator of $\lambda(F)$ is $\hat{\lambda}(F)$ given by

$$n(n-1)(n-2)k(k-1)(k-2)\hat{\lambda}(F) = \sum_{(i,j,l)\in C_n} \sum_{(u,v,w)\in C_k} \eta(X_{iv} - X_{iu} + X_{ju} - X_{jv})\eta(X_{iw} - X_{iu} + X_{lu} - X_{lw}) \tag{3.3}$$

where $\eta(t) = 1$ as $t \geq 0$ and is otherwise 0. The sets $C_n(C_k)$ are defined as the collection of all permutations of three integers chosen from the first $n(k)$ integers.

a) *Normal:*

n	5	10	11	12	13	15
α \k	3	4	8	5	7	6
.010	.010	.006	.010	.002	.002	.010
.025	.018	.020	.018	.024	.016	.030
.050	.052	.038	.038	.040	.042	.052
.100	.100	.078	.096	.082	.082	.092
w	161	377	499	463	497	497

b) *Cauchy:*

n	5	10	11	12	13	15
α \k	3	4	8	5	7	6
.010	.018	.016	.008	.008	.004	.004
.025	.042	.032	.024	.024	.016	.016
.050	.064	.054	.052	.052	.056	.034
.100	.122	.118	.110	.084	.090	.078
w	177	483	500	500	500	500

c) *Exponential:*

n	5	10	11	12	13	15
α \k	3	4	8	5	7	6
.010	.008	.004	.006	.012	.006	.006
.025	.028	.022	.020	.022	.022	.016
.050	.060	.044	.052	.048	.050	.048
.100	.114	.094	.094	.090	.118	.108
w	164	438	500	494	500	500

d) *Uniform:*

n	5	10	11	12	13	15
α \k	3	4	8	5	7	6
.010	.018	.010	.010	.014	.004	.010
.025	.038	.030	.024	.034	.020	.020
.050	.058	.052	.058	.044	.048	.038
.100	.120	.128	.104	.078	.090	.078
w	172	322	463	396	454	462

e) *Logistic:*

n	5	10	11	12	13	15
α \k	3	4	8	5	7	6
.010	.020	.014	.008	.008	.004	.006
.025	.040	.030	.024	.028	.016	.018
.050	.070	.054	.048	.048	.046	.034
.100	.118	.116	.108	.082	.096	.072
w	185	415	500	488	499	499

Table 4. Observed levels of Hollander's test

For the simulations of Sections 3.3 and 3.4, we generated independent samples from the following populations : (a) normal, (b) Cauchy, (c) exponential, (d) uniform, (e) logistic. Samples from the normal and exponential populations were obtained using the Zigurrat method as discussed in Marsaglia and Tsang (1984), samples from the uniform were obtained using a random number generator that combines, with subtraction mod 1, element c in arithmetic sequence generated by $c = c - cdmod(16777213./16777216.)$, period $2^{24} - 3$. All these are available from the Statistics Laboratory of the Florida State University. Samples from the remaining populations were obtained by transforming the numbers generated from the uniform distribution. All codes were written in FORTRAN and ran using an f77 compiler on the Sun Network system. All results were based on 500 iterations. Finally, Lehmann's estimator $\hat{\lambda}(F)$ was computed using the algorithm proposed by Mann and Pirie (1982).

3.4. *Observed Coverage Probabilities of Hsu's Procedure*

Hsu's (1982) procedure involves simultaneous inference with respect to a so called "best" treatment. Consider again model (1.1) where θ_j is the effect of treatment π_j. The treatment corresponding to the largest θ_j is said to be the "best" treatment. If there is more than one such treatment, then exactly one is arbitrarily designated the "best" treatment.

Let $\theta_{[k]} = \max_{1 \leq i \leq k} \theta_i$ and denote by (k) the unknown index of the "best" treatment, i.e., $\pi_{(k)}$ is the unique "best" treatment. Hsu's procedure gives a confidence set C for $\pi_{(k)}$, and simultaneously a set of simultaneous upper bounds $D = (D_1, \ldots, D_k)$ for $\phi = (\theta_{[k]} - \theta_1, \theta_{[k]} - \theta_2, \ldots, \theta_{[k]} - \theta_k)$.

Let $m = n(n+1)/2$ and let

$$A_{uv}^{[1]} \leq A_{uv}^{[2]} \leq \cdots \leq A_{uv}^{[m]}$$

denote the ordered averages $(X_{iu} - X_{iv} + X_{ju} - X_{jv})/2, 1 \leq i \leq j \leq n$. Then Hsu's procedure is outlined as follows:

For all $u \neq v$, *let*

$$\begin{aligned} T_{uv} &= A_{uv}^{[(m+1)/2]} && \text{if } m \text{ is odd} \\ &= (A_{uv}^{[m/2]} + A_{uv}^{[m/2+1]})/2 && \text{if } m \text{ is even} \end{aligned}$$

For $u = 1, \ldots, k$ *,calculate*

$$T_{u.} = \sum_{v \neq u} T_{uv}/k$$

Choose c_α *such that*

$$P\{\ c_\alpha \leq W \leq m - c_\alpha\ \} = 1 - \alpha$$

where W *is the Wilcoxon signed-rank statistic on* n *observations and* $0 < \alpha < 1$. *For all* $u \neq v$, *calculate*

$$B_{uv} = (3n)^{1/2}(A_{uv}^{[m+1-c_\alpha]} - A_{uv}^{[c_\alpha]})/z_{\alpha/2}$$

where $z_{\alpha/2}$ *is the upper* $\alpha/2$ *percentile point of a N(0,1) distribution. Let*

$$B = \binom{k}{2}^{-1} \sum_{u<v} B_{uv}$$

and

$$\hat{\tau}^2 = [1/12 + (k-2)(\hat{\lambda}(F) - 1/4)]B^2/k$$

where $\hat{\lambda}(F)$ *is Lehmann's estimator of* $\lambda(F)$ *modified so that* $\hat{\lambda}(F)$ *is replaced by* $89/315$ *if it is less than this lower bound and by* $7/24$ *if it exceeds this upper bound. Take*

$$\begin{aligned} C &= \{\pi_i : \min_{j \neq i} T_{i.} - T_{j.} \geq -(2/n)^{1/2} d(k-1, 1/2, P^*)\hat{\tau}\} \\ D_i &= \max\{\max_{j \neq i} T_{j.} - T_{i.} + (2/n)^{1/2} d(k-1, 1/2, P^*)\hat{\tau}, 0\} \\ & \quad i = 1, \ldots, k \end{aligned}$$

where $0 < P^* < 1$ *and* $d(h, \rho, P^*)$ *denotes the number such that*

$$P(Z_i \geq -d(h, \rho, P^*) \ \textit{for}\ i = 1, \ldots, h) = P^*$$

where $Z_1, \ldots, Z_h$ *are equally correlated N(0,1) random variables with correlation* ρ. *Values for* $d(k-1, 1/2, P^*)$ *were obtained from Gupta, Nagel, and Panchapakesan* (1973).

Now define the coverage probability of a procedure R with confidence set C for $\pi_{(k)}$ and bounds $D = (D_1, \ldots, D_k)$ for $\phi = (\theta_{[k]} - \theta_1, \ldots, \theta_{[k]} - \theta_k)$ by

$$P\{\ \texttt{coverage|R}\ \} \equiv P\{\pi_{(k)} \in C \text{and } \theta_{[k]} - \theta_i \leq D_i \text{ for } i = 1, \ldots, k\}$$

Hsu showed that

$$\lim_{n\to\infty} \inf_{\theta} P_\theta\{\ \texttt{coverage|R}\ \} = P^* \tag{3.4}$$

where R is the procedure we described above. For our Monte Carlo study we determined the observed coverage probability for various choices of P^* and model structures for θ_j's. Results of the simulation are given in Table 5. In most cases the observed probability is higher, as is to be expected, than the nominal P^* and in only a few cases are the observed levels less than P^*, a direction change that can be attributed to the Monte Carlo sampling rather than the theoretical requirement (3.4) not being met.

a) *Normal:*

Model 1: $\theta_1 = \theta_2 = \cdots = \theta_{k-1} = 0\ ,\theta_k = 0.1$

n	5	5	6
P^* \k	4	5	4
.990	.990	.996	.998
.975	.982	.994	.992
.950	.968	.980	.982
.900	.942	.938	.958
.750	.806	.848	.826

Model 2: $\theta_1 = 0,\ \theta_k = \theta_{k-1} + 0.1$

n	5	5	6
P^* \k	4	5	4
.990	.992	.996	1.000
.975	.988	.994	.992
.950	.974	.986	.990
.900	.948	.958	.968
.750	.826	.862	.844

Model 3: $\theta_1 = 1,\ \theta_k = 2\theta_{k-1}$

n	5	5	6
P^* \k	4	5	4
.990	.988	.996	.996
.975	.976	.986	.990
.950	.960	.964	.972
.900	.928	.920	.940
.750	.758	.810	.784

Table 5(a). Observed coverage probabilities of Hsu's procedure (with $c_\alpha = 2$)

b) *Cauchy:*

Model 1: $\theta_1 = \theta_2 = \cdots = \theta_{k-1} = 0, \theta_k = 0.1$

n	5	5	6
P^* \k	4	5	4
.990	.996	.996	.996
.975	.992	.996	.992
.950	.984	.990	.982
.900	.964	.978	.974
.750	.886	.922	.906

Model 2: $\theta_1 = 0$, $\theta_k = \theta_{k-1} + 0.1$

n	5	5	6
P^* \k	4	5	4
.990	.996	.998	.996
.975	.992	.996	.992
.950	.984	.992	.984
.900	.968	.980	.976
.750	.898	.926	.916

Model 3: $\theta_1 = 1, \theta_k = 2\theta_{k-1}$

n	5	5	6
P^* \k	4	5	4
.990	.996	.996	.996
.975	.992	.996	.992
.950	.982	.990	.984
.900	.958	.976	.968
.750	.882	.916	.904

Table 5(b). Observed coverage probabilities of Hsu's procedure (with $c_\alpha = 2$)

c) *Exponential:*

Model 1: $\theta_1 = \theta_2 = \cdots = \theta_{k-1} = 0, \theta_k = 0.1$

n	5	5	6
P^* \k	4	5	4
.990	.998	.990	.996
.975	.992	.982	.990
.950	.978	.968	.980
.900	.938	.940	.954
.750	.826	.832	.832

Model 2: $\theta_1 = 0,\ \theta_k = \theta_{k-1} + 0.1$

n	5	5	6
P^* \k	4	5	4
.990	1.000	.990	.996
.975	.992	.988	.990
.950	.988	.972	.982
.900	.944	.952	.968
.750	.836	.852	.846

Model 3: $\theta_1 = 1,\ \theta_k = 2\theta_{k-1}$

n	5	5	6
P^* \k	4	5	4
.990	.996	.990	.996
.975	.986	.976	.984
.950	.972	.962	.968
.900	.916	.922	.940
.750	.794	.792	.790

Table 5(c). Observed coverage probabilities of Hsu's procedure (with $c_\alpha = 2$)

d) *Uniform:*

Model 1: $\theta_1 = \theta_2 = \cdots = \theta_{k-1} = 0, \theta_k = 0.1$

n P^* \k	5 4	5 5	6 4
.990	.994	1.000	1.000
.975	.988	.996	1.000
.950	.982	.988	.994
.900	.962	.964	.976
.750	.854	.884	.878

Model 2: $\theta_1 = 0, \theta_k = \theta_{k-1} + 0.1$

n P^* \k	5 4	5 5	6 4
.990	.996	.998	1.000
.975	.990	.988	.998
.950	.978	.984	.994
.900	.956	.962	.978
.750	.816	.866	.856

Model 3: $\theta_1 = 1, \theta_k = 2\theta_{k-1}$

n P^* \k	5 4	5 5	6 4
.990	.986	.990	.996
.975	.972	.976	.986
.950	.958	.960	.978
.900	.914	.912	.944
.750	.786	.814	.818

Table 5(d). Observed coverage probabilities of Hsu's procedure (with $c_\alpha = 2$)

e) *Logistic:*

Model 1: $\theta_1 = \theta_2 = \cdots = \theta_{k-1} = 0, \theta_k = 0.1$

n	5	5	6
P^* \k	4	5	4
.990	.992	.994	1.000
.975	.986	.986	.992
.950	.972	.976	.976
.900	.936	.942	.944
.750	.818	.818	.850

Model 2: $\theta_1 = 0$, $\theta_k = \theta_{k-1} + 0.1$

n	5	5	6
P^* \k	4	5	4
.990	.994	.994	1.000
.975	.988	.992	.994
.950	.974	.976	.982
.900	.948	.952	.954
.750	.844	.860	.876

Model 3: $\theta_1 = 1, \theta_k = 2\theta_{k-1}$

n	5	5	6
P^* \k	4	5	4
.990	.988	.994	1.000
.975	.982	.986	.990
.950	.966	.964	.976
.900	.922	.936	.940
.750	.796	.800	.842

Table 5(e). Observed coverage probabilities of Hsu's procedure (with $c_\alpha = 2$)

References

Gupta, S., Nagel, K. and Panchapakesan, S. (1973). On the order statistics from statistics from equally correlated normal random variables. *Biometrika* **60** 403-413.

Hodges, J.L., Jr. and Lehmann, E.L. (1961). Comparison of the normal scores and Wilcoxon tests. In *Proc. Fourth Berkeley Symp. Math. Statist. Probab.* **1**, J. Neyman, ed., University of California Press, Berkeley, CA. 307-317.

HOLLANDER, M. (1966). An asymptotically distribution-free multiple comparisons procedure-treatments vs. control. *Ann. Math. Statist.* **37** 735-738.

HOLLANDER, M. (1967). Rank tests for randomized blocks when the alternatives have an a priori ordering. *Ann. Math. Statist.* **38** 867-877.

HSU, J. C. (1982). Simultaneous inference with respect to the best treatment in block designs. *J. Amer. Statist. Assoc.* **77** 461-467.

LEHMANN, E. L. (1964). Asymptotically nonparametric inference in some linear models with one observation per cell. *Ann. Math. Statist.* **35** 726-734.

MANN, B. L. AND PIRIE, W. R. (1982). Tighter bounds and simplified estimation for moments of some rank statistics. *Commun. Statist.-Theor. Meth.* **11** 1107-1117.

MARSAGLIA, G. AND TSANG, W. (1984). A fast, easily implemented method for sampling from decreasing or symmetric unimodal density functions. *Siam J. Sci. Statist. Comput.* **5** 349-359.

SPURRIER, J. D. (1991). Improved bounds for moments of some rank statistics. *Commun. Statist.-Theor. Meth.* **20** 2603–2608.

DEPARTMENT OF STATISTICS
FLORIDA STATE UNIVERSITY
TALLAHASSEE, FL 32306–3033

Stochastic Inequalities
IMS Lecture Notes – Monograph Series
Volume 22 (1993)

ORDERINGS ARISING FROM EXPECTED EXTREMES, WITH AN APPLICATION

By PETER J. DOWNEY and ROBERT S. MAIER[1]

University of Arizona

We bound the expected maximum order statistics $\{EX_{(n)}\}_{n=1}^{\infty}$ of a d.f. F_X both above and below. Our results have an interpretation in terms of stochastic orderings $\leq_e$ and $\leq_{we}$ defined as follows: $F_X \leq_e F_Y$ iff $EX_{(n)} \leq EY_{(n)}$ for all n, and $F_X \leq_{we} F_Y$ iff $EX_{(n)} \leq EY_{(n)}$ for n sufficiently large. We apply our results on $\leq_{we}$ to the end-to-end delay in a resequencing $M/G/\infty$ queue.

1. Introduction

If $X_1, \ldots, X_n$ are i.i.d. random variables with parent distribution F_X, let $X_{(n)}$ denote the maximum order statistic $\max(X_1, \ldots, X_n)$. We are interested in the case when F_X has nonnegative lower endpoint, and upper endpoint $+\infty$. In this case we wish to control the behavior of $X_{(n)}$ as $n \to \infty$; in particular, to bound it above and below in expectation or in related senses. The bounds should be as free of assumptions on the distribution F_X as possible.

Our original motivation for investigating this question was the study of stochastic models arising in computing (Downey and Maier (1990)). There the X_i are interpreted as time delays. (See Section 3 for a typical example, a resequencing $M/G/\infty$ queueing model.) But our results have a more general interpretation, in terms of stochastic inequalities. If a relation $\leq_e$ and its weak counterpart $\leq_{we}$ are defined on the class of finite-mean distributions of nonnegative r.v.'s by

$$F_X \leq_e F_Y \iff EX_{(n)} \leq EY_{(n)}, \quad n \geq 1 \tag{1}$$

$$F_X \leq_{we} F_Y \iff EX_{(n)} \leq EY_{(n)}, \quad n \text{ suff. large} \tag{2}$$

then our results have implications for $\leq_e$ and $\leq_{we}$.

The orderings $\leq_e$ and $\leq_{we}$ are very natural, but seem never to have been studied before. Chan (1967) showed that a distribution is uniquely

[1]Partially supported by National Science Foundation Grant NCR-9016211.

AMS 1991 *subject classifications.* Primary 60G70; secondary 60E05, 60K25.

Key words and phrases. Extreme order statistics, stochastic orderings, stochastic inequalities, resequencing delay, heavy traffic limit.

determined by its expected extreme order statistics, a result that has been considerably generalized (Huang (1987)). In fact F_X is uniquely determined by the sequence $\{EX_{(n)}\}_{n=N}^{\infty}$, for any $N \geq 1$. So both $\leq_e$ and $\leq_{we}$ are antisymmetric relations, and are therefore partial orders. We shall see that they are related to the increasing convex order $\leq_{icx}$.

Several different lines of research have yielded upper and lower bounds on $EX_{(n)}$. Arnold (1985) showed that if $EX^p < \infty$, then $EX^p = O(n^{1/p})$, $n \to \infty$. The precise statement is

$$EX_{(n)} \leq EX + \|X - EX\|_p n^{1/p} \tag{3}$$

with $\|Z\|_p$ signifying the L^p norm $(E|Z|^p)^{1/p}$; this result was rediscovered by Downey (1990). This is an example of a distribution-free result. Other results follow from the classical theory of the convergence in distribution of $X_{(n)}$, suitably normalized, as $n \to \infty$. It is well known that many distributions F_X lie in the domain of attraction $\mathcal{D}(\Lambda)$ of $\Lambda(t) \stackrel{\text{def}}{=} \exp(-e^{-t})$, the double exponential distribution. For them we have $(X_{(n)} - b_n)/a_n \Longrightarrow Y$ with Y distributed according to the law Λ, if a_n and b_n are appropriately chosen. Gnedenko (1943) showed that one may take $b_n = \bar{F}_X^{\leftarrow}(n^{-1})$ and $a_n = \bar{F}_X^{\leftarrow}(e^{-1}n^{-1}) - \bar{F}_X^{\leftarrow}(n^{-1})$; here $\bar{F}_X^{\leftarrow}$ is the right-continuous inverse of the complementary d.f. $\bar{F}_X \stackrel{\text{def}}{=} 1 - F_X$.

De Haan (1975) showed that if $F_X \in \mathcal{D}(\Lambda)$, convergence in distribution also obtains if a_n is chosen to equal $\mu_X(\bar{F}_X^{\leftarrow}(n^{-1}))$. Here $\mu_X(t)$ signifies the mean residual life after time t, $\bar{F}_X(t)^{-1}\int_t^\infty \bar{F}_X(s)\,ds$. Pickands (1968) showed that moments converge as well. So if $F_X \in \mathcal{D}(\Lambda)$,

$$EX_{(n)} \sim \bar{F}_X^{\leftarrow}(n^{-1}) + \gamma\mu_X(\bar{F}_X^{\leftarrow}(n^{-1})), \qquad n \to \infty \tag{4}$$

since the Euler-Mascheroni constant γ is the first moment of the double exponential distribution. In general one expects that even if $F \notin \mathcal{D}(\Lambda)$, if F_X has a sufficiently thin and well-behaved tail then $X_{(n)}$ is not likely to differ from $\bar{F}_X(n^{-1})$ by much more than $\mu_X(\bar{F}_X^{\leftarrow}(n^{-1}))$ in the $n \to \infty$ limit. However the question of which distributions F_X have the property that for all $\epsilon > 0$, there is an M such that

$$\limsup_{n\to\infty} P\left\{\left|\left(X_{(n)} - \bar{F}_X^{\leftarrow}(n^{-1})\right)/\mu_X(\bar{F}_X^{\leftarrow}(n^{-1}))\right| > M\right\} \leq \epsilon \tag{5}$$

seems not to have been resolved. This property defines a larger class than $\mathcal{D}(\Lambda)$. Geometric distributions, for example, satisfy it but are not attracted to Λ.

It is known however (Gnedenko (1943)) that if $\bar{F}_X \in R_{-\infty}$, *i.e.*, the complementary d.f. is regularly varying with index $-\infty$, then

$$X_{(n)}/\bar{F}_X^{\leftarrow}(n^{-1}) \to 1 \tag{6}$$

in probability; the converse also holds. (In fact by the work of Lai and Robbins (1978) and Pickands (1968)) we may substitute for (6) the statement that for all $p > 0$, $E\left|X_{(n)}/\bar{F}_X^{\leftarrow}(n^{-1}) - 1\right|^p \to 0$.) Recall that $\bar{F} \in R_{-\infty}$ means that for all $c > 1$

$$\lim_{t\to\infty} \bar{F}(ct)/\bar{F}(t) = 0. \tag{7}$$

It is known (Resnick (1987)) that if F has upper endpoint $+\infty$, then $F \in \mathcal{D}(\Lambda) \Rightarrow \bar{F} \in R_{-\infty}$. So $\bar{F} \in R_{-\infty}$ is another natural weakening of the condition $F \in \mathcal{D}(\Lambda)$.

In general imposing such regularity conditions as the property (5), $F_X \in \mathcal{D}(\Lambda)$, or $\bar{F}_X \in R_{-\infty}$ will facilitate the control of the sequence $\{EX_{(n)}\}_{n=1}^{\infty}$. But as we sketch in the next section, the large-n asymptotics of this sequence can be usefully bounded in terms of $\bar{F}_X^{\leftarrow}(n^{-1})$ and $\mu_X(\bar{F}_X^{\leftarrow}(n^{-1}))$ even if *no* regularity assumptions are imposed on F_X.

2. Recent Results

Suppose that F_X has upper endpoint t_X^* and is the distribution of an r.v. with finite mean. Since $X_{(n)}$ has distribution F_X^n, we have $\bar{F}_{X_{(n)}} = h_n(\bar{F}_X)$ with $h_n(u) \stackrel{\text{def}}{=} 1 - (1-u)^n$. So

$$EX_{(n)} = \int_0^\infty \bar{F}_{X_{(n)}}(t)\,dt = \int_0^\infty h_n(\bar{F}_X(t))\,dt. \tag{8}$$

It is natural to extend this statement to noninteger values of n; indeed, to all $n \in [0,\infty)$. With this definition $EX_{(n)}$, as a function of n, will be increasing and concave; in fact, its derivative is completely monotone in the sense of Widder (1971). For the remainder of this paper we allow n to take on noninteger values.

THEOREM 2.1 (Downey and Maier (1990)) *We have the following bounds on $EX_{(n)}$. For all $t \in [0,\infty)$ and $n \in [1,\infty)$*

$$EX_{(n)} \le t + n\int_t^\infty \bar{F}_X(s)\,ds \tag{9}$$

and for all $t \in [0, t_X^*)$

$$EX_{(n)} > (1-e^{-1})\left(t + n\int_t^\infty \bar{F}_X(s)\,ds\right) \tag{10}$$

in which $n \stackrel{\text{def}}{=} \bar{F}_X(t)^{-1}$. The same lower bound holds for arbitrary $n \in [1,\infty)$ if t is defined to equal $\bar{F}_X^{\leftarrow}(n^{-1})$.

REMARK The upper bound of the theorem is well known (Lai and Robbins (1978)); the lower bound follows from (8), and is a refinement of Chebyshev's inequality.

COROLLARY *In general*

$$(11)\qquad \begin{aligned}(1-e^{-1})\left[\bar{F}_X^{\leftarrow}(n^{-1})+\mu_X(\bar{F}_X^{\leftarrow}(n^{-1})-)\right] &< EX_{(n)}\\ &\le \bar{F}_X^{\leftarrow}(n^{-1})+\mu_X(\bar{F}_X^{\leftarrow}(n^{-1}))\end{aligned}$$

for all $n \ge 1$. *So if the distribution* F_X *is continuous,*

$$(12)\qquad \frac{EX_{(n)}}{\bar{F}_X^{\leftarrow}(n^{-1})+\mu_X(\bar{F}_X^{\leftarrow}(n^{-1}))} \in (1-e^{-1},1]$$

for all $n \ge 1$.

PROOF (of Corollary). To obtain the upper bound we set $t = \bar{F}_X^{\leftarrow}(n^{-1})$. This implies $n \le \bar{F}_X(t)^{-1}$, so the upper bound follows. It also implies $n \ge \bar{F}_X(t-)^{-1}$, so the lower bound follows as well. □

The corollary provides the desired distribution-free bound on $EX_{(n)}$ in terms of $\bar{F}_X^{\leftarrow}(n^{-1})$ and $\mu_X(\bar{F}_X^{\leftarrow}(n^{-1}))$, or rather $\mu_X(\bar{F}_X^{\leftarrow}(n^{-1})-)$. Due to the presence of the $1-e^{-1}$ factor, for general distributions $X_{(n)}$ is allowed to differ in expectation from $\bar{F}^{\leftarrow}(n^{-1})$ by much more than $O(\mu_X(\bar{F}_X^{\leftarrow}(n^{-1})))$ in the large-n limit. The deviation may only be in the negative direction however. So for continuous distributions the inequality (5) may be replaced by

$$(13)\qquad \limsup_{n\to\infty} P\left\{\left(X_{(n)}-\bar{F}_X^{\leftarrow}(n^{-1})\right)/\mu_X(\bar{F}_X^{\leftarrow}(n^{-1})) < -M\right\} \le \epsilon$$

without any loss of generality.

Another consequence of Theorem 2.1 is the abovementioned relation between $\le_e$ and $\le_{icx}$. Recall that $F_X \le_{icx} F_Y$ iff $\int_t^\infty \bar{F}_X\,ds \le \int_t^\infty \bar{F}_Y\,ds$ for all $t \ge 0$. Equivalently, $Ef(X) \le Ef(Y)$ for all increasing convex functions f on $[0,\infty)$. So $\le_{icx}$ is a weaker ordering than $\le_d$, the standard stochastic ordering.

THEOREM 2.2 (Downey and Maier (1990)) $\le_e$ *and* $\le_{icx}$ *are related as follows.*

1. $F_X \le_{icx} F_Y \Rightarrow F_X \le_e F_Y$.

2. $F_X \le_e F_Y \Rightarrow F_X \le_{icx} F_{\kappa Y}$ *for some universal constant* κ, *which may be taken to equal* $(1-e^{-1})^{-1}$.

PROOF (1) It is well known (Ross (1983)) that if $X_1,\dots,X_n$ and $Y_1,\dots,Y_n$ are independent and $X_i \leq_{icx} Y_i$ for all i, then

$$g(X_1,\dots,X_n) \leq_{icx} g(Y_1,\dots,Y_n) \tag{14}$$

for all increasing convex functions g on $\mathbb{R}^n$. Since max is an increasing convex function of its arguments, $X_{(n)} \leq_{icx} Y_{(n)}$. So $EX_{(n)} \leq EY_{(n)}$.

(2) We shall prove the contrapositive of the claim. Assume that $F_X \not\leq_{icx} F_Y$, *i.e.*, that $\int_t^\infty \bar{F}_X(s)\,ds > \int_t^\infty \bar{F}_Y(s)\,ds$ for some $t \in [0, t_X^*)$. The lower bound of Theorem 2.1, applied to F_X, says that

$$EX_{(n)} > (1-e)^{-1}\left(t + n\int_t^\infty \bar{F}_X(s)\,ds\right) \tag{15}$$

with $n \stackrel{\text{def}}{=} \bar{F}_X(t)^{-1}$. The upper bound of Theorem 2.1, applied to F_Y, says that

$$EY_{(n)} \leq t + n\int_t^\infty \bar{F}_Y(s)\,ds. \tag{16}$$

Combining the bounds (15) and (16) yields $EX_{(n)} > (1-e^{-1})EY_{(n)}$. That is, if $\kappa \stackrel{\text{def}}{=} (1-e^{-1})^{-1}$ then $EX_{(n)} > E\kappa^{-1}Y_{(n)}$. So $F_X \not\leq_e F_{\kappa^{-1}Y}$. □

Theorem 2.2 implies that if distributions which differ only by a change of scale are identified, $\leq_e$ and $\leq_{icx}$ become identical. This is a very curious result, and suggests that it may prove profitable to explore the ways in which stochastic orderings relate such 'scaling equivalence classes' of distributions. Scaling equivalence classes have been considered by Barlow and Proschan (1975).

For the queueing theory application of the next section we need a variant form of Theorem 2.2, which characterizes $\leq_{we}$ rather than $\leq_e$. Theorem 2.3, the proof of which is almost identical, relates $\leq_{we}$ to the *weak* increasing convex ordering $\leq_{wicx}$, defined as follows:

$$F_X \leq_{wicx} F_Y \iff \int_t^\infty \bar{F}_X\,ds \leq \int_t^\infty \bar{F}_Y\,ds, \quad t \text{ suff. large.} \tag{17}$$

Equivalently, $F_X \leq_{wicx} F_Y$ if and only if $Ef(X) \leq Ef(Y)$ for all increasing convex f supported sufficiently far away from zero. $\leq_{wicx}$, unlike $\leq_{icx}$, $\leq_e$ and $\leq_{we}$, is not a partial order: it is merely a pre-order.

THEOREM 2.3 *$\leq_{we}$ and $\leq_{wicx}$ are related as follows. For any $\gamma > 1$*

1. *$F_X \leq_{wicx} F_Y \Rightarrow F_X \leq_{we} F_{\gamma Y}$.*
2. *$F_X \leq_{we} F_Y \Rightarrow F_X \leq_{wicx} F_{\gamma\kappa Y}$, for κ the universal constant of* Theorem 2.2.

3. A Queueing Application

We now show how the above results yield useful bounds on a stochastic model introduced by Harrus and Plateau (1982) and pursued by Baccelli, Gelenbe and Plateau (1984). The model is based on an $M/G/\infty$ queue. Arrivals to the queue are Poisson; that is, interarrival times are distributed according to the law $EXP(\lambda)$, with λ some specified arrival rate. Since there are an infinite number of servers available, customers are processed immediately upon arrival; service time has some finite-mean distribution F_X. We write $\mu \stackrel{\text{def}}{=} (EX)^{-1}$ for the processing rate.

This $M/G/\infty$ queue will be recurrent, irrespective of the traffic intensity $\rho \stackrel{\text{def}}{=} \lambda/\mu$, and the stationary distribution of the number of busy servers will be Poisson with parameter ρ. However we require that for a customer to depart, all its predecessors must have departed. In other words the processing must not be allowed to alter the order of the arriving customers; they are released only in sequence. This introduces an additional *resequencing delay*: a customer's total delay time Y, the 'end-to-end' delay, will be the sum of the processing time X and (possibly) some additional holding time.

A formally stationary distribution for Y was worked out by Harrus and Plateau. Baccelli, Gelenbe and Plateau showed that if the queue begins empty, the distribution of the end-to-end delay of the jth customer does indeed converge, as $j \to \infty$, to the formula given by Harrus and Plateau. Their formula is equivalent to the following (Downey (1992a)):

$$F_Y(t) = \sum_{n=0}^{\infty} \frac{e^{-\rho}\rho^n}{n!} F_X(t) F_{X^*}^n(t) \tag{18}$$

in which F_{X^*} is the distribution of the equilibrium excess of the renewal process with renewal period distribution F_X. That is,

$$\bar{F}_{X^*}(t) = (EX)^{-1} \int_t^{\infty} \bar{F}_X(s)\, ds. \tag{19}$$

The interpretation of formula (18) is simple. If we condition on n servers being busy with previous arrivals when a new customer arrives, since the arrival time is random the time to completion of the kth server, $k = 1, \ldots, n$, will have distribution F_{X^*}. So the end-to-end delay of the new arrival will necessarily be $\max(X, X_1^*, \ldots, X_n^*)$, in which $X_1^*, \ldots, X_n^*$ are i.i.d. with parent distribution F_{X^*}. Since n is Poisson, removing the conditioning yields (18).

We wish to study how the end-to-end delay of this system, in the heavy traffic limit, depends on characteristics of the service time distribution other than its expectation. So we fix μ, and restrict ourselves to distributions with expectation μ^{-1}. We equip this class with a pre-order $\prec$ defined as follows:

if F_{X_1} and F_{X_2} are two service time distributions, we say that $F_{X_1} \prec F_{X_2}$ iff $EY_1 \le EY_2$ for all sufficiently large ρ. Here Y_1 and Y_2 are the corresponding end-to-end delay times, whose distributions are computed from F_{X_1} and F_{X_2} by (18).

It follows from (18) that

$$EY = \sum_{n=0}^{\infty} \frac{e^{-\rho}\rho^n}{n!} E\max(X, X_1^*, \ldots, X_n^*) \tag{20}$$

$$= \left(\sum_{n=0}^{\infty} \frac{e^{-\rho}\rho^n}{n!} EX_{(n)}^* \right) + \frac{1 - e^{-\rho}}{\lambda}. \tag{21}$$

The expression (21) follows from (20) by noting that

$$E\max(X, X_1^*, \ldots, X_n^*) = EX_{(n)}^* + (n+1)^{-1} EX. \tag{22}$$

This is easily verified by integration by parts.

The formula (21) allows us to prove Theorem 3.1 below. The statement of the theorem relies on an ordering $\le_3$ and its weak counterpart $\le_{w3}$, defined as follows. We say that

$$F_{X_1} \le_3 F_{X_2} \iff \int_t^{\infty}\int_s^{\infty} \bar{F}_{X_1}(u)\,du\,ds \le \int_t^{\infty}\int_s^{\infty} \bar{F}_{X_2}(u)\,du\,ds, \quad t \ge 0 \tag{23}$$

Equivalently, $F_{X_1} \le_3 F_{X_2}$ iff $Ef(X_1) \le Ef(X_2)$ for all functions f on $[0, \infty)$ that are increasing, convex and have nonnegative third derivative. $\le_{w3}$ is the corresponding weak pre-order:

$$F_{X_1} \le_{w3} F_{X_2} \iff \int_t^{\infty}\int_s^{\infty} \bar{F}_{X_1}(u)\,du\,ds$$

$$\le \int_t^{\infty}\int_s^{\infty} \bar{F}_{X_2}(u)\,du\,ds, \quad t \text{ suff. large.} \tag{24}$$

Equivalently, $F_{X_1} \le_{w3} F_{X_2}$ iff $Ef(X_1) \le Ef(X_2)$ for all functions f on $[0, \infty)$ that are increasing, convex and have nonnegative third derivative, and are supported sufficiently far away from zero. The definitions (23) and (24) serve to define $\le_3$ and $\le_{w3}$ on the class of d.f.'s that have finite mean and variance.

THEOREM 3.1 *If F_{X_1} and F_{X_2} are two service time distributions with finite variance and the same (finite) mean, then for any $\gamma > 1$*

1. $F_{X_1} \le_{w3} F_{\gamma^{-1}X_2} \Rightarrow F_{X_1} \prec F_{\gamma X_2}$.
2. $F_{X_1} \prec F_{X_2} \Rightarrow F_{X_1} \le_{w3} F_{\kappa\gamma^2 X_2}$, *for κ the universal constant of* Theorem 2.2.

PROOF Since $EX_1 = EX_2$, the parameter ρ is the same for both service time distributions. Also, since $EX_1{}^2$, $EX_2{}^2 < \infty$ we have by examination that EX_1^*, $EX_2^* < \infty$. But if Z is any nonnegative r.v. with finite expectation, it is easily verified that

$$(25) \qquad \sum_{n=0}^{\infty} \frac{e^{-\rho}\rho^n}{n!} EZ_{(n)} \sim EZ_{(\rho)}, \quad \rho \to \infty.$$

(This is a special case of an Abelian theorem for completely monotone functions (Downey (1992b)).) It follows by (21) that $EY_i \sim E(X_i^*)_{(\rho)}$, $\rho \to \infty$. Accordingly for any $\gamma > 1$

$$(26) \qquad F_{X_1^*} \leq_{we} F_{X_2^*} \quad \Rightarrow \quad F_{X_1} \prec F_{\gamma X_2}$$

$$(27) \qquad F_{X_1} \prec F_{X_2} \quad \Rightarrow \quad F_{X_1^*} \leq_{we} F_{\gamma X_2^*}$$

These two implications may be extended by applying Theorem 2.3; we get

$$(28) \qquad F_{X_1^*} \leq_{wicx} F_{\gamma^{-1} X_2^*} \Rightarrow F_{X_1^*} \leq_{we} F_{X_2^*} \Rightarrow F_{X_1} \prec F_{\gamma X_2}$$

$$(29) \qquad F_{X_1} \prec F_{X_2} \Rightarrow F_{X_1^*} \leq_{we} F_{\gamma X_2^*} \Rightarrow F_{X_1^*} \leq_{wicx} F_{\kappa\gamma^2 X_2^*}$$

By (19), the hypothesis of (28) may be written as

$$(30) \qquad (\forall t \geq 0) \qquad \int_t^\infty \int_s^\infty \bar{F}_{X_1}(u)\, du\, ds \leq \gamma \int_t^\infty \int_s^\infty \bar{F}_{\gamma^{-1} X_2}(u)\, du\, ds,$$

and the conclusion of (29) as

$$(31) \quad (\forall t \geq 0) \qquad \int_t^\infty \int_s^\infty \bar{F}_{X_1}(u)\, du\, ds \leq \kappa^{-1}\gamma^{-2} \int_t^\infty \int_s^\infty \bar{F}_{\kappa\gamma^2 X_2}(u)\, du\, ds.$$

But (30) is implied by $F_{X_1} \leq_{w3} F_{\gamma^{-1} X_2}$, and similarly (31) implies $F_{X_1} \leq_{w3} F_{\kappa\gamma^2 X_2}$. So we are finished. □

Theorem 3.1 makes it clear that in analysing the effects of the service time distribution on the expected end-to-end delay in the heavy traffic limit, the ordering $\leq_{w3}$ on service time distributions will prove useful. It is difficult to see how this could have been deduced without the aid of Theorem 2.2.

It would of course be desirable to reduce the constant κ toward unity. Theorem 3.1 is a distribution-free result, and we expect substantial strengthening will be possible if regularity conditions are imposed on the service time distributions.

4. Conclusions

We have seen that for any finite-mean distribution F_X, $EX_{(n)}$ may be bounded for any n above and below in terms of $\bar{F}_X^{\leftarrow}(n^{-1})$ and the mean residual life $\mu_X(\bar{F}_X^{\leftarrow}(n^{-1}))$. $\mu_X(t)$ is expressible in terms of an integral of $\bar{F}_X(t)$, so it proved possible to relate $\leq_e$ to the 'integrated' stochastic ordering $\leq_{icx}$.

Our result Theorem 2.2, and its weak counterpart Theorem 2.3, are expressed in terms of a universal constant κ. It is not clear that our bounds, when the choice $\kappa = (1 - e^{-1})^{-1}$ of Section 2 are used, are tight. It would be desirable either to prove this or to compute the minimal value of κ, particularly from the point of view of applications such as that of Section 3. Moreover the classes of d.f.'s for which $X_{(n)} - \bar{F}_X^{\leftarrow}(n^{-1})$ is $O(\mu_X(\bar{F}_X^{\leftarrow}(n^{-1}))$, in expectation or in other senses, remain to be characterized.

References

ARNOLD, B. C. (1985). p-norm bounds on the expectation of the maximum of a possibly dependent sample. *J. Multivariate Anal.* **17** 316–332.

BACCELLI, F., GELENBE, E., AND PLATEAU, B. (1984). An end-to-end approach to the resequencing problem. *J. Assoc. Comput. Mach.* **31** 474–485.

BARLOW, R. E. AND PROSCHAN, F. (1975). *Statistical Theory of Reliability and Life Testing: Probability Models.* Holt, Rinehart and Winston, New York.

CHAN, L. K. (1967). On a characterization of distributions by expected values of extreme order statistics. *Amer. Math. Monthly* **74** 950–951.

DE HAAN, L. (1975). *On Regular Variation and Its Application to the Weak Convergence of Sample Extremes.* Mathematical Centre Tracts No. 32, Mathematisch Centrum, Amsterdam.

DOWNEY, P. J. (1990). Distribution-free bounds on the expectation of the maximum, with scheduling applications. *Oper. Res. Lett.* **9** 189–201.

DOWNEY, P. J. (1992a). Bounds and approximations for overheads in the time to join parallel forks. Tech. Report 92-10, Department of Computer Science, University of Arizona.

DOWNEY, P. J. (1992b). An Abelian theorem for completely monotone functions. In preparation.

DOWNEY, P. J. AND MAIER, R. S. (1990). Stochastic orderings and the growth of expected extremes. Tech. Report 90-9, Department of Computer Science, University of Arizona.

GNEDENKO, B. V. (1943). Sur la distribution limité du terme maximum d'une série aléatoire. *Ann. Math.* (2) **44** 423–453.

HARRUS, G. AND PLATEAU, B. (1982). Queueing analysis of a reordering issue. *IEEE Trans. Software Engrg.* **SE-8** 113–123.

HUANG, J. S. (1987). Moment problems of order statistics: A review. *Internat. Statist. Review* **57** 59–66.

LAI, T. L. AND ROBBINS, H. (1978). A class of dependent random variables and their maxima. *Z. Wahrsch. Verw. Gebiete* **42** 89–111.

PICKANDS, III., J. (1968). Moment convergence of sample extremes. *Ann. Math. Statist.* **39** 881–889.

RESNICK, S. I. (1987). *Extreme Values, Regular Variation, and Point Processes.* Springer-Verlag, New York.

ROSS, S. M. (1983). *Stochastic Processes.* Wiley, New York.

WIDDER, D. V. (1971). *An Introduction to Transform Theory.* Academic Press, New York.

DEPARTMENT OF COMPUTER SCIENCE
UNIVERSITY OF ARIZONA
TUCSON, AZ 85721

DEPT. OF MATHEMATICS
UNIVERSITY OF ARIZONA
TUCSON, AZ 85721

Stochastic Inequalities
IMS Lecture Notes – Monograph Series
Volume 22 (1993)

A GROUP ACTION ON COVARIANCES WITH APPLICATIONS TO THE COMPARISON OF LINEAR NORMAL EXPERIMENTS[1]

By MORRIS L. EATON

University of Minnesota

Consider a linear normal experiment with a fixed regression subspace and a known covariance matrix Σ. A classical method for comparing such experiments involves the covariance matrix of the Gauss–Markov estimator of the regression coefficients, say $V(\Sigma)$. We introduce a group action on covariance matrices and show that a maximal invariant is $V(\Sigma)$. The concavity of $V(\Sigma)$ in the Loewner ordering shows that $V(\Sigma)$ is monotone in the natural group induced ordering on covariances. In addition, the group structure is used to provide an easy proof of a main theorem in the comparison of linear normal experiments. A related problem concerns the behavior of $V(\Sigma)$ as a function of the elements of Σ. Some results related to positive dependence ideas are presented via examples.

1. Introduction

In simple linear model problems, the covariance matrix of the Gauss–Markov estimator of the vector of regression coefficients is often used to choose between competing linear models with the same regression coefficients. Given an $n \times k$ design matrix X of rank k with $1 \leq k < n$ and a known non-singular covariance matrix Σ, let $\mathcal{E}(X, \Sigma)$ denote the experiment with an observation vector Y whose distribution is multivariate normal $N(X\beta, \Sigma)$ where β is the k–vector of regression coefficients. The reason for the assumption that $k < n$ is explained at the end of Section 4.

Now, the covariance matrix of $\widehat{\beta}$, the Gauss Markov estimator of β, is

$$\text{Cov}(\widehat{\beta}) = (X'\Sigma^{-1}X)^{-1}. \tag{1.1}$$

For two experiments with the same $\beta \in R^k$, say $\mathcal{E}(X_i, \Sigma_i)$, $i = 1, 2$, it is well known that experiment $\mathcal{E}(X_1, \Sigma_1)$ is sufficient for $\mathcal{E}(X_2, \Sigma_2)$ iff

$$(X_1'\Sigma_1^{-1}X_1)^{-1} \leq (X_2'\Sigma_2^{-1}X_2)^{-1} \tag{1.2}$$

[1]Work supported in part by National Science Foundation Grant DMS–89–22607.
AMS 1991 *subject classifications.* Primary 62B15; Secondary 62J05.
Key words and phrases. Comparison of experiments, positive correlation, linear normal experiments, group induced orderings.

where "$\leq$" is the standard Loewner ordering on symmetric matrices ($A \leq B$ means $B - A$ is non-negative definite). [Here, we are using sufficiency in the sense discussed in Blackwell (1951, 1953) — that is, an experiment $\mathcal{E}_1$ is sufficient for $\mathcal{E}_2$ if for every decision problem and prior distribution, the Bayes risk from $\mathcal{E}_1$ is not greater than that from $\mathcal{E}_2$.] A proof of this and related results can be found in Hansen and Torgersen (1974). A few other relevant references include Ehrenfeld (1955), Torgersen (1972), Goel and De Groot (1979), Stepniak, Wang and Wu (1984), Torgersen (1984), Shaked and Tong (1990), and Torgersen (1991).

In the simple case when $k = 1$ and X is the vector of ones in $\mathbb{R}^n$, say $X = e$, replace β by θ. Thus the data vector Y is $N(\theta e, \Sigma)$ and

$$\text{var}(\widehat{\theta}) = (e'\Sigma^{-1}e)^{-1}. \tag{1.3}$$

With the further assumption that Σ is a correlation matrix, say $\Sigma = R$, (1.3) becomes

$$\phi(R) = \text{var}(\widehat{\theta}) = (e'R^{-1}e)^{-1}. \tag{1.4}$$

When R is an intraclass correlation matrix with off diagonal elements equal to $\rho \in (-(n-1)^{-1}, 1)$ then $\phi(R) = n^{-1}[(n-1)\rho + 1]$ which is increasing in ρ. For this case, then, experiments can be ordered by sufficiency in terms of ρ. In a recent paper, Shaked and Tong (1990) studied this and other problems by comparing experiments with i.i.d. observations to those with exchangeable variables. They showed that under certain conditions, positive dependence (corresponding to $\rho > 0$ in the example above) tends to decrease information (increase $\phi(R)$). This raises the rather natural question of how $\phi(R)$ behaves as a function of R when all the elements of R are non-negative. For example, when is it true that $\phi(R) \geq \phi(I_n)$ so that an experiment with i.i.d. observations is preferred to one with correlation matrix R?

In Section 2 of this paper, we present a number of examples — all of which concern the behavior of $\phi(R)$ when the elements of R are non-negative. For $n = 3$, the examples show that for some R's with non-negative elements, perfect estimation of θ is possible — that is, $\widehat{\theta}$ has variance zero. In other cases, $\phi(R)$ first increases and then decreases as certain elements of R increase. However, when R has some special structure, such as the circular correlation structure of Olkin and Press (1969) or the special correlation structure described in Tong (1990, p. 129), $\phi(R)$ increases as certain elements of R increase. These examples show that our rather vague intuitive feeling that "positive correlation tends to decrease information content in an experiment" is very far from the truth, even for rather simple normal experiments with three observations. But, when R has some special structure, our intuition may be correct.

For a general linear normal experiment $\mathcal{E}(X, \Sigma)$, the quantity

$$\psi(\Sigma) = (X'\Sigma^{-1}X)^{-1} \tag{1.5}$$

is used to order experiments in terms of sufficiency. For X fixed, this induces a natural equivalence relationship on Σ's — namely, Σ_1 and Σ_2 are equivalent iff $\psi(\Sigma_1) = \psi(\Sigma_2)$. An alternative way to induce an equivalence relationship on Σ's is to consider $n \times n$ non-singular linear transformations A which fix the regression space (the column space of X), and then define a group action on the set $\mathcal{S}_n^+$ of all $n \times n$ positive definite Σ's. More precisely, suppose the observation vector Y satisfies

$$\mathcal{L}(Y) = N(X\beta, \Sigma), \tag{1.6}$$

where $\mathcal{L}(\bullet)$ denotes "the law of $\bullet$". Let $G(X)$ be the group of all $n \times n$ non-singular matrices which satisfy $AX = X$ (A fixes the regression subspace). In terms of sufficiency, Y and AY are equivalent since they are 1–1 transformations of each other. Further,

$$\begin{aligned} \mathcal{L}(Y) &= N(X\beta, \Sigma) \\ \mathcal{L}(AY) &= N(X\beta, A\Sigma A'). \end{aligned}$$

Thus, the group action $\Sigma \to A\Sigma A'$, $A \in G(X)$, also induces an equivalence relationship on $\mathcal{S}_n^+$ which is clearly relevant for the comparison of experiments. A main result in Section 3 shows that the two equivalence relations are the same. This is accomplished by showing that $\psi(\Sigma)$ is, in fact, a maximal invariant under the action of $G(X)$ on $\mathcal{S}_n^+$. A basic lemma which is used to prove that $\psi(\Sigma)$ is a maximal invariant is also used to give a very easy proof that (1.2) implies that $\mathcal{E}(X_1, \Sigma_1)$ is sufficient for $\mathcal{E}(X_2, \Sigma_2)$.

Finally, the action of $G(X)$ on $\mathcal{S}_n^+$ induces a natural partial ordering — namely $\Sigma_1 < \Sigma_2$ iff Σ_1 is in the convex hull of the $G(X)$ orbit of Σ_2. This type of ordering arises in a number of problems in probability and statistics — see Eaton and Perlman (1977) or Eaton (1987). However, unlike the case here, only compact groups have arisen naturally in the examples familiar to me. Because the function

$$\Sigma \to \psi(\Sigma)$$

is concave (in the Loewner ordering — see Ylvisaker (1964)), it follows immediately that ψ is decreasing in this $G(X)$–induced ordering on $\mathcal{S}_n^+$. A main result in Section 4 relates the ordering induced by $G(X)$ to the ordering induced by ψ.

2. Examples with Non–Negative Correlations

Throughout this section, R denotes an $n \times n$ correlation matrix with non–negative elements. With e denoting the vector of ones in R^n, the examples below concern the behavior of

$$\phi(R) = (e'R^{-1}e)^{-1} \tag{2.1}$$

when R is non–singular. Cases where R is singular are also of interest and will be discussed separately. When Y is $N_n(\theta e, R)$ and R is non–singular, then $\phi(R)$ is the variance of the Gauss–Markov estimator of θ. Marginally, each coordinate of Y is $N(\theta, 1)$ so it is natural to ask for conditions on R which imply that

$$\phi(R) \geq \phi(I_n). \tag{2.2}$$

In other words, when is the experiment $\mathcal{E}_1(e, I_n)$ consisting of i.i.d. observations sufficient for $\mathcal{E}_2(e, R)$? Clearly (2.2) holds for $n = 2$, since R has non–negative elements. However, the following example for $n = 3$ exhibits, what is to some, a rather counter–intuitive result.

EXAMPLE 2.1 For $n = 3$ consider

$$R = \begin{pmatrix} 1 & 0 & a \\ 0 & 1 & a \\ a & a & 1 \end{pmatrix}, \qquad 0 \leq a < 2^{-1/2}.$$

Given that a is non–negative, the condition on a is necessary and sufficient that R be positive definite. An easy calculation shows that

$$h(a) = (e'R^{-1}e)^{-1} \tag{2.3}$$

is given by

$$h(a) = \frac{1 - 2a^2}{3 - 4a}, \qquad 0 \leq a < 2^{-1/2}. \tag{2.4}$$

Differentiation shows that h is concave, increases in $[0,1/2)$ and decreases in $(1/2, 2^{-1/2})$. The maximum value of $1/2$ is for $a = 1/2$, and

$$h(2^{-1/2}) = 0. \tag{2.5}$$

In fact, when the correlation matrix is

$$R_0 = \begin{pmatrix} 1 & 0 & 2^{-1/2} \\ 0 & 1 & 2^{-1/2} \\ 2^{-1/2} & 2^{-1/2} & 1 \end{pmatrix}, \tag{2.6}$$

R_0 is singular. When

$$\mathcal{L}(Y) = N_3(\theta e, R_0), \tag{2.7}$$

let $c_3 = (1 - 2^{1/2})^{-1}$, $c_1 = c_2 = -2^{-1/2}c_3$, and let $c \in R^3$ have coordinates c_1, c_2, c_3. Then $c'Y$ is an unbiased estimator of θ which has variance zero. Thus perfect estimation of θ is possible when the covariance matrix is R_0.

This example has the following implication. Given Y_1 and Y_2 which are i.i.d. $N(\theta, 1)$, suppose we can further select a Y_3 which is marginally $N(\theta, 1)$, but which has correlation ρ with Y_1 and Y_2, $\rho \in [0, 2^{-1/2}]$. From a design point of view, some intuition suggests that $\rho = 0$ may be the preferred value of ρ for inferential problems concerning θ. However, the example shows that $\rho = 2^{-1/2}$ is the preferred value for all inferential problems since perfect estimation of θ is then possible. □

REMARK 2.1 The above example is easily extended to correlation matrices of the form

$$R = \begin{pmatrix} 1 & \delta & a \\ \delta & 1 & a \\ a & a & 1 \end{pmatrix}$$

where $0 \le \delta < 1$, $0 \le a < 1$, and $1 + \delta - 2a^2 > 0$. In this case,

(1) for a fixed, $a \in [0, 2^{-1/2})$, the function $\delta \to (e'R^{-1}e)^{-1}$ is increasing in $\delta, \delta \in [0, 1)$

(2) for $\delta \in [0, 1)$ fixed, the function $a \to (e'Re)^{-1}$ is concave on $[0, ((1 + \delta)/2)^{1/2})$, has a maximum at a point strictly between the two endpoints, and converges to zero as a converges to the right endpoint. When $a = ((1 + \delta)/2)^{1/2}$, perfect estimation of θ is again possible. □

REMARK 2.2 Extensions of Example 2.1 to higher dimensions is easy. For dimension n, let u be a fixed n–vector of length one with non–negative coordinates and let b be a real number in [0,1). Then the $(n + 1) \times (n + 1)$ correlation matrix

$$R = \begin{pmatrix} I_n & bu \\ bu' & 1 \end{pmatrix}$$

has non–negative elements and is non–singular. When $u'e \neq 1$, the behavior of

$$b \to (e'R^{-1}e)^{-1}$$

is similar to that of h defined in (1.3). When $u'e \neq 1$ and $b = 1$, then R is singular and again, perfect estimation is possible. □

The following proposition gives a sufficient condition for (2.2) to hold.

PROPOSITION 2.1 *Let R be an $n \times n$ positive definite correlation matrix with non–negative entries such that $Re = ce$ for some real number c. Then*

$$(e'R^{-1}e)^{-1} \ge \frac{1}{n}. \tag{2.8}$$

PROOF Since $Re = ce$ and R has non-negative elements, $c \geq 1$. Thus,

$$(e'R^{-1}e)^{-1} = \left(\frac{e'e}{c}\right)^{-1} = \frac{c}{n} \geq \frac{1}{n}. \qquad \square$$

Examples of R's which satisfy the assumptions of Proposition 2.1 include intraclass correlation matrices with non-negative entries and the circular stationary correlation matrices (see Olkin and Press (1969)) with non-negative entries.

The final example of this section deals with a class of correlation matrices which might be called intra-inter-class correlation matrices as described in Tong (1990, p. 129). For a positive integer n, let $n_1 \geq n_2 \geq \cdots \geq n_r \geq 1$ be a partition of n — that is, each n_i is a positive integer and

$$\sum_1^r n_i = n. \tag{2.9}$$

Partition a correlation matrix R into $n_i \times n_j$ blocks, $1 \leq i, j \leq r$ and assume

(2.10)
- (i) R_{ii} has intraclass correlation structure with correlation coefficient $\rho_2 \in [0, 1)$
- (ii) for $i \neq j$, the block R_{ij} has all entries equal to $\rho_1 \in [0, \rho_2]$.

PROPOSITION 2.2 *If R has the structure given in* (2.10), *then*

$$e'R^{-1}e = \frac{\sum_1^r n_i(1 - \rho_2 + (\rho_2 - \rho_1)n_i)^{-1}}{1 + \rho_1 \sum_1^r n_i(1 - \rho_2 + (\rho_2 - \rho_1)n_i)^{-1}} \tag{2.11}$$

PROOF This is proved in Appendix I. $\square$

Using the expression (2.11) it is easy to show that

(i) for fixed $\rho_2, e'R^{-1}e$ is decreasing in ρ_1 for $\rho_1 \in [0, \rho_2]$

(ii) for fixed $\rho_1, e'R^{-1}e$ is decreasing in ρ_2 for $\rho_2 \in (\rho_1, 1)$.

Thus, $\phi(R)$ in (2.1) is increasing in ρ_1 for $\rho_1 \in [0, \rho_2]$ and is increasing in ρ_2 for $\rho_2 \in [\rho_1, 1)$. Hence we have

COROLLARY 2.3 *A correlation matrix with the structure* (2.10) *satisfies* $\phi(R) \geq \phi(I_n)$. $\square$

REMARK 2.3 Using (2.11) it is possible to give an alternative proof of a recent result of Shaked and Tong (1992). To describe this result, fix ρ_1 and ρ_2 with $0 \leq \rho_1 < \rho_2 < 1$ and regard $\phi(R)$ as a function of the partition $\mathbf{n} =$

$(n_1, n_2, \ldots, n_r, 0, \ldots, 0)$ — an n–dimensional vector. Let $\mathbf{n}$ and $\mathbf{n}^*$ be two partitions such that $\mathbf{n}$ majorizes $\mathbf{n}^*$ (see Marshall and Olkin (1979) for the relevant definitions) and let R and R^* be correlation matrices corresponding to these two partitions (with the same ρ_1 and ρ_2). Then, the experiment based on $Y^* \sim N(\theta e, R^*)$ is sufficient for the experiment based on $Y \sim N(\theta e, R)$. The proof of this in Shaked and Tong (1992) is based on results in Torgersen (1984). However, using (2.11), a direct verification that the function ϕ is a Schur convex function of partitions $\mathbf{n}$ is not difficult. Thus if $\mathbf{n}$ majorizes $\mathbf{n}^*$, then $\phi(R^*) \le \phi(R)$ so the Shaked and Tong result follows. □

REMARK 2.4 It is well known that for Σ for positive definite and x fixed, the function

$$\Sigma \to (x'\Sigma^{-1}x)^{-1}$$

is a concave function of Σ (see Section 4 for a more general result and further discussion). Thus, on any line segment contained in the set of positive definite matrices, the function $(x'\Sigma^{-1}x)^{-1}$ is concave on that line segment. In particular, $(x'\Sigma^{-1}x)^{-1}$ is concave in each element of Σ, as long as Σ remains positive definite. These remarks explain the concavity property in Example 2.1. □

3. A Group Action on Covariances

Consider an observation vector Y in R^n which has a $N(X\beta, \Sigma)$ distribution where the design matrix X is $n \times k$ of rank k, the known covariance matrix Σ is non–singular, and $\beta \in R^k$ is the vector of regression coefficients. As described in Section 1, such experiments can be compared via the function

$$\psi(\Sigma) = (X'\Sigma^{-1}X)^{-1} \tag{3.1}$$

which is the covariance matrix of the Gauss–Markov estimator $\widehat{\beta}$ of β. Let $G(X)$ be the group of $n \times n$ non–singular matrices A which satisfy $AX = X$. Such A's fix the elements of the regression subspace. Then $G(X)$ acts on the set $\mathcal{S}_n^+$ of $n \times n$ positive definite matrices via the group action

$$S \to ASA'. \tag{3.2}$$

Clearly, ψ in (3.1) is invariant under this group action. A main result in this section shows that ψ is a maximal invariant. To establish this, some preliminaries are needed. First, each $n \times k$ X of rank k can be written

$$X = \Gamma X_0 M \tag{3.3}$$

where Γ is an $n \times n$ orthogonal matrix, M is a $k \times k$ non-singular matrix, and X_0 is the special design matrix

$$X_0 = \begin{pmatrix} I_k \\ O \end{pmatrix}. \tag{3.4}$$

Also note that

$$X = \Gamma V X_0 \tag{3.5}$$

where

$$V = \begin{pmatrix} M & O \\ O & I_{n-k} \end{pmatrix} \tag{3.6}$$

is $n \times n$ and non-singular.

LEMMA 3.1 *Given $\Sigma \in \mathcal{S}_n^+$, there exists an $A \in G(X_0)$ such that*

$$A\Sigma A' = \begin{pmatrix} (X_0\Sigma^{-1}X_0')^{-1} & O \\ O & I_{n-k} \end{pmatrix}. \tag{3.7}$$

PROOF Partition Σ as

$$\Sigma = \begin{pmatrix} \Sigma_{11} & \Sigma_{12} \\ \Sigma_{21} & \Sigma_{22} \end{pmatrix}$$

where Σ_{11} is $k \times k$ and Σ_{22} is $(n-k) \times (n-k)$. It is clear that $A \in G(X_0)$ iff A has the form

$$A = \begin{pmatrix} I_k & A_{12} \\ O & A_{22} \end{pmatrix}$$

where A_{22} is $(n-k) \times (n-k)$ and non-singular. Now, pick $A_{22} = \Sigma_{22}^{-1/2}$ and $A_{12} = -\Sigma_{12}\Sigma_{22}^{-1}$. With this choice of A, some algebra yields

$$A\Sigma A' = \begin{pmatrix} \Sigma_{11\cdot 2} & O \\ O & I_{n-k} \end{pmatrix}$$

where $\Sigma_{11\cdot 2} = \Sigma_{11} - \Sigma_{12}\Sigma_{22}^{-1}\Sigma_{21}$. But it is well known that

$$\Sigma_{11\cdot 2} = (X_0'\Sigma^{-1}X_0)^{-1}.$$

□

COROLLARY 3.2 *The function*

$$\psi_0(\Sigma) = (X_0'\Sigma^{-1}X_0)^{-1}$$

is a maximal invariant under the action of $G(X_0)$.

PROOF The invariance of ψ_0 is obvious. If $\psi_0(\Sigma_1) = \psi_0(\Sigma_2)$, use Lemma 3.1 to find A_1 and A_2 so that

$$A_1\Sigma_1A_1' = \begin{pmatrix} \psi_0(\Sigma_1) & O \\ O & I_{n-k} \end{pmatrix} = \begin{pmatrix} \psi_0(\Sigma_2) & O \\ O & I_{n-k} \end{pmatrix} = A_2\Sigma_2A_2'.$$

Thus Σ_1 and Σ_2 are in the same orbit so ψ_0 is maximal. □

THEOREM 3.3 *A maximal invariant under the action of* $G(X)$ *on* $\mathcal{S}_n^+$ *is* $\psi(\Sigma)$ *in* (3.1).

PROOF The invariance of ψ is clear. If $\psi(\Sigma_1) = \psi(\Sigma_2)$, then writing X in the form (3.3) yields

$$X_0'(\Gamma'\Sigma_1\Gamma)^{-1}X_0 = X_0'(\Gamma'\Sigma_2\Gamma)^{-1}X_0. \tag{3.8}$$

From Corollary 3.2, there then exists an $A \in G(X_0)$ such that

$$A\Gamma'\Sigma_1\Gamma A' = \Gamma'\Sigma_2\Gamma. \tag{3.9}$$

Setting $B = \Gamma A\Gamma'$, B is in $G(X)$ and $B\Sigma_1 B' = \Sigma_2$. □

Using Lemma 3.1 it is fairly straightforward to give a proof from first principles that $\mathcal{E}(X_1, \Sigma_1)$ is sufficient for $\mathcal{E}(X_2, \Sigma_2)$ iff (1.2) holds. Here, we just sketch the details. Let Y_i be the data vector for $\mathcal{E}(X_i, \Sigma_i)$, $i = 1, 2$.

CLAIM 1 *Without loss of generality,* $X_1 = X_2 = X_0$. To see this, use (3.5) to find an $n \times n$ non-singular matrix C_i such that $C_iX_i = X_0$, $i = 1, 2$. Then

$$\mathcal{L}(C_iY_i) = N(X_0\beta, C_i\Sigma_iC_i'), \qquad i = 1, 2.$$

Because C_i is non-singular, the experiments $\mathcal{E}(X_i, \Sigma_i)$ and $\mathcal{E}(C_iX_i, C_i\Sigma_iC_i')$ are equivalent. Further, (1.2) holds iff

$$(X_0'(C_1\Sigma_1C_1')^{-1}X_0)^{-1} \le (X_0'(C_2\Sigma_2C_2')^{-1}X_0)^{-1}.$$

This establishes Claim 1. □

Now take $X_1 = X_2 = X_0$ so we want to prove

THEOREM 3.4 *The experiment* $\mathcal{E}(X_0, \Sigma_1)$ *is sufficient for* $\mathcal{E}(X_0, \Sigma_2)$ *iff*

$$(X_0'\Sigma_1^{-1}X_0)^{-1} \le (X_0'\Sigma_2^{-1}X_0)^{-1}. \tag{3.10}$$

PROOF Assume (3.10) holds. Using Lemma 3.1, find $A_i \in G(X_0)$ such that

$$\mathcal{L}(A_iY_i) = N(X_0\beta, A_i\Sigma_iA_i'), \qquad i = 1, 2$$

where

$$A_i\Sigma_iA_i' = \begin{pmatrix} (X_0'\Sigma_i^{-1}X_0)^{-1} & O \\ O & I \end{pmatrix}.$$

Since the A_i's are non-singular, the experiments $\mathcal{E}(X_0, \Sigma_i)$ and $\mathcal{E}(X_0, A_i\Sigma_iA_i)$ are equivalent. But, when (3.10) holds, we can then find a random vector Z which is $N(0, \Delta)$ and is independent of Y_1 such that for all β,

$$\mathcal{L}(A_1Y_1 + Z) = \mathcal{L}(A_2Y_2). \tag{3.11}$$

Of course,

$$\Delta = A_2\Sigma_2 A_2' - A_1\Sigma_1 A_1'$$

is non-negative definite. But (3.11) clearly implies that $\mathcal{E}(X_0, A_1\Sigma_1 A_1')$ is sufficient for $\mathcal{E}(X_0, A_2\Sigma_2 A_2')$ (see Lehmann (1959, p. 75)).

For the converse, let $T_i = A_i\Sigma_i A_i'$ so the experiment $\mathcal{E}(X_0, T_i)$ has data vector $A_i Y_i$, $i = 1, 2$. Consider the decision problem of estimating β with a loss function

$$L(a, \beta) = (a - \beta)' D(a - \beta)$$

where D is a fixed non-negative definite matrix. When β has a $N(0, \alpha I)$ prior distribution with $\alpha \in (0, \infty)$ standard calculations yield a Bayes risk for experiment $\mathcal{E}(X_0, T_i)$ of

$$\begin{aligned} r_i(\alpha) &= \alpha^2 \mathrm{tr} T_i(T_i + \alpha I)^{-1} D(T_i + \alpha I)^{-1} \\ &\quad + \alpha \mathrm{tr}[\alpha(T_i + \alpha I)^{-1} - I] D[\alpha(T_i + \alpha I)^{-1} - I]. \end{aligned}$$

Letting $\alpha \to \infty$ produces the limit

$$r_i(\infty) = \mathrm{tr} T_i D, \qquad i = 1, 2.$$

When $\mathcal{E}(X_0, T_1)$ is sufficient for $\mathcal{E}(X_0, T_2)$, we then must have

$$\mathrm{tr} T_1 D \le \mathrm{tr} T_2 D \tag{3.12}$$

for all non-negative definite D. This is clearly equivalent to (3.10). □

REMARK 3.1 When comparing $\mathcal{E}(X_1, \Sigma_1)$ and $\mathcal{E}(X_2, \Sigma_2)$, the argument in Claim 1 above shows that it is sufficient to consider the case $X_1 = X_2 = X_0$ where X_0 is given in (3.4). For comparing $\mathcal{E}(X_0, \Sigma_1)$ and $\mathcal{E}(X_0, \Sigma_2)$, the group $G(X_0)$ is obviously relevant since $\psi(\Sigma) = (X_0\Sigma^{-1}X_0')^{-1}$ is maximal invariant and characterizes sufficiency. □

4. A Group Induced Ordering on Covariances

Again consider an experiment $\mathcal{E}(X, \Sigma)$ corresponding to a random vector Y with $\mathcal{L}(Y) = N(X\beta, \Sigma)$, $\beta \in \mathbb{R}^k$. The function

$$\psi(\Sigma) = (X'\Sigma^{-1}X)^{-1} \tag{4.1}$$

is a maximal invariant under the action of $G(X)$ on $\mathcal{S}_n^+$, and is clearly relevant for the comparison of experiments. An important property of ψ, established in Ylvisaker (1964), is that ψ is concave in the Loewner ordering on $\mathcal{S}_n^+$. That is, for $S_1, S_2 \in \mathcal{S}_n^+$ and $\alpha \in [0, 1]$, the matrix

$$\psi(\alpha S_1 + (1 - \alpha) S_2) - \alpha\psi(S_1) - (1 - \alpha)\psi(S_2)$$

is non–negative definite. This we write as

$$\psi(\alpha S_1 + (1-\alpha)S_2) \geq \alpha\psi(S_1) + (1-\alpha)\psi(S_2). \tag{4.2}$$

To define an ordering on $\mathcal{S}_n^+$, let $C(S)$ denote the convex hull of the $G(X)$–orbit of $S \in \mathcal{S}_n^+$. In other words, $C(S)$ is the convex hull of

$$\{ASA' \mid A \in G(X)\}.$$

DEFINITION 4.1 Write $S_1 \leq_G S_2$ if $S_1 \in C(S_2)$.

Since $C(S_2)$ is the convex hull of the $G(X)$–orbit of S_2, each point in $C(S_2)$ is a finite convex combination of points in the orbit. Thus S_1 is in $C(S_2)$ iff S_1 can be represented as

$$S_1 = \sum_1^r \alpha_i A_i S_2 A_i'$$

for some integer $r \geq 1$ where $\alpha_i \geq 0$ and $\Sigma\alpha_i = 1$. In other words, $S_1 \leq_G S_2$ iff S_1 has the above representation. It is easily checked that $S_1 \leq_G S_2$ iff for any $A, B \in G(X)$,

$$AS_1A' \leq_G BS_2B'.$$

Using this, it follows that $S_1 \leq_G S_2$ iff $C(S_1) \subseteq C(S_2)$. Thus, $\leq_G$ is a pre–order in the sense described in Marshall and Olkin (1979, p. 13). Group induced orderings of this type have arisen in a number of contexts related to both probability inequalities and inequalities more generally. The classical majorization ordering is a group induced ordering, as is one version of the submajorization ordering. Some relevant references are Eaton and Perlman (1977), Marshall and Olkin (1979), Eaton (1982) and Eaton (1987). It should be noted that in all of the examples I know, except the current one, the underlying groups are compact.

The main result of this section shows that $S_1 \leq_G S_2$ iff $\psi(S_1) \geq \psi(S_2)$. Therefore, the $G(X)$ induced ordering on covariances is the same as the comparison of experiments ordering given by ψ in (4.1). The implication in one direction is easy.

THEOREM 4.1 *If $S_1 \leq_G S_2$, then $\psi(S_1) \geq \psi(S_2)$ (in the Loewner ordering).*

PROOF Since $S_1 \leq_G S_2$ we can find $A_1, \ldots, A_r$ in $G(X)$ and non–negative numbers $\alpha_1, \ldots, \alpha_r$ satisfying $\Sigma\alpha_i = 1$ such that

$$S_1 = \sum_1^r \alpha_i A_i S_2 A_i'. \tag{4.3}$$

Using the concavity of ψ in the Loewner ordering, we have

$$\psi(S_1) = \psi\left(\sum_1^r \alpha_i A_i S_2 A_i'\right) \geq \sum_1^r \alpha_i \psi(A_i S_2 A_i') = \psi(S_2).$$

The final equality follows from the $G(X)$–invariance of ψ. □

For the converse, we first establish a special case.

LEMMA 4.2 *Consider the ordering $\leq_G$ induced on $\mathcal{S}_n^+$ by $G(X_0)$. If*

$$(X_0' S_2^{-1} X_0)^{-1} \leq (X_0' S_1^{-1} X_0)^{-1} \tag{4.4}$$

then there exists a discrete probability measure μ on $G(X_0)$ such that

$$S_1 = \int A S_2 A' \mu(dA). \tag{4.5}$$

PROOF Let

$$T_i = (X_0' S_i^{-1} X_0)^{-1}, \qquad i = 1, 2.$$

From Lemma 3.1, there exists $A_i \in G(X_0)$ such that

$$A_i S_i A_i' = \begin{pmatrix} T_i & O \\ O & I_{n-k} \end{pmatrix}, \qquad i = 1, 2.$$

By assumption, $\Delta = T_1 - T_2$ is non–negative definite. Write $\Delta = \Sigma_1^k v_i v_i'$ where $v_i, \ldots, v_k$ are vectors in R^k. Fix $u_0 \in R^{n-k}$ such that $u_0' u_0 = 1$. Let B be the random $k \times (n-k)$ matrix which takes on the values $\pm(2k)^{1/2} v_i u_0'$ with probabilities $1/2k$. Denote by μ_0 the distribution (on $G(X_0)$) of the random matrix

$$A = \begin{pmatrix} I_k & B \\ O & I_{n-k} \end{pmatrix} \in G(X_0).$$

Because $\mathcal{E}B = 0$ and $\mathcal{E}BB' = \Delta$, it is easy to verify that

$$\mathcal{E}_{\mu_0} A(A_2 S_2 A_2')A' = A_1 S_1 A_1'. \tag{4.6}$$

Setting $\tilde{A} = A_1^{-1} A A_2$, let μ denote the distribution of $\tilde{A}$ on $G(X_0)$. Then (4.6) can be written

$$\mathcal{E}_\mu \tilde{A} S_2 \tilde{A}' = S_1 \tag{4.7}$$

which is just (4.5). □

Of course, (4.5) is just the assertion that S_1 is in $C(S_2)$ — that is, $S_1 \leq_G S_2$ when $G = G(X_0)$. The general case is now easy.

THEOREM 4.3 *Consider the ordering $\leq_G$ induced on $\mathcal{S}_n^+$ by $G(X)$. If*

$$(X' S_2^{-1} X)^{-1} \leq (X' S_1^{-1} X)^{-1}, \tag{4.8}$$

then $S_1 \leq_G S_2$.

PROOF Using (3.3), write $X = \Gamma X_0 M$ so (4.8) is equivalent to

$$(X_0'(\Gamma' S_2 \Gamma)^{-1} X_0)^{-1} \leq (X_0'(\Gamma' S_1 \Gamma)^{-1} X_0)^{-1}. \tag{4.9}$$

From Lemma (4.2), $\Gamma' S_1 \Gamma$ is in the convex hull of the $G(X_0)$–orbit of $\Gamma' S_2 \Gamma$. Thus, we can write

$$\Gamma' S_1 \Gamma = \Sigma \alpha_i A_i (\Gamma' S_2 \Gamma) A_i'$$

where $A_i \in G(X_0)$, $\alpha_i \geq 0$ and $\Sigma \alpha_i = 1$. Therefore

$$S_1 = \Sigma \alpha_i (\Gamma A_i \Gamma') S_2 (\Gamma A_i \Gamma')'.$$

Since $A_i \in G(X_0)$, $\Gamma A_i \Gamma' \in G(X)$ so S_1 is in the $G(X)$–orbit of S_2. Thus $S_1 \leq_G S_2$. □

Now, consider experiments $\mathcal{E}(X_1, \Sigma_1)$ and $\mathcal{E}(X_2, \Sigma_2)$. To compare these experiments, we can take $X_1 = X_2 = X_0$ (see Remark 3.1). The results in this section show that $\mathcal{E}(X_0, \Sigma_1)$ is sufficient for $\mathcal{E}(X_0, \Sigma_2)$ iff $\Sigma_2 \leq_G \Sigma_1$ where $G = G(X_0)$. Note that this result is not correct when $k = n$ since in this case $X_0 = I_n$ and $G = \{I_n\}$; so the convex hull of the G–orbit of Σ_1 is just $\{\Sigma_1\}$. However, the characterization (1.2) of sufficiency does continue to hold when $X_1 = X_2 = X_0$ and $k = n$. Thus the assumption that $k < n$ is necessary.

Appendix I

In this appendix, we establish a result which verifies equation (2.11) of Proposition 2.2. Let $x_1, \ldots, x_r$ be vectors in R^n which satisfy

(i) $x_i' x_j = 0$ if $i \neq j$

(ii) $x_i' x_i = a_i^2$ with $a_i > 0$.

Also, let $x = \Sigma_1^r x_i$. For $\alpha \geq 0$ and $\beta \geq 0$, consider the matrix

$$A = I + \alpha x x' + \beta \sum_1^r x_i x_i'. \tag{A.1}$$

The following proposition gives a formula for $x' A^{-1} x$.

PROPOSITION A.1 *With x and A as defined above,*

$$x' A^{-1} x = \frac{\sum_1^r a_i^2 (1 + \beta a_i^2)^{-1}}{1 + \alpha \sum_1^r a_i^2 (1 + \beta a_i^2)^{-1}}. \tag{A.2}$$

PROOF Without loss of generality (just make an orthogonal transformation of coordinates), we can assume that $n = r$ and $x_i = a_i \epsilon_i$ where ϵ_i is the ith

standard basis vector in R^r. Under this assumption, $\sum_1^r x_i x_i'$ is a diagonal matrix D with diagonal entries a_i^2, $i = 1, \ldots, r$. Further

$$x = \sum_1^r x_i = \begin{pmatrix} a_1 \\ a_2 \\ \vdots \\ a_r \end{pmatrix} = a \in R^r$$

and

$$A = I + \alpha aa' + \beta D.$$

Therefore,

$$x'A^{-1}x = a'(I + \beta D + \alpha aa')^{-1}a = v'(I + \alpha vv')^{-1}v$$

where v is given by

$$v = (I + \beta D)^{-1/2}a.$$

But, it is easy to show that

(A.3)
$$v'(I + \alpha vv')^{-1}v = \frac{vv'}{1 + \alpha vv'}.$$

From this, we have

$$x'A^{-1}x = \frac{a'(I + \beta D)^{-1}a}{1 + \alpha a'(I + \beta)^{-1}a}$$

from which the result follows by noting that

$$a'(I + \beta D)^{-1}a = \sum_1^r a_i^2(1 + \beta a_i^2)^{-1}. \qquad \square$$

To apply this result to correlation matrices R satisfying (2.10), first let $n_1 \geq \cdots \geq n_r \geq 1$ be a partition of n. Then, let $e^{(i)} \in R^n$ be the vector whose first $n_1 + \cdots + n_{i-1}$ coordinates are zero, whose next n_i coordinates are one, and whose remaining coordinates are zero. Then it is clear that

(A.4)
$$\begin{cases} \text{(i)} \ e^{(i)\prime}e^{(j)} = 0 \text{ if } i \neq j \\ \text{(ii)} \ e^{(i)\prime}e^{(i)} = n_i \\ \text{(iii)} \ \sum_1^r e^{(i)} = e \end{cases}$$

where e is the vector of ones in R^n. An easy calculation shows that

$$R = (1 - \rho_2)I + \rho_1 ee' + (\rho_2 - \rho_1)\sum_1^r e^{(i)}e^{(i)\prime}.$$

A direct application of Proposition A.1 and a bit of algebra yields Proposition 2.2.

References

Blackwell, D. (1951). Comparison of experiments. In *Proc. Second Berkeley Symp. Math. Statist. Probab.*, J. Neyman, ed., University of California Press, Berkeley, CA. 93–102.

Blackwell, D. (1953). Equivalent comparison of experiments. *Ann. Math. Statist.* **24** 265–272.

Eaton, M. L. (1982). A review of selected topics in probability inequalities. *Ann. Statist.* **10** 11–43.

Eaton, M. L. (1987). *Lectures on Topics in Probability Inequalities.* Centrum voor Wiskunde en Informatica, Amsterdam.

Eaton, M. L. and Perlman, M. (1977). Reflection groups, generalized Schur functions and the geometry of majorization. *Ann. Probab.* **5** 829–860.

Ehrenfeld, S. (1955). Complete class theorem in experimental design. In *Proc. Third Berkeley Symp. Math. Statist. Probab.* **1**, J. Neyman, ed., University of California Press, Berkeley, CA. 69–75.

Goel, P. K. and De Groot, M. H. (1979). Comparison of experiments and information measures. *Ann. Statist.* **7** 1066–1077.

Hansen, O. H. and Torgersen, E. N. (1974). Comparison of linear normal experiments. *Ann. Statist.* **2** 367–373.

Lehmann, E. L. (1959). *Testing Statistical Hypotheses.* Wiley, New York.

Marshall, A. W. and Olkin, I. (1979). *Inequalities. Theory of Majorization and its Applications*, Academic Press, New York.

Olkin, I. and Press, S. J. (1969). Testing and estimation for a circular stationary model. *Ann. Math. Statist.* **40** 1358–1373.

Shaked, M. and Tong, Y. L. (1990). Comparison of experiments for a class of positively dependent random variables. *Canadian J. Statist.* **18** 79–86.

Shaked, M. and Tong, Y. L. (1992). Comparison of experiments via dependence of normal variables with a common marginal distribution. *Ann. Statist.* **20** 614–618.

Stepniak, C., Wang, S.-G. and Wu, C. F. J. (1984). Comparison of linear experiments with known covariances. *Ann. Statist.* **12** 358–365.

Tong, Y. L. (1990). *The Multivariate Normal Distribution.* Springer–Verlag, New York.

Torgersen, E. (1972). Comparison of translation experiments. *Ann. Math. Statist.* **43** 1383–1399.

Torgersen, E. N. (1984). Orderings of linear models. *J. Statist. Plan. Inf.* **9** 1–17.

Torgersen, E. N. (1991). *Comparison of Experiments.* Cambridge University Press, Cambridge.

Ylvisaker, D. (1964). Lower bounds for minimum covariance matrices in time series regression problems. *Ann. Math. Statist.* **35** 362–368.

School of Statistics
University of Minnesota
Minneapolis, MN 55455

Stochastic Inequalities
IMS Lecture Notes – Monograph Series
Volume 22 (1993)

THE ROLE OF A MODULE IN THE FAILURE OF A SYSTEM[1]

By EMAD EL–NEWEIHI and JAYARAM SETHURAMAN

University of Illinois at Chicago Circle and Florida State University

Arrangement increasing and Schur functions play a central role in establishing stochastic inequalities in several areas of statistics and reliability. The role of a module in the failure of a system measures the importance of the module. We define the role to be the probability that this module is among the modules that failed before the failure of the system. A system is called a second order r-out-of-k system if it is a r-out-of-k system based on k modules, without common components, and where each module is an a_i-out-of-n_i system. For such systems, we show that the role of a module is an arrangement increasing or Schur function of parameters that describe the system. These results allow us to compare the role of a module under different values of the parameters of the system.

1. Introduction

In Reliability Theory, after answering questions concerning the **reliability** of a system, the **importance of a component** in a system becomes the next natural question to study. The importance of a component may be measured in many ways. It may be measured by the increment in reliability of the system per unit increase in the reliability of the component. This view is taken in the pioneering paper of Birnbaum (1969). Boland, El-Neweihi and Proschan (1988) and Natvig (1985) have built upon this concept of importance.

The probability that a component is among the components that failed before the failure of a system provides another measure of the importance of the component. This view can be found in Fussell and Vesely (1972) and Barlow and Proschan (1975).

A general summary of many different ways to measure the importance of a component may be found in the expository paper of Boland and El-Neweihi (1990).

---

[1]Research supported by U.S. Army Research Office Grant DAAL03–90–G–0103.
AMS 1991 *subject classifications.* 60E15, 62N05.
Key words and phrases. Schur functions, importance of components, role of module.

A system generally consists of modules which themselves are subsystems of individual components. In this work we will talk about the role of a module in the failure of a system. There can be several notions of the **role** of a **module**. In this paper, we define the **role** of a module to be the probability that the the module is among the modules that caused the failure of the system.

We will compare the role of a module with the role of another module, or compare the role of several modules simultaneously, or compare the role of a module under several values of other parameters of the system. Each of these comparisons can be made by showing that the role of a module is an arrangement increasing or Schur function of the appropriate arguments. In this expository paper we describe such results without proof. The complete proofs are given in the cited references.

The theory of arrangement increasing (AI) and Schur functions plays a central role in establishing stochastic inequalities in several areas of statistics and reliability. This theory is well established, for instance see Proschan and Sethuraman (1977), Hollander, Proschan and Sethuraman (1977). A comprehensive treatment of these functions is given in Marshall and Olkin (1979). We therefore do not give the definitions and known facts concerning arrangement increasing and Schur functions.

2. Series-Parallel System

Consider a system S which is a series system based on modules $C_0, C_1, \ldots, C_k$ where C_i is a parallel system based on n_i components, $i = 1, \ldots, k$. We assume that the lifetimes of $n = n_0 + n_1 + \cdots + n_k$ components are independent with a common continuous distribution. This system was studied in El-Neweihi, Proschan and Sethuraman (1978). In the following $\mathbf{n}$ will stand for the vector $(n_1, n_2, \cdots, n_k)$.

The probability that the failure of the cut set C_0 causes the failure of the system S, *i.e.* the role of C_0, will be denoted by $P(n_0; \mathbf{n})$. It is easy to see that $P(n_0; \mathbf{n})$ is decreasing in n_0 and increasing in $\mathbf{n}$. Theorem 2.1 below gives a compact expression to evaluate $P(n_0; \mathbf{n})$. (See El-Neweihi, Proschan and Sethuraman (1978)).

THEOREM 2.1

$$P(n_0; \mathbf{n}) = \int_0^1 \prod_{i=1}^{n} (1 - x)^{n_i} n_0 x^{n_0 - 1} dx.$$

From this it follows that $P(n_0; \mathbf{n})$ is a Schur-concave function of $\mathbf{n}$. The implication of this statement is that C_0 is more likely to fail first if the remaining cut sets are homogeneous in size than if they are more heterogeneous.

Let $n_0 \leq n_1 \leq \cdots \leq n_k$. The order in which the cut sets will fail is another quantity of interest. This will compare the relative roles of all the cut sets. Let

$$Q(i_0, i_1, \ldots, i_k) = P(C_{i_0} < C_{i_1} < \cdots < C_{i_k}),$$

where $C_{i_0} < C_{i_1} < \cdots < C_{i_k}$ stands for the event that cut set C_{i_0} fails first, cut set C_{i_1} fails next, and so on.

The following theorem can be found in El-Neweihi, Proschan and Sethuraman (1978).

THEOREM 2.2

$$P(C_0 < C_1 < \cdots < C_k) = \prod_i^k n_i \prod_{i=1}^k (\sum_{j=0}^i n_j)^{-1}.$$

This shows that $Q(i_0, i_1, \ldots, i_k)$ is a AI function of $(i_0, i_1, \ldots, i_k)$ and thus the modules C_i are more likely to fail in the order of their sizes.

Let $L(\mathbf{n})$ be the number of components that have failed in all the modules at the time of the failure of the system S. The following were proved in El-Neweihi, Proschan and Sethuraman, (1978):

1 : $L(\mathbf{n}) \overset{st}{\geq} L(\mathbf{n}^*)$ if $\mathbf{n}^* \overset{m}{\geq} \mathbf{n}$.

2 : The distribution of $L(\mathbf{n})$ is NBU.

It was also conjectured in that paper that the distribution of $L(\mathbf{n})$ is IFR; this was later proved in Ross, Shahshahani and Weiss (1980).

3. A (k+1−r+1)-out-of-(k+1) System Based on Parallel Modules

Consider a system S constructed from $k+1$ modules $P_0, P_1, \ldots, P_k$. Assume that P_i contains n_i components whose lifetimes have a common continuous distribution $F_i(x)$, $i = 0, \ldots, k$. Assume that the $n_0 + \cdots + n_k$ components are independent. Let $\mathbf{n}$ denote $(n_1, \ldots, n_k)$. Consider the following structure **(A)** for S:

A1 : The modules $P_0, P_1, \ldots, P_k$ are all parallel systems, and

A2 : the system S is a $(k+1-r+1)$-out-of-$(k+1)$ system based on the $k+1$ modules $P_0, P_1, \ldots, P_k$.

This means that the system S fails as soon as r modules fail.

Denote the lifetimes of the modules P_i by T_i, $i = 0, \ldots, k$ and let $R_0, R_1, \ldots, R_k$ be the ranks of $T_0, T_1, \ldots, T_k$. Denote the probability that P_0 is among the r modules that failed first and caused the failure of the system by

$$P_r(n_0, F_0; \mathbf{n}, \mathbf{F}) = P\{R_0 \leq r\}.$$

A study of properties of the quantity $P_r(n_0, F_0; \mathbf{n}, \mathbf{F})$ is useful to determine the contribution of the module P_0 towards the failure of S. This quantity may be viewed as a measure of importance of the module P_0.

The system considered in this section reduces to the series-parallel system considered in Section 2 when $r = 1$ and $F_1 = F_2 = \cdots = F_k = F$.

Let $h_{r|k}(p_1, \ldots, p_k) = P\{\sum_i^k Y_i \geq r\}$ where $Y_1, \ldots, Y_k$ are k independent Bernoulli random variables with parameters $p_1, \ldots, p_k$. The quantity $h_{r|k}(p_1, \ldots, p_k)$ represents the reliability of an r-out-of-k system with k independent components having reliabilities $p_1, \ldots, p_k$.

A compact expression for $P_r(n_0, F_0; \mathbf{n}, \mathbf{F})$ is given by the following theorem.

THEOREM 3.1

$$P_r(n_0, F_0; \mathbf{n}, \mathbf{F}) = 1 - \int h_{r|k}((F_1(x))^{n_1}, \ldots, (F_k(x))^{n_k}) dF_{T_0}(x).$$

The following theorem can be shown by using Theorem 3.1 and a result on order statistics from heterogeneous distributions found in Pledger and Proschan (1971).

THEOREM 3.2 *For each n_0, F_0 and F, $P_r(n_0, F_0; \mathbf{n}, F)$ is Schur-concave in* $\mathbf{n}$.

This theorem states that the module P_0 is more likely to be among the modules that fail before the failure of the system S when the sizes of the modules $P_1, \ldots, P_k$ are more homogeneous. This fact is intuitively more obvious when $r = 1$, the case considered in El-Neweihi, Proschan and Sethuraman (1978). Theorem 3.2 shows that this is true for all values of r.

Let $P_r^*(n_0, F_0; \mathbf{n}, F)$ be the probability that module P_0 is the rth module to fail among the modules $P_0, P_1, \ldots, P_k$. Clearly, $P_r^*(n_0, F_0; \mathbf{n}, F) = P_r(n_0, F_0; \mathbf{n}, F) - P_{r-1}(n_0, F_0; \mathbf{n}, F)$ and is therefore the difference of two Schur functions. It is not true that $P_r^*(n_0, F_0; \mathbf{n}, F)$ is Schur-concave in $\mathbf{n}$. For instance when $k = 2, r = 2$ and $F_0 = F_1 = F_2 = F$, we have

$P_r^*(n_0, F_0; \mathbf{n}, F) = \int_0^1 (x^{n_1} + x^{n_2} - 2x^{n_1+n_2}) n_0 x^{n_0-1} dx$, which is Schur-convex in $\mathbf{n}$, for each n_0. This remark shows that the claim in Theorem 3.8 in El-Neweihi (1980) is false.

Assume that $n_1 = \cdots = n_k = n$ and that the life distribution F_i of the components of the module P_i have proportional hazards, *i.e.*, $\bar{F}_i(x) = \exp(-\lambda_i R(x)), i = 1, \ldots, k$. In this case, $P_r(n_0, F_0; n, \mathbf{F})$ is a function which depends on $\mathbf{F}$ only through $\boldsymbol{\lambda}$ and therefore may be denoted by $P_r^+(n_0, F_0; n, \boldsymbol{\lambda})$. Theorem 3.3 below shows that $P_r^+(n_0, F_0; n, \boldsymbol{\lambda})$ is Schur-concave in $\boldsymbol{\lambda}$ when $r = 1$. We do not know whether this result will extend to other cases of r.

THEOREM 3.3 *$P_1^+(n_0, F_0; n, \boldsymbol{\lambda})$ is Schur-concave in $\boldsymbol{\lambda}$.*

We can give more complete results if we assume that the distributions F_i have proportional left-hazards. Assume that $F_i(x) = \exp(-\lambda_i A(x)), i = 1, \ldots, k$. In this case, $P_r(n_0, F_0; n, \mathbf{F})$ is a function which depends on $\mathbf{F}$ only through $\boldsymbol{\lambda}$ and therefore may be denoted by $P_r^-(n_0, F_0; n, \boldsymbol{\lambda})$. In Theorem 3.4 below we show that $P_r^-(n_0, F_0; n, \boldsymbol{\lambda})$ is Schur-concave in $\boldsymbol{\lambda}$.

THEOREM 3.4 *$P_r^-(n_0, F_0; n, \boldsymbol{\lambda})$ is Schur-concave in $\boldsymbol{\lambda}$.*

El-Neweihi (1980) studied the joint monotonicity properties of $P_r(n_0, F_0; \mathbf{n}, \mathbf{F})$ in $\mathbf{n}, \mathbf{F}$. He considered the case $r = 1$ and showed that $P_1(n_0, F_0; \mathbf{n}, \mathbf{F})$ is an AI function of $(\mathbf{n}, \mathbf{F})$. Example 2.8 of El-Neweihi and Sethuraman (1991) shows that this AI property is not generally true for other values of r.

4. Series System Based on $a_i + 1$-out-of-n_i Systems

Consider an alternate structure **(B)** for the system S.

B1 : The module P_i is an $a_i + 1$-out-of-n_i system, $i = 0, \ldots, k$, and

B2 : the system S is a series system based on $P_0, P_1, \ldots, P_k$.

The system considered in this section reduces to the series-parallel system considered in Section 2 when $a_i = 0, i = 0, 1, \ldots, k$ and $F_1 = F_2 = \cdots = F_k = F$. This system allows for more general modules than the system considered in Section 3 and requires the modules to be connected in series.

The probability that the module P_0 causes the system to fail, $P_1(n_0, F_0; \mathbf{n}, \mathbf{F})$, will now be denoted by $P(a_0, n_0, F_0; \mathbf{a}, \mathbf{n}, \mathbf{F})$. We will say that $F \leq G$ if $F(x) \leq G(x)$ for all x. The following theorem gives an AI property using this ordering on distribution functions.

THEOREM 4.1 *$P(a_0, n_0, F_0; a, \mathbf{n}, \mathbf{F})$ is AI in $\mathbf{n}, \mathbf{F}$, for each a_0, n_0, F_0, and a.*

Theorem 4.8 of El-Neweihi (1980) treats the special case of the above when $a = a_0 = 0$.

We now give an application of the above results to an optimal allocation problem. Suppose that the sizes $n_1, \dots, n_k$ of the modules $P_1, \dots, P_k$ are in increasing order. Suppose that we have collections of components with reliabilities $p_1 \geq \cdots \geq p_k$ at a particular time t. Theorem 4.1 shows that the reliability of S at time t is maximized by allocating components of reliability p_i to the module $P_i, i = 1, \dots, k$.

The following theorem considers the case $n_i = n, F_i = F, i = 1, 2, \dots, k$.

THEOREM 4.2 *$P(a_0, n_0, F_0; \mathbf{a}, n, F)$ is Schur-concave in $\mathbf{a}$.*

The case when $a_i = a, F_i = F, i = 1, 2, \dots, k$ was treated in El-Neweihi, Proschan and Sethuraman (1978) where the following theorem was established.

THEOREM 4.3 *$P(a_0, n_0, F_0; a, \mathbf{n}, F)$ is Schur-concave in $\mathbf{n}$.*

Theorem 4.3 shows that the probability that module P_0 fails first is Schur-concave in $\mathbf{n}$. We can ask the question whether the probability that module P_0 is among the first r modules to fail is also Schur-concave. The following example shows that this is not so for $r = 2$.

EXAMPLE 4.4 Let $k = 2, a_1 = 1, a_2 = 1, F_0 = F_1 = F_2 = F$ where F is the uniform distribution on $[0, 1]$. Then
The probability that module P_0 is among the first two modules to fail

$$= 1 - \int [t^{n_1+n_2} + (n_1 + n_2)(1-t)t^{n_1+n_2-1} + n_1 n_2 (1-t)^2 t^{n_1+n_2-2}] \times \binom{n_0}{a_0} (n_0 - a_0) t^{n_0 - a_0 - 1} (1-t)^{a_0} \, dt.$$

The integrand is Schur-concave in $\mathbf{n}$ and hence this probability is Schur-convex.

Theorems 4.2 and 4.3 have obvious applications to optimal allocation along the lines of the remark following Theorem 4.1.

5. Dual Systems

Every coherent structure possesses a dual structure. The dual of a parallel structure is a series structure. The dual of a k-out-of-n structure is an $(n-k+1)$-out-of-n structure, and is a structure of the same type. Consider the system S with structure **A** based on the modules $P_0, P_1, \ldots, P_k$ as in Section 3. The dual of this is a system S' based on the modules $P'_0, P'_1, \ldots, P'_k$, consisting of $n_0, n_1, \ldots, n_k$ components, and possessing the structure $\mathbf{A}'$ as follows:

A′1 The modules $P'_0, P'_1, \ldots, P'_k$ are all series systems, and

A′2 the system S' is an r-out-of-$(k+1)$ system based on the $k+1$ modules $P'_0, P'_1, \ldots, P'_k$.

This means that the system S' fails as soon as $k-r+1$ modules fail. Let T_i be the lifetime of the modules P_i, $i = 0, \ldots, k$ and let $R_0, R_1, \ldots, R_k$ be the ranks of $T_0, T_1, \ldots, T_k$. Let T'_i be the lifetime of the modules P'_i, $i = 0, \ldots, k$ and let $R'_0, R'_1, \ldots, R'_k$ be the ranks of $T'_0, T'_1, \ldots, T'_k$. Suppose that $T'_i = f(T_i)$ where f is a positive, strictly decreasing and continuous function. This happens when the lifetimes of the components in S' are the same function f of the lifetimes of the corresponding components of S. Let $P'_r(n_0, F'_0; \mathbf{n}, \mathbf{F}')$ be the probability that R'_0 is less than or equal to r, that is P'_0 is among the first r modules to fail in S'.

It is easy to see that $P'_{k-r+1}(n_0, F'_0; \mathbf{n}, \mathbf{F}') = 1 - P_r(n_0, F_0; \mathbf{n}, \mathbf{F})$, that is, the probability that P'_0 is among the modules that caused the failure of the system S' is the complement of the probability that P_0 is among the modules that caused the failure of the system S.

Theorems 5.1 to 5.3 below follow directly from the above relationship between dual structures, see El-Neweihi and Sethuraman (1991).

THEOREM 5.1 *For each* n_0, F'_0, F', $P'_r(n_0, F'_0; \mathbf{n}, F')$ *is Schur-convex in* $\mathbf{n}$.

THEOREM 5.2 *The probability that* P'_0 *fails last among all the* $k+1$ *modules is* $1-$ $P'_{k'}(n_0, F'_0; \mathbf{n}, \mathbf{F}')$ *and is arrangement decreasing in* $\mathbf{n}, \mathbf{F}'$.

THEOREM 5.3 *Let* $\bar{F}'_i(x) = \exp(-\lambda_i R(x))$, $i = 1, \ldots, k$ *(the proportional hazards case). Then* $P'_r(n_0, F'_0; n, \mathbf{F}')$ *is Schur-convex in* $\boldsymbol{\lambda}$.

We will now consider the dual of the system S with the structure **B** defined in Section 4. This is a system S' with modules $P'_0, P'_1, \ldots, P'_k$ satisfying the following structure.

B′1 The module P_i in an $(n_i - a_i)$-out-of-n_i system , $i = 0, \ldots, k$, and

B′2 the system S' is a parallel system based on the modules $P_0', P_1', \ldots, P_k'$.

We will denote the probability that P_0' fails last by $P'(a_0, n_0, F_0'; \mathbf{a}, \mathbf{n}, \mathbf{F}')$. The following theorems follow by using the relation between dual structures.

THEOREM 5.4 *For each a_0, n_0, F_0' and a, $P'(a_0, n_0, F_0'; a, \mathbf{n}, \mathbf{F}')$ is arrangement decreasing in $\mathbf{n}, \mathbf{F}'$.*

THEOREM 5.5 *For each a_0, n_0, F_0' and F', $P'(a_0, n_0, F_0'; \mathbf{a}, n, F')$ is Schur-concave in $\mathbf{a}$.*

6. Further Extensions

The structures that have been considered in this paper are special cases of second order r-out-of-k systems. A definition of such a system S is as follows. Let $P_1, \ldots, P_k$ be k modules with no common components where each module is a a_i-out-of-n_i system. The system S fails as soon as $k - r + 1$ of the modules $P_1, \ldots, P_k$ fail.

We need to investigate questions similar to those considered in this paper for such second order systems in general. This would be a first step. New kinds of questions also arise for these systems. One can study the role of groups of modules rather than that of a single module. The role of a group of modules can be defined to be the probability that at least m of the modules in the group have failed prior to the failure of the system, where m can vary from 1 to the size of the group.

REFERENCES

BARLOW, R. E. AND PROSCHAN F. (1975). Importance of system components and failure tree events. *Stochastic Proc. Appl.* **3** 153–173.

BIRNBAUM, Z. W. (1969). On the importance of different components in a multi-component system. *Multivariate Analysis II*, P. R. Krishnaiah, ed. Academic Press, New York, 581–592.

BOLAND P.J. AND EL-NEWEIHI, E. (1990). Measures of component importance in reliability theory. Tech. Report, Department of Statistics, University College, Dublin.

BOLAND, P.J., EL-NEWEIHI, E. AND PROSCHAN, F. (1988). Active redundancy allocation in coherent systems. *Prob. Eng. Inform. Sci.* **2** 343–353.

EL-NEWEIHI, E. (1980). Extensions of a simple model with applications in structural reliability, extinction of species, inventory depletion and urn sampling. *Commun. Statist.-Theor. Meth.* **A9** 399–414.

EL-NEWEIHI, E., PROSCHAN, F. AND SETHURAMAN, J. (1978). A simple model with applications in structural reliability, extinction of species, inventory depletion and urn sampling. *Adv. Appl. Prob.* **10** 232–254.

EL-NEWEIHI, E. AND SETHURAMAN, J. (1991). A study of the role of modules in the failure of systems. *Prob. Eng. Inform. Sci.* **5** 221–227.

FUSSELL, J. B. AND VESELY, W. E. (1972). A new methodology for obtaining cut sets for fault trees. *Amer. Nuclear Soc. Trans.* **15(1)** 262–263.

HOLLANDER, M., PROSCHAN, F. AND SETHURAMAN, J. (1977). Functions decreasing in transposition and their applications in ranking problems. *Ann. Statist.* **5** 722–733.

MARSHALL, A. W. AND OLKIN, I. (1979). *Theory of Majorization and its Applications.* Academic Press, New York.

NATVIG, B. (1985). New light on measures of importance of system components. *Scand. J. Statist.* **12** 43–54.

PLEDGER, G. AND PROSCHAN, F. (1971). Comparisons of order statistics and of spacings from heterogeneous distributions. *Optimization Methods in Statistics,* J. Rustagi, ed. Academic Press, New York 89–113.

PROSCHAN, F. AND SETHURAMAN, J. (1977). Schur functions in statistics - I: the preservation theorem. *Ann. Statist.* **5** 256–262.

ROSS, S. M., SHAHSHAHANI, M. AND WEISS, G. (1980). On the number of component failures in systems whose component lives are exchangeable. *Math. Oper. Res.* **5(3)** 358–365.

DEPARTMENT OF MATHEMATICS, STATISTICS
AND COMPUTER SCIENCE
UNIVERSITY OF ILLINOIS AT CHICAGO CIRCLE
CHICAGO, IL 60680

DEPARTMENT OF STATISTICS
FLORIDA STATE UNIVERSITY
TALLAHASSEE, FL 32306-3033

Stochastic Inequalities
IMS Lecture Notes – Monograph Series
Volume 22 (1993)

EXTREME ORDER STATISTICS FOR A SEQUENCE OF DEPENDENT RANDOM VARIABLES[1]

By JOSEPH GLAZ

University of Connecticut

Let $X_1, \dots, X_n$ be a sequence of dependent random variables from a continuous density function. Denote by $X_{(1)} = \min(X_1, \dots, X_n)$ and $X_{(n)} = \max(X_1, \dots, X_n)$ the extreme order statistics. In this article Bonferroni–type inequalities and product–type approximations of order $k \geq 1$ are derived for the distribution and the moments of extreme order statistics for a sequence of stationary random variables. These results are particularized to m–spacings from a uniform distribution and moving sums of size m for independent normal random variables. These inequalities and approximations are compared with approximations and asymptotic results that have been previously derived.

From the numerical results it is evident that there is merit in studying higher order Bonferroni–type inequalities and product–type approximations. The product–type approximations appear to be the most accurate approximations for the distribution and the moments of extreme order statistics.

1. Introduction

Let $X_1, \dots, X_n$ be a sequence of dependent random variables from a continuous density function. Denote the extreme order statistics by

$$X_{(1)} = \min(X_1, \dots, X_n) \quad \text{and} \quad X_{(n)} = \max(X_1, \dots, X_n).$$

The distribution and the moments of extreme order statistics have been studied extensively in the iid case (David (1981) and Leadbetter, Lindgren and Rootzen (1983)). A major part of these studies focuses on the elegant asymptotic theory that has been developed for the iid case or in the dependent case for stationary sequences of random variables that satisfy the strong mixing condition, including the stationary m–dependent sequences (Leadbetter, Lindgren and Rootzen (1983) and Reiss (1989)).

[1]Work supported in part by the Research Foundation of the University of Connecticut.
AMS 1991 *subject classifications.* Primary 60E15; Secondary 62G30.
Key words and phrases. Bonferroni–type inequalities, moving sums, order statistics, product–type approximations, quasi–stationary, uniform spacings.

As the asymptotic results are not always accurate, the problem of approximating the distribution and the mean of extreme order statistics for dependent random variables has attracted many researchers. Arnold (1980, 1985), Arnold and Groeneveld (1979), Aven (1985), Gravey (1985) and Hoover (1989) have derived general bounds for the mean of the extreme order statistics. The use of the Boole–Bonferroni inequality and the Sidak inequality is outlined in David (1981, Section 5.3). Tong (1982, 1990, Chapter 6) has used the Sidak inequality to approximate the distribution and the mean of extreme order statistics for dependent normal random variables.

The methods that have been used in the articles listed above, only lightly utilize the dependence structure inherent in the distribution of $X_1, \ldots, X_n$. In this article I will consider only stationary sequences of dependent random variables. For this case, Bonferroni–type inequalities and product–type approximations of order $k \geq 1$ will be discussed in Section 2. The performance of Bonferroni–type and product–type inequalities for positively and negatively dependent random variables is being investigated in Glaz, Kuo and Yiannoutsos (1991).

In Section 3 these inequalities and approximations are applied to the distribution and the mean of the smallest m–spacing from a uniform distribution. A comparison with approximations and asymptotic results that have been previously derived will be presented in Tables 1 and 2 in Section 3.

In Section 4 the distribution of the maximum of a moving sum of m independent and identically distributed normal random variables is considered. Bonferroni–type inequalities and product–type approximations are derived and compared in Tables 3 and 4 with the Poisson approximation using the Chen–Stein method.

2. Product–Type and Bonferroni–Type Approximations and Inequalities

The first occurrence of using a *product–type inequality* to approximate a multivariate cumulative distribution function is recorded in Kimball's (1951) article. The following result is proved there:

THEOREM 2.1 *Let Y be a random variable with the density function $f(y)$ and let $g_i(y)$, $i = 1, \ldots, n$ be nonnegative monotone functions of the same type. Then,*

$$E\left[\prod_{i=1}^{n} g_i(Y)\right] \geq \prod_{i=1}^{n} E[g_i(Y)]. \tag{2.1}$$

An immediate consequence of this result is the following:

COROLLARY 2.2 *Let Y be a random variable with the density function $f(y)$ and $X_i = g_i(Y)$, where $g_i(y)$ are nonnegative monotone functions of the same type, $i = 1, \ldots, n$. Then for any sequence of nonnegative constants $c_1, \ldots, c_n$*

$$P[X_1 \leq c_1, \ldots, X_n \leq c_n] \geq \prod_{i=1}^{n} P[X_i \leq c_i] \tag{2.2}$$

and

$$P[X_1 > c_1, \ldots, X_n > c_n] \geq \prod_{i=1}^{n} P[X_i > c_i]. \tag{2.3}$$

To extend the product-type inequalities (2.2)–(2.3) to a larger class of distributions, Esary, Proschan and Walkup (1967) introduced the following concept of positive dependence.

DEFINITION 2.1 The random vector $\mathbf{X} = (X_1, \ldots, X_n)'$ is *associated* if for all coordinatewise nondecreasing functions f and g

$$\text{Cov}[f(\mathbf{X}), g(\mathbf{X})] = E[f(\mathbf{X})g(\mathbf{X})] - E[f(\mathbf{X})]E[g(\mathbf{X})] \geq 0.$$

If $\mathbf{X}$ is associated we will say that the random variables $X_1, \ldots, X_n$ are associated.

THEOREM 2.3 (Esary, Proschan and Walkup (1967)). *If $\mathbf{X} = (X_1, \ldots, X_n)'$ is associated then for all $c_1, \ldots, c_n$ inequalities* (2.2) *and* (2.3) *hold.*

REMARK Inequalities (2.2) and (2.3) are often called, in the context of simultaneous confidence intervals, Sidak inequalities (Sidak (1967, 1971)).

Inequalities (2.2)–(2.3) are referred to as *first order product–type inequalities,* since only one dimensional marginal distributions have been used. We say that an approximation or an inequality for the probability of an intersection or union of n events is of *order* k, if j dimensional marginal distributions are used in computing it, where $j \leq k$. While the first order product–type approximations and inequalities have the advantage of ease of computation, they are often quite inaccurate (Glaz and Johnson (1984)), the reason being that the dependence structure inherent in the random process is exploited only to a minimal degree. Therefore, Glaz and Johnson (1984, 1986) proposed to study product–type inequalities and approximations of degree $k \geq 2$. The following representation has motivated the study of these inequalities and approximations.

Let $A_j = (X_j \in I_j)$, $j = 1, \ldots, n$. Then for $k \geq 2$,

$$P\left(\bigcap_{j=1}^{n} A_j\right) = P\left(\bigcap_{j=1}^{k} A_j\right) \prod_{j=k+1}^{n} P\left(A_j \mid A_{j-1} \cap \cdots \cap A_{j-k+1} \bigcap_{s=1}^{j-k} A_s\right). \tag{2.4}$$

If one is able to evaluate marginal probabilities up to dimension k, then it is of interest to study when

$$(2.5) \qquad \gamma_k = P\left(\bigcap_{j=1}^{k} A_j\right) \prod_{j=k+1}^{n} P(A_j \mid A_{j-1} \cap \cdots \cap A_{j-k+1})$$

is an accurate (upper or lower) bound or an accurate approximation.

If the X_j's are stationary and $I_j = I$ then the above equation simplifies to:

$$(2.6) \qquad \gamma_k = P(X_j \in I; j = 1, \ldots, k) \\ [P(X_k \in I \mid X_j \in I; j = 1, \ldots, k-1)]^{n-k}.$$

A thorough discussion of the higher order product–type inequalities, conditions for their validity and the applications to various areas in probability and statistics are presented in Block, Costigan and Sampson (1988a, 1988b) and Glaz (1990a, 1991b).

In some applications (Glaz (1983, 1989), Glaz and Johnson (1986), Glaz and Naus (1991), Kenyon (1990) and Ravishanker, Wu and Glaz (1991)) the dependence structure of the distribution does not support the product–type inequalities for $k > 1$. Instead one can sometimes assert that

$$(2.7) \qquad \lim_{j \to \infty} P(X_j \in I_j \mid X_i \in I_i; i = 1, \ldots, j-1) = \theta,$$

where $0 < \theta < 1$ (Glaz (1989) and Glaz and Johnson (1986)), a property referred to in the statistical literature as *quasi–stationarity* (Darroch and Seneta (1965) and Tweedie (1974)). In this case, if X_j are stationary and $I_j = I$, one can use γ_k given in equation (2.6) as an approximation for $P(X_j \in I; j = 1, \ldots, n)$.

In Sections 3 and 4 the performance of product–type approximations will be evaluated for the distribution of extreme order statistics of two stationary sequences of random variables. In both cases the concept of quasi–stationarity will be utilized to support the use of these approximations.

The classical *Bonferroni inequalities* for the probability of a union of n events have been introduced in Bonferroni (1937). Let $A_1, \ldots, A_n$ be a sequence of events and define the event $A = \bigcup_{i=1}^{n} A_i$. Then for $2 \leq k \leq n$,

$$\sum_{j=1}^{k} (-1)^{j-1} S_j \leq P(A) \leq \sum_{j=1}^{k-1} (-1)^{j-1} S_j$$

where k is an even integer and for $j = 1, \ldots, n$

$$(2.8) \qquad S_j = \sum_{1 \leq i_1 < \cdots < i_j \leq n} P\left(\bigcap_{m=1}^{j} A_{i_m}\right).$$

The first order Bonferroni upper bound is referred to as Boole's inequality and was introduced earlier in Boole (1854). Boole's inequality has been used in approximating the tail distribution of extreme order statistics (David (1981, Section 5.3)).

The classical Bonferroni inequalities for order $k > 1$ are computationally complex and can be quite inaccurate (Prékopa (1988)). Therefore, attempts have been made to improve them. The improved inequalities are referred to as Bonferroni–type. In this article I will concentrate mainly on one special class of Bonferroni–type inequalities proposed by Hunter (1976) and Worsley (1982). Let $A_j = (X_j \in I_j)$, $j = 1, \ldots, n$. The basic idea of this approach is to express

$$(2.9) \qquad \begin{aligned} A = A_1 &\cup (A_2 \cap A_1^c) \cup (A_3 \cap A_2^c \cap A_1^c) \\ &\cup \cdots \cup (A_n \cap A_{n-1}^c \cap \cdots \cap A_1^c) \end{aligned}$$

and obtain the following inequality

$$(2.10) \qquad \begin{aligned} P(A) &\leq P(A_1) + \sum_{j=2}^{n} P(A_j \cap A_{j-1}^c) \\ &= \sum_{j=1}^{n} P(A_j) - \sum_{j=2}^{n} P(A_j \cap A_{j-1}), \end{aligned}$$

which is a special case of a more general second order inequality discussed in Hunter (1976) and Worsley (1982).

Recently, Hoover (1990) extended this class of Bonferroni–type inequalities to order $k \geq 3$. Consider again the identity (2.9). Then,

$$(2.11) \qquad P(A) \leq S_1 - \sum_{i=1}^{n-1} p_{i,i+1} - \sum_{j=2}^{k-1} \sum_{i=1}^{n-j} p^*_{i,i+1,\ldots,i+j}$$

where S_1 is defined in equation (2.8),

$$p_{i,i+1} = P(A_i \cap A_{i+1})$$

and

$$p^*_{i,i+1,\ldots,i+j} = P(A_i \cap A_{i+1}^c \cap \cdots \cap A_{i+j-1}^c \cap A_{i+j}).$$

The inequality (2.11) is an improvement over the inequality (2.10) and is a member of a more general class of inequalities discussed in Hoover (1990).

In Sections 3 and 4, X_j are stationary and $I_j = I$ for $j = 1, \ldots, n$ and hence $A_1, \ldots, A_n$ are stationary events. It is tedious but routine to show that inequality (2.11) implies for the problem at hand:

$$(2.12) \qquad P\left(\bigcap_{j=1}^{n} A_j\right) \geq (n-k+1)P\left(\bigcap_{j=1}^{k} A_j\right) - (n-k)P\left(\bigcap_{j=1}^{k-1} A_j\right) = \beta_k.$$

The advantage in using the Bonferroni–type inequalities (2.12) is that they are valid without any assumptions on the distribution of the random vector $\mathbf{X}$. On the other hand, they are usually inaccurate and quite often give a negative value. The product–type inequalities or approximations always produce a value between 0 and 1. The following result supports the use of product–type approximations and inequalities.

THEOREM 2.4 (Glaz (1990a)). *Let* $\mathbf{X} = (X_1, \ldots, X_n)'$ *be a random vector and* $A_j = (X_j \in I_j)$, $j = 1, \ldots, n$. *If* γ_k *and* β_k *are given by equations* (2.6) *and* (2.12) *respectively, then*

$$\gamma_k > \beta_k.$$

REMARK Other interesting approaches to obtain improved Bonferroni–type inequalities are discussed in Hoppe and Seneta (1990), Prékopa (1988), Seneta (1988) and Tomescu (1986).

In Sections 3 and 4 of this article I will illustrate the performance of the Bonferroni–type inequalities and the product–type approximations that were discussed above for the problem of evaluating the distribution and the mean of extreme order statistics for two sequences of stationary dependent random variables.

3. Extreme Spacings

Let $X_{(1)}, \ldots, X_{(n)}$ be the order statistics of iid observations from the uniform distribution on the interval (0,1]. Consider the m–spacings defined by

$$(3.1) \qquad X_{(m+i-1)} - X_{(i)}, \qquad i = 1, \ldots, n-m+1.$$

The distribution of the smallest of the m–spacings,

$$(3.2) \qquad M_n^{(m)} = \min_{1 \leq i \leq n-m+1} \left\{X_{(m+i-1)} - X_{(i)}\right\},$$

has been studied extensively (Barton and David (1956), Berman and Eagleson (1983, 1985), Cressie (1977a, 1977b, 1980, 1984), Darling (1953), Glaz

(1989, 1991a), Huntington and Naus (1975), Naus (1965, 1966, 1982), Neff and Naus (1980), Newell (1963), Pyke (1965), and Wallenstein and Neff (1987)). The distribution of $M_n^{(m)}$ is closely related to the distribution of *scan statistic*,

$$N_d = \sup_{0 \leq x \leq 1-d} N_{x,x+d}, \tag{3.3}$$

where $N_{x,x+d}$ is the number of observations that are in the scanning interval $(x, x+d]$ and $0 \leq d \leq 1$. $M_n^{(m)}$ is the smallest interval containing m points. The following relation is true:

$$P\{N_d \geq m\} = P\{M_n^{(m)} \leq d\}. \tag{3.4}$$

The exact distribution of $M_n^{(m)}$ is derived for $m = 2$ in Barton and David (1956) and Darling (1953), and for $m > 2$, under certain restriction in Naus (1965, 1966a, 1966b). A thorough discussion about the exact result for the case $m > 2$ (without any restrictions) is presented in Neff and Naus (1980), who tabulated $P\{M_n^{(m)} \leq d\}$ for $0 < d < .5$ and $3 \leq m < n \leq 20$. The formulas for evaluating the distribution of $M_n^{(m)}$ are complicated and computationally impractical for large value of n, moderate value of m and small value of d. Therefore, there has been an interest in evaluating asymptotic results: Berman and Eagleson (1983), Cressie (1977a, 1980), and McClure (1976). One can also employ the Chen–Stein method (Arratia, Goldstein and Gordon (1989, 1990), Chen (1975) and Stein, (1972, 1986, Chapter VIII)) and obtain Poisson approximation for the distribution of $M_n^{(m)}$.

In what follows an mth order product–type approximation and an mth order Bonferroni–type inequality are derived for $P\{M_n^{(m)} \leq d\}$. Similar results can be obtained for the distribution of the largest of the m–spacings.

For $3 \leq m \leq n/2$, $0 < d < .5$ and $0 \leq i \leq n - m + 1$ define the events

$$A_i = \{X_{(m+i-1)} - X_{(i)} \leq d\},$$

where $X_{(0)} = 0$. It follows that

$$P\left\{M_n^{(m)} \leq d\right\} = P\left\{\bigcup_{i=1}^{n-m+1} A_i\right\} = 1 - P\left\{\bigcap_{i=1}^{n-m+1} A_i^c\right\}.$$

The following notation will be used throughout this article:

$$Q_1^* = P\{A_0\}, Q_k^* = P\left\{A_0 \cap \left(\bigcap_{j=1}^{k-1} A_j^c\right)\right\}, 2 \leq k \leq n - m + 1 \tag{3.5}$$

and

$$(3.6) \qquad Q_k = P\left\{\bigcap_{j=0}^{k-1} A_j^c\right\}, \qquad 1 \le k \le n-m+1.$$

For $i = 1,\dots,n-m$, $Q_{i+1} = Q_i - Q_{i+1}^*$ and, therefore, for $i \ge 2$, $Q_i = Q_1 - \sum_{j=2}^{i} Q_j^*$. It follows from Glaz (1991a, Section 2) that for $1 \le k \le n-m$

$$(3.7) \qquad P\left\{M_n^{(m)} \le d\right\} \le 1 - Q_k + (n-m+1-k)Q_k^*.$$

Equation (3.7) is the Bonferroni–type inequality given in equation (2.11). The following result is used in evaluating the inequality (3.7) for $3 \le k \le m$:

THEOREM 3.1 (Glaz (1991a)). *Let $X_{(1)},\dots,X_{(n)}$ be the order statistics of iid observations from the uniform distribution on the interval $(0,1]$. Then for $3 \le k \le m \le n/2$ and $0 < d < .5$*

$$(3.8) \qquad Q_k^* = b(m-1;n,d) - b(m;n,d) + \sum_{j=k}^{n-m+1} (-1)^j \prod_{i=1}^{k-2}\left[1 - \frac{j(j-1)}{i(i+1)}\right] b(m+j-1;,n,d),$$

where

$$b(j;n,d) = \binom{n}{j} d^j(1-d)^{n-j}.$$

The above result is also useful in evaluating the following product–type approximation for $P\{M_n^{(m)} \le d\}$. For $1 \le i \le n-m$, write

$$(3.9) \qquad P\{M_n^{(m)} \le d\} = 1 - Q_{n-m+1} = 1 - Q_i \prod_{j=i+1}^{n-m+1} (Q_j/Q_{j-1}),$$

which can be approximated by

$$(3.10) \qquad P^{(k)}\{M_n^{(m)} \le d\} = 1 - Q_k(Q_k/Q_{k-1})^{n-m+1-k},$$

where $1 \le k \le n-m$. Approximation (3.10), referred to as the kth order product–type approximation, has been studied in Glaz (1989, 1991a). The product–type approximation (3.10) for $k = m$ can be viewed as an $(m-1)$ order Markov like approximation, where the terms Q_k/Q_{k-1}, for $m+1 \le k \le n-m+1$, are approximated by Q_m/Q_{m-1}. This approximation is supported by the asymptotic result stating that as $n \to \infty$ and $k \to \infty$ and $nd = O(1)$, $Q_k/Q_{k-1} \to \theta$, where $0 < \theta < 1$ is a constant (Glaz (1989, Theorem 3.1)).

In Table 1 the performance of the product–type approximation, the Bonferroni–type inequality and the asymptotic approximations mentioned above are compared for $n = 500$ and selected values of m, and d. From the numerical results it is evident that the product–type approximations are the most accurate ones. The Poisson asymptotic approximation using the Chen–Stein method, denoted by CS, outperforms all the other asymptotic approximations.

					Asymptotic Approximations		
d	m	Simulation	$P^{(m)}\{M_n^{(m)} < d\}$ (3.10)	$UB(m)$ (3.7)	Berman and Eagleson (1983)	Cressie (1977a)	CS
.001	4	.960	.997	> 1	.999	1.000	.999
	5	.507	.506	.700	.587	.728	.577
	6	.069	.068	.071	.083	.122	.080
.005	8	.977	.971	> 1	.999	1.000	.999
	9	.680	.670	> 1	.883	1.000	.869
	10	.271	.267	.309	.440	.995	.417
	12	.019	.018	.018	.031	.258	.028
.01	12	.935	.916	> 1	.999	1.000	.999
	13	.665	.645	> 1	.938	1.000	.921
	14	.341	.331	.397	.643	1.000	.604
	15	.135	.135	.144	.300	1.000	.270
	16	.048	.048	.049	.109	.999	.095
.05	38	.618	.582	.807	.999	1.000	.997
	40	.314	.304	.351	.947	1.000	.880
	44	.046	.045	.046	.293	1.000	.194
	46	.014	.014	.014	.099	1.000	.058

Table 1. Comparison of Seven Approximations to $P\{M_n^{(m)} < d\}$ for $n = 500$.

Note: This simulation is based on 20,000 trials.

We now turn to the problem of evaluating $E[M_n^{(m)}]$. Exact results are available for $m = 2$ (Parzen (1960)) and for $n/2 < m \leq n$ (Naus (1966)). For $2 < m \leq n/2$ the following approximation is evaluated in Glaz (1991a). First, note that in this case $P\{M_n^{(m)} > x\} = 0$ for $x > .5$ and therefore

$$E[M_n^{(m)}] = \int_0^{.5} \bar{F}(x)dx, \tag{3.11}$$

where

$$\bar{F}(x) = P\{M_n^{(m)} > x\}.$$

Using the extended Simpson's rule for $2N$ points (Davis and Polonsky (1972, p. 886)) to evaluate numerically the integral in equation (3.11) we get that

$$E[M_n^{(m)}] \approx \frac{1}{12N}\left[1 + 4\sum_{i=1}^{N} \bar{F}(x_{2i-1}) + \sum_{i=1}^{N-1} \bar{F}(x_{2i})\right], \tag{3.12}$$

where $x_i = i/4N$. The numerical procedure is set up as follows. Start with $N = 25$ (50 points) and proceed to double the number of points in the interval $[0, .5]$ until the difference between successive approximations for $E[M_n^{(m)}]$ is less than 10^{-6}. In Table 2 below the approximation (3.12) is evaluated for $n = 100$ and selected values of m. These approximations are compared with simulated values and inequalities derived by Arnold and Groeneveld (1979) and Aven (1985, Corollary 2.1). The approximation that one can obtain from the Bonferroni–type inequalities will not be presented here as they have not produced accurate results. Related Bonferroni–type inequalities for expected values of order statistics have been studied in Hoover (1989).

m	Simulation $E[M_n^{(m)}]$	$E[M_n^{(m)}]$ (3.12)	LB Arnold &	UB Groeneveld (1979)	LB Aven (1985, Corr. 2.1)
10	.039	.039	—	.089	—
20	.120	.122	—	.188	—
30	.213	.215	—	.287	—
40	.312	.313	—	.386	—
50	.415	.416	—	.485	—

Table 2. A Comparison of Five Approximations for $E[M_n^{(m)}]$, $n = 100$.

Note: $E[M_n^{(m)}]$ was estimated from a simulation with 20,000 trials. — denotes negative values for the lower bounds.

From Table 2 it is evident that the product–type approximation provides an accurate approximation for the expected length of the smallest m^{th} order spacing. For large m the Arnold and Groeneveld (1979) upper bound appear to be quite good.

4. Moving Sums of Normal Random Variables

Let $Z_1, \ldots, Z_n$ be iid standard normal random variables. Define the sequence of moving sums of order m:

$$X_i = Z_i + \cdots + Z_{i+m-1}, \qquad i = 1, \ldots, n - m + 1.$$

Approximations for the distribution of $X_{(n)} = \max(X_1, \ldots, X_n)$ have been discussed in Lai (1974), Bauer and Hackl (1980), Glaz and Johnson (1986) and Glaz (1990a). In this section I will evaluate the product–type and the Bonferroni–type approximations for the distribution of $X_{(n)}$ and compare them with the Poisson approximation (Aldous (1989, p. 50) and Holst and Janson (1990)). If the Z_i's have a discrete distribution the problem of approximating the distribution of the extreme order statistics of moving sums has been discussed in Glaz (1983), Glaz and Naus (1991), Naus (1982) and Samuel–Cahn (1983).

To evaluate the product–type approximation γ_k in equation (2.6) and the Bonferroni–type inequality β_k in equation (2.12) for

$$P\{X_{(n)} \leq a\} = P(X_1 \leq a, \ldots, X_n \leq a)$$

we have to evaluate the $k-1$ and k dimensional multivariate normal probabilities. In this paper I will use the algorithm developed in Schervish (1984). The inequalities and the approximations will be computed for $1 \leq k \leq 5$ only since for $k = 6$ the evaluation of the multivariate probabilities becomes time consuming. The use of product–type approximations for approximating the distribution of extreme order statistics for moving sums of normal random variables is supported by the quasi–stationarity property for moving sums of iid random variables in Glaz and Johnson (1986). For moving sums of iid discrete random variables the quasi–stationarity has been established in Samuel–Cahn (1983).

The Poisson approximation using the Chen–Stein method (Aldous (1989, p. 48–52)) is given by

$$\lim_{n\to\infty} P\{X_{(n)} \leq a\} = \exp\{-nx\}, \tag{4.1}$$

where a is large and

$$x = P\{X_1 + \cdots + X_m > a\}. \tag{4.2}$$

In Tables 3 and 4 the performance of the product–type approximations, Bonferroni–type inequalities and the Poisson asymptotic approximation using the Chen–Stein method (denoted in the tables by CS) is compared for $m = 10$ and $a = 10$.

n	P^*	γ_5	β_5	γ_4	β_4	γ_3	β_3	γ_1	β_1	CS
200	.926	.926	.923	.925	.923	.923	.920	.861	.850	.861
400	.856	.855	.844	.854	.842	.849	.836	.736	.694	.736
600	.792	.789	.764	.788	.762	.781	.752	.629	.537	.630
800	.732	.729	.684	.726	.681	.718	.669	.538	.381	.538
1000	.675	.673	.605	.670	.601	.660	.585	.460	.224	.460
2000	.455	.451	.206	.448	.198	.434	.168	.210	—	.210
3000	.305	.303	—	.299	—	.286	—	.096	—	.096
4000	.202	.203	—	.200	—	.188	—	.044	—	.044
5000	.133	.136	—	.133	—	.128	—	.020	—	.020

Table 3. A Comparison of Ten Approximations for $P[X_{(n)} \leq 10]$.

Note: P^* was estimated from a simulation with 10,000 trials. — denotes a negative value for the approximations.

n	P^*	γ_5	γ_4	γ_3	γ_2	CS
6000	.090	.091	.089	.081	.069	.0092
7000	.060	.061	060	.054	.045	.0042
8000	.041	.041	.040	.035	.029	.0019
9000	.028	.028	.027	.023	.018	.0009
10000	.020	.019	.018	.015	.012	.0004

Table 4. A Comparison of Six Approximations for $P[X_{(n)} \leq 10]$.

Note: P^* was estimated from a simulation with 10,000 trials.

From Tables 3 and 4 it is evident that the product–type approximations are more accurate than the Bonferroni–type inequalities and the Poisson approximation. The product–type approximation γ_5 is remarkably accurate throughout the entire range. It is also interesting to note that the Poisson approximation (4.1) is equal to the value of

$$\gamma_1 = (1 - x)^n$$

where x is defined in equation (4.2). It is routine to verify that the Poisson approximation always exceeds γ_1, but for a value of x that is close to 0 (which is the case in this example that was chosen to enhance the performance of the Poisson approximation) both approximations are equivalent.

At this point, I would like to note that Prékopa (1988) has used a linear programming approach to derive optimal k^{th} order Bonferroni–type inequalities for the probability of a union or intersection of n events. The difficulty in applying this approach is the need to evaluate the terms S_j given in equation (2.8) for large j. For a special case of the problem considered in this section ($m = 2$) Glaz (1990b) compares the performance of the product–type approximation with these Bonferroni–type inequalities. Again, the product–type approximation appears to be a more accurate approximation.

Acknowledgements

The author wishes to thank the editors and the referees for their comments which improved the presentation of the results in this article.

REFERENCES

ALDOUS, D. (1989). *Probability Approximations via the Poisson Clumping Heuristic.* Springer–Verlag, New York.

ARNOLD, B. C. (1980). Distribution–free bounds on the mean of a dependent sample. *SIAM J. Appl. Math.* **38** 163–167.

ARNOLD, B. C. (1985). p–Norm bounds on the expectation of the maximum of possibly dependent sample. *J. Multivariate Anal.* **17** 316–332.

ARNOLD, B. C. AND BALAKRISHNAN, N. (1989). *Relations, Bounds and Approximations for Order Statistics.* Lecture Notes in Statistics Vol. 53, Springer–Verlag, New York.

ARNOLD, B. AND GROENEVELD, R. (1979). Bound on expectations of linear systematic statistics based on dependent samples. *Ann. Statist.* **7** 220–223. Correction **8** 1401.

ARRATIA, R., GOLDSTEIN, L. AND GORDON, L. (1989). Two moments suffice for Poisson approximations: The Chen–Stein method. *Ann. Probab.* **17** 9–25.

ARRATIA, R., GOLDSTEIN, L. AND GORDON, L. (1990). Poisson approximation and the Chen–Stein method. *Statist. Sci.* **5** 403–434.

AVEN, T. (1985). Upper (lower) bounds on the mean of the maximum (minimum) of a number of random variables. *J. Appl. Prob.* **22** 723–728.

BALKRISHNAN, N. (1990). Improving the Hartley–David–Gumbel bound for the mean of extreme order statistics. *Statist. Prob. Lett.* **9** 291–294.

BARTON, D. E. AND DAVID, F. N. (1956). Some notes on ordered random intervals. *J. Royal Stat. Soc. Ser. A* **18** 79–94.

BAUER, P. AND HACKL, P. (1980). An extension of the mosum technique for quality control. *Technometrics* **22** 1–7.

BERMAN, M. AND EAGLESON, G. K. (1983). A Poisson limit theorem incomplete symmetric statistics. *J. Appl. Prob.* **20** 47–60.

BERMAN, M. AND EAGLESON, G. K. (1985). A useful upper bound for the tail probabilities when the sample size is large. *J. Amer. Statist. Assoc.* **80** 886–889.

BLOCK, H. W., COSTIGAN, T. AND SAMPSON, A. R. (1988a). Product–type probability bounds of higher order. Tech. Report No. 88–08, Department of Mathematics and Statistics, University of Pittsburgh.

BLOCK, H. W., COSTIGAN, T. AND SAMPSON, A. R. (1988b). Optimal second order product–type probability bounds. Tech. Report No. 88–07, Department of Mathematics and Statistics, University of Pittsburgh.

BONFERRONI, C. E. (1937). Teoria statistica delle classi e calcolo delle probabilita. *Volume in onore di Riccardo Dalla Volta,* Universita di Firenze, 1–62.

BOOLE, G. (1854). *An Investigation of the Laws of Thought on which are Founded the Mathematical Theories of Logic and Probabilities.* Macmillan, London (Repr. Dover, New York, 1958).

CHEN, L. H. Y. (1975). Poisson approximations for dependent trails. *Ann. Probab.* **3** 534–545.

CRESSIE, N. (1977a). The minimum of higher order gaps. *Austral. J. Statist.* **19** 132–143.

CRESSIE, N. (1977b). On some properties of the scan statistic on the circle and the line. *J. Appl. Prob.* **14** 272–283.

CRESSIE, N. (1980). The asymptotic distribution of the scan under uniformity. *Ann. Probab.* **8** 828–840.

CRESSIE, N. (1984). Using the scan statistic to test uniformity. *Colloq. Math. Soc. Janos Bolayai* **45** Goodness of Fit, Debrecen, Hungary, p. 87–100.

DARLING, D. A. (1953). On a class of problems related to the random division of an interval. *Ann. Math. Statist.* **24** 239–253.

DARROCH, J. N. AND SENETA, E. (1965). On quasi-stationary distributions in absorbing discrete time finite Markov chains. *J. Appl. Prob.* **2** 88–100.

DAVID, H. A. (1981). *Order Statistics*, 2nd ed., Wiley, New York.

ESARY, J. D., PROSCHAN, F. AND WALKUP, D. W. (1967). Association of random variables with applications. *Ann. Math. Statist.* **38** 1466–1474.

DAVIS, P. J. AND POLONKSY, I. (1972). Numerical interpolation, differentiation and integration. In *Handbook of Mathematical Functions with Formulas, Graphs, and Mathematical Tables*, M. Abramowitz and L. E. Stegun, eds., National Bureau of Standards AMS **55** Washington, DC, 875–924.

GLAZ, J. (1983). Moving window detection for discrete data. *IEEE Trans. Inform. Theory IT*-29, 457–462.

GLAZ, J. (1989). Approximations and bounds for the distribution of the scan statistic. *J. Amer. Statist. Assoc.* **84** 560–566.

GLAZ, J. (1990a). A comparison of Bonferroni–type and product–type inequalities in the presence of dependence. In *Topics in Statistical Dependence,* H. W. Block, A. R. Sampson and T. H. Savits, eds. Institute of Mathematical Statistics, Hayward, CA. 223–235.

GLAZ, J. (1990b). Product–type approximations and inequalities. *Appl. Probab. News* Vol. 14, 3–6.

GLAZ, J. (1991a). Approximations for tail probabilities and moments of the scan statistic. (to appear in *Comput. Stat. Data Anal.*)

GLAZ, J. (1991b). Approximate simultaneous confidence intervals. Tech. Report No. 11, Department of Statistics, University of Connecticut.

GLAZ, J. AND JOHNSON, B. MCK. (1984). Probability inequalities for multivariate distributions with dependence structures. *J. Amer. Stat. Assoc.* **79** 436–440.

GLAZ, J. AND JOHNSON, B. MCK. (1986). Boundary crossing for moving sums. *J. Appl. Prob.* **25** 81–88.

GLAZ, J., KUO, L. AND YIANNOUTSOS, C. (1991). Approximations and inequalities for the distribution and the moments of extreme order statistics for positive and negative dependent random variables. *Preliminary Report.*

GLAZ, J. AND NAUS, J. (1986). Approximating probabilities of first passage in a particular Gaussian process. *Comm. Statist.-Theor. Meth. Ser. A* **15** 1709–1722.

GLAZ, J. AND NAUS, J. (1991). Tight bounds and approximations for scan statistic probabilities for discrete data. *Ann. Appl. Probab.* **1** 306–318.

GRAVEY, A. (1985). A simple construction of an upper bound for the mean of the maximum of n identically distributed random variables. *J. Appl. Prob.* **22** 844–851.

HOLST, L. AND JANSON, S. (1990). Poisson approximations using the Chen–Stein method and coupling: Number of exceedances of Gaussian random variables. *Ann. Probab.* **18** 713–723.

HOOVER, D. R. (1989). Bounds on expectations of order statistics for dependent samples. *Statist. Prob. Lett.* **8** 261–264.

HOOVER, D. R. (1990). Subset complement addition upper bounds — an improved inclusion–exclusion method. *J. Statist. Plann. Inf.* **24** 195–202.

HOPPE, F. M. AND SENETA, E. (1990). A Bonferroni–type identity and permutation bounds. *Inter. Statist. Rev.* **58** (3), 253–261.

HUNTER, D. (1976). An upper bound for the probability of a union. *J. Appl. Prob.* **13** 597–603.

HUNTINGTON, R. J. AND NAUS, J. I. (1975). A simpler expression for k^{th} nearest neighbour coincidence probabilities. *Ann. Probab.* **3** 894–896.

KENYON, J. R. (1990). Bounds and approximations for pairwise comparisons of independent multinormal means. *Commun. Statist. Simul.* **19** 555-590.

KIMBALL, A. W. (1951). On dependent tests of significance in the analysis of variance. *Ann. Math. Statist.* **22** 600–602.

LAI, T. L. (1974). Control charts based on weighted sums. *Ann. Statist.* **2** 134–147.

LEADBETTER, M. R., LINDGREN, G. AND ROOTZEN, H. (1983). *Extremes and Related Properties of Random Sequences and Processes.* Springer–Verlag, New York.

MCCLURE, D. E. (1976). Extreme non uniform spacings. *Report No.* 44 *in Pattern Analysis*, Brown University, Providence, RI.

NAUS, J. I. (1965). The distribution of the size of the maximum cluster of points on a line. *J. Amer. Statist. Assoc.* **60** 532–538.

NAUS, J. I. (1966). Some probabilities, expectations and variances for the size of the largest clusters and smallest intervals. *J. Amer. Statist. Assoc.* **61** 1191–1199.

NAUS, J. I. (1982). Approximations for distributions of scan statistics. *J. Amer. Statist. Assoc.* **77** 177–183.

NEFF, N. D. AND NAUS, J. I. (1980). The distribution of the size of the maximum cluster of points on a line. *Vol.* 6 *in IMS Series of Selected Tables in Mathematical Statistics*, American Mathematical Society, Providence, RI.

NEWELL, G. F. (1963). Distribution for the smallest distance between any pair of kth nearest neighbor random points on a line. In *Proc. Symp. Time Series Analysis,* Wiley, New York, 89–103.

PARZEN, E. (1960). *Modern Probability Theory and its Applications.* Wiley, New York.

PRÉKOPA, A. (1988). Boole–Bonferroni inequalities and linear programming. *Oper. Res.* **36** 146–162.

PYKE, R. (1965). Spacings. *J. Royal Stat. Soc. Ser. B* **27** 395–449.

RAVISHANKER, N., WU, L. S. Y. AND GLAZ, J. (1991). Multiple prediction intervals for time series: comparison of simultaneous and marginal intervals. *J. Forecast.* **10** 445–463.

REISS, R. D. (1989). *Approximate Distributions of Order Statistics.* Springer–Verlag, New York.

SAMUEL–CAHN, E. (1983). Some approximations to the expected waiting time for a cluster of any given size for point processes. *Adv. Appl. Prob.* **15** 21–38.

SCHERVISH, M. J. (1984). Algorithm AS 195, multivariate normal probabilities with error bound. *Appl. Statist.* **33** 81–94.

SENETA, E. (1988). Degree, iteration and permutation in improving Bonferroni–type bounds. *Australian J. Statist.* **30A** 27–38.

SIDAK, Z. (1967). Rectangular confidence regions for means of multivariate normal distributions. *J. Amer. Stat. Assoc.* **62** 626–633.

SIDAK, Z. (1971). On probabilities on rectangles in multivariate Student distributions: their dependence on correlations. *Ann. Math. Statist.* **42** 169–175.

STEIN, C. (1972). A bound for the error in the normal approximations to the distribution of a sum of dependent random variables. In *Proc. Sixth Berkeley Symp. Math. Statist. Probab.* **2**, L. M. LeCam, J. Neyman and E. L. Scott, eds., University of California Press, Berkeley, CA. 583–602.

STEIN, C. (1986). *Approximate Computation of Expectations.* Institute of Mathematical Statistics, Hayward, CA.

TOMESCU, I. (1986). Hypertrees and Bonferroni inequalities. *J. Combina. Theory* **B41** 209–217.

TONG, Y. L. (1982). Some applications of inequalities for extreme order statistics to a genetic selection problem. *Biometrics* **38** 333–339.

TONG, Y. L. (1990). *The Multivariate Normal Distribution.* Springer–Verlag, New York.

TWEEDIE, R. L. (1974). Quasi-stationary distributions for Markov chains on a general state space. *J. Appl. Prob.* **11** 726–741.

WALLENSTEIN, S. AND NEFF, N. (1987). An approximation for the distribution of the scan statistic. *Statist. Medicine* **6** 197–207.

WORSLEY, K. J. (1982). An improved Bonferroni inequality and applications. *Biometrika* **69** 297–302.

DEPARTMENT OF STATISTICS
UNIVERSITY OF CONNECTICUT
STORRS, CT 06269-3120

Stochastic Inequalities
IMS Lecture Notes – Monograph Series
Volume 22 (1993)

PARTITIONING INEQUALITIES IN PROBABILITY AND STATISTICS[1]

By THEODORE P. HILL

Georgia Institute of Technology

This article surveys fair–division or cake–cutting inequalities in probability and statistics, including bisection inequalities, basic fairness inequalities, convexity tools, superfairness inequalities, and partitioning inequalities in hypotheses testing and optimal stopping theory. The emphasis is measure theoretic, as opposed to game theoretic or economic, and a number of open problems in the area are mentioned.

1. Introduction

The main purpose of this article is to present a unified study of a class of partitioning inequalities in the theories of probability and statistics; it is not meant to be a complete review of the subject. The emphasis is measure theoretic with emphasis on both constructive (algorithmic) and non–constructive techniques, including generalizations of classical "cake–cutting" inequalities, the ham sandwich theorem, and classical statistical problems such as Fisher's Problem of the Nile, the problem of smiliar regions, and the classification problem.

The overall framework is as follows. There are a finite number of (countably additive) probability measures $\boldsymbol{\mu} = (\mu_1, \ldots, \mu_n)$ defined on the *same* measurable space $(\Omega, \mathcal{F})$, and a class Π of $\mathcal{F}$–measurable partitions of Ω is specified. (Recall that $(A_i)_1^k$ is an (ordered) $\mathcal{F}$–partition of Ω if $\cup_1^k A_i = \Omega, A_i \cap A_j = \emptyset$ for $i \neq j$, and $A_i \in \mathcal{F}$ for all i.) From this collection of partitions Π a single partition is sought which will satisfy some objective such as bisection or minimax–risk property. It may help the reader to keep in mind either a cake–cutting or a hypotheses–testing interpretation of this setting throughout the following sections.

In the cake–cutting interpretation, Ω is a cake which must be divided among n people having values $\mu_1, \ldots, \mu_n$ (that is, $\mu_j(A)$ is the relative value

[1]Research supported in part by National Science Foundation Grant DMS 89–01267.

AMS 1991 *subject classifications.* Primary 60E15, 62C20, 28B05; Secondary 62H15, 60A10, 28A99, 90A05.

Key words and phrases. Partitioning inequalities, fair–division problems, cake–cutting theorems, bisection, ham–sandwich theorems, Lyapounov convexity, hypothesis testing, classification problem, optimal stopping.

of piece (measurable subset) A to person j), and Π describes the permissible divisions (e.g., into parallel slices, or convex connected pieces, or general Borel sets). The basic measure–theoretic assumptions of nonnegativity and (countable) additivity seem natural in this setting: the value of any piece is at least zero (otherwise the piece could be discarded); and the value of the union of several disjoint pieces is the sum of the values of the individual pieces.

In the hypotheses testing interpretation, a single observation is made of a random variable X taking values in Ω, and it must then be decided from which of n known distributions $\mu_1, \ldots, \mu_n$ the observation came; this is known as the *classification problem.* In this case, the decision rule "if $X \in A_i$, guess that the distribution of X is μ_i" corresponds to a partition of Ω, and the expected *risk* associated with this decision given that the true distribution is μ_i is given by $1 - \mu_i(A_i) = P(X \notin A_i \mid \text{true distribution of } X \text{ is } \mu_i)$.

The organization of this article is as follows: Section 2 addresses bisection results including ham sandwich theorems and medians; Section 3 the basic fairness inequalities; Section 4 the convexity tools, with special emphasis on Lyapounov's theorem and extensions; Section 5 various superfairness inequalities; Section 6 inequalities in statistical decision theory related to the classification problem; Section 7 partitioning inequalities in optimal–stopping theory; and Section 8 a list of open problems.

It should be emphasized that the results in this article are focused on probabilistic and statistical partitioning inequalities, and do not include discussion of related fair–division results in other areas such as combinatorics (cf. Alon and West (1986)), economics (cf. Kirman (1981), Samuelson (1980), Svensson (1983), Weller (1985), Young (1987)) or game theory (cf. Kuhn (1973), Legut (1990)).

2. Bisection

The Bisection Problem is the question of the existence of a single ($\mathcal{F}$–measurable) subset A of Ω which bisects Ω simultaneously with respect to each of the measures $\mu_1, \ldots, \mu_n$, that is

$$\mu_i(A) = 1/2 \quad \text{for all } i = 1, \ldots, n. \tag{1}$$

In general such a set does not exist (e.g., if $\mu_1 = \mu_2$ is a Dirac measure assigning mass 1 to a single point ω in Ω), but the ham sandwich theorem of Steinhaus (cf. Stone and Tukey (1942)) guarantees that there is even a *half–space* simultaneous bisection in certain cases, namely

(2) if $\mu_1, \ldots, \mu_n$ are uniformly distributed (probability measures) on bounded Borel subsets of $\Omega = \mathbb{R}^n$, there is halfspace H^+ satisfying $\mu_i(H^+) = 1/2$ for all $i = 1, \ldots, n$.

The classical proof of (2) uses the Borsuk–Ulam fixed point theorem, and relies heavily on the fact that the number of measures is no more than the dimension of the space. The hypotheses of (2) can be weakened (Stone and Tukey (1942)) to "$\mu_1, \ldots, \mu_n$ are probability measures on $\mathbb{R}^n$ satisfying $\mu_i(H) = 0$ for every hyperplane H and all i," but the conclusion may fail if the measures have atoms.

On the other hand, hyperplane bisection in a *median* sense is always possible for arbitrary (including atomic) probability measures in this setting. Say that a hyperplane $H = \sum_{i=1}^n a_i x_i - b$ in $\mathbb{R}^n$ is a *hyperplane median* for μ if $\mu(H^+) \geq 1/2$ and $\mu(H^-) \geq 1/2$, where $H^+ = \{\mathbf{r} = (r_1, \ldots, r_n) \in \mathbb{R}^n : \sum_{i=1}^n a_i r_i \geq b\}$ and $H^- = \{\mathbf{r} \in \mathbb{R}^n : \sum_{i=1}^n a_i r_i \leq b\}$. Using the Borsuk-Ulam theorem applied to a "midpoint-median" function, it was shown in Hill (1988a) that

(3) every collection $\mu_1, \ldots, \mu_n$ of arbitrary Borel probability measures on $\Omega = \mathbb{R}^n$ always has a common hyperplane median.

Using countable additivity, it can be seen that if $\boldsymbol{\mu}$–bisection of every measurable set is possible, then the range of $\boldsymbol{\mu}$ is convex (and conversely); cf. Dubins and Spanier (1961) and Section 4 below. Stone and Tukey (1942) have shown that for any two nonatomic Borel measures on the closed unit circle S^1 there is always an *interval* which bisects each of the measures simultaneously. For an interesting example of combinatorial bisection also based on the Borsuk-Ulam theorem, the reader is referred to Alon and West (1986).

For general $\alpha \in [0,1]$ the question of the existence of a set A satisfying $\mu_i(A) = \alpha$ for all i (as opposed to the exact bisection in (1)), is called the Problem of Similar Regions (Feller (1938)), as such a set A is in some gross sense a smaller copy of Ω itself, and this question has been related to the efficiency of tests of statistical hypotheses by Neyman and Pearson (1933). In contrast to (2), it is not always possible to find a hyperplane H satisfying $\mu_i(H^+) = \alpha$ for $\alpha \neq 1/2$, even if the measures are continuous (Hill (1988a)). On the other hand, it follows from convexity (Section 4 below) that if the measures are all nonatomic, then for each $\alpha \in [0,1]$ there is a measurable set A_α satisfying $\mu_i(A_\alpha) = \alpha$ for all $i \leq n$.

3. Basic Fairness Inequalities

The basic question of the existence of a *fair division* is that of the existence of an ($\mathcal{F}$–measurable) n–partition $(A_i)_{i=1}^n$ of Ω satisfying

$$\mu_i(A_i) \geq 1/n \quad \text{for all } i = 1, \ldots, n. \tag{4}$$

The cake–cutting interpretation of (4) says that cutting the cake into pieces $A_1, \ldots, A_n$ and distributing it so that the ith piece is given to person i guarantees that each person receives a portion which he considers, by his own measure, to be at least one n^{th} of the total. The hypotheses–testing interpretation of (4) is that the decision rule corresponding to the partition $(A_i)_1^n$ has expected risk at most $(n-1)/n$. Just as bisections (1) do not exist in general, neither do fair divisions in the sense of (4). They do, however, if all the measures are nonatomic (Steinhaus (1949)), and more generally if at most one of the measures has atoms (Hill and Kennedy (1990)):

(5) if $\mu_1, \ldots, \mu_{n-1}$ are nonatomic, then there is a measurable partition $(A_i)_1^n$ with $\mu_i(A_i) \geq 1/n$ for all $i \leq n$.

For $n = 2$, the demonstration of (5) is the classical "cut–and–choose" algorithm: the first person (the person with the nonatomic measure) identifies a subset A which bisects his own measure of Ω (i.e., $\mu_1(A) = 1/2$), and the second person chooses between A and the complement of A. Kirby (1988) has recently applied this idea to obtain an algorithm for nuclear arms reduction.

For $n > 2$, there are several well known algorithms to demonstrate (5). One algorithm, called a "sliding knife" solution, is a modification of an algorithm of Knaster and Steinhaus (1946, 1953) by Dubins and Spanier (1961). Although stated under the hypothesis that *all n* measures are continuous (that is, absolutely continuous with respect to Lebesgue measure on $\Omega \subset \mathbb{R}^n$), the procedure also works if at most one of the measures has atoms. In this algorithm, a long knife is passed slowly parallel to itself over the cake Ω until one of the participants feels that the increasing portion under the knife is exactly one n^{th} the total value, at which point he says "stop," and the cake is cut at that point and that slice is given to the person who said stop (ties are broken in any manner), and the remaining $n-1$ participants continue the process. For continuous measures, any starting orientation of the knife will suffice, and for the more general nonatomic case, it follows from Jones (1989) that almost all starting angles will suffice (a "sufficient" starting angle is one in which at most one measure of the corresponding increasing slice at that angle is discontinuous).

Fink (1964) has given an algorithm demonstrating (5) which generalizes the cut–and–choose algorithm and which has two advantages over the sliding–knife solution: first, implementation of the algorithm does not require *a priori* knowledge of the number of participants; and second, the algorithm is essentially finite, as opposed to the continuous–evaluation method of the sliding–knife solution. In Fink's algorithm the first person bisects the cake Ω according to his own measure. The second person arriving chooses between the two pieces cut by the first player, and if a third person arrives then each of these first two players *trisects* his own portion, and the third person selects one portion from each. The algorithm continues in this manner (e.g., quadrisection at the next stage), until no new arrivals appear and the algorithm termminates. (Note that the single measure with atoms must "arrive" last in order to guarantee the solution in (5).) A variation of this algorithm requiring at most $\mathcal{O}(n \log n)$ cuts for parallel slices in $\mathbb{R}^n$ is described in Even and Paz (1984).

The bound $1/n$ in (5) is easily seen to be best possible (taking $\mu_1 = \mu_2 = \ldots = \mu_n$), and the corresponding best possible bound for nonatomic finite (e.g., non–probability) measures is one–n^{th} the harmonic mean of the total masses (Hill (1985)), a probabilistic analog of which is

(6) if $X_1, X_2, \ldots, X_n$ are nonnegative continuous random variables on $(\Omega, \mathcal{F}, P)$ with finite means, then there is a measurable partition $(A_i)_1^n$ of Ω satisfying $\int_{A_i} X_i \geq ((EX_1)^{-1} + \ldots + (EX_n)^{-1})^{-1}$ for all i, and this bound is best possible.

Although an algorithmic proof of (6) is possible in some special cases, such as when the $\{EX_i\}$ are all rational numbers, proofs of the general case seem to rely on non–constructive results such as the convexity conclusions in Section 4 below.

If more than one of the measures $\mu_1, \ldots, \mu_n$ has atoms, then the conclusion of (5) may fail, but if an upper bound is known for the maximum atom size of the measures, the following best possible fairness bound is known (Hill (1987a)); taking limits as $\alpha \to 0$ yields (5) as a corollary.

(7) If $\mu_1, \ldots, \mu_n$ each have atoms at most $\alpha > 0$, then there exists a measurable partition $(A_i)_1^n$ of Ω satisfying $\mu_i(A_i) \geq V_n(\alpha)$ for all $i = 1, \ldots, n$, where $V_n : [0,1] \to [0, n^{-1}]$ is the unique nonincreasing function satisfying $V_n(\alpha) = 1 - k(n-1)\alpha$ for $\alpha \in [(k+1)k^{-1}((k+1)n-1)^{-1}, (kn-1)^{-1}]$. Moreover, this bound is attained for all n and α.

The function V_n is piecewise linear and satisfies $V_n(\alpha) \nearrow n^{-1}$ as $\alpha \searrow 0$; see Hill (1987a) for the graphs of V_2 and V_3. The proof of (7) is largely combinatorial and nonconstructive in nature.

Analogs of the fairness inequalities in this section for game theory and economics can be found in Crawford (1977), Crawford and Heller (1979), Demko and Hill (1988), Kuhn (1973), and Legut (1985).

4. Convexity Tools

For nonatomic measures, the basic partitioning–inequality tool is the celebrated convexity theorem of Lyapounov (1940) which states that the range of every countably–additive, finite–dimensional, vector–valued measure is closed and convex. Many proofs of this theorem and various generalizations have appeared (e.g. Armstrong and Prikry (1981), Artstein (1990), Blackwell (1951a), Dvoretzky, Wald and Wolfowitz (1951), Elton and Hill (1987), Gouweleeuw (1991), Halmos (1948), Karlin (1953), Lindenstrauss (1966), Margolies (1978)); that of Lindenstrauss (1966) based on the Krein–Milman Theorem being perhaps the shortest, and that of Artstein (1990) perhaps the most elementary. Since $\mathbf{0} = \boldsymbol{\mu}(\emptyset)$ and $\mathbf{1} = (1, 1, \ldots, 1) = \boldsymbol{\mu}(\Omega)$ are in the range of $\boldsymbol{\mu}$, Lyapounov's theorem immediately guarantees that if the measures are all nonatomic, measurable bisecting sets (1) and fair divisions ((4) with *equality*) always exist; a little extra effort yields (6) (also with equality).

In non–probabilistic applications, the convexity theorem has had widespread application (see Akemann and Anderson (1990)) in combinatorics, control theory (to prove the basic bang–bang principle; LaSalle (1960)) differential equations, economics, functional analysis, graph theory, and logic. Another curious probabilistic implication of the convexity theorem apparently first obtained by Blackwell (1951b) is: given any finite collection of continuous (Borel) probability distributions on the real line, there is a sub–σ–algebra $\mathcal{G}$ of the Borels with the property that, restricted to $\mathcal{G}$, those measures are *identical* non–atomic probability measures.

A generalization of Lyapounov's convexity theorem due to Dvoretzky, Wald, and Wolfowitz (1951) (see also Dubins and Spanier (1961)) which is particularly useful in the present setting is the following:

(8) if $\mu_1, \ldots, \mu_n$ are nonatomic, then for each k,

$$\{(\mu_i(A_j))_{i=1}^{n}{}_{j=1}^{k} : (A_j)_1^k \text{ is a measurable } k\text{–partition of } \Omega\}$$

is a compact and convex subset of $n \times k$ real matrices.

The proof of (8) is based upon Lyapounov's theorem and an idea of stringing together measures attributed to Blackwell (1951a). Application of (8) yields an affirmative solution to R. A. Fisher's "Problem of the Nile" (1936) for nonatomic measures $\mu_1, \ldots, \mu_n$; namely, the existence for each natural number k of a measurable k–partition $(A_j)_1^k$ satisfying

$$\mu_i(A_j) = 1/k \text{ for all } i \le n \text{ and all } j \le k,$$

and more generally, for each k and each set of positive numbers $\alpha_1, \ldots, \alpha_k$ with $\sum \alpha_i = 1$, a k–partition satisfying $\mu_i(A_j) = \alpha_j$ for all $i \le n$ and all

$j \le k$. Some information concerning the minimal number of "cuts" required to obtain such a partition (in the case $k = n$; $\alpha_j = 1/k$) is contained in Legut (1991) and in Stromquist and Woodall (1985); Legut (1991) gives a qualitative characterization of the partitions for continuous measures.

The "closed" conclusion of Lyapounov's theorem holds even if the measures have atoms, but the "convexity" conclusion fails in general. On the other hand, if a bound on the maximum atom size is known, then the following generalization by Elton and Hill (1987) of the convexity theorem gives a bound on how non-convex the range may be; intuitively, if the atoms are all very small, the range of $\boldsymbol{\mu}$ will be close to convex:

(9) if $\mu_1, \ldots, \mu_n$ each have atoms at most $\alpha > 0$, then the Hausdorff distance from the range $\boldsymbol{\mu}$ to its convex hull is at most $\alpha n/2$.

(Recall that the range of $\boldsymbol{\mu}$ is the subset of $\mathbb{R}^n$ given by $\{\boldsymbol{\mu}(A) : A \in \mathcal{F}\}$, and the Hausdorff distance between $S_1 \subset S_2$ is $d(S_1, S_2) = \sup_{x \in S_2} \inf_{y \in S_1} |x - y|$.)

Thus (9) affords approximate solutions to the bisection and fair–division problems (as well as the Problem of the Nile), in the case of measures with atoms no bigger than α. For example, it implies that if μ_1, μ_2, μ_3 each have atoms no larger than $1/100$, then there is an "almost bisecting" set A satisfying $97/200 \le \mu_i(A) \le 103/200$ for all $i \le 3$.

The convexity conclusion of Lyapounov's theorem may fail if the number of measures is infinite (Feller (1938)). If the hypothesis of countable additivity is weakened to finite additivity, the convexity conclusion still holds (Margolies (1978), Armstrong and Prikry (1981)).

5. Superfairness Inequalities

If $\mu_1, \ldots, \mu_n$ are *identical* measures, then any partition $(A_i)_1^n$ satisfying the fairness inequality (4) necessarily holds with equality for all i. On the other hand,

(10) if $\mu_1, \ldots, \mu_n$ are nonatomic and $\mu_i \ne \mu_j$ for some $i \ne j$, then there is a measurable partition $(A_i)_1^n$ of Ω satisfying $\mu_i(A_i) > 1/n$ for all $i \le n$.

The result (10) was apparently first stated by Knaster and Steinhaus (1953), proved by Urbanik (1955) for the case the measures all have the same null sets, and proved independently by Dubins and Spanier (1961) for the general nonatomic case. Although both these proofs used Lyapounov's Convexity Theorem (note that the strict inequality conclusion $> 1/n$ requires a much more subtle argument than the weak inequality $\ge 1/n$), Woodall

(1980) has modified Fink's (1964) fair–division algorithm to yield an algorithm to generate the superfair partition appearing in (10). Woodall's algorithm, however, requires more information than just $\mu_i \neq \mu_j$ for some $i \neq j$; it requires knowledge of a set A, real numbers $\alpha \neq \beta$, and exact indices i and j satisfying $\mu_i(A) = \alpha \neq \beta = \mu_j(A)$.

Woodall's algorithm proving (10) does not give any bound strictly greater than $1/n$, but such bounds are possible if the total masses of the supremum or infimum of the μ_i's are known. (Here $\bigvee_{i=1}^{n} \mu_i$ is the smallest measure dominating each μ_i, and $\bigwedge_{i=1}^{n} \mu_i$ is the largest measure dominated by each μ_i; it is an easy exercise to show that such measures always exist.) The superfair inequality of Elton, Hill and Kertz (1986)

(11) if $\mu_1, \ldots, \mu_n$ are nonatomic, then there is a measurable partition $(A_i)_1^n$ of Ω satisfying $\mu_i(A_i) \geq (n - M + 1)^{-1}$, where M is the total mass of $\bigvee_{i=1}^{n} \mu_i$,

is best possible, and improves (10) since $M \geq 1$, with equality if and only if $\mu_1 = \ldots = \mu_n$. The proof in Elton, Hill and Kertz (1986) is partly constructive, and Legut (1988) contains an easy nonconstructive proof based on Lyapounov's convexity theorem. Using the convexity theorem and an "inversion principle," an analog of (11) for the infimum was obtained by Hill (1987b):

(12) if $\mu_1, \ldots, \mu_n$ are nonatomic, then there exists a measurable partition $(A_i)_1^n$ of Ω satisfying $\mu_i(A_i) \geq (n + m - 1)^{-1}$, where m is the total mass of $\bigwedge_{i=1}^{n} \mu_i$.

The bound in (12) is also best possible, and improves (10) since $m \leq 1$, with equality if and only if $\mu_1 = \ldots = \mu_n$. If $n = 2$, (11) and (12) are equivalent since $m + M = 2$, but for $n > 2$ neither (11) nor (12) implies the other.

The superfairness inequality (10) can also be generalized in another direction. If $\{\alpha_i\}_1^n$ are nonnegative numbers with $\alpha_1 + \ldots + \alpha_n = 1$, then the same hypothesis as in (10) guarantees the existence of measurable partition $(A_i)_1^n$ of Ω satisfying $\mu_i(A_i) > \alpha_i$ for all $i \leq n$ (cf. Dubins and Spanier (1961)). The weights $\{\alpha_i\}$ can be viewed in the cake–cutting framework as representing non–uniform shares to which each participant is entitled and in the hypotheses–testing framework as non–uniform loss functions.

6. Inequalities in Hypotheses Testing

In the classification problem setting, using decision rule (partition) $(A_i)_1^n$ against $\mu_1,\ldots,\mu_n$ results in a maximum expected risk of $\max_{i\le n} P(X \notin A_i \mid \text{dist}(X) = \mu_i)$, and the objective is to minimize this risk. That is, a partition $(A_i)_1^n$ of Ω is sought which will attain the *minimax risk*

$$R(\boldsymbol{\mu}) = \inf\{\max_{i\le n} P(X \notin A_i \mid \text{dist}(X) = \mu_i) : (A_i)_1^n \text{ is an } \mathcal{F}\text{–partition of } \Omega\}.$$

Since $\inf\max P(X \notin A_i \mid \text{dist}X = \mu_i) = \inf\max(1-\mu_i(A_i)) = 1 - \sup\min \mu_i(A_i)$, it follows that $R(\boldsymbol{\mu}) = 1 - \max\{\min_{i\le n}\mu_i(A_i) : (A_i)_1^n$ is an $\mathcal{F}$–partition of $\Omega\}$, so the fairness and superfairness inequalities above can all be translated immediately into minimax risk inequalities. For example, the analog of (11) (cf. Elton, Hill and Kertz (1986)) is

(13) if $\mu_1,\ldots,\mu_n$ are nonatomic, then the minimax risk R satisfies $n^{-1}(n-M) \le R(\boldsymbol{\mu}) \le (n-M+1)^{-1}(n-M)$, where M is the total mass of $\bigvee_{i=1}^{n} \mu_i$.

Both bounds in (13) are sharp (the lower bound is easy, and the upper bound follows from (11)), and are attained.

In a similar application of the convexity theorem to the classification problem, Dvoretzky, Wald, and Wolfowitz (1951) showed that if $\mu_1,\ldots,\mu_n$ are nonatomic, then given any *randomized* decision function and loss functions $L_{i,j}$ there exists a non–randomized decision function (i.e., partition) with exactly the same expected risks for each i.

In a special case of the classification problem called the *location–parameter* problem, $\Omega = \mathbb{R}^1$ and the measures μ_i are translates of one another (for example μ_i is $N(a_i,1)$ for each i), and the classification problem is now that of guessing which parameter was underlying the observation X. Based on concentration parameters of the (location–parameter) distributions, sharp bounds for the minimax risk were obtained by Hill and Tong (1989) using the convexity theorem. For example, letting $\rho(\mu_1,d)$ denote the *tail–d concentration* of μ_1 (see Hill and Tong (1989)),

(14) if μ_1 is continuous and $\mu_i(A) = \mu_i(A-(i-1)d)$ for all A and $i = 1,\ldots,n$, then $R(\boldsymbol{\mu}) \le \left(\sum_{j=1}^{n-1} q^j\right) / \left(\sum_{j=0}^{n-1} q^j\right)$, where $q = 1-\rho(\mu_1,d)$. Moreover this bound is best possible and is attained for all n, all d and all $q<1$.

And, letting $\lambda(\mu,d)$ denote the *Levy d–concentration* $\sup_x \mu([x,x+d])$ of μ,

(15) if μ_1 is continuous and $\mu_2(A) = \mu_1(A - d)$ for all A, then there is a test for testing $H_1 : \text{dist}(X) = \mu_1$ versus $H_2 : \text{dist}(X) = \mu_2$ which satifies

$$\max\{\alpha, \beta\} \leq (1 - \lambda)/(2 - \lambda)$$

(where α and β are the type I and type II errors, respectively, and $\lambda = \lambda(\mu_1, d)$.) Moreover, this bound is attained for all d and all λ.

Again, a key element in the proof of (15) is the convexity theorem; Legut and Wilczynski (1991) have found improvements of (14) using a similar argument.

The convexity theorem was also used by Hill (1987a) to establish a proportionality principle for partitioning problems, which essentially says that in a general class of partitioning problems, the worst case is when the measures are proportional. One corollary of that principle related to the classification problem is

(16) if $X_1, X_2, \ldots, X_n$ are independent continuous random variables on $(\Omega, \mathcal{F}, P)$, then for each positive integer $k \leq n$ and each set of k distinct integers $K = \{1 \leq i_1 < \ldots < i_k \leq n\}$, there is a real Borel set B satisfying

$$P(X_i \in B \text{ if and only if } i \in K) \geq \left(\frac{k}{n}\right)^k \left(\frac{n-k}{n}\right)^{n-k},$$

and this bound is best possible.

7. Partitioning Inequalities in Optimal–Stopping Theory

The classical problem in optimal–stopping theory is: given a sequence of integrable random variables $\mathbf{X} = (X_1, X_2, \ldots, X_n)$ on $(\Omega, \mathcal{F}, P)$, find a stopping time t which maximizes EX_t. Here the stopping times are required to be adapted to an increasing filtration of σ–algebras $\mathcal{F}_1 \subseteq \mathcal{F}_2 \subseteq \ldots \subseteq \mathcal{F}_n \subseteq \mathcal{F}$, where typically $\mathcal{F}_j$ is the σ–algebra $\sigma(X_1, \ldots, X_j)$ generated by $X_1, \ldots, X_j$. In other words, a stopping time t corresponds to an n–partition $(A_i)_1^n$ of Ω satisfying $A_j \in \mathcal{F}_j$ for each $j \leq 1$; the correspondence is simply $\{t = j\} = A_j$. Thus the set Π of allowable partitions of Ω for optimal stopping is more restricted, by these σ–algebra constraints, than for the above fair–division problems. Still, many of the convexity tools apply in optimal stopping, where for example it was shown by Hill and Pestien (1983) using Lyapounov's theorem and an idea of Blackwell (1951a) that even in a finitely additive setting, the function $t \mapsto EX_t$ has convex range on the nonatomic components of the distributions.

A useful analog to the Lyapounov convexity result for optimal stopping is (Hill and Kennedy (1990))

(17) if $X_1, \ldots, X_n$ are integrable random variables on $(\Omega, \mathcal{F}, P)$, and X_1 is continuous, then the stopping time range of $\mathbf{X}$

$$\left\{ \left(\int_{t=1} X_1, \ldots, \int_{t=n} X_n \right) : t \text{ is a stopping time for } \mathbf{X} \right\}$$

is a closed convex subset of $\mathbb{R}^n$.

With *no* restrictions on X_1, the stopping time range may fail to be convex, although the *randomized*-stopping rule range is always convex. Note the contrast between (17), where the nonatomic assumption is only required for the first distribution, and (5), where nonatomicity is required for all but one of the distributions.

Using (17), the separating hyperplane theorem and the classical prophet inequality of Krengel and Sucheston (1978), the following sharp minimax partitioning inequality (18) in a prophet problem (optimal stopping) setting was proved in Hill and Kennedy (1990). Here $\mathcal{T}$ is the set of stopping times for $X_1, \ldots, X_n$, and $\mathcal{S}$ is the set of stopping functions which are $\mathcal{F}_n$ measurable, that is, $s \in \mathcal{S}$ has the property that the decision to stop at time j may depend on the *whole* sequence $X_1, \ldots, X_n$ as opposed to just the first j variables. In this sense, a player allowed to use stopping times from $\mathcal{S}$ is like a prophet, in that he can use information about future variables to decide when to stop (cf. Hill and Kennedy (1990) for the formal definition of $\mathcal{S}$).

(18) If $X_1, X_2, \ldots, X_n$ are integrable nonnegative random variables on $(\Omega, \mathcal{F}, P)$ and X_1 is continuous, then

$$\sup_{s \in \mathcal{S}} \min_i \int_{s=i} X_i \leq 2 \sup_{t \in \mathcal{T}} \min_i \int_{t=i} X_i,$$

and the bound 2 is best possible. If X_1 is not continuous, the inequality may fail, but does hold (and is best possible) if $\mathcal{T}$ is replaced with the collection of *randomized* stopping times.

A probabilistic interpretation of (18) is that if the objective is to maximize the minimum expected reward of stopping at $i = 1, \ldots, n$ then a prophet (or player with complete foresight) may never do better than twice that of an ordinary player restricted to using non–anticipatory stopping times. The convexity result (17) is also used in Hill and Kennedy (1990) to prove a stopping time analog of the partitioning principle in Hill (1988b); typical corollaries of which are (6) and

(19) if $X_1, \ldots, X_n$ are integrable nonnegative random variables on $(\Omega, \mathcal{F}, P)$, then there is a randomized stopping time t satisfying

$$\prod_{i=1}^{n} \int_{t=i} X_i \geq n^{-n} \prod_{i=1}^{n} EX_i,$$

and this bound is best possible.

8. Open Problems

The main purpose of this section is to record a number of open problems related to the partitioning inequalities mentioned above.

PROBLEM 1. Find a finite algorithm for generating a hyperplane median guaranteed by (3) for general distributions.

PROBLEM 2. Find an *efficient* algorithm for generating a hyperplane median based on a finite set of data points. (If there are k n–dimensional data points in $\mathbb{R}^n$, then it is easy to see that at least one of the $\binom{k}{n}$ hyperplanes will be a median, but checking all possible hyperplanes is certainly not optimal.)

The algorithms of Steinhaus, Banach and Knaster and of Fink guarantee a fair solution (i.e., $\mu_i(A_i) \geq 1/n$ for all i), but do not guarantee a *first-choice* solution in which each participant gets the piece he values most highly, i.e., satisfying

$$\mu_i(A_i) \geq \mu_i(A_j) \quad \text{for all } i \leq n \text{ and } j \leq n. \tag{20}$$

Of course convexity (e.g., (8)) guarantees the existence of a partition satisfying $\mu_i(A_j) = 1/n$ for all $i \leq n$ and all $j \leq n$ (and hence satisfying (20)) but Gamow and Stern (1958) raised the question of finding an *algorithm* generating a first–choice solution. Stromquist (1980) and Woodall (1980) independently found an algorithm for $n = 3$.

PROBLEM 3. Find (or demonstrate non–existence of) a finite algorithm yielding a first–choice partition (20) for $n \geq 4$.

PROBLEM 4. Find the best possible inequality generalizing (6) and (7), that is, find the largest constant $k = k(n, \alpha, \|\mu_1\|, \ldots, \|\mu_n\|)$ so that if $\mu_1, \ldots, \mu_n$ have atoms at most α, then there is a partition $(A_i)_1^n$ of Ω satisfying $\min_i \mu_i(A_i) \geq k$.

PROBLEM 5. Find necessary and sufficient geometric conditions on a set $S \subset \mathbb{R}^n$ so that S is the range of a nonatomic vector (probability) measure. (By

Lyapounov's theorem, S must be convex and compact; by general principles, S must be centrally symmetric, contain the origin, and lie in the positive orthant. For $n = 2$, these five conditions are also sufficient, but they are not sufficient for $n > 2$. Bolker (1969, 1971) attributes this question to Blaschke, and proves non-geometric characterizations.)

PROBLEM 6. Find necessary and sufficient conditions on $\boldsymbol{\mu} = (\mu_1, \ldots, \mu_n)$ so that the range of $\boldsymbol{\mu}$ is convex. (By Lyapounov's theorem, nonatomicity of the $\{\mu_i\}$ suffices, but even for $n = 1$ it is not necessary, as can be seen by looking at a purely atomic measure with atoms of size $\frac{1}{2^k}$ for all $k \geq 1$. Gouweleeuw (1991) has partial results in this direction.)

PROBLEM 7. Find natural topological and geometric generalizations of the convexity theorem; for example, if Ω is a polyhedron in $\mathbb{R}^n$, is the range of nonatomic measures over subpolyhedra convex? What about open simply connected sets? (The problem seems to be that taking limits is not always possible here; the limit of polygons need not be a polygon, nor must the limit of connected sets be connected. Approximate convexity is known in some cases, such as the fair–border results in Hill (1983) and Beck (1987), and Samuelson (1980) has a geometric fair–division scheme for coastal waterways. Note that if Ω is a convex subset of $\mathbb{R}^n$, the sliding–knife fair–division algorithm, as well as algorithms of Stromquist (1980) and Woodall (1980) generate partitions consisting of convex pieces. Gardner's problem (1978) is purely geometric.)

PROBLEM 8. Find the best possible constant in the generalization of Lyapounov's theorem to measures with atoms (9). (The bound $\alpha n/2$ is not sharp for small n, but is known to be of the correct order in n; the best possible bound is at least $n/8$ for general n and at least $n/4$ if n is a power of 2 (Elton and Hill (1987).)

PROBLEM 9. Find the best possible bound generalizing both superfairness inequalities (11) and (12); that is, find the largest $k = k(n, m, M)$ depending on the number of measures n, and the masses m and M of the infimum and supremum, respectively, so that if $\mu_1, \ldots, \mu_n$ are nonatomic with $\| \vee \mu_i \| = M$ and $\| \wedge \mu_i \| = m$, then there is a partition $(A_i)_1^n$ satisfying $\mu_i(A_i) \geq k(n, m, M)$.

PROBLEM 10. Find finite algorithms for generating the superfair partitions guaranteed by (11) and (12).

PROBLEM 11. Prove (or give a counterexample) that the minimax risk result (14) holds if ρ is replaced by the Levy concentration λ.

PROBLEM 12. For a general class of problems, find a proportionality principle for general measures (the proportionality principle in Hill (1988b) was for nonatomic measures, but in many partitioning problems for general measures (e.g., Hill (1987a)), the worst case is also known to be when all the measures are proportional, i.e., *equal*, in the case of probability measures. A basic superfairness property generalizing (10) should hold.)

Acknowledgements

The author is grateful to Z. Thomas for assistance in researching these fair–division inequalities, and to Professors M. Shaked and Y. L. Tong for an invitation to present these ideas at the AMS–IMS–SIAM Conference on Stochastic Inequalities in Seattle in July 1991.

References

AKEMANN, C. A. AND ANDERSON, J. (1990). Lyapounov theorems for operator algebras, preprint. University of California at Santa Barbara.

ALON, N. AND WEST, D. B. (1986). The Borsuk–Ulam theorem and bisection of necklaces. *Proc. Amer. Math. Soc.* **98** 623–628.

ARMSTRONG, T. AND PRIKRY, K. (1981). Liapounoff's theorem for nonatomic finitely–additive, bounded, finite–dimensional, vector–valued measures. *Trans. Amer. Math. Soc.* **266** 499–514.

ARTSTEIN, Z. (1990). Yet another proof of the Lyapounov convexity theorem. *Proc. Amer. Math. Soc.* **108** 89–91.

BECK, A. (1987). Constructing a fair border. *Amer. Math. Monthly* **94** 157–162.

BLACKWELL, D. (1951a). On a theorem of Lyapounov. *Ann. Math. Statist.* **22** 112–114.

BLACKWELL, D. (1951b). The range of certain vector integrals. *Proc. Amer. Math. Soc.* **2** 390–395.

BOLKER, E. D. (1969). A class of convex bodies. *Trans. Amer. Math. Soc.* **145** 323–345.

BOLKER, E. D. (1971). The zonoid problem. *Amer. Math. Monthly* **78** 529–531.

CRAWFORD, V. P. (1977). A game of fair division. *Rev. Econ. Studies* **44** 235–247.

CRAWFORD, V. P. AND HELLER, W. P. (1979). Fair division with indivisible commodities. *J. Econ. Th.* **21** 10–27.

DEMKO, S. AND HILL, T. P. (1988). Equitable distribution of indivisible objects. *Math. Soc. Sci.* **16** 1–14.

DUBINS, L. E. AND SPANIER, E. H. (1961). How to cut a cake fairly. *Amer. Math. Monthly* **68** 1–17.

DVORETZKY, A., WALD, A. AND WOLFOWITZ, J. (1951). Relations among certain ranges of vector measures. *Pacific J. Math.* **1** 59–74.

ELTON, J. AND HILL, T. P. (1987). A generalization of Lyapounov's convexity theorem to measures with atoms. *Proc. Amer. Math. Soc.* **99** 297-304.

ELTON, J., HILL, T. P. AND KERTZ, R. P. (1986). Optimal–partitioning inequalities for nonatomic probability measures. *Trans. Amer. Math. Soc.* **296** 703–725.

EVEN, S. AND PAZ, A. (1984). A note on cake–cutting. *Discrete Appl. Math.* **7** 285–296.

FELLER, W. (1938). Note on regions similar to the sample space. *Statistical Research Memoirs.*

FINK, A. M. (1964). A note on the fair division problem. *Math. Magazine* **37** 341–342.

FISHER, R. A. (1936). Uncertain inference. *Proc. Amer. Acad. Arts Sci.* **71** 245–258.

GAMOW, G. AND STERN, M. (1958). *Puzzle Math.* Viking, New York.

GARDNER, M. (1978). The curious cake cut. *Aha! Insight!* 56–57, W. H. Freeman, New York.

GOUWELEEUW, J. (1991). A generalization of Lyapounov's theorem. Tech. Report No. WS–381, Free University, Amsterdam.

HALMOS, P. R. (1948). The range of a vector measure. *Bull. Amer. Math. Soc.* **54** 416–421.

HILL, T. P. (1983). Determining a fair border. *Amer. Math. Mo.* **90** 438–442.

HILL, T. P. (1985). Equipartitioning common domains of non–atomic measures. *Math. Z.* **189** 415–419.

HILL, T. P. (1987a). Partitioning general probability measures. *Ann. Probab.* **15** 804–813.

HILL, T. P. (1987b). A sharp partitioning–inequality for non–atomic probability measures based on the mass of the infimum of the measures. *Prob. Th. and Rel. Fields* **75** 143–147.

HILL, T. P. (1988a). Common hyperplane medians for random vectors. *Amer. Math. Mo.* **95** 437–441.

HILL, T. P. (1988b). A proportionality principle for partitioning problems. *Proc. Amer. Math. Soc.* **103** 288–293.

HILL, T. P. AND KENNEDY, D. P. (1990). Optimal stopping problems with generalized objective functions. *J. Appl. Prob.* **28** 828–838.

HILL, T. P. AND PESTIEN, V. (1983). The advantage of using non–measurable stop rules. *Ann. Probab.* **11** 442–450.

HILL, T. P. AND TONG, Y. L. (1989). Optimal–partitioning inequalities in classification and multi–hypotheses testing. *Ann. Statist.* **17** 1325–1334.

JONES, M. (1989). *Universal constants in optimal stopping theory.* Ph.D. Dissertation, Georgia Institute of Technology.

KARLIN, S. (1953). Extreme points of vector functions. *Proc. Amer. Math. Soc.* **4** 603–610.

KARLIN, S. AND STUDDEN, W. (1966). *Tchebycheff Systems with Applications in Analysis and Statistics.* Wiley, New York.

KIRBY, R. (1988). A new proposal for nuclear arms negotiation, preprint. Department of Mathematics, University of California at Berkeley.

KIRMAN, A. (1981). Measure theory with applications to economics. In *Handbook of Mathematical Economics.* Vol. I, Ch. 5, K. Arrow and M. Intrilgator, eds., North–Holland, Amsterdam.

KNASTER, B. (1946). Sur le probleme du partage pragmatique de H. Steinhaus. *Comptes Rendus de la Societé Polonaise de Mathématique, Annales de la Societé Polonaise de Mathematique* **19** 228–230.

KNASTER, B. AND STEINHAUS, H. (1953). Sur le partage pragmatique. *Comptes Rendus de la Societé des Sciences et des Lettres de Wroclaw.* Communication No. 1, **2**.

KRENGEL, U. AND SUCHESTON, L. (1978). On semiamarts, amarts, and processes with finite value. *Probability on Banach Spaces.* J. Kuelbs, ed., Marcel Dekker, New York.

KUHN, H. W. (1973). On games of fair division. In *Essays in Modern Math. Econ.* M. Shubik, ed., Princeton University Press, Princeton, NJ. 29–37.

LASALLE, J. P. (1960). The time optimal control problem. In *Contributions to the Theory of Nonlinear Oscillators V.* Princeton University Press, Princeton, NJ.

LEGUT, J. (1985). The problem of fair division for countably many participants. *J. Math. Anal. Appl.* **109** 83–89.

LEGUT, J. (1987). A game of fair division with continuum of players. *Colloq. Math.* **53** 323-331.

LEGUT, J. (1988). Inequalities for α–optimal partitioning of a measurable space. *Proc. Amer. Math. Soc.* **104** 1249–1251.

LEGUT, J. (1990). On totally balanced games arising from cooperation in fair division. *Games and Econ. Behavior* **2** 47–60.

LEGUT, J. (1991). On dividing the unit interval. Preprint, Technical University of Wroclaw.

LEGUT, J. AND WILCZYNSKI, M. (1988). Optimal partitioning of a measurable space. *Proc. Amer. Math. Soc.* **104** 262–264.

LEGUT, J. AND WILCZYNSKI, M. (1991). Inequalities for the minimax risk in multihypotheses testing. Preprint, Technical University of Wroclaw.

LINDENSTRAUSS, J. (1966). A short proof of Liapounoff's convexity theorem. *J. Math. Mech.* **15** 971–972.

LYAPOUNOV, A. (1940). Sur les fonctions–vecteurs complètement additives. *Bull. Acad. Sci. URSS* **4** 465–478.

MARGOLIES, D. (1978). *A study of finitely additive measures as regards amenable groups, Liapounov's theorem, and the elimination of infinite integrals via non–standard real numbers.* Ph.D. Dissertation, University of California at Berkeley.

NEYMAN, J. (1946). Un theorémè d'existence. *C. R. Acad. Sci. Paris* **222** 843–845.

NEYMAN, J. AND PEARSON, E. (1933). On the problem of most efficient tests of statistical hypotheses. *Philos. Trans. Roy. Soc. London Ser. A* **231** 289–337.

REBMAN, K. (1979). How to get (at least) a fair share of the cake. *Mathematical Plums, Dolciani Math. Expositions* #4. R. Honsberger, ed., 22–37.

SAMUELSON, W. (1980). The object distribution problem revisited. *Quart. J. Econ.* 85–90.

SAMUELSON, W. (1985). Dividing coastal waterways. *J. Conflict Resolution* **29** 83–111.

STEINHAUS, H. (1948). The problem of fair division. *Econometrika* **16** 101–104.

STEINHAUS, H. (1949). Sur la division pragmatique. *Econometrika (Supplement)* **17** 315–319.

STEINHAUS, H. (1960). *Mathematical Snapshots.* Oxford University Press, New York.

STONE, A. H. AND TUKEY, J. W. (1942). Generalized "sandwich" theorems. *Duke Math. J.* **9** 356–359.

STROMQUIST, W. (1980). How to cut a cake fairly. *Amer. Math. Monthly* **87** 640-644.

STROMQUIST, W. AND WOODALL, D. R. (1985). Sets on which several measures agree. *J. Math. Anal. Appl.* **108** 241–248.

SVENSSON, L. G. (1983). On the existence of fair allocations. *J. Econ.* **43** 301–308.

URBANIK, K. (1955). Quelques theorémès sur les mesures. *Fund. Math.* **41** 150–162.

WELLER, W. (1985). Fair division of a measurable space. *J. Math. Econ.* **14** 5–17.

WOODALL, D. R. (1980). Dividing a cake fairly. *J. Math. Anal. Appl.* **78** 233–247.

WOODALL, D. R. (1985). A note on the cake–division problem. Preprint, University of Nottingham.

YOUNG, H. P. (1987). On dividing an amount according to individual claims or liabilities. *Math. Oper. Res.* **12** 398–414.

SCHOOL OF MATHEMATICS
GEORGIA INSTITUTE OF TECHNOLOGY
ATLANTA, GA 30332–0160

Stochastic Inequalities
IMS Lecture Notes – Monograph Series
Volume 22 (1993)

MATRIX EXTREMES AND RELATED STOCHASTIC BOUNDS

By D. R. JENSEN

Virginia Polytechnic Institute and State University

The spaces $F_{n\times k}$ and S_k^+ consisting of rectangular and positive definite matrices are developed as partially ordered sets having lower and upper bounds. Under Loewner (1934) ordering, spectral lower and upper bounds are constructed for pairs $\mathbf{A}, \mathbf{B} \in (S_k^+, \succeq_L)$ and are shown to be tight. Similar bounds are given for pairs in $(F_{n\times k}, \succeq)$ in terms of singular decompositions under an induced ordering. Applications pertaining to $(S_k^+, \succeq_L)$ include stochastic bounds for distributions of quadratic forms, minimal dispersion bounds in certain regular ensembles, and bounds on the peakedness of certain weighted vector sums. Applications to $(F_{n\times k}, \succeq)$ support the uniform improvement of any pair of first-order experimental designs.

1. Introduction

Extremal problems persist throughout applied probability and statistics. Their solutions often shed new light on structural aspects of the system at hand.

To fix ideas, we reexamine the concentration properties of measures $\mu(\cdot;\mathbf{p})$ induced by weighted sums $\sum_{i=1}^n p_i X_i$ of *iid* random scalars $\{X_1, \ldots, X_n\}$ having a symmetric log-concave density. Here $\mathbf{p} = [p_1, \ldots, p_n]$ satisfies $\{0 \le p_i \le 1, p_1 + \cdots + p_n = 1\}$, and we let $F(t;\mathbf{p}) = \mu([-t,t];\mathbf{p})$ with $t > 0$. Proschan (1965) has shown for each $t > 0$ that $F(t;\mathbf{p})$ is order-reversing under majorization, *i.e.*, if $\mathbf{p}$ majorizes $\mathbf{q}$, then $F(t;\mathbf{q}) \ge F(t;\mathbf{p})$ and thus $\mu(\cdot;\mathbf{q})$ is more peaked than $\mu(\cdot;\mathbf{p})$ in the sense of Birnbaum (1948).

Since linear functions arise in a variety of contexts not entailing ordered weights, we pose the further question: If neither $\mathbf{p}$ majorizes $\mathbf{q}$ nor $\mathbf{q}$ majorizes $\mathbf{p}$, what then may be said regarding the concentration properties of $\mu(\cdot;\mathbf{p})$ and $\mu(\cdot;\mathbf{q})$? One answer follows immediately on observing that the ordered simplex supporting majorization is a lattice with greatest lower

AMS 1991 *subject classifications.* Primary 15A45, 60E15.
Key words and phrases. Matrix orderings, positive definite and rectangular matrices, monotone functions, stochastic bounds, applications.

bound *(glb)* and least upper bound *(lub)* given respectively by $\mathbf{p} \wedge \mathbf{q}$ and $\mathbf{p} \vee \mathbf{q}$. Proschan's (1965) result immediately gives the bounds

$$F(t; \mathbf{p} \vee \mathbf{q}) \leq \{F(t; \mathbf{p}), F(t; \mathbf{q})\} \leq F(t; \mathbf{p} \wedge \mathbf{q}) \tag{1.1}$$

for each $t > 0$. Accordingly, we may refer to $F(t; \mathbf{p} \vee \mathbf{q})$ as the *stochastic minorant*, and to $F(t; \mathbf{p} \wedge \mathbf{q})$ as the *stochastic majorant*, of measures $\{\mu(\cdot; \mathbf{p}), \mu(\cdot; \mathbf{q})\}$ when evaluated over symmetric sets $[-t, t]$. We return to this topic later.

In this paper we extend the foregoing concepts to include other spaces and other orderings. It is seen that lattice properties need not carry forward. Nonetheless, these spaces are developed as partially ordered sets having lower and upper bounds, and these bounds are shown to be tight. An outline of the paper follows.

Preliminary developments occupy Section 2. Our main results are set forth in Section 3, first with regard to an ordering for symmetric matrices due to Loewner (1934), then with regard to an induced ordering on the space of real rectangular matrices. Applications are developed in Section 4. For the positive semidefinite ordering these include stochastic bounds for distributions of quadratic forms, minimal dispersion bounds on vector estimators in certain regular ensembles, and bounds on the peakedness of certain weighted vector sums. Applications to rectangular matrices support the uniform improvement of any pair of first-order experimental designs.

2. Preliminaries

We establish conventions for notation and review basic properties of some ordered spaces and functions monotone on them.

2.1. *Notation*

Symbols include $\mathbb{R}^n$ as Euclidean n-space, $F_{n \times k}$ as the real $(n \times k)$ matrices with $n \geq k$, S_k as the real symmetric $(k \times k)$ matrices, and S_k^0, S_k^+ and D_k as their positive semidefinite, positive definite, and diagonal varieties, respectively. The simplex $R_n(c)$ in $\mathbb{R}^n$ is given by $R_n(c) = \{\mathbf{x} \in \mathbb{R}^n : x_1 \geq \cdots \geq x_n, x_1 + \cdots + x_n = c\}$, and the transpose of $\mathbf{x} \in \mathbb{R}^n$ is $\mathbf{x}' = [x_1, \ldots, x_n]$. Special arrays include the unit vector $\mathbf{1}_n = [1, \ldots, 1]' \in \mathbb{R}^n$, the unit matrix $\mathbf{I}_n$, and a typical diagonal matrix $\mathbf{D}_\alpha = \text{Diag}(\alpha_1, \ldots, \alpha_k) \in D_k$. Groups of transformations on $\mathbb{R}^n$ include the general linear group $Gl(n)$ and the real orthogonal group $O(n)$.

The *spectral decomposition* $\mathbf{A} = \sum_{i=1}^k \alpha_i \mathbf{q}_i \mathbf{q}_i'$ of $\mathbf{A} \in S_k^+$ yields its *symmetric root* $\mathbf{A}^{1/2} = \sum_{i=1}^k \alpha_i^{1/2} \mathbf{q}_i \mathbf{q}_i'$. The *singular decomposition* of $\mathbf{X} \in F_{n \times k}$

is $\mathbf{X} = \sum_{i=1}^{k} \xi_i \mathbf{p}_i \mathbf{q}_i' = \mathbf{P}\mathbf{D}_\xi \mathbf{Q}'$ in which $\mathbf{P} = [\mathbf{p}_1, \ldots, \mathbf{p}_k]$ is semiorthogonal containing the *left singular vectors*, $\mathbf{Q} = [\mathbf{q}_1, \ldots, \mathbf{q}_k]$ is orthonormal containing the *right singular vectors*, and $\mathbf{D}_\xi = \text{Diag}(\xi_1, \ldots, \xi_k)$ contains the ordered *singular values* of $\mathbf{X}$.

Standard usage refers to independent, identically distributed *(iid)* variates and their cumulative distribution function *(cdf)*. $\mathcal{L}(\mathbf{X})$ denotes the distribution of $\mathbf{X}$, with $N_k(\boldsymbol{\mu}, \boldsymbol{\Sigma})$ as the Gaussian law on $\mathbb{R}^k$ having some mean $\boldsymbol{\mu}$ and dispersion matrix $\boldsymbol{\Sigma}$.

2.2. *Ordered Spaces*

A set $\mathcal{H}$ together with a binary relation $\succeq_0$ is said to be *linearly ordered* if the relation is reflexive, transitive, antisymmetric and complete. A *partial ordering* is reflexive, transitive and antisymmetric, and a *preordering* is reflexive and transitive. A partially ordered set $(\mathcal{H}, \succeq_0)$ is a *lower semi-lattice* if for any two elements x, y in $\mathcal{H}$, there is a greatest lower bound $x \wedge y$ in $\mathcal{H}$; an *upper semi-lattice* if there is a least upper bound $x \vee y$ in $\mathcal{H}$; and a *lattice* if it is both a lower and upper semi-lattice.

Ordered spaces of note include $(\mathbb{R}^k, \geq_k)$, with $\mathbf{x} \geq_k \mathbf{y}$ in $\mathbb{R}^k$ if and only if $\{x_i \geq y_i; 1 \leq i \leq k\}$, and the simplex $(R_n(c), \succeq_M)$ ordered by majorization (cf. Marshall and Olkin (1979)). The space $(S_k, \succeq_L)$ is ordered as in Loewner (1934) such that $\mathbf{A} \succeq_L \mathbf{B}$ if and only if $\mathbf{A} - \mathbf{B} \in S_k^0$, with $\mathbf{A} \succ_L \mathbf{B}$ whenever $\mathbf{A} - \mathbf{B} \in S_k^+$. The space $(F_{n \times k}, \succeq)$ has an induced ordering in which $\mathbf{X} \succeq \mathbf{Z}$ if and only if $\mathbf{X}'\mathbf{X} \succeq_L \mathbf{Z}'\mathbf{Z}$; see Jensen (1984). This ordering is invariant in the sense that $\mathbf{X} \succeq \mathbf{Z}$ if and only if $\mathbf{PXB} \succeq \mathbf{QZB}$ for any $\mathbf{P}, \mathbf{Q} \in O(n)$ and $\mathbf{B} \in Gl(k)$, and antisymmetry holds up to equivalence under $O(n)$ acting from the left.

Spaces with lower and upper bounds are germane to our studies. Clearly $(\mathbb{R}^k, \geq_k)$ is a lattice with $\mathbf{a} \wedge \mathbf{b} = [a_1 \wedge b_1, \ldots, a_k \wedge b_k]'$ and $\mathbf{a} \vee \mathbf{b} = [a_1 \vee b_1, \ldots, a_k \vee b_k]'$, where $\{a_i \wedge b_i = \min(a_i, b_i)$ and $a_i \vee b_i = \max(a_i, b_i); 1 \leq i \leq k\}$; see Vulikh (1967), for example. On working backwards from

$$
\begin{aligned}
v_1 &= x_1 \wedge y_1 \\
v_1 + v_2 &= (x_1 + x_2) \wedge (y_1 + y_2) \\
&\cdots \quad \cdots \\
v_1 + \cdots + v_k &= (x_1 + \cdots + x_k) \wedge (y_1 + \cdots + y_k) = c
\end{aligned}
\tag{2.1}
$$

then from

$$
\begin{aligned}
u_1 &= x_1 \vee y_1 \\
u_1 + u_2 &= (x_1 + x_2) \vee (y_1 + y_2) \\
&\cdots \quad \cdots \\
u_1 + \cdots + u_k &= (x_1 + \cdots + x_k) \vee (y_1 + \cdots + y_k) = c,
\end{aligned}
\tag{2.2}
$$

we conclude that $(R_k(c), \succeq_M)$ is a lattice with $\mathbf{x} \wedge \mathbf{y} = \mathbf{v}$ and $\mathbf{x} \vee \mathbf{y} = \mathbf{u}$. The space $(S_k, \succeq_L)$ is not a lattice (cf. Halmos (1958, p. 142)), nor can $(F_{n\times k}, \succeq)$ inherit lattice properties through its induced ordering. Nonetheless, both $(S_k^+, \succeq_L)$ and $(F_{n\times k}, \succeq)$ are shown subsequently to have lower and upper bounds that are tight.

2.3. *Monotone Functions*

A real-valued function $f(\cdot)$ on $(\mathcal{H}, \succeq_0)$ is said to be *order-preserving* if $x \succeq_0 y$ on $\mathcal{H}$ implies $f(x) \geq f(y)$ on $\mathbb{R}^1$, and to be *order-reversing* if $x \succeq_0 y$ on $\mathcal{H}$ implies $f(x) \leq f(y)$ on $\mathbb{R}^1$. Denote by $\Phi(\mathcal{H}, \succeq_0)$ the class of order-preserving functions, and by $\Phi^-(\mathcal{H}, \succeq_0)$ the order-reversing functions, on $(\mathcal{H}, \succeq_0)$. Specifically, $\Phi(\mathbb{R}^k, \geq_k)$ consists of functions $f(x_1, \ldots, x_k)$ non-decreasing in each argument, and functions in $\Phi(R_k(c), \succeq_M)$ comprise the Schur convex functions (cf. Marshall and Olkin (1979)). The class $\Phi(S_k, \succeq_L)$ is characterized in Marshall, Walkup and Wets (1967), and $\Phi(F_{n\times k}, \succeq)$ may be characterized through compositions as

$$\Phi(F_{n\times k}, \succeq) = \{\phi(\mathbf{X}) = \psi(\mathbf{X}'\mathbf{X}) : \psi \in \Phi(S_k, \succeq_L)\}; \tag{2.3}$$

for further details see Jensen (1984).

3. Matrix Extremes in S_k^+ and $F_{n\times k}$

Our principal findings are developed here. We first characterize the sets of lower and upper bounds for pairs of matrices in $(S_k^+, \succeq_L)$.

3.1. *Bounds in* $(S_k^+, \succeq_L)$

Given $(\mathbf{A}, \mathbf{B})$ in $(S_k^+, \succeq_L)$ we study first the lower bounds $H_L(\mathbf{A}, \mathbf{B}) = \{\mathbf{S} \in S_k^+ : \mathbf{S} \preceq_L \mathbf{A} \text{ and } \mathbf{S} \preceq_L \mathbf{B}\}$, and then the upper bounds $\mathbf{H}_U(\mathbf{A}, \mathbf{B}) = \{\mathbf{T} \in S_k^+ : \mathbf{T} \succeq_L \mathbf{A} \text{ and } \mathbf{T} \succeq_L \mathbf{B}\}$. The ordering $\mathbf{L} \preceq_L \{\mathbf{A}, \mathbf{B}\} \preceq_L \mathbf{U}$ always holds with $\mathbf{L} = \mathbf{0}$ and $\mathbf{U} = \mathbf{A} + \mathbf{B}$, and if $\mathbf{A} \preceq_L \mathbf{B}$, then $\mathbf{L} = \mathbf{A}$ and $\mathbf{U} = \mathbf{B}$. Since $\mathbf{A} \succeq_L \mathbf{S}$ if and only if $\mathbf{GAG}' \succeq_L \mathbf{GSG}'$ for any $\mathbf{G} \in Gl(k)$, it suffices to consider a canonical form in which $(\mathbf{A}, \mathbf{B}) \rightarrow (\mathbf{B}^{-1/2}\mathbf{A}\mathbf{B}^{-1/2}, \mathbf{I}_k) \rightarrow (\mathbf{GAG}', \mathbf{GBG}') \rightarrow (\mathbf{D}_\gamma, \mathbf{I}_k)$, where $\mathbf{B}^{-1/2}\mathbf{A}\mathbf{B}^{-1/2} = \sum_{i=1}^k \gamma_i \mathbf{q}_i \mathbf{q}_i'$ is its spectral decomposition and $\mathbf{D}_\gamma = \text{Diag}(\gamma_1, \ldots, \gamma_k)$ contains the ordered roots of $|\mathbf{A} - \gamma\mathbf{B}| = 0$. We thus seek $\mathbf{E} = \mathbf{GLG}'$ and $\mathbf{F} = \mathbf{GUG}'$ such that $\mathbf{E} \preceq_L \{\mathbf{D}_\gamma, \mathbf{I}_k\} \preceq_L \mathbf{F}$ or, equivalently, the classes $\mathbf{H}_L(\mathbf{D}_\gamma, \mathbf{I}_k)$ and $\mathbf{H}_U(\mathbf{D}_\gamma, \mathbf{I}_k)$.

First note that $\mathbf{A} \succeq_L \mathbf{B}$ on S_k^+ if and only if $\{\gamma_1 \geq \cdots \geq \gamma_k \geq 1\}$. If neither $\mathbf{A} \succeq_L \mathbf{B}$ nor $\mathbf{B} \succeq_L \mathbf{A}$, then at least one of two integers (r, s) can be found such that

$$\text{(3.1)} \qquad \{\gamma_1 \geq \cdots \geq \gamma_r > \gamma_{r+1} = 1 = \cdots = \gamma_{r+s} > \gamma_{r+s+1} \geq \cdots \geq \gamma_k > 0\}.$$

Now let $t = k - r - s$, and let $\{\epsilon_1 \geq \cdots \geq \epsilon_k > 0\}$ be the ordered eigenvalues of $\mathbf{E} \in S_k^+$. Essential properties of the lower bounds $\mathbf{H}_L(\mathbf{D}_\gamma, \mathbf{I}_k)$ are summarized in the following lemma.

LEMMA 1 *Let* $\mathbf{E} = [e_{ij}]$ *have eigenvalues* $\{\epsilon_1 \geq \cdots \geq \epsilon_k > 0\}$, *and consider the class* $\mathbf{H}_L(\mathbf{D}_\gamma, \mathbf{I}_k)$ *with* $\mathbf{D}_\gamma$ *fixed. In order that* $\mathbf{E} \in \mathbf{H}_L(\mathbf{D}_\gamma, \mathbf{I}_k)$ *it is necessary that*

(i) $\{\epsilon_i \leq 1; 1 \leq i \leq k\}$, *and that*

(ii) $\{e_{ii} \leq 1; 1 \leq i \leq r+s\}$ *and* $\{e_{ii} \leq \gamma_i; r+s+1 \leq i \leq k\}$.

Moreover, if $\{e_{ii}; 1 \leq i \leq k\}$ *are assigned their maximal values, a necessary and sufficient condition that* $\mathbf{E} \in \mathbf{H}_L(\mathbf{D}_\gamma, \mathbf{I}_k)$ *is that* $\mathbf{E}$ *should take the form*

(iii) $\mathbf{E}_M = \text{Diag}(\mathbf{I}_r, \mathbf{I}_s, \gamma_{r+s+1}, \ldots, \gamma_k)$.

PROOF The equivalence of $\mathbf{I}_k \succeq_L \mathbf{E}$ and $\mathbf{I}_k \succeq_L \text{Diag}(\epsilon_1, \ldots, \epsilon_k)$ gives conclusion (i). Conclusion (ii) follows on noting that the diagonal elements of the positive semidefinite matrices $\mathbf{D}_\gamma - \mathbf{E}$ and $\mathbf{I}_k - \mathbf{E}$ are necessarily nonnegative. To see necessity in conclusion (iii), assume first that $\mathbf{D}_\gamma - \mathbf{E} \in S_k^0$, so that $\mathbf{E}$ when assigned its maximal diagonal elements takes the form $\mathbf{E}_0 = \text{Diag}(\mathbf{H}, \mathbf{I}_s, \gamma_{r+s+1}, \ldots, \gamma_k)$ such that $\mathbf{H} = [h_{ij}]$ with $\{h_{ii} = 1; 1 \leq i \leq r\}$, and $\text{Diag}(\gamma_1, \ldots, \gamma_r) - \mathbf{H} \in S_k^0$. Other off-diagonal elements vanish since $\mathbf{D}_\gamma - \mathbf{E} \in S_k^0$ and the corresponding diagonal elements vanish. Now invoking the assumption $\mathbf{I}_k - \mathbf{E}_0 \in S_k^0$ stipulates further that $\mathbf{I}_r - \mathbf{H} \in S_r^0$, hence the off-diagonal elements of $\mathbf{H}$ must vanish also, giving $\mathbf{E}_M$ as in conclusion (iii). Sufficiency of the form (iii) follows since the diagonals of $\mathbf{E}_M$ take their maximal values, and both $\mathbf{D}_\gamma - \mathbf{E}_M$ and $\mathbf{I}_k - \mathbf{E}_M$ are positive semidefinite by construction. □

Turning to upper bounds for the pair $(\mathbf{D}_\gamma, \mathbf{I}_k)$, and thereby for $(\mathbf{A}, \mathbf{B})$, let $\{\eta_1 \geq \cdots \geq \eta_k > 0\}$ be the ordered eigenvalues of $\mathbf{F} \in S_k^+$. Without further proof, essential properties of $\mathbf{H}_U(\mathbf{D}_\gamma, \mathbf{I}_k)$ are as summarized in the following.

LEMMA 2 *Let* $\mathbf{F} = [f_{ij}]$ *have eigenvalues* $\{\eta_1 \geq \cdots \geq \eta_k > 0\}$, *and consider the class* $\mathbf{H}_U(\mathbf{D}_\gamma, \mathbf{I}_k)$ *with* $\mathbf{D}_\gamma$ *fixed. In order that* $\mathbf{F} \in \mathbf{H}_U(\mathbf{D}_\gamma, \mathbf{I}_k)$ *it is necessary that*

(i) $\{1 \leq \eta_i < \infty; 1 \leq i \leq k\}$, *and that*

(ii) $\{f_{ii} \geq \gamma_i; 1 \leq i \leq r\}$ *and* $\{f_{ii} \geq 1; r+1 \leq i \leq k\}$.

Moreover, if $\{f_{ii}; 1 \leq i \leq k\}$ *are assigned their minimal values, a necessary and sufficient condition that* $\mathbf{F} \in \mathbf{H}_U(\mathbf{D}_\gamma, \mathbf{I}_k)$ *is that* $\mathbf{F}$ *should take the form*

(iii) $\mathbf{F}_m = \mathrm{Diag}(\gamma_1, \ldots, \gamma_r, \mathbf{I}_s, \mathbf{I}_t)$ *with* $t = k - r - s$.

All lower and upper bounds for $(\mathbf{A}, \mathbf{B})$ in $(S_k^+, \succeq_L)$ follow on mapping back to the original space. Since $\mathbf{A} = \mathbf{B}^{1/2}\mathbf{Q}\mathbf{D}_\gamma\mathbf{Q}'\mathbf{B}^{1/2}$ and $\mathbf{B} = \mathbf{B}^{1/2}\mathbf{Q}\mathbf{I}_k\mathbf{Q}'\mathbf{B}^{1/2}$, we conclude that $\mathbf{L} \in \mathbf{H}_L(\mathbf{A}, \mathbf{B})$ if and only if $\mathbf{L} = \mathbf{B}^{1/2}\mathbf{Q}\mathbf{E}\mathbf{Q}'\mathbf{B}^{1/2}$ for some $\mathbf{E} \in \mathbf{H}_L(\mathbf{D}_\gamma, \mathbf{I}_k)$. Similarly, $\mathbf{U} \in \mathbf{H}_U(\mathbf{A}, \mathbf{B})$ if and only if $\mathbf{U} = \mathbf{B}^{1/2}\mathbf{Q}\mathbf{F}\mathbf{Q}'\mathbf{B}^{1/2}$ for some $\mathbf{F} \in \mathbf{H}_U(\mathbf{D}_\gamma, \mathbf{I}_k)$.

3.2. *Spectral Bounds in* $(S_k^+, \succeq_L)$

If we require that lower and upper bounds for diagonal matrices should be diagonal, then $(D_k, \succeq_L)$ is seen to be a lattice on imbedding it in $(\mathbb{R}^k, \geq_k)$. Then the *glb* and *lub* of $(\mathbf{D}_\gamma, \mathbf{I}_k)$ are $\mathbf{D}_\gamma \wedge \mathbf{I}_k = \mathbf{E}_M$ and $\mathbf{D}_\gamma \vee \mathbf{I}_k = \mathbf{F}_m$, precisely as defined in Lemmas 1 and 2. This in turn prompts the following.

DEFINITION 1 The matrices given by $\mathbf{A} \wedge \mathbf{B} = \mathbf{B}^{1/2}\mathbf{Q}(\mathbf{D}_\gamma \wedge \mathbf{I}_k)\mathbf{Q}'\mathbf{B}^{1/2}$ and $\mathbf{A} \vee \mathbf{B} = \mathbf{B}^{1/2}\mathbf{Q}(\mathbf{D}_\gamma \vee \mathbf{I}_k)\mathbf{Q}'\mathbf{B}^{1/2}$ are called the *spectral glb* and the *spectral lub* for $(\mathbf{A}, \mathbf{B})$ in $(S_k^+, \succeq_L)$.

Properties of these spectral extremes are studied next. The main issues include the possible interchangeability of $\mathbf{A}$ and $\mathbf{B}$, and whether the spectral bounds are tight. Both are answered affirmatively in developments culminating in Theorem 1.

With regard to the reduction $(\mathbf{A}, \mathbf{B}) \to (\mathbf{D}_\gamma, \mathbf{I}_k)$, we may take instead $(\mathbf{A}, \mathbf{B}) \to (\mathbf{I}_k, \mathbf{A}^{-1/2}\mathbf{B}\mathbf{A}^{-1/2}) \to (\mathbf{I}_k, \mathbf{D}_\theta)$. Here $\mathbf{D}_\theta$ is the diagonal matrix $\mathbf{D}_\theta = \mathrm{Diag}(\theta_1, \ldots, \theta_k)$ with $\{0 < \theta_1 \leq \cdots \leq \theta_k\}$ as the reverse-ordered roots of $|\mathbf{B} - \theta\mathbf{A}| = 0$, where $\mathbf{A}^{-1/2}\mathbf{B}\mathbf{A}^{-1/2} = \sum_{i=1}^k \theta_i \mathbf{p}_i \mathbf{p}_i'$ is the spectral decomposition with $\mathbf{P} = [\mathbf{p}_1, \ldots, \mathbf{p}_k] \in O(k)$. Proceeding as before, define $\mathbf{B} \wedge \mathbf{A} = \mathbf{A}^{1/2}\mathbf{P}(\mathbf{I}_k \wedge \mathbf{D}_\theta)\mathbf{P}'\mathbf{A}^{1/2}$ and $\mathbf{B} \vee \mathbf{A} = \mathbf{A}^{1/2}\mathbf{P}(\mathbf{I}_k \vee \mathbf{D}_\theta)\mathbf{P}'\mathbf{A}^{1/2}$. To investigate whether the spectral bounds are invariant with regard to decomposition, *i.e.*, whether $\mathbf{A} \wedge \mathbf{B} = \mathbf{B} \wedge \mathbf{A}$ and $\mathbf{A} \vee \mathbf{B} = \mathbf{B} \vee \mathbf{A}$, we first note several duality relations. These are $\mathbf{D}_\theta = \mathbf{D}_\gamma^{-1}$, $\mathbf{I}_k \wedge \mathbf{D}_\theta = (\mathbf{D}_\gamma \vee \mathbf{I}_k)^{-1}$, $\mathbf{I}_k \vee \mathbf{D}_\theta = (\mathbf{D}_\gamma \wedge \mathbf{I}_k)^{-1}$, $(\mathbf{D}_\gamma \wedge \mathbf{I}_k)(\mathbf{D}_\gamma \vee \mathbf{I}_k) = \mathbf{D}_\gamma$, and $\mathbf{B}^{1/2}\mathbf{Q} = \mathbf{A}^{1/2}\mathbf{P}\mathbf{D}_\gamma^{-1/2}$ and $\mathbf{A}^{-1/2}\mathbf{P}\mathbf{D}_\gamma^{1/2} = \mathbf{B}^{-1/2}\mathbf{Q}$. The latter expressions follow on establishing relationships between the normalized eigenvectors of $\{\mathbf{B}^{-1/2}\mathbf{A}\mathbf{B}^{-1/2}\mathbf{q}_i = \gamma_i\mathbf{q}_i; 1 \leq i \leq k\}$ and $\{\mathbf{A}^{-1/2}\mathbf{B}\mathbf{A}^{-1/2}\mathbf{p}_i = \theta_i\mathbf{p}_i; 1 \leq i \leq k\}$. Our principal findings with regard to the lower and upper spectral bounds are as follow.

THEOREM 1 *Let* $\{\mathbf{A} \wedge \mathbf{B}, \mathbf{A} \vee \mathbf{B}\}$ *and* $\{\mathbf{B} \wedge \mathbf{A}, \mathbf{B} \vee \mathbf{A}\}$ *be spectral glb's and lub's as defined. Then for any* $(\mathbf{A}, \mathbf{B})$ *in* $(S_k^+, \succeq_L)$, *(i)* $\mathbf{A} \wedge \mathbf{B} \preceq_L \{\mathbf{A},\mathbf{B}\} \preceq_L \mathbf{A} \vee \mathbf{B}$,

(ii) $\phi(\mathbf{A} \wedge \mathbf{B}) \le \{\phi(\mathbf{A}), \phi(\mathbf{B})\} \le \phi(\mathbf{A} \vee \mathbf{B})$ *for each* $\phi \in \Phi(S_k^+, \succeq_L)$*, and*

(iii) $\mathbf{A} \wedge \mathbf{B} = \mathbf{B} \wedge \mathbf{A}$ *and* $\mathbf{A} \vee \mathbf{B} = \mathbf{B} \vee \mathbf{A}$.

Moreover, the bounds are tight in the sense that

(iv) if $\{\mathbf{A},\mathbf{B}\} \preceq_L \mathbf{T}$ *and* $\mathbf{T} \preceq_L \mathbf{A} \vee \mathbf{B}$*, then* $\mathbf{T} = \mathbf{A} \vee \mathbf{B}$*, and*

(v) if $\{\mathbf{A},\mathbf{B}\} \succeq_L \mathbf{S}$ *and* $\mathbf{S} \succeq_L \mathbf{A} \wedge \mathbf{B}$*, then* $\mathbf{S} = \mathbf{A} \wedge \mathbf{B}$.

PROOF Conclusion (i) follows from Lemmas 1 and 2, and this implies (ii) in view of the monotonicity of $\Phi(S_k^+, \succeq_L)$. To see conclusion (iii), note that if $\mathbf{A} \preceq_L \mathbf{B}$, then $\mathbf{A}\wedge\mathbf{B} = \mathbf{A} = \mathbf{B}\wedge\mathbf{A}$ and $\mathbf{A}\vee\mathbf{B} = \mathbf{B} = \mathbf{B}\vee\mathbf{A}$. Otherwise let $\mathbf{B}_1' = \mathbf{B}^{1/2}\mathbf{Q}$ and $\mathbf{B}_2' = \mathbf{A}^{1/2}\mathbf{P}\mathbf{D}_\gamma^{-1/2}$; observe that $\mathbf{A} \wedge \mathbf{B} = \mathbf{B}_1'(\mathbf{D}_\gamma \wedge \mathbf{I}_k)\mathbf{B}_1$; and recall that $\mathbf{B} \wedge \mathbf{A} = \mathbf{A}^{1/2}\mathbf{P}(\mathbf{I}_k \wedge \mathbf{D}_\theta)\mathbf{P}'\mathbf{A}^{1/2}$. From the duality relations cited earlier it follows that $\mathbf{I}_k \wedge \mathbf{D}_\theta = (\mathbf{D}_\gamma \vee \mathbf{I}_k)^{-1} = (\mathbf{D}_\gamma \wedge \mathbf{I}_k)\mathbf{D}_\gamma^{-1}$, so that $\mathbf{B}\wedge\mathbf{A} = \mathbf{A}^{1/2}\mathbf{P}\mathbf{D}_\gamma^{-1/2}(\mathbf{D}_\gamma\wedge\mathbf{I}_k)\mathbf{D}_\gamma^{-1/2}\mathbf{P}'\mathbf{A}^{1/2} = \mathbf{B}_2'(\mathbf{D}_\gamma\wedge\mathbf{I}_k)\mathbf{B}_2$. But another duality asserts that $\mathbf{B}_1' = \mathbf{B}_2'$, so that $\mathbf{A} \wedge \mathbf{B} = \mathbf{B} \wedge \mathbf{A}$ as claimed. The assertion $\mathbf{A} \vee \mathbf{B} = \mathbf{B} \vee \mathbf{A}$ follows similarly. To establish (iv), first suppose that $\{\mathbf{A},\mathbf{B}\} \preceq_L \mathbf{T}$ and $\mathbf{T} \preceq_L \mathbf{A} \vee \mathbf{B}$. Then since $\mathbf{D}_\gamma = \mathbf{Q}'\mathbf{B}^{-1/2}\mathbf{A}\mathbf{B}^{-1/2}\mathbf{Q}$ and $\mathbf{D}_\gamma \vee \mathbf{I}_k = \mathbf{Q}'\mathbf{B}^{-1/2}(\mathbf{A} \vee \mathbf{B})\mathbf{B}^{-1/2}\mathbf{Q}$, the ordering $\mathbf{A} \preceq_L \mathbf{T} \preceq_L \mathbf{A} \vee \mathbf{B}$ implies

$$\mathbf{D}_\gamma \preceq_L \mathbf{Q}'\mathbf{B}^{-1/2}\mathbf{T}\mathbf{B}^{-1/2}\mathbf{Q} \preceq_L \mathbf{D}_\gamma \vee \mathbf{I}_k \tag{3.2}$$

whereas $\mathbf{B} \preceq_L \mathbf{T} \preceq_L \mathbf{A} \vee \mathbf{B}$ gives

$$\mathbf{I}_k \preceq_L \mathbf{Q}'\mathbf{B}^{-1/2}\mathbf{T}\mathbf{B}^{-1/2}\mathbf{Q} \preceq_L \mathbf{D}_\gamma \vee \mathbf{I}_k. \tag{3.3}$$

Letting $\mathbf{c}_i' = [0,\ldots,0,1,0,\ldots,0]$ have unity in the ith coordinate and zeros elsewhere, we infer from (3.2) and (3.3) that

$$\gamma_i \le \mathbf{c}_i'\mathbf{Q}'\mathbf{B}^{-1/2}\mathbf{T}\mathbf{B}^{-1/2}\mathbf{Q}\mathbf{c}_i \le \gamma_i, \qquad 1 \le i \le r \tag{3.4}$$

$$1 \le \mathbf{c}_i'\mathbf{Q}'\mathbf{B}^{-1/2}\mathbf{T}\mathbf{B}^{-1/2}\mathbf{Q}\mathbf{c}_i \le 1, \qquad r+1 \le i \le k. \tag{3.5}$$

Combining these and letting $\mathbf{W} = [(\mathbf{D}_\gamma \vee \mathbf{I}_k) - \mathbf{Q}'\mathbf{B}^{-1/2}\mathbf{T}\mathbf{B}^{-1/2}\mathbf{Q}]$ we find that $\mathbf{c}_i'\mathbf{W}\mathbf{c}_i = 0$, so that the diagonal elements of $\mathbf{W}$ are zero. But since $\mathbf{W} \succeq_L \mathbf{0}$, this implies that $\mathbf{W} = \mathbf{0}$ and thus $\mathbf{T} = \mathbf{B}^{1/2}\mathbf{Q}(\mathbf{D}_\gamma \vee \mathbf{I}_k)\mathbf{Q}'\mathbf{B}^{1/2} = \mathbf{A} \vee \mathbf{B}$ as claimed in conclusion (iv). The proof for (v) proceeds similarly, to complete our proof. □

3.3. *Bounds on* $(F_{n\times k}, \succeq)$

We seek lower and upper bounds on $(F_{n\times k}, \succeq)$, now requiring that matrices in $F_{n\times k}$ should be of full rank $k \le n$. We proceed constructively as follows, starting with $(\mathbf{X},\mathbf{Z})$ in $(F_{n\times k}, \succeq)$.

First transform $(\mathbf{X},\mathbf{Z}) \to [\mathbf{X}(\mathbf{Z}'\mathbf{Z})^{-1/2}, \mathbf{Z}(\mathbf{Z}'\mathbf{Z})^{-1/2}]$, observing that $\mathbf{H} = \mathbf{Z}(\mathbf{Z}'\mathbf{Z})^{-1/2}$ is semiorthogonal such that $\mathbf{H}'\mathbf{H} = \mathbf{I}_k$ and $\mathbf{H}\mathbf{H}'$ is idempotent of rank k. We next undertake the singular decompositions $\mathbf{X}(\mathbf{Z}'\mathbf{Z})^{-1/2} = \sum_{i=1}^{k} \lambda_i \mathbf{p}_i \mathbf{q}_i' = \mathbf{P}\mathbf{D}_\lambda\mathbf{Q}'$ and $\mathbf{Z}(\mathbf{Z}'\mathbf{Z})^{-1/2} = \sum_{i=1}^{k} \mathbf{u}_i\mathbf{q}_i' = \mathbf{U}\mathbf{I}_k\mathbf{Q}'$ such that $\mathbf{D}_\lambda = \mathrm{Diag}(\lambda_1,\ldots,\lambda_k)$, and $\mathbf{P}$ and $\mathbf{U}$ are semiorthogonal, whereas $\mathbf{Q} \in O(k)$. It is essential to note that $\mathbf{X}$ and $\mathbf{Z}$ may be recovered as $\mathbf{X} = \mathbf{P}\mathbf{D}_\lambda\mathbf{Q}'(\mathbf{Z}'\mathbf{Z})^{1/2}$ and $\mathbf{Z} = \mathbf{U}\mathbf{I}_k\mathbf{Q}'(\mathbf{Z}'\mathbf{Z})^{1/2}$. Corresponding to earlier usage we now define provisional lower and upper bounds, to be verified subsequently, as

$$\mathbf{X} \wedge \mathbf{Z} = \mathbf{P}(\mathbf{D}_\lambda \wedge \mathbf{I}_k)\mathbf{Q}'(\mathbf{Z}'\mathbf{Z})^{1/2} \tag{3.6}$$

$$\mathbf{X} \vee \mathbf{Z} = \mathbf{P}(\mathbf{D}_\lambda \vee \mathbf{I}_k)\mathbf{Q}'(\mathbf{Z}'\mathbf{Z})^{1/2}. \tag{3.7}$$

Subject to later justification we set forth the following.

DEFINITION 2 The matrix $\mathbf{X} \wedge \mathbf{Z}$ in (3.6) is called a *singular lower bound* for $(\mathbf{X},\mathbf{Z})$ in $(F_{n\times k}, \succeq)$, and $\mathbf{X} \vee \mathbf{Z}$ in (3.7) is called a *singular upper bound.*

Basic properties of the singular bounds are given in the following.

THEOREM 2 *Consider matrices* $\mathbf{X}\wedge\mathbf{Z}$ *and* $\mathbf{X}\vee\mathbf{Z}$ *as constructed from* $(\mathbf{X},\mathbf{Z})$ *in* $(F_{n\times k}, \succeq)$. *Then*

(i) $\mathbf{X} \wedge \mathbf{Z} \preceq \{\mathbf{X},\mathbf{Z}\} \preceq \mathbf{X} \vee \mathbf{Z}$;

(ii) $\phi(\mathbf{X} \wedge \mathbf{Z}) \le \{\phi(\mathbf{X}), \phi(\mathbf{Z})\} \le \phi(\mathbf{X} \vee \mathbf{Z})$ *for each* $\phi \in \Phi(F_{n\times k}, \succeq)$; *and*

(iii) $\mathbf{X} \wedge \mathbf{Z}$ *and* $\mathbf{X} \vee \mathbf{Z}$ *are determined up to equivalence under* $O(n)$ *acting from the left. Moreover, the bounds are tight in the sense that*

(iv) if $\{\mathbf{X},\mathbf{Z}\} \preceq \mathbf{T}$ *and* $\mathbf{T} \preceq \mathbf{X} \vee \mathbf{Z}$, *then* $\mathbf{T}$ *is equivalent to* $\mathbf{X} \vee \mathbf{Z}$, *and*

(v) if $\{\mathbf{X},\mathbf{Z}\} \succeq \mathbf{S}$ *and* $\mathbf{S} \succeq \mathbf{X} \wedge \mathbf{Z}$, *then* $\mathbf{S}$ *is equivalent to* $\mathbf{X} \wedge \mathbf{Z}$.

PROOF The conclusions follow from Theorem 1 since the orderings "$\mathbf{X} \succeq \mathbf{Y}$ on $(F_{n\times k}, \succeq)$" and "$\mathbf{X}'\mathbf{X} \succeq_L \mathbf{Y}'\mathbf{Y}$ on $(S_k^+, \succeq_L)$" are equivalent. In particular, with $\mathbf{X}'\mathbf{X} = \mathbf{A}$, $\mathbf{Z}'\mathbf{Z} = \mathbf{B}$ and $\mathbf{D}_\gamma = \mathbf{D}_\lambda^2$, conclusion (i) follows from its counterpart in Theorem 1. Conclusion (ii) follows from (i) and the monotonicity of $\Phi(F_{n\times k}, \succeq)$. Conclusion (iii) is apparent since $\mathbf{X} \succeq \mathbf{Y}$ and $\mathbf{P}\mathbf{X} \succeq \mathbf{U}\mathbf{Y}$ are equivalent for any $\mathbf{P},\mathbf{U}$ in $O(n)$. Conclusions (iv) and (v) follow from their counterparts in Theorem 1 together with the foregoing conclusion (iii). □

In view of Theorem 2 we now see that there are equivalence classes of singular lower and upper bounds for $(\mathbf{X},\mathbf{Z})$ in $(F_{n\times k}, \succeq)$. Thus $\mathbf{X} \wedge \mathbf{Z}$ and $\mathbf{X} \vee \mathbf{Z}$ in (3.6) and (3.7) may be replaced by their equivalents

$$\{\mathbf{X} \wedge \mathbf{Z}\} = \{\mathbf{R}(\mathbf{D}_\lambda \wedge \mathbf{I}_k)\mathbf{Q}'(\mathbf{Z}'\mathbf{Z})^{1/2};\ \mathbf{R} \in O(n)\} \tag{3.8}$$

$$\{\mathbf{X} \vee \mathbf{Z}\} = \{\mathbf{R}(\mathbf{D}_\lambda \vee \mathbf{I}_k)\mathbf{Q}'(\mathbf{Z}'\mathbf{Z})^{1/2};\ \mathbf{R} \in O(n)\}. \tag{3.9}$$

4. Some Applications

We illustrate stochastic bounds arising from Theorems 1 and 2.

4.1. *Distributions of Quadratic Forms*

Quadratic forms in Gaussian variates arise in a variety of settings, often under one of several alternative models. Specifically, with $\mathbf{A} \in S_k^0$, let $G_A(t;\mathbf{\Sigma})$ be the *cdf* of $V = \mathbf{X}'\mathbf{A}\mathbf{X}$ under $\mathcal{L}(\mathbf{X}) = N_k(\mathbf{0},\mathbf{\Sigma})$. Standard results include the following. (i) $\mathcal{L}(\mathbf{X}'\mathbf{A}\mathbf{X}) = \mathcal{L}\left(\sum_{i=1}^k \gamma_i Z_i^2\right)$, where $\{Z_1,\ldots,Z_k\}$ are *iid* $N_1(0,1)$ and $\{\gamma_1,\ldots,\gamma_k\}$ are the roots of $|\mathbf{A} - \gamma\mathbf{\Sigma}^{-1}| = 0$; see Johnson and Kotz (1970), including series expansions for $G_A(t;\mathbf{\Sigma})$. (ii) For fixed $\mathbf{A} \in S_k^0$ and $t > 0$, it is clear that $G_A(t,\mathbf{\Sigma}) \in \Phi^-(S_k^+,\succeq_L)$ when considered as a function on S_k^+, as may be seen on applying a result of Anderson (1955). Given two models, $N_k(\mathbf{0},\mathbf{\Sigma})$ and $N_k(\mathbf{0},\mathbf{\Omega})$, our earlier developments support the bounds

$$(4.1) \qquad G_A(t;\mathbf{\Sigma}\vee\mathbf{\Omega}) \leq \{G_A(t;\mathbf{\Sigma}), G_A(t;\mathbf{\Omega})\} \leq G_A(t;\mathbf{\Sigma}\wedge\mathbf{\Omega})$$

for each $t > 0$. As t varies, this provides an envelope bounding the two *cdf's* $G_A(t,\mathbf{\Sigma})$ and $G_A(t;\mathbf{\Omega})$. Moreover, these bounds may be evaluated numerically in particular cases using known series expansions; see Johnson and Kotz (1970).

To continue, consider an ensemble $\{N_k(\mathbf{0},\Xi);\Xi \in K_0\}$, for which $K_0 \subset S_k^+$ contains a minimal element Ξ_m and a maximal element Ξ_M under the ordering $\succeq_L$. Corresponding to (4.1) we have stochastic bounds given by

$$(4.2) \qquad G_A(t;\Xi_M) \leq \{G_A(t;\Xi);\Xi \in K_0\} \leq G_A(t;\Xi_m)$$

for each $t > 0$. For fixed $\mathbf{A}$, this gives an envelope of curves for all such *cdf's* as t varies over $[0,\infty)$.

4.2. *Minimal Dispersion Bounds*

We consider the regular estimation of vector parameters. Specifically, consider a family of dominated probability measures having density functions $\{f(\mathbf{X};\theta);\theta \in \Theta\}$ with $\Theta \subset \mathbb{R}^r$; let $\mathbf{G}(\theta) = [g_1(\theta),\ldots,g_k(\theta)]'$ be estimable functions on Θ having the partial derivatives $\Delta(\theta) = [\delta_{ij}(\theta)] = [\partial g_i(\theta)/\partial\theta_j]$ for all $\theta \in \Theta$; and let $\widehat{\mathbf{G}}(\mathbf{X}) = [\widehat{g}_1(\mathbf{X}),\ldots,\widehat{g}_k(\mathbf{X})]'$ be any unbiased estimator for $\mathbf{G}(\theta)$ having some dispersion matrix $V(\widehat{\mathbf{G}}(\mathbf{X}))$. Under regularity conditions $\mathcal{R}$, given as (i)–(vi) on page 194 of Zacks (1971), for example, a standard result is a minimal dispersion bound given by $V(\widehat{\mathbf{G}}(\mathbf{X})) \succeq_L$

$\Delta(\theta)[\mathcal{I}(\theta)]^{-1}\Delta'(\theta)$, where $\mathcal{I}(\theta) = [\mathcal{I}_{ij}(\theta)]$ is the Fisher information matrix. Moreover, this bound holds for every unbiased estimator $\widehat{\mathbf{G}}(\mathbf{X})$.

If two such regular families have information matrices $\mathcal{I}_1(\theta)$ and $\mathcal{I}_2(\theta)$, then our earlier construction applies directly. The resulting bound, namely,

$$V(\widehat{\mathbf{G}}(\mathbf{X})) \succeq_L \Delta(\theta)[\mathcal{I}_1(\theta) \vee \mathcal{I}_2(\theta)]^{-1}\Delta'(\theta) \tag{4.3}$$

applies to any unbiased estimator for $\mathbf{G}(\theta)$ taken from either model.

These developments extend to any regular ensemble $\{f_\tau(\mathbf{X};\theta); \theta \in \Theta, \tau \in T\}$ for which there is a maximal Fisher information matrix $\mathcal{I}_M(\theta)$. Corresponding to (4.3) we now have the bound

$$V(\widehat{\mathbf{G}}(\mathbf{X})) \succeq_L \Delta(\theta)[\mathcal{I}_M(\theta)]^{-1}\Delta'(\theta) \tag{4.4}$$

uniformly for every unbiased estimator $\widehat{\mathbf{G}}(\mathbf{X})$ for $\mathbf{G}(\theta)$ taken from any family in the ensemble.

4.3. *Peakedness of Vector Sums*

Consider the measures $\mu(\cdot;\mathbf{p})$ on $\mathbb{R}^k$ induced by weighted sums $\sum_{i=1}^n p_i\mathbf{X}_i$ of *iid* random vectors $\{\mathbf{X}_1,\ldots,\mathbf{X}_n\}$ having a symmetric log-concave density. Let $\mathbf{C}_0^k$ be the class of all compact convex subsets of $\mathbb{R}^k$ symmetric under reflection through $\mathbf{0} \in \mathbb{R}^k$, *i.e.* $\mathbf{x} \in A$ implies $-\mathbf{x} \in A$ for each $A \in \mathbf{C}_0^k$. It is known for each $A \in \mathbf{C}_0^k$ that $\mu(A;\mathbf{p})$ is order-reversing when considered as a function on $(R_n(c), \succeq_M)$, *i.e.*, $\mu(A;\mathbf{p}) \in \Phi^-(R_n(c), \succeq_M)$; see Olkin and Tong (1988) and Chan, Park and Proschan (1989). Thus if $\mathbf{p} \succeq_M \mathbf{q}$, then $\mu(\cdot;\mathbf{q})$ is more peaked about $\mathbf{0} \in \mathbb{R}^k$ than $\mu(\cdot;\mathbf{p})$ in the sense of Sherman (1955). Corresponding to (1.1) we therefore have the bounds

$$\mu(A;\mathbf{p} \vee \mathbf{q}) \le \{\mu(A;\mathbf{p}), \mu(A;\mathbf{q})\} \le \mu(A;\mathbf{p} \wedge \mathbf{q}) \tag{4.5}$$

for each $A \in \mathbf{C}_0^k$. Accordingly, we may refer to $\mu(\cdot;\mathbf{p} \vee \mathbf{q})$ as the stochastic minorant, and to $\mu(\cdot;\mathbf{p} \wedge \mathbf{q})$ as the stochastic majorant, of measures $\{\mu(\cdot;\mathbf{p}), \mu(\cdot;\mathbf{q})\}$ when evaluated over sets in $\mathbf{C}_0^k$.

To continue, we consider a possibly contaminated Gaussian sample as follows. Let $\{\mathbf{Y}_1,\ldots,\mathbf{Y}_n\}$ be *iid* $N_k(\mathbf{0},\mathbf{\Sigma})$ and let $\{\mathbf{Z}_1,\ldots,\mathbf{Z}_n\}$ be *iid* $N_k(\mathbf{0},\mathbf{\Omega})$. Now consider a possibly contaminated sample $\{\mathbf{X}_1,\ldots,\mathbf{X}_n\}$ in which $\{\mathbf{X}_i = \delta_i\mathbf{Y}_i + (1-\delta_i)\mathbf{Z}_i; 1 \le i \le n\}$ with $\{\delta_i \in \{0,1\}; 1 \le i \le n\}$. Further let $\mu(\cdot;\mathbf{p},\mathbf{\Sigma})$ be the measure induced by $\sum_{i=1}^n p_i\mathbf{Y}_i$. A basic result of Anderson (1955) shows that for each fixed $A \in \mathbf{C}_0^k$ and $\mathbf{p} \in R_n(c)$, the measure $\mu(A;\mathbf{p},\mathbf{\Sigma})$ is order-reversing when considered as a function on $(S_k^+, \succeq_L)$.

On combining the foregoing developments with Theorem 1, we now have the following bounds on the measure $\mu_X(\cdot;\mathbf{p})$ induced by $\sum_{i=1}^n p_i\mathbf{X}_i$ in a possibly contaminated sample, as

$$\mu(A;\mathbf{p} \vee \mathbf{q};\mathbf{\Sigma} \vee \mathbf{\Omega}) \le \{\mu_X(A;\mathbf{p}), \mu_X(A;\mathbf{q})\} \le \mu(A;\mathbf{p} \wedge \mathbf{q},\mathbf{\Sigma} \wedge \mathbf{\Omega}) \tag{4.6}$$

for each $A \in \mathbf{C}_0^k$, where the brackets in the center encompass all 2^n possible choices for $\{\delta_1, \ldots, \delta_n\}$ in the contaminated model. Equality is achieved on the left when $\mathbf{p} \succeq_M \mathbf{q}$ and $\mathbf{\Sigma} \succeq_L \mathbf{\Omega}$ at $\delta_1 = \cdots = \delta_n = 1$, and on the right at $\delta_1 = \cdots = \delta_n = 0$.

With regard to the peakedness of vector sums, the results of Olkin and Tong (1988) and of Chan, Park and Proschan (1989) have been extended by Eaton (1988) to include concentration properties in Gauss-Markov estimation. Our developments here, applied in Eaton's setting, would appear to provide lower and upper bounds for concentration probabilities in two or more linear models having dispersion parameters not ordered in $(S_k^+, \succeq_L)$. Earlier work by the author (Jensen (1979)) complements Eaton's work in demonstrating that Gauss-Markov estimators are most peaked among median-unbiased linear estimators, even without first or second moments as assumed by Eaton (1988).

4.4. *First-Order Experimental Designs*

Consider models $\mathbf{Y} = \alpha_1 \mathbf{1}_n + \mathbf{X}\boldsymbol{\beta} + \mathbf{e}$ and $\mathbf{Y} = \alpha_2 \mathbf{1}_n + \mathbf{Z}\boldsymbol{\beta} + \mathbf{e}$ having design matrices $\mathbf{X}, \mathbf{Z} \in F_{n \times k}$ in centered form as deviations from their column means. We are concerned with inferences regarding the elements of $\boldsymbol{\beta} = [\beta_1, \ldots, \beta_k]'$ in the two models, based on the Gauss-Markov estimators $\widehat{\boldsymbol{\beta}}(\mathbf{X}) = (\mathbf{X}'\mathbf{X})^{-1}\mathbf{X}'\mathbf{Y}$, and similarly $\widehat{\boldsymbol{\beta}}(\mathbf{Z}) = (\mathbf{Z}'\mathbf{Z})^{-1}\mathbf{Z}'\mathbf{Y}$, when $\mathcal{L}(\mathbf{e}) = N_n(\mathbf{0}, \sigma^2 \mathbf{I}_n)$. The directed Fisher efficiency of design $\mathbf{Z}$ relative to $\mathbf{X}$ in estimating $\mathbf{a}'\boldsymbol{\beta}$ is given by

$$E_F(\mathbf{Z}, \mathbf{X}; \mathbf{a}) = \mathbf{a}'(\mathbf{X}'\mathbf{X})^{-1}\mathbf{a} / \mathbf{a}'(\mathbf{Z}'\mathbf{Z})^{-1}\mathbf{a}. \tag{4.7}$$

The Pitman efficiency of $\mathbf{Z}$ relative to $\mathbf{X}$, in normal-theory tests for H : $\mathbf{A}\boldsymbol{\beta} = \delta_0$ against $K : \mathbf{A}\boldsymbol{\beta} \neq \boldsymbol{\delta}_0$ is given by

$$E_P(\mathbf{Z}, \mathbf{X} \mid \mathbf{A}) = \frac{(\mathbf{A}\boldsymbol{\beta} - \boldsymbol{\delta}_0)'[\mathbf{A}(\mathbf{Z}'\mathbf{Z})^{-1}\mathbf{A}']^{-1}(\mathbf{A}\boldsymbol{\beta} - \boldsymbol{\delta}_0)}{(\mathbf{A}\boldsymbol{\beta} - \boldsymbol{\delta}_0)'[\mathbf{A}(\mathbf{X}'\mathbf{X})^{-1}\mathbf{A}']^{-1}(\mathbf{A}\boldsymbol{\beta} - \boldsymbol{\delta}_0)}. \tag{4.8}$$

It is clear that design $\mathbf{Z}$ is more efficient than $\mathbf{X}$, uniformly for all $\mathbf{a} \in \mathbb{R}^k$ under Fisher efficiency, and for all $\{\mathbf{A} \in F_{r \times k}; 1 \leq r \leq k\}$ under Pitman efficiency, if and only if $\mathbf{Z} \succeq \mathbf{X}$ on $(F_{n \times k}, \succeq)$.

If neither $\mathbf{Z} \succeq \mathbf{X}$ nor $\mathbf{X} \succeq \mathbf{Z}$, then Theorem 2 supports the construction of a new design dominating both $\mathbf{X}$ and $\mathbf{Z}$ in efficiency. This is precisely the design $\mathbf{X} \vee \mathbf{Z}$ given in expression (3.7), or any equivalent design from expression (3.9). In summary, given any specified pair of first-order experimental designs in $F_{n \times k}$, we may construct numerically a new design dominating both designs in its efficiency for inferences regarding $\boldsymbol{\beta}$.

References

ANDERSON, T. W. (1955). The integral of a symmetric unimodal function over a symmetric convex set and some probability inequalities. *Proc. Amer. Math. Soc.* **6** 170-176.

BIRNBAUM, Z. (1948). On random variables with comparable peakedness. *Ann. Math. Statist.* **19** 76-81.

CHAN, W. T., PARK, D. H. AND PROSCHAN, F. (1989). Peakedness of weighted averages of jointly distributed random variables. In *Contributions to Probability and Statistics. Essays in Honor of Ingram Olkin.* L. J. Gleser, M. D. Perlman, S. J. Press and A. R. Sampson, eds., Springer–Verlag, New York, 58-62.

EATON, M. L. (1988). Concentration inequalities for Gauss-Markov estimators. *J. Multivariate Anal.* **19** 76-81.

HALMOS, P. R. (1958). *Finite-Dimensional Vector Spaces.* Second edition. Van Nostrand, New York.

JENSEN, D. R. (1979). Linear models without moments. *Biometrika* **66** 611-617.

JENSEN, D. R. (1984). Invariant ordering and order preservation. In *Inequalities in Statistics and Probability.* Y. L. Tong, ed., Institute of Mathematical Statistics, Hayward, CA. 26–34.

JOHNSON, N. L. AND KOTZ, S. (1970). *Distributions in Statistics: Continuous Univariate Distributions-2.* Houghton-Mifflin, Boston, MA.

LOEWNER, C. (1934). Uber monotone Matrixfunktionen. *Math. Z.* **38** 177-216.

MARSHALL, A. W. AND OLKIN, I. (1979). *Inequalities: Theory of Majorization and Its Applications.* Academic Press, New York.

MARSHALL, A. W., WALKUP, D. W. AND WETS, R. J. B. (1967). Order preserving functions: Applications to majorization and order statistics. *Pacific J. Math.* **23** 569-584.

OLKIN, I. AND TONG, Y. L. (1988). Peakedness in multivariate distributions. In *Statistical Decision Theory and Related Topics IV, Volume* 2. S. S. Gupta and J. O. Berger, eds., Springer–Verlag, New York, 373-383.

PROSCHAN, F. (1965). Peakedness of distributions of convex combinations. *Ann. Math. Statist.* **36** 1703-1706.

SHERMAN, S. (1955). A theorem on convex sets with applications. *Ann. Math. Statist.* **26** 763-766.

VULIKH, B. Z. (1967). *Introduction to the Theory of Partially Ordered Spaces.* Wolters-Noordhoff, Groningen.

ZACKS, S. (1971). *The Theory of Statistical Inference.* Wiley, New York.

DEPARTMENT OF STATISTICS
VIRGINIA POLYTECHNIC INSTITUTE AND STATE UNIVERSITY
BLACKSBURG, VA 24061

Stochastic Inequalities
IMS Lecture Notes – Monograph Series
Volume 22 (1993)

GENERALIZED MAJORIZATION ORDERINGS AND APPLICATIONS

By HARRY JOE[1]

University of British Columbia

Orderings that are special cases of or related to the majorization ordering in Joe (1987a) for functions on a measure space are reviewed. Applications in probability and statistics that have motivated the orderings are briefly discussed and some new applications are given. Also some new links are made between results of previous papers.

1. Introduction

This article reviews a class of majorization orderings that generalize vector majorization and some applications motivating or coming from the orderings. The emphasis is on work that has come after the publication of Marshall and Olkin (1979). The class fits within the majorization ordering in Joe (1987a) for functions on a measure space and includes most generalized majorization orderings. Exceptions are group majorization (see Eaton (1987), Giovagnoli and Wynn (1985)) and stochastic majorization (see Shanthikumar (1987)).

In Section 2, the definition of Joe (1987a) is given and then it is shown how other orderings are either special cases or are related in some way. A diversity of applications are discussed or summarized in Section 3. Marshall and Olkin (1979) unified inequalities through majorization, and although generalized majorization leads to inequalities, they have not always been the motivation for extensions. It is hoped that the results in this paper will lead readers to discover further applications and extensions.

2. Generalized Majorization Orderings

The goal in this section is to show that results of various authors fit within a unified framework. These authors have often not cross-referenced each other. We start with the definition of Joe (1987a).

[1]Research supported by NSERC Canada Grant A–8698.
AMS 1991 *subject classifications.* Primary 06A06, 62A99.
Key words and phrases. Entropy, majorization, ordering, probability distribution.

Let $(\mathcal{X}, \Lambda, \nu)$ be a measure space. For most applications, $\mathcal{X}$ will be a subset of a Euclidean space, and ν will be Lebesgue measure or counting measure. For a nonnegative integrable function h on $(\mathcal{X}, \Lambda, \nu)$, let $m_h(t) = \nu(\{x : h(x) > t\})$, $t \geq 0$, and let $h^*(u) = m_h^{-1}(u) = \sup\{t : m_h(t) > u\}$, $0 \leq u \leq \nu(\mathcal{X})$; h^* is the (left–continuous) decreasing rearrangement of h.

DEFINITION 2.1 Let a and b be nonnegative integrable functions on $(\mathcal{X}, \Lambda, \nu)$ such that $\int a\, d\nu = \int b\, d\nu$. Then a is majorized by b, written $a \prec b$, if one of the following four equivalent conditions hold.

(a) $\int [a - t]^+ d\nu \leq \int [b - t]^+ d\nu$ for all $t \geq 0$, where $[y]^+ = \max\{y, 0\}$.

(b) $\int \psi(a) d\nu \leq \int \psi(b) d\nu$ for all convex, continuous real-valued functions ψ such that $\psi(0) = 0$ and the integrals exist.

(c) $\int_t^\infty m_a(s)\, ds \leq \int_t^\infty m_b(s)\, ds$ for all $t \geq 0$.

(d) $\int_0^t a^*(u)\, du \leq \int_0^t b^*(u)\, du$ for all $0 \leq t < \nu(\mathcal{X})$.

ASIDE 2.2 The ideas in Definition 2.1 go back to Hardy, Littlewood and Pólya (1929) for $\mathcal{X}$ being a finite interval of the real line and to Chong (1974) for the general case. Except in some cases such as Case 1 and Case 2 (with $-\infty < C < D < \infty$) when $\mathcal{X}$ is a bounded subset of the real line, it is not the same as the dilation ordering of measures given in Chapter 13 of Phelps (1966) and references therein.

The above definition is suitable for all stochastic applications except one in this article; in the exception, a, b can be partly negative. If one wants to compare a, b that can have negative parts, then it appears necessary to have $\nu(\mathcal{X})$ finite. Note that $\int [a - t]^+ d\nu$ is not defined for an integrable function a if $t < 0$ and $\nu(\mathcal{X}) = \infty$.

If the $\nu(\mathcal{X})$ is finite and the nonnegativity condition for a, b is removed, then (i) m_a, m_b can be defined on $\mathbb{R}$, (ii) in condition (a), the inequality holds for all $t \in \mathbb{R}$, and (iii) $\psi(0) = 0$ in (b) is not required. In this case, the equivalence of parts (a), (c) and (d) still follows from Chong (1974). The main new contribution in Joe (1987a) is stating together the equivalent conditions as a generalized majorization ordering, and including condition (b) which is crucial to the applications in that paper. For $\nu(\mathcal{X})$ finite, another equivalent condition is

(a') $\int (a - t)^- d\nu \leq \int (b - t)^- d\nu$ for all real t, where $(y)^- = \max\{0, -y\}$.

The proof of condition (b) from (a') and (a) then follows with a few steps. The general case will follow after proving condition (b) for ψ such that $\psi'(0+) = 0 = \psi(0)$. The case where ψ has domain on $(-\infty, 0]$ can be handled similarly to when the domain is $[0, \infty)$ (see the proof of Theorem 2.1 in Joe (1987a)). If ψ has domain $[C, D]$ with $C < 0 < D$, then $\psi(x)$ can be approximated from below by functions of the form $\sum_{s_m < \cdots < s_1 < 0} (s_i - x)^+ + \sum_{0 < t_1 < \cdots < t_n} (x - t_i)^+$ and then the monotone convergence theorem can be used.

The following are specific cases (of Definition 2.1) that have been studied.

Case 1. The usual vector majorization results if $\mathcal{X} = \{1,\ldots,n\}$ and $a = (a_1,\ldots,a_n)$, $b = (b_1,\ldots,b_n)$ and ν is counting measure. Condition (d) in Definition 2.1 is the usual definition with decreasing components of the vectors a, b. The generalization to Case 2 below comes from thinking of a, b as probability vectors; i.e., a_i, b_i are probability masses at a point x_i. The generalization to Case 3 below comes from considering random variables X and Y with masses n^{-1} at points a_i and b_i respectively.

Case 2. Continuous majorization for densities on an interval $[C, D]$ (possibly unbounded) results if $a = f$, $b = g$, where f, g are densities on $[C, D]$ with respect to Lebesgue measure. In this case, $\prec$ is an ordering of closeness to uniformity of the density, and it has been studied in Hickey (1984). This ordering on densities can be extended to higher dimensional Euclidean space. In related work, Chan, Proschan and Sethuraman (1987), following up on Ryff (1963), study the majorization ordering for integrable functions (not necessarily nonnegative) on [0,1].

Case 3. The Lorenz ordering in Arnold (1987) and Das Gupta and Bhansali (1989) results if $\mathcal{X} = [0,1]$, ν is Lebesgue measure and $a = F^{-1}$, $b = G^{-1}$, where F, G are the respective distribution functions of nonnegative random variables X, Y (X and Y have a common finite mean) and F^{-1}, G^{-1} are the corresponding quantile functions. Since a, b are monotone increasing, the decreasing rearrangements are respectively $a^*(u) = F^{-1}(1-u)$ and $b^*(u) = G^{-1}(1-u)$. Condition (d) of Definition 2.1 is $\int_0^t F^{-1}(1-p)\,dp \le \int_0^t G^{-1}(1-p)\,dp$, $0 < t < 1$ or

$$\int_0^t F^{-1}(p)\,dp \ge \int_0^t G^{-1}(p)\,dp, \qquad 0 < t < 1; \tag{2.1}$$

the latter being the definition in Arnold (1987) and Das Gupta and Bhansali (1989). Condition (b) becomes

$$\mathrm{E}\,\psi(X) \le \mathrm{E}\,\psi(Y) \qquad \forall \text{ convex continuous functions } \psi \tag{2.2}$$

and condition (a) becomes

$$\mathrm{E}\,(X-t)^+ \le \mathrm{E}\,(Y-t)^+, \qquad \forall t \ge 0. \tag{2.3}$$

The Lorenz ordering is known by other names in earlier work. Let $\overline{F} = 1 - F, \overline{G} - 1 - G$ be the survival functions of F, G. In the form of condition (c) of Definition 2.1, that is,

$$\int_u^\infty \overline{F}(x)dx \le \int_u^\infty \overline{G}(x)dx \qquad \forall \text{ real } u, \tag{2.4}$$

the ordering is referred to as the "more variable" ordering in Ross (1983) and as a majorization ordering in Boland and Proschan (1986).

If the random variables X, Y have support on a bounded interval, say [0,1], then the doubly stochastic condition of vector majorization generalizes in two ways. From Ryff (1965), Definition 2.1 for $F^{-1} \prec G^{-1}$ in this case is equivalent to the condition:

(e) there exists a doubly stochastic operator T (a positive, contraction operator such that $\int_0^1 (TI_E)d\nu = \nu(E)$, where ν is Lebesgue measure and I_E is the indicator function of the measureable set E) from L^1 to L^1 such that $F^{-1} = TG^{-1}$, where L^1 is the space of Lebesgue integrable functions on [0,1].

Also, from Theorem 10 of Blackwell (1951), it is equivalent to the condition:

(f) there is a stochastic transformation $H(x|y)$ ($H(\cdot|y)$ is a distribution for each y in the support of G) such that $\int H(x|y)\,dG(y) = F(x)$ for all x and $E(Y|X) = X$.

Note that if X, Y have support on the points $x_1, \ldots, x_n$ and $y_1, \ldots, y_n$ respectively with masses n^{-1} at each support point, then both conditions (e) and (f) are equivalent to the existence of a doubly stochastic matrix P such that $(x_1, \ldots, x_n) = (y_1, \ldots, y_n)P$.

Case 4. Simonis (1988) defines a "spectral order" which is the Lorenz ordering in Case 3 without the constraint of nonnegativity on the random variables X, Y (see also Aside 2.2). Let X, Y have respective distribution functions F, G, and corresponding survival functions $\overline{F}, \overline{G}$. Simonis defines $X \prec Y$ if $\mathrm{E}X = \mathrm{E}Y$ and $\int_0^t \overline{F}^{-1}(u)du \le \int_0^t \overline{G}^{-1}(u)du$ for all $0 < t < 1$ This is the same as (2.1) and condition (d) of Definition 2.1 but without the nonnegativity requirement for $a = F^{-1}$ and $b = G^{-1}$. Simonis proves the (a), (b) and (c) are equivalent, that is, $X \prec Y$ if and only if (2.3), (2.2) or (2.4) hold.

This ordering is also used in Stoyan (1983), where it is called a convex ordering or an ordering of "mean residual life" (with (2.3) as the definition).

Case 5. Non-uniform weighted majorization results if a measure other than Lebesgue measure or counting measure is used. For the vector case, let $\mathcal{X} = \{1, \ldots, n\}$ and ν be a measure with positive mass q_i at the point i, $i = 1, \ldots, n$. Cheng (1977) defined this ordering for vectors that are similarly ordered and called it p–majorization. Joe (1990) uses this ordering without the constraint of similarly ordered and in addition used the continuous version, that is, with $\mathcal{X} = [C, D]$ being an interval of the real line and ν corresponding to a positive density $q(\cdot)$ on $[C, D]$. Both of these orderings will be referred to as majorization with respect to q and denoted by $\prec_q$. For this case, the various forms of Definition 2.1 and other equivalent conditions are given and discussed in Joe (1990).

Case 6. The r–majorization ordering with respect to q of Joe (1990), denoted by $\prec_q^r$, follows from Case 5, with ratios of densities with respect to q. That is, for n–dimensional probability vectors $p_1 = (p_{11}, \ldots, p_{1n})$ and

$p_2 = (p_{21}, \ldots, p_{2n})$, $p_1 \prec_q^r p_2$ if $(p_{11}/q_1, \ldots, p_{1n}/q_n) \prec_q (p_{21}/q_1, \ldots, p_{2n}/q_n)$, and for densities p_1, p_2 of random variables on $[C, D]$, $p_1 \prec_q^r p_2$ if $p_1/q \prec_q p_2/q$. R–majorization puts the probability vector or density q at the lower end of the ordering instead the the uniform vector or density, and can be interpreted as an ordering of divergence or distance from q. With applications to thermodynamics, this ordering is called a mixing distance in Ruch and Mead (1976) and Ruch, Schranner and Seligman (1978).

3. Applications

In this section, we summarize some recent applications of the majorization orderings in Section 2, and also we give some new applications (in 3.5 and 3.9 and part of 3.2). In some cases, the application motivated the study of the majorization ordering. One goal is to show a diversity of applications so they are mainly brief (in which case details can be found in the papers that are referred to). The applications taking up more space are the new ones and the one in 3.8 which is more detailed in order to mention an open problem.

3.1. *Ordering on Random Variables or Cumulative Distribution Functions*

The ordering in Case 3 of Section 2 has been applied in diverse areas. Applications in reliability are given in Boland and Proschan (1986) and Ross (1983) with earlier such applications going back to Marshall and Proschan (1970). Applications to queueing models are given in Ross (1983) and Stoyan (1983). Arnold (1987) has applications to distributions of wealth; the idea is that with a given mean wealth μ, a distribution of wealth F with a constant F^{-1} is most equitable (this corresponds to a mass of 1 at μ) and a distribution F which is larger in the (Lorenz) ordering is less equitable. A further, more detailed application is given next.

3.2. *Probability Forecasting*

Conditions (e) and (f) of Section 2 are used in this application. DeGroot and Fienberg (1982) and DeGroot and Erikkson (1985) consider an ordering of forecasters with possible forecasts in the set $\{x_0, x_1, \ldots, x_m\}$, where $x_0 = 0, x_m = 1$ and $x_0 < x_1 < \cdots < x_m$. Their ordering can be generalized to allow forecasts in [0,1], in which case, their ordering becomes that in Application 3.1 or Case 3 of Section 2 with support of random variables in the interval [0,1].

The framework is that forecasters give a probability each day for an event like occurrence of rain. A forecast of 1 means a prediction that the event will happen and a forecast of 0 means a prediction that the event will not happen. A forecaster is well–calibrated if the conditional probability of the event, given that the forecaster's prediction is x, is x. Let two well–calibrated forecasters A, B have distributions F, G for their forecasts. Then the means of F, G must be the same, both being the probability or relative frequency of the event. Forecaster B is at least as refined as A if condition (f) holds. In well–behaved situations, the doubly stochastic operator in conditions (e) and (f) becomes an integral operator, that is, there is a function k on $[0,1]^2$ such that $\int k(u,v)du = 1$ for all v, $\int k(u,v)dv = 1$ for all u, and $\int k(u,v)G^{-1}(v)\,dv = F^{-1}(u)$. In fact, $k(u,v) = h(F^{-1}(u)|G^{-1}(v))/f(F^{-1}(u))$, where $h(\cdot|y)$ is the density of $H(\cdot|y)$ and f is the density of F; both densities are with respect to a measure that dominates F (for simple cases, Lebesgue measure, counting measure or a combination).

Let p be the relative frequency of the event. It is intuitively true, and not difficult to show from Definition 2.1, that the best or most refined well–calibrated forecaster has a forecast distribution G that has mass of p at 1 and mass of $1-p$ at 0. The least refined well–calibrated forecaster has a forecast distribution F that has a mass of 1 at p. With the restriction that forecasts are in the set $\{x_0, x_1, \ldots, x_m\}$, referred to above, DeGroot and Fienberg (1982) proved that the least refined well–calibrated forecaster has a distribution F that puts mass α at x_i and mass $1-\alpha$ at x_{i+1}, where i is such that $x_i < p \leq x_{i+1}$ and $\alpha = (x_{i+1} - p)/(x_{i+1} - x_i)$.

3.3. *Constrained Majorization*

The author has used constrained majorization in several papers: Joe (1985, 1987a,b, 1988a, 1990). This comes about when there are additional constraints on a, b (in Definition 2.1) of the form

$$\int h_\tau a\,d\nu = \int h_\tau b\,d\nu, \quad \forall \tau \in \Upsilon,$$

where Υ is an index set. Only functions that satisfy the additional constraints are comparable. With constrained majorization, maximal and minimal functions in the orderings can be of interest.

3.4. *Orderings of Dependence*

If the h_τ are taken to be appropriate indicator functions and $a = f$, $b = g$ are m–dimensional multivariate densities with respect to a measure ν, the constraints can be that f, g have the same set of univariate margins, say $f_1, \ldots, f_m$. Then the constrained majorization ordering is an ordering of dependence among densities in the set $\Gamma(f_1, \ldots, f_m)$ of multivariate densities with univariate margins $f_1, \ldots, f_m$. This and generalizations are studied in Joe (1987a) and Joe (1985), with the latter mainly concerned with the bivariate discrete case with counts from a two–way contingency table (see also Application 3.6) as well as bivariate distributions. With this ordering, the density $f_I = \prod_j f_j$ is among those which are minimal. If it is desired to have f_I as the "unique" minimal density, then ratios relative to f_I can be used with Case 6 of Section 2 (that is, r–majorization with respect to f_I). The ordering of dependence in Scarsini (1990) is r–majorization for the bivariate case. However Scarsini also allows for the pair of univariate margins not being identical for comparison of two densities.

3.5. *Exploratory Data Analysis for Two–Way Tables*

There is a benefit to having more than just f_I as the minimal density in Application 3.4, especially for two–way tables of counts or sample proportions. For this special case, Joe (1985) proves that a necessary condition for minimal tables is that each row is similarly ordered with the column sum margin and each column is similarly ordered with the row sum margin. Since minimal tables with respect to the constrained majorization ordering can be interpreted as those "closest" to independence, this result provides a quick way to check whether two categorical variables are approximately independent. This can be done in one's head, unlike computation of the expected counts under the assumption of independence.

An example illustrating this is given below; the data are from students at the University of British Columbia in a recent year who took a first year calculus course. The two–way table below is constructed from the two variables, with grade in calculus (A, B or $\leq C$) as the column variable and type of high school (Vancouver, rest of Greater Vancouver Regional District, rest of British Columbia, private) as the row variable. The last row and the last column are the marginal totals by grade and by type of high school.

	A	B	$\leq C$	Total
1	198	143	201	542
2	186	169	284	639
3	80	102	159	341
4	42	40	83	165
Total	506	456	727	1687

By checking for similarly ordered rows and columns, one can see that the two variables are close to being independent, with the main discrepancies being that more A's than expected under independence are from "Vancouver" (compare first and fourth columns) and less A's than expected under independence are from "rest of B.C." (compare third and fifth rows). A measure of dependence given in Joe (1987a) is $\delta^* = (1 - e^{-2\delta})^{1/2}$, where

$\delta = \sum_{i,j} p_{ij} \log[p_{ij}/(p_{i+}p_{+j})]$, p_{ij} is the proportion simultaneously in category i of the row variable and category j of the column variable, $p_{i+} = \sum_j p_{ij}$ and $p_{+j} = \sum_i p_{ij}$. The value of δ^* is 0.13 for the above table, and this suggests very little dependence as δ^* can take values between 0 and 1.

3.6. *Fisher's Exact Test*

The maximal and minimal tables from the constrained majorization ordering in Applications 3.3 and 3.4 were used in the network algorithm of Mehta and Patel (1983) for the computation of the P–value of Fisher's exact test for two–way contingency tables. This allowed the use of maxima and minima of certain functions instead of the use of the bounds in Mehta and Patel. The theorems from Joe (1985) were programmed into a Fortran routine, and the improvement in computational time reported in Joe (1988b) and Clarkson, Fan and Joe (1990). The routine has now been adapted into IMSL and Splus.

3.7. *Ordering of Transitivity*

Depending on what $\mathcal{X}$ is, constrained majorization can lead to interpretations other than an ordering of dependence. An example is for paired comparison matrices $a = \{p_{ij} : 1 \le i, j \le n, i \ne j\}$, where there are n items and p_{ij} is the probability that item i is preferred to item j. The ordering on a with the constraints $\sum_{j \ne i} p_{ij} = m_i$, $i = 1, \ldots, n$, is interpreted as an ordering of transitivity in Joe (1988b), in that matrices at the lower end of the ordering are such that there is a preference transitivity among the items.

3.8. *Conjugate Priors and Majorization with Moment Constraints*

The ordering Case 6 was partly motivated with the aim to justify the use of some common conjugate priors. This goal was only partly reached and the remaining step to be proved is posed as a problem here. This application also shows the connection between majorization and entropy, and illustrates what constrained majorization results are like.

Consider the problem of choosing a prior distribution for a random (continuous) quantity after having elicited the mean and/or variance. We compare the use of r–majorization with the maximum entropy principle and the principle of minimum cross entropy (Jaynes 1983, Shore and Johnson 1980). The following results from Joe (1990) are needed; they concern minimal densities relative to r–majorization with respect to q when there are first and/or second moment constraints.

THEOREM 3.8.1 *Let $q(x)$ be a positive continuous function on the interval $[C,D]$. Let $\mathcal{P} = \mathcal{P}(C,D;\mu)$ be the class of densities f (with respect to Lebesgue measure) on the interval $[C,D]$ satisfying $\int_C^D x f(x)dx = \int_C^D x\, r(x)q(x)\,dx = \mu$, where $r(x) = f(x)/q(x)$. Then $f \in \mathcal{P}$ is minimal with respect to q if and only if $r(x)$ is monotone.*

THEOREM 3.8.2 *Let $q(x)$ be a positive continuous function on the interval $[C,D]$. Let $\mathcal{P} = \mathcal{P}(C,D;\mu_1,\mu_2)$ be the class of densities f (with respect to Lebesgue measure) on the interval $[C,D]$ satisfying $\int_C^D x f(x)dx = \mu_1$, $\int_C^D x^2 f(x)dx = \mu_2$. Then $f \in \mathcal{P}$ is minimal with respect to q only if $r(x) = f(x)/q(x)$ is monotone, U-shaped or unimodal.*

By taking a limit, Theorems 3.8.1 and 3.8.2 are valid for open (and possibly infinite) intervals and q can approach ∞ at one or both of the endpoints.

One approach to eliciting a prior distribution might be to first use the invariance principle to obtain a "non–informative", possibly improper, prior q (Cox and Hinkley (1974, Chapter 10)) and then use knowledge (possibly subjective) of first and/or second order moments to choose a prior close to q satisfying the moment constraints. The minimum cross entropy principle (and the maximum entropy principle as a special case) lead to a very small class of priors – for example, only the exponential and normal, possibly truncated, result as prior distributions from the maximum entropy principle. Although several common conjugate priors are maximum entropy based on other constraints, a stronger justification is based on first and/or second moment constraints since these are more easily elicited than something like the expected value of the logarithm of the quantity.

It is known (see, for example, Berger (1980)) that if the minimum cross entropy distribution with respect to q exists, then it has the form

$$f(x) = C(\lambda_1,\ldots,\lambda_m)q(x)\exp[-\sum_j \lambda_j h_j(x)], x \in \mathcal{X},$$

when the constraints are $\int_{\mathcal{X}} h_j(x)f(x)d\nu = \mu_j$, $j = 1,\ldots,m$. When q is not a constant function and the constraints are moments, no commonly used conjugate prior probability distribution has this form. It is shown below that several conjugate prior distributions of an unknown scale parameter σ satisfy the necessary conditions for minimality with respect to the invariant (improper) prior which is proportional to $q(\sigma) = 1/\sigma$. That is, the constrained r–majorization results lead to a larger class of distributions that are "near" q, subject to the moment constraints. Nearness of a density to q here means that for any closed subinterval of $(0,\infty)$ the density is (relatively) near c/σ, where c is the normalizing constant for the interval.

Conjugate priors for scale parameters include (a) the inverse gamma density, $f(\sigma) \propto \sigma^{-\alpha-1}\exp\{-\beta/\sigma\}$, $\sigma > 0$, for a gamma distibution with known shape parameter, (b) the density, $f(\sigma) \propto \sigma^{-2\alpha-1}\exp\{-\lambda/\sigma^2\}, \sigma > 0$, for a normal distribution with known mean, and (c) the hyperbolic or Pareto density (Raiffa and Schlaifer (1961)), $f(\sigma) \propto \sigma^{-\alpha-1}$, $\sigma \geq M$ $(M > 0)$, for a uniform distribution on 0 to an unknown upper bound. It is easy to check that $\sigma f(\sigma)$ is unimodal over $(0, \infty)$ for (a) and (b) and $\sigma f(\sigma)$ is monotone over $[M, \infty)$ for (c) when $\alpha > 0$. Hence the necessary condition for minimality in Theorem 3.8.2 is satisfied for (a) and (b), and the necessary condition for minimality in Theorem 3.8.1 is satisfied for (c). Note that $C(\lambda_1, \lambda_2)\sigma^{-1}\exp\{\lambda_1\sigma + \lambda_2\sigma^2\}$ from (3.1) is not a proper density on $(0, \infty)$.

For an example not involving a scale parameter, consider the probability parameter θ $(0 < \theta < 1)$ of a binomial distribution. "Noninformative" priors that have been proposed are $q(\theta) = [\theta(1-\theta)]^{-0.5}$ (Jeffreys (1961)), $q(\theta) = [\theta(1-\theta)]^{-1}$ (Haldane (1948)) and $q(\theta) \equiv 1$. The conjugate prior for the parameter θ is the Beta density $f(\theta)$, which is proportional to $\theta^{\alpha-1}(1-\theta)^{\beta-1}$. Hence, for all three of these q, $f(\theta)/q(\theta)$ is monotone, U–shaped or unimodal, and the necessary condition for minimality when there are two moment constraints is satisfied. This is better than saying that the beta distribution is a maximum entropy distribution subject to knowing the expectation of $\log\theta$ and $\log(1-\theta)$. It would be nice to prove in addition that the beta density is minimal relative to r–majorization with respect to q for any of the above q (since satisfying the necessary condition for minimality need not imply minimality). This however is an open problem (a similar comment holds for cases (a) and (b) in the preceding paragraph). The techniques in Joe (1990) do not work but numerical comparisons with the maximum entropy density on (0,1) show that the beta density with the same first two moments does not majorize the maximum entropy density.

3.9. *Updating Subjective Probability*

This is another application of r–majorization.

For simplicity of presentation, we think of a random quantity which has a finite number of possible outcomes, labelled as $1, 2, \ldots, n$; however results do generalize. Suppose our initial prior distribution is $q = (q_1, \ldots, q_n)$, with q_i being the probability of outcome i. If we then get further information that cause us to revise our probabilities for some pairwise mutually exclusive events, how should we update our probability distribution? Diaconis and Zabell (1982) study this problem using Jeffrey's conditionization rule and a divergence or distance approach; the latter was called mechanical updating. We look at this updating problem using majorization as an ordering of divergence instead of using several different measures of divergence. This is, in a sense, a simpler way of obtaining and viewing the results in Section

5 of Diaconis and Zabell.

Let $E_1, \ldots, E_e$ be pairwise mutually exclusive events (subsets of $S = \{1, \ldots, n\}$). Without loss of generality, we suppose that these form a partition of S. Let $q^* = (q_1^*, \ldots, q_n^*)$ be the updated subjective probability distribution, where our further information causes us to decide

$$q^*(E_j) = \sum_{i \in E_j} q_i^* = \eta_j, \qquad j = 1, \ldots, e, \tag{3.1}$$

where $\sum_j \eta_j = 1$. The constrained r–majorization approach is to consider the r–majorization ordering with respect to q ($\prec_q^r$) for the class of q^* that satisfy (3.1).

The minimal q^* is such that q^*/q is as close to uniform as possible, and therefore, from Joe (1990), it is piecewise uniform and satisfies $q_i^*/q_i = c_j$, $i \in E_j$, $j = 1, \ldots, e$, for some constants c_j, and from (3.1), clearly, $c_j = \eta_j / \sum_{i \in E_j} q_i$. Hence Jeffrey's rule,

$$q^*(A|E_j) = \frac{\sum_{i \in A \cap E_j} q_i^*}{\eta_j} = \frac{c_j \sum_{i \in A \cap E_j} q_i}{\eta_j} = \frac{\sum_{i \in A \cap E_j} q_i}{\sum_{i \in E_j} q_i} = q(A|E_j), \tag{3.2}$$

for the conditional probabilities holds. Theorem 5.1 of Diaconis and Zabell (1982) makes the conclusion (3.2) for the q^* that minimizes the Hellinger or cross entropy divergence from q subject to (3.1). This follows as a corollary of the r–majorization result since both of these divergence measures are increasing in the ordering $\prec_q^r$ (Section 3.1 of Joe, 1990).

Diaconis and Zabell (1982) also study the case of (compatible) updated probabilities for two partitions $E_1, \ldots, E_e$ and $D_1, \ldots, D_d$. Let q^* be the updated probability distribution. Suppose $q^*(E_j) = \eta_j$ and $q^*(D_k) = \theta_k$. Let $A_{jk} = E_j \cap D_k$. Note that A_{jk}, $j = 1, \ldots, e$, $k = 1, \ldots, d$, form a partition of S. Let $q(A_{jk}) = \zeta_{jk} = \sum_{i \in A_{jk}} q_i$ and let $q^*(A_{jk}) = \zeta_{jk}^* = \sum_{i \in A_{jk}} q_i^*$. Then the revised probabilites are specified up to

$$\sum_k \zeta_{jk}^* = \eta_j, \qquad j = 1, \ldots, e, \qquad \sum_j \zeta_{jk}^* = \theta_k, \qquad k = 1, \ldots, d.$$

Consider the set Q^* of q^* which satisfy these constraints and put the r–majorization ordering with respect to q on this set. Now there is not a unique minimal q^* in Q^* for this ordering. This explains why Diaconis and Zabell obtained different q^* for different divergence measures (cross entropy and variation distance) However, in an argument similar to the above, it can be concluded that minimal distributions in Q^* are piecewise uniform and must satisfy $q_i^*/q_i = \zeta_{jk}^*/\zeta_{jk} = c_{jk}$ for $i \in A_{jk}$ for some constants c_{jk}. If the ζ_{jk}'s are all the same, then similar to Theorem 1 of Joe (1985), a minimal q^* must be such that the matrix (c_{jk}) has rows which are similarly ordered to $(\theta_1, \ldots, \theta_d)$ and columns which are similarly ordered to $(\eta_1, \ldots, \eta_e)$. In general there is no simple characterization of the minimal q^*. But a sufficient condition for q^* in Q^* to be minimal is that it minimizes

$$\sum_i q_i \psi(q_i^*/q_i) = \sum_j \sum_k \zeta_{jk} \psi(\zeta_{jk}^*/\zeta_{jk})$$

for a strictly convex real–valued function ψ. If $\psi(u) = u \log u$, the minimum cross entropy distribution results and its form is $\zeta_{jk}^* = \zeta_{jk}\alpha_j\beta_k$ for some positive constants α_j, β_k. Hence $\zeta_{jk}^* = \eta_j\theta_k$ is a minimum cross entropy distribution only if $\zeta_{jk} = \zeta_{j+}\zeta_{+k}$ where $\zeta_{j+} = \sum_k \zeta_{jk}$ and $\zeta_{+k} = \sum_j \zeta_{jk}$. The conclusion here is not always the suggestion of updated probabilities, $\zeta_{jk}^* = \eta_j\theta_k$, in Section 4.2 of Diaconis and Zabell (1982).

Acknowledgements. I am grateful to the referees for comments leading to an improved presentation.

REFERENCES

ARNOLD, B. C. (1987). *Majorization and the Lorenz Order.* Springer–Verlag, New York.

BERGER, J. O. (1980). *Statistical Decision Theory.* Springer–Verlag, New York.

BLACKWELL, D. (1951). Comparison of experiments. In *Proceedings of the Second Berkeley Symposium on Math. Statist. Prob.*, J. Neyman, ed. University of California Press, Berkeley, CA. 93–102.

BOLAND, P. J. AND PROSCHAN, F. (1986). An integral inequality with applications to order statistics. In *Reliability and Quality Control*, A.P. Basu, ed. North Holland, Amsterdam, 107–116.

CHAN, W. T., PROSCHAN, F. AND SETHURAMAN, J. (1987). Schur–Ostrowski theorems for functionals on $L_1(0,1)$. *SIAM J. Math. Anal.* **18** 566–578.

CHONG, K.-M. (1974). Some extensions of a theorem of Hardy, Littlewood and Pólya and their applications. *Canad. J. Math.* **26** 1321–1340.

CLARKSON, D. B., FAN, Y.-A. AND JOE, H. (1990). A remark on algorithm 643: FEXACT: An algorithm for performing Fisher's exact test in $r \times c$ contingency tables. Tech. Report, Department of Statistics, University of British Columbia.

COX, D. R. AND HINKLEY, D. V. (1974). *Theoretical Statistics.* Chapman and Hall, London.

DAS GUPTA, S. AND BHANDARI, S.K. (1989). Multivariate majorization. In *Contributions to Probability and Statistics, Essays in honour of Ingram Olkin.* L. J. Gleser, M. D. Perlman, S. J. Press and A. R. Sampson, eds. Springer–Verlag, New York, 63–74.

DEGROOT, M. H. AND ERIKKSON, E. A. (1985). Probability forecasting, stochastic dominance, and the Lorenz curve. In *Bayesian Statistics* **2** J. M. Bernardo, M. H. DeGroot, D. V. Lindley and A. F. Smith, eds. North-Holland, Amsterdam, 99–118.

DEGROOT, M. H. AND FIENBERG, S. E. (1982). Assessing probability assessors: calibration and refinement. In *Statistical Decision Theory and Related Topics III*, **Vol. 1**. S. S. Gupta and J. O. Berger, eds. Academic Press, New York, 291–314.

DIACONIS, P. AND ZABELL, S. A. (1982). Updating subjective probability. *J. Amer. Statist. Assoc.* **77** 822–831.

EATON, M. L. (1987). *Lecture on Topics in Probability Inequalities.* Centrum voor Wiskunde en Informatica, Amsterdam.

GIOVAGNOLI, A. AND WYNN, H. P. (1985). G–majorization with applications to matrix orderings. *Linear Alg. Appl.* **67** 111–135.

HALDANE, J. B. S. (1948). The precision of observed values of small frequencies. *Biometrika* **35** 297–303.

HARDY, G. H., LITTLEWOOD, J. E. AND PÓLYA, G. (1929). Some simple inequalities satisfied by convex functions. *Messenger Math.* **58** 145–152.

HICKEY, R. J. (1984). Continuous majorisation and randomness. *J. Appl. Prob.* **21** 924–929.

JAYNES, E. T. (1983). *Papers on Probability, Statistics and Statistical Physics*, R. D. Rosenkrantz, ed. Reidel, Dordrecht.

JEFFREYS, H. (1961). *Theory of Probability,* 3rd edition. Oxford University Press, New York.

JOE, H. (1985). An ordering of dependence for contingency tables. *Linear Alg. Appl.* **70** 89–103.

JOE, H. (1987a). Majorization, randomness and dependence for multivariate distributions. *Ann. Probab.* **15** 1217–1225.

JOE, H. (1987b). An ordering of dependence for distributions of k–tuples, with applications to lotto games. *Canad. J. Statist.* **15** 227–238.

JOE, H. (1988a). Majorization, entropy and paired comparisons. *Ann. Statist.* **16** 915–925.

JOE, H. (1988b). Extreme probabilities for contingency tables under row and column independence with application to Fisher's exact test. *Commun. Statist.-Theor. Meth.* **17** 3677–3685.

JOE, H. (1990). Majorization and divergence. *J. Math. Anal. Appl.* **148** 287–305.

MARSHALL, A. W. AND OLKIN, I. (1979). *Inequalities: Theory of Majorization and its Applications.* Academic Press, New York.

MARSHALL, A. W. AND PROSCHAN, F. (1970). Mean life of series and parallel systems. *J. Appl. Prob.* **7** 165–174.

MEHTA, C. R. AND PATEL, N. R. (1983). A network algorithm for performing Fiesher's exact test in $r \times c$ contingency tables. *J. Amer. Statist. Assoc.* **78** 427–434.

PHELPS, R. R. (1966). *Lectures on Choquet's Theorem.* Van Nostrand, Princeton, NJ.

RAIFFA, H. AND SCHLAIFER, R. (1961). *Applied Statistical Decision Theory.* Colonial Press, Clinton, MA.

ROSS, S. M. (1983). *Stochastic Processes.* Wiley, New York.

RUCH, E. AND MEAD, C. A. (1976). The principle of increasing mixing character and some of its consequences. *Theor. Chim. Acta.* **41** 95–117.

RUCH, E., SCHRANNER, R. AND SELIGMAN, T. H. (1978). The mixing distance. *J. Chem. Phys.* **69** 386–392.

RYFF, J. V. (1963). On the representation of doubly stochastic operators. *Pacific J. Math.* **13** 1379–1386.

RYFF, J. V. (1965). Orbits of L^1 –functions under doubly stochastic transformations. *Trans. Amer. Math. Soc.* **117** 92–100.

SCARSINI, M. (1990). An ordering of dependence. In *Topics in Statistical Dependence.* H. W. Block, A. R. Sampson, and T. H. Savits, eds. Institute of Mathematical Statistics, Hayward, CA. 403–414.

SHANTHIKUMAR, J. G. (1987). Stochastic majorization of random variables with proportional equilibrium rates. *Adv. Appl. Prob.* **19** 854–872.

SHORE, J. E. AND JOHNSON, R. W. (1980). Axiomatic derivation of the principle of maximum entropy and the principle of minimum cross–entropy. *IEEE Trans. Inform. Theory* **IT-26** 26–37.

SIMONIS, A. (1988). Ordem Espectral e Schur-Convexidade em Espacos de Probabilidade. M.Sc. Thesis, Instituto de Mathematica e Estatistica, Universidade de Sao Paulo.

STOYAN, D. (1983). *Comparison Methods for Queues and Other Stochastic Models.* Wiley, New York.

DEPARTMENT OF STATISTICS
UNIVERSITY OF BRITISH COLUMBIA
VANCOUVER, B.C. CANADA V6T 1Z2

Stochastic Inequalities
IMS Lecture Notes – Monograph Series
Volume 22 (1993)

MULTIVARIATE MAJORIZATION BY POSITIVE COMBINATIONS[1]

By HARRY JOE and JOSEPH VERDUCCI

University of British Columbia and Ohio State University

Multivariate majorization orderings are used to compare matrices according to their dispersiveness. When applied to matrices whose rows represent distributions of different resources, the ordering that appears to be most useful is called majorization by positive linear combinations (PC-majorization). Two matrices are PC-majorized if all positive linear combinations of the rows are ordered by ordinary vector majorization. Properties of PC-majorization are derived; an algorithm is given to determine whether or not one matrix is PC-majorized by another; and elementary operations that reduce a matrix in the PC-ordering are explained.

1. Introduction

The key idea of majorization is to pre-order vectors according to a universal standard of dispersiveness. That is, any reasonable measure of dispersiveness of the components of a vector should imply an ordering that is consistent with the pre-ordering of majorization. The universality of the majorization ordering is well illustrated by the hundreds of applications mentioned in Marshall and Olkin (1979), and many other sources.

Several attempts have been made to extend majorization to a pre-ordering of matrices. However, there appears to be no 'universal' extension, but rather several different extensions that are useful for different purposes. For example, Joe (1985) uses a 'vectorized' generalization to describe association in contingency tables, and Tong (1989) uses uniform majorization (described below) to obtain probability inequalities for rectangles. Several other multivariate majorization orderings may be found in the books by Marshall and Olkin (1979) and Arnold (1987).

In this paper, we study multivariate majorization orderings that can be interpreted as orderings of distribution of wealth of several resources, with lower in the ordering meaning closer to equal division of the resources. Our

[1]Research supported by an NSERC Canada grant and a U. S. National Science Foundation grant.

AMS 1991 *subject classifications.* Primary: 15A45; Secondary: 15A39, 15A48.

Key words and phrases. Convexity, dispersion, linear programming, matrix inequalities.

orderings are on $m \times n$ matrices (m resources and n individuals) of real numbers. One potential area of application is to the monitoring and management of economic and ecological systems, where interest focuses on interventions that lead to more equitable consumption of resources, especially when some resources may become scarce. In such a situation the total value of the resources consumed (called *wealth*) by an individual or species may fluctuate, depending on current availability of each resource. The essential criterion for judging if a redistribution of resources leads to something 'universally' more equitable is that the vector describing the distribution of wealth changes to something smaller in the majorization pre-ordering, *no matter what values are assigned to each resource.* This idea is described in more detail in Arnold (1987, pp. 60-61).

The formal definition and basic properties of our proposed PC-majorization ordering are given in Section 2. Section 3 contains an algorithm to determine whether two given $m \times n$ matrices are ordered or not. The theory behind elementary methods for reducing matrices in this ordering is developed in Section 4. Section 5 contains a preliminary study of the set of matrices that are PC-majorized by a given matrix, and suggests how this leads to a more general method of reduction.

2. Definitions and Basic Properties

We first define vector majorization and give some of its equivalent forms (see Marshall and Olkin (1979) and Arnold (1987) for details) that we will use. Then we define multivariate majorization by positive comparisons (PC-majorization), and relate it to other forms of multivariate majorization. Through examples and results, we motivate PC-majorization as providing the most useful interpretion for distributions of several resources. Our notation follows Arnold (1987).

DEFINITION 2.1 Let $\mathbf{x} = (x_1, \ldots, x_n)$ and $\mathbf{y} = (y_1, \ldots, y_n)$ be n–dimensional row vectors. Let the ordered x_i and y_i be denoted by $x_{[1]} \geq \cdots \geq x_{[n]}$ and $y_{[1]} \geq \cdots \geq y_{[n]}$. Then $\mathbf{x}$ is majorized by $\mathbf{y}$ (written $\mathbf{x} \prec \mathbf{y}$) if

$$\sum_{i=1}^{k} x_{[i]} \leq \sum_{i=1}^{k} y_{[i]}, \quad k = 1, \ldots, n-1$$

and $\sum_{i=1}^{n} x_i = \sum_{i=1}^{n} y_i$.

Equivalent definitions of $\mathbf{x} \prec \mathbf{y}$ are: (a) $\mathbf{x} = \mathbf{y}D$, where D is a doubly stochastic $n \times n$ matrix (each row and column having nonnegative entries that sum to 1), and (b) $\mathbf{x}$ is in the convex hull of the vectors which are permutations of $\mathbf{y}$.

We now go on to matrices. An $m \times n$ matrix of reals is interpreted as a distribution of m resources among n people; the ith row is the distribution of the ith resource among n individuals and the jth column is the vector of amounts of the m resources for the jth individual. Negative values correspond to debts. Arnold (1987) gives an interpretation of in terms of n individuals with money in m different currencies.

Throughout this paper, X and Y are real $m \times n$ matrices, and $\mathbb{R}_+^m$ is the set of m–dimensional row vectors with nonnegative elements. We emphasize the following pre-ordering of these matrices.

DEFINITION 2.2 $X \prec^{\mathrm{PC}} Y$ or X is PC-majorized by Y if $\mathbf{a}X \prec \mathbf{a}Y$ for all $\mathbf{a} \in \mathbb{R}_+^m$.

We shall compare PC-majorization with three other kinds of multivariate majorization, defined as follows.

DEFINITIONS 2.3 $X \prec^{\mathrm{UM}} Y$ or X is *uniformly majorized* by Y if $X = YD$ for a $n \times n$ doubly stochastic matrix D.

$X \prec^{\mathrm{LC}} Y$ or X is majorized by Y through *linear combinations* if $\mathbf{a}X \prec \mathbf{a}Y$ for all $\mathbf{a} \in \mathbb{R}^m$.

$X \prec^{\mathrm{MM}} Y$ or X is *marginally majorized* by Y if $\mathbf{x}_i \prec \mathbf{y}_i$, $i = 1, \ldots, m$, where $\mathbf{x}_i, \mathbf{y}_i$ are the ith rows of X and Y respectively.

Arnold refers to our Definition 2.2 as $\prec^{\mathrm{MO}}$ for Marshall–Olkin, but Marshall and Olkin (1979) have $\mathbf{a} \in \mathbb{R}^m$, which is a stronger condition. To avoid confusion, we refer to the Marshall–Olkin ordering by $\prec^{\mathrm{LC}}$.

The following simple example illustrates PC-majorization and indicates why it is more appropriate for ordering distributions of resources than are the other three forms of multivariate majorization.

EXAMPLE 2.4 Let $X = \begin{bmatrix} 1 & 3 \\ 4 & 2 \end{bmatrix}$ and let $Y = \begin{bmatrix} 1 & 3 \\ 2 & 4 \end{bmatrix}$.

First we verify that $X \prec^{\mathrm{PC}} Y$. To do this we must show that for any $\mathbf{a} = (a_1, a_2) \in \mathbb{R}_+^m$,

$$(a_1 + 4a_2, 3a_1 + 2a_2) \prec (a_1 + 2a_2, 3a_1 + 4a_2);$$

but this follows immediately, since $3a_1 + 4a_2 \geq \max\{a_1 + 4a_2, 3a_1 + 2a_2\}$.

On the other hand, X is not uniformly majorized by Y, because the only doubly stochastic matrix that would leave the first row of Y unaltered is the identity matrix. Neither is X LC-majorized by Y, as can be seen by taking $\mathbf{a} = (1, -1)$. Most people would consider X to represent a more equitable distribution of resources than Y, since the second individual in Y is clearly the richer. Uniform and LC-majorization do not make such desirable distinctions.

It should also be clear that X and Y are equivalent with respect to marginal majorization; that is, $X \prec^{\text{MM}} Y$ and $Y \prec^{\text{MM}} X$. The fact that Y can be considered as 'smaller' than X, in the sense of marginal majorization, makes marginal majorization unsuitable as a pre-ordering for joint distributions of resources. □

The point of this example is that uniform and LC-majorization are too restrictive in determining which matrices should be ordered, and that marginal majorization is not restrictive enough. This is made clear in the next theorem, which formally relates these four types of multivariate majorization.

THEOREM 2.5 *PC-majorization possesses the following properties:*

Invariance by permutations. If $X \prec^{\text{PC}} Y$, *then* $PXQ_1 \prec^{\text{PC}} PYQ_2$, *for any choice of permutation matrices* P, Q_1, *and* Q_2.

Invariance by addition. If $X \prec^{\text{PC}} Y$, *then* $X + (c_1, \ldots, c_m)^T\mathbf{e} \prec^{\text{PC}} Y + (c_1, \ldots, c_m)^T\mathbf{e}$ *for real constants* c_i, *where* T *stands for transpose and* $\mathbf{e}$ *is a* $1 \times n$ *vector of ones.*

Marginal Majorization. If $X \prec^{\text{PC}} Y$, *then* $X \prec^{\text{MM}} Y$.

Uniform Reduction. If $X \prec^{\text{UM}} Y$, *then* $X \prec^{\text{LC}} Y$, *which in turn implies* $X \prec^{\text{PC}} Y$.

PROOF Invariance with respect to P follows from the invariance of the domain $\mathbb{R}_+^m$ of the weighting vector $\mathbf{a}$. Invariance with respect to Q_1 and Q_2 follows from the permutation invariance of ordinary vector majorization. Also, vector majorization is invariant under addition of the same constant to all components, which makes PC-majorization invariant under the addition of constant rows.

Suppose $X \prec^{\text{PC}} Y$. Let $\mathbf{a} \in \mathbb{R}_+^m$ be a vector with a 1 in the ith position and 0 elsewhere. Then $\mathbf{a}X \prec \mathbf{a}Y$ is equivalent to $\mathbf{x}_i \prec \mathbf{y}_i$. By letting i go from 1 to m, $X \prec^{\text{MM}} Y$.

Finally, suppose $X \prec^{\text{UM}} Y$. Then there exists a doubly stochastic matrix D such that $X = YD$, and $\mathbf{a}X = (\mathbf{a}Y)D$ or $\mathbf{a}X \prec \mathbf{a}Y$ for all $\mathbf{a} \in \mathbb{R}^m$. Hence $X \prec^{\text{LC}} Y$. Restricting $\mathbf{a}$ to $\mathbb{R}_+^m$ gives $X \prec^{\text{PC}} Y$. □

We note also that the $\prec^{\text{MM}}$ and $\prec^{\text{UM}}$ pre-orderings possess the same invariance properties as $\prec^{\text{PC}}$. Further relationships among these these multivariate majorization pre-orderings are developed in Section 4, in the context of finding elementary operations that reduce a matrix Y to something smaller in the PC-majorization ordering. In particular, we will give conditions under which the orderings $\prec^{\text{MM}}$ and $\prec^{\text{PC}}$ become equivalent, in which case a reduction can be obtained by reducing one row of Y in the usual vector majorization ordering. First, however, it is useful to know how each of the three multivariate majorization relations is verified.

3. Determining Whether Two given Matrices are Ordered

In this section we show for given matrices X, Y how to check whether one or more of the multivariate majorization orderings in Section 2 holds. Marginal majorization is easily checked since it is rowwise majorization. If $X \prec^{\mathrm{MM}} Y$, it is logical next to check if $X \prec^{\mathrm{PC}} Y$. PC-majorization would seem to require checking vector majorization for an infinite number of $\mathbf{a}$ but we show that a finite number of $\mathbf{a}$'s will do. Finally, if $X \prec^{\mathrm{PC}} Y$, the check for uniform majorization also requires some work because there could be zero, one, or more doubly stochastic matrices D such that $X = YD$.

The main difficulty in checking for PC-majorization is that the ordering of the components of $\mathbf{a}X$ and $\mathbf{a}Y$ change as $\mathbf{a}$ varies over $\mathbb{R}^m_+$. The following example illustrates the difficulty, offers some geometric intuition, and suggests the general solution.

EXAMPLE 3.1 Suppose that

$$X = \begin{bmatrix} 4 & 3 & 3 \\ 3 & 3 & 4 \end{bmatrix}, \quad Y = \begin{bmatrix} 5 & 4 & 1 \\ 2 & 3 & 5 \end{bmatrix}.$$

For any $\mathbf{a} = (a_1, a_2) \in \mathbb{R}^m_+$, we must check if

$$(4a_1 + 3a_2, 3a_1 + 3a_2, 3a_1 + 4a_2) \prec (5a_1 + 2a_2, 4a_1 + 3a_2, a_1 + 5a_2).$$

Let $\mathbf{A}$, $\mathbf{B}$, and $\mathbf{C}$ denote the columns of Y, so that $\mathbf{A} = \begin{bmatrix} 5 \\ 2 \end{bmatrix}, \mathbf{B} = \begin{bmatrix} 4 \\ 3 \end{bmatrix}$, and $\mathbf{C} = \begin{bmatrix} 1 \\ 5 \end{bmatrix}$ correspond to individuals A, B, and C. As the first resource, whose worth is measured by a_1, decreases in value relative to the second resource, the ranking of individuals in decreasing order of total wealth goes from (ABC) to (BAC) to (BCA) to (CBA). Assuming that we have already checked for marginal majorization, we claim that we need to check $\mathbf{a}X \prec \mathbf{a}Y$ only for 3 values of $\mathbf{a}$ corresponding to the three transitions in the rankings of the relative wealth of individuals A, B, and C.

Some geometric intuition can be obtained by visualizing the column vectors $\mathbf{A}$, $\mathbf{B}$, and $\mathbf{C}$ as points in the euclidean plane, and $\mathbf{a}$ as determining a ray from the origin through the point $\mathbf{a}$. For any given $\mathbf{a}$, the wealth of individuals A, B, and C is proportional to the orthogonal projection of the points $\mathbf{A}$, $\mathbf{B}$, and $\mathbf{C}$ onto the ray determined by $\mathbf{a}$. A transition in rankings occurs when the ray is orthogonal to a line connecting some pair of points among $\mathbf{A}$, $\mathbf{B}$, and $\mathbf{C}$.

The transition from (ABC) to (BAC) occurs at $\mathbf{a} = (1,1)$; from (BAC) to (BCA) at $\mathbf{a} = (3,4)$; and from (BCA) to (CBA) at $\mathbf{a} = (2,3)$. In particular, for $\mathbf{a} = (3,4), \mathbf{a}X = (24, 21, 25)$ and $\mathbf{a}Y = (23, 24, 23)$. Therefore Y does not PC-majorize X.

The argument why attention can be restricted to the above three values of $\mathbf{a}$ is best given in the general setting that we now develop.

THEOREM 3.2 *Suppose that $X \prec^{\mathrm{MM}} Y$, where the columns of Y are denoted by $\mathbf{Y}_1, \ldots, \mathbf{Y}_n$. For each (i,j) with $1 \le i \le j \le n$, let*

$$T_{ij} = \{\mathbf{a} \in \mathbb{R}_+^m : \sum_k a_k = 1 \quad \text{and} \quad \mathbf{a}(\mathbf{Y}_i - \mathbf{Y}_j) = 0\}$$

and

$$T = \cup_{i<j} T_{ij}.$$

Then $X \prec^{\mathrm{PC}} Y$ if and only if $\mathbf{a}X \prec \mathbf{a}Y$ for every $\mathbf{a} \in T$.

PROOF First note that $\mathbf{a}X \prec \mathbf{a}Y$ if and only if $c\mathbf{a}X \prec c\mathbf{a}Y$ for every $c > 0$. Thus it is enough to check the definition of PC-majorization for all $\mathbf{a}$ such that $\sum_j a_j = 1$.

For each $k = 1, \ldots, m$, define

$$f_k(\mathbf{a}) = \sum_{i=1}^{k} (\mathbf{a}Y)_{[i]}$$

to be the sum of the k largest components of $\mathbf{a}Y$. We may also express $f_k(\mathbf{a})$ as

$$f_k(\mathbf{a}) = \max_Q \sum_{i=1}^{k} (\mathbf{a}YQ)_i$$

where the maximum is taken over all $n \times n$ permutation matrices. Expressed as a maximum of linear functions, it is clear that $f_k(\mathbf{a})$ is convex and piecewise linear. Transitions from one linear section to another can occur only when at least two of the components $\mathbf{a}\mathbf{Y}_i$ and $\mathbf{a}\mathbf{Y}_j$ are equal. A similar argument shows that

$$g_k(\mathbf{a}) = \sum_{i=1}^{k} (\mathbf{a}X)_{[i]},$$

which is the sum of the k largest components of $\mathbf{a}X$, is also convex.

To establish the PC-majorization $X \prec^{\mathrm{PC}} Y$, it is enough to show that $f_k(\mathbf{a}) \ge g_k(\mathbf{a})$ for each $\mathbf{a} \in \mathcal{H} \equiv \{\mathbf{a} \in \mathbb{R}_+^m : \sum_j a_j = 1\}$. The assumption that $X \prec^{\mathrm{MM}} Y$ implies that $f_k(\mathbf{a}) \ge g_k(\mathbf{a})$ at the extreme points $(1,\ldots,0),\ldots,(0,\ldots,1)$ of $\mathcal{H}$. Now suppose that $g_k(\mathbf{a}) > f_k(\mathbf{a})$ at some point $\mathbf{a} \in \mathcal{H}$. We may assume that the components of $\mathbf{a}Y$ are distinct, since otherwise $\mathbf{a} \in T$. Let $\mathcal{H}_{\mathbf{a}}$ denote the subset of $\mathcal{H}$ that contains $\mathbf{a}$ and over which $f_k(\cdot)$ is linear. This is a convex set, defined by

$$\mathcal{H}_{\mathbf{a}} = \{\mathbf{b} \in \mathcal{H} : \mathbf{b}_{\pi(i)} \ge \mathbf{b}_{\pi(i+1)}, i = 1, \ldots, m-1\},$$

where π is the permutation that puts the components of $\mathbf{a}Y$ in decreasing order; that is, $(\mathbf{a}Y)_{\pi(i)} = (\mathbf{a}Y)_{[i]}, i = 1, \ldots, m$. Since $g_k(\cdot)$ is a convex function, if it exceeds $f_k(\cdot)$ anywhere in the region $\mathcal{H}_\mathbf{a}$, it must also exceed $f_k(\cdot)$ at one of the extreme points of $\mathcal{H}_\mathbf{a}$. Each of these extreme points is contained either in T or in the set $\{(1, \ldots, 0), \ldots, (0, \ldots, 1)\}$ of extreme points of $\mathcal{H}$. It thus follows that if $g_k(\cdot)$ exceeds $f_k(\cdot)$ anywhere in $\mathcal{H}$, it must exceed $f_k(\cdot)$ on T. Hence, majorization needs to be checked only for $\mathbf{a} \in T$. □

In the case when $m = 2, T$ is finite, and Theorem 3.2 is all that is needed to check for PC-majorization. For $m > 2$, the sets T_{ij} are either empty or are convex polytopes. In this case, it suffices to check the extreme points of each nonempty T_{ij}. Hence overall only a finite number of $\mathbf{a}$'s need to be checked. A precise algorithm that checks for both marginal and PC-majorization in a finite number of steps is now described.

ALGORITHM 3.3. CHECK FOR PC-MAJORIZATION. Consider the hyperplanes

$$\{\mathbf{a} : \sum_k a_k(y_{ki} - y_{kj}) = 0,\} \qquad i < j.$$

The intersection of these hyperplanes with $\mathcal{H} = \{\mathbf{a} \in \mathbb{R}_+^m : \sum_j a_j = 1\}$ are the convex polytopes T_{ij}, but notice that these hyperplanes need not intersect $\mathcal{H}$ at all, as happens, for example, when the two column vectors $\mathbf{Y}_i$ and $\mathbf{Y}_j$ are similarly ordered. In general, the above hyperplanes divide A into at most $n!$ convex polytopes (with faces T_{ij}) such that for each $\mathbf{a}$ in a particular polytope, $\mathbf{a}Y$ always has a certain ordering. The vertices of the convex polytopes may be found by taking $m-1$ of the hyperplane equations at a time and solving for a root $\mathbf{a}$ in $\mathcal{H}$ (if any). Since all these vertices are similarly ordered, the following lemma (with $\sum_j \lambda_j = 1$) proves that it is enough to check $\mathbf{a}X \prec \mathbf{a}Y$ for $\mathbf{a}$'s being one of these vertices. □

The next lemma is equivalent to result 5.A.6 on page 121 of Marshall and Olkin (1979).

LEMMA 3.4 *Suppose that the vectors* $\mathbf{z}^{(j)} = (z_1^{(j)}, \ldots, z_n^{(j)})$, $j = 1, \ldots, L$, *are similarly ordered and that there are corresponding vectors* $\mathbf{x}^{(j)}$ *such that* $\mathbf{x}^{(j)} \prec \mathbf{z}^{(j)}$. *Then* $\sum_j \lambda_j \mathbf{x}^{(j)} \prec \sum_j \lambda_j \mathbf{z}^{(j)}$ *if* $\lambda_j \geq 0$, $j = 1, \ldots, L$.

In the next section, we discuss some elementary ways to transform a matrix Y into something smaller with respect to PC-majorization. The simplest way is to post multiply Y by a doubly stochastic matrix D, which leads to something smaller in uniform majorization. For a certain class of matrices Y, which we have not yet been able to specify completely, this is the only way to obtain something that is PC-majorized by Y.

Also, in our study of elementary operations that PC-reduce Y, we are interested in how a targeted X, which is known to be PC-majorized, might be obtained through a series of elementary operations. That is, given $X \prec^{PC} Y$, we would like to construct a sequence $Y \to Y_1 \to \ldots \to Y_k \to X$ such that each step $Y_{j-1} \to Y_j$ is an elementary operation.

A valuable tool for studying each of these problems is the following check for uniform majorization.

ALGORITHM 3.5. CHECK FOR UNIFORM MAJORIZATION ($\prec^{UM}$). First note that if $X = YD$ for a doubly stochastic matrix D, and if certain rows of Y are linearly dependent, say $\mathbf{b}Y = 0$ for a $\mathbf{b} \in \mathbb{R}^m$, then $\mathbf{b}X = 0$. However $X = YD$ and $\mathbf{b}X = 0$ need not imply $\mathbf{b}Y = 0$ unless D is invertible.

Given X, Y a procedure to use is the following. First check whether $X \prec^{MM} Y$. If not, then X, Y are not ordered by $\prec^{UM}$. Assuming $X \prec^{MM} Y$, there are a few cases to consider.

(a) If Y has at least n linearly independent rows, let Y_0 be a $n \times n$ submatrix of Y with n linearly independent rows, say rows $i_1 < \cdots < i_n$ of Y. Let X_0 be a $n \times n$ submatrix of X consisting of rows $i_1, \ldots, i_n$ of X. Y_0 is invertible. Let $D = Y_0^{-1} X_0$ and $Z = YD$. If D is doubly stochastic and $Z = X$, then $X \prec^{UM} Y$. Otherwise X and Y are not ordered by $\prec^{UM}$.

(b) If the number of linearly independent rows of Y is less than n, check whether each linear dependency in rows of Y implies the same linear dependency in rows of X. If not, then X and Y are not ordered by $\prec^{UM}$.

(c) Suppose the number of linearly independent rows of Y is exactly $n-1$. Let Y_0 be a (sub)matrix of Y consisting of $n-1$ independent rows, and let X_0 be the corresponding (sub)matrix of X. Then $X \prec^{MM} Y$ implies that the row sum vectors corresponding to X and Y are the same. Therefore $X_0 = Y_0 D$ reduces to solving for $(n-1)^2$ linear equations in $(n-1)^2$ unknowns d_{jk}, $j = 1, \ldots, n-1$, $k = 1, \ldots, n-1$. Note that the doubly stochastic requirement means that d_{nk} and d_{jn} can be substituted for. If $D = (d_{jk})_{1 \le j,k \le n}$ is nonnegative and $X = YD$, then $X \prec^{UM} Y$.

(d) If the number of linearly independent rows of Y is less than $n-1$, then potentially more than one D exists. Using $X = YD$ and D doubly stochastic as linear constraints on the d_{ij}, the simplex method of linear programming, for example, can be used to see if a feasible solution exists (any linear function of the d_{ij} can be used as an objective function). If a feasible solution exists, then $X \prec^{UM} Y$.

4. PC-Reduction of Matrices

The preceding results enable us to recognize when two matrices are related by any of the $\prec^{MM}, \prec^{PC}$ or $\prec^{UM}$ pre-orderings for dispersiveness.

We now consider interventions that reduce dispersiveness according to PC-majorization. These interventions take the mathematical form of special types of linear operations on an initial matrix Y. We describe three different types of operations.

DEFINITION 4.1 *Type (A) (Uniform reduction) operations.*
A type (A) operation is defined as a linear transformation $Y \to YD$ where D is a doubly stochastic matrix.

That type A operations always produce a PC-smaller matrix is an immediate consequence of Theorem 2.5; namely, if D is doubly stochastic, then $YD \prec^{\mathrm{PC}} Y$. We further conjecture that any doubly stochastic matrix can be written as a product of matrices of the form $\lambda Q + (1-\lambda)I$ where Q is an $n \times n$ permutation matrix, I is the $n \times n$ identity matrix, and $0 \leq \lambda \leq 1$. This conjectured decomposition would allow an arbitrary uniform reduction to be achieved as a sequence of 'rearrangement transfers' $\lambda Q + (1-\lambda)I$. An exercise in Arnold (1987, p. 75) demonstrates that the conjecture is false if the permutation matrices Q are restricted to transpositions of two coordinates, in which case $\lambda Q + (1-\lambda)I$ is called a 'Robin Hood transfer.'

Recall Example 2.4, where a PC-reduction was obtained by permuting elements in the second row of Y. It was shown there that such a reduction is not obtainable via type A operations. We now devise an elementary operation, based on majorization reduction of individual rows of certain submatrices of Y. Toward this end, consider the following definitions.

DEFINITION 4.2 Let J be a subset of $\{1,\ldots,n\}$ with cardinality n_0 where $n_0 \geq 2$. X is said to have *similarly ordered rows in* J, if there is a permutation $j_1,\ldots,j_{n_0}$ of the indices in J such that $x_{ij_1} \geq \cdots \geq x_{ij_{n_0}}$, for each $i = 1,\ldots,m$. If $J = \{1,\ldots,n\}$ then we will say that X has similarly ordered rows. In the case that X has two rows, the notion of *oppositely ordered* rows can be well defined, in the obvious way.

DEFINITION 4.3 *Type (B) (Marginal reduction) operations.*
A type (B) operation $Y \to X$ is described as follows: Let Y^* be an $m \times n_0$ submatrix of Y $(2 \leq n_0 \leq n)$ with similarly ordered rows; let the ith row of Y^* be denoted by $\mathbf{y}_i^*$. For each i, replace $\mathbf{y}_i^*$ by a vector $\mathbf{x}_i^*$ which is smaller with respect to $\prec$. Assuming that the jth column of Y^* corresponds to the kth column of Y, replace y_{ik} by x_{ij}^*. The result is defined to be X.

We need to show that any X resulting from a type B operation satisfies $X \prec^{\mathrm{PC}} Y$. We begin with a simple lemma whose proof follows directly from elementary properties of vector majorization.

LEMMA 4.4 *Suppose X and Y differ only in the columns $j_1,\ldots,j_{n_0}$ where $2 \leq n_0 \leq n$. Then $X \prec^{\mathrm{PC}} Y$ if and only if $X^* \prec^{\mathrm{PC}} Y^*$, where X^*, Y^* are the* *submatrices obtained from the $j_1, \ldots, j_{n_0}$ columns of X and Y respectively. A similar argument shows that $\prec^{\mathrm{PC}}$ can be replaced by $\prec^{\mathrm{UM}}$, $\prec^{\mathrm{LC}}$ or $\prec^{\mathrm{MM}}$.*

The key set of conditions under which PC- and marginal majorization become equivalent are described formally as follows:

THEOREM 4.5 *If Y has similarly ordered rows and $X \prec^{\mathrm{MM}} Y$, then $X \prec^{\mathrm{PC}} Y$.*

PROOF Because $\prec^{\mathrm{PC}}$ and $\prec^{\mathrm{MM}}$ share the same invariance property given in Theorem 2.5, we can assume without loss of generality that $Y = (y_{ij})$ satisfies $y_{i1} \geq \cdots \geq y_{in}$, $i = 1, \ldots, m$. Let $\mathbf{a}$ be an arbitrary vector in $\mathbb{R}_+^m$. Then $\mathbf{z} = \mathbf{a}Y$ satisfies $z_1 \geq \cdots \geq z_n$ and the sum of the k largest elements of $\mathbf{z}$ is $\sum_{i=1}^m a_i \sum_{j=1}^k y_{ij}$. Let j_ℓ be the index of the ℓth largest element of $\mathbf{a}X$ and let $x_{i[j]}$ be the jth largest element of the ith row of X. Then the sum of the k largest elements of $\mathbf{a}X$ is $\sum_{i=1}^m a_i \sum_{\ell=1}^k x_{ij_\ell}$, and

$$\sum_{i=1}^m a_i \sum_{\ell=1}^k x_{ij_\ell} \leq \sum_{i=1}^m a_i \sum_{j=1}^k x_{i[j]} \leq \sum_{i=1}^m a_i \sum_{j=1}^k y_{ij},$$

where the last inequality follows from the assumption $X \prec^{\mathrm{MM}} Y$. □

Together, Lemma 4.4 and Theorem 4.5 prove that a type B operation on Y leads to something smaller with respect to $\prec^{\mathrm{PC}}$. Reduction via marginal majorization is easy, and we have so far shown that reduction, via marginal majorization, of an $m \times n_0$ submatrix of Y with similarly ordered rows also leads to something smaller with respect to PC-majorization.

We wish to investigate the class of matrices that can be obtained by applying sequences of type (A) and type (B) to a given matrix Y. For this purpose, the following definition is convenient.

DEFINITION 4.6 X is majorized by Y via *simple transfers* ($X \prec^{\mathrm{ST}} Y$) if X can be obtained from Y via a finite number of operations of type (A) or (B) above.

By the above results, $X \prec^{\mathrm{ST}} Y$ implies $X \prec^{\mathrm{PC}} Y$. The next theorem shows that these two orderings are, in fact, equivalent when $n = 2$. Section 5 contains some other conditions under which equivalence is obtained.

THEOREM 4.7 *If $n = 2$, then $X \prec^{\mathrm{ST}} Y$ if and only if $X \prec^{\mathrm{PC}} Y$.*

PROOF Suppose $X \prec^{\mathrm{PC}} Y$. By Theorem 2.5, $X \prec^{\mathrm{MM}} Y$. If Y has similarly ordered, then by Definition 4.6, X can be obtained from Y via one operation of type (B). Next suppose that Y does not have similarly ordered rows. Since $X \prec^{\mathrm{MM}} Y$, for each $i = 1, \ldots, m$, there exist constants θ_i in the interval [0,1] such that $x_{i1} = \theta_i y_{i1} + (1-\theta_i)y_{i2}$, $x_{i2} = (1-\theta_i)y_{i1} + \theta_i y_{i2}$. Suppose that rows i and i' of Y are oppositely ordered. Let $\mathbf{a}$ be a vector with $a_i = 1$, $a_{i'} = (y_{i1} - y_{i2})/(y_{i'2} - y_{i'1}) > 0$ and $a_k = 0$, $k \neq i, i'$. Then $\mathbf{a}Y$ is a 2–vector of the constant $y_{i1} + a_{i'}y_{i'1}$; and $\mathbf{a}X \prec \mathbf{a}Y$ implies that $\mathbf{a}X = \mathbf{a}Y$, which can happen for this $\mathbf{a}$ only if $\theta_i = \theta_{i'}$. Since each non–constant row must be oppositely ordered to either row i or i' and θ can be anything (in [0,1]) for a constant row, it follows that $X = YD$, where $D = \begin{bmatrix} \theta_i & 1-\theta_i \\ 1-\theta_i & \theta_i \end{bmatrix}$. □

EXAMPLES 4.8 In general, there can be matrices that are $\prec^{\mathrm{PC}}$ ordered but not $\prec^{\mathrm{ST}}$ ordered. The methods of Section 3 can be used to verify that

$$X = \begin{bmatrix} 1 & 2 & 3 & 4 & 5 \\ 2 & 3 & 4 & 5 & 1 \end{bmatrix} \prec^{\mathrm{PC}} \begin{bmatrix} 1 & 2 & 3 & 4 & 5 \\ 1 & 4 & 2 & 3 & 5 \end{bmatrix} = Y.$$

For this example, it is easily checked that the matrices X and Y are not $\prec^{\mathrm{ST}}$ ordered.

The example on page 59 of Arnold (1987) with $Y = \begin{bmatrix} 1 & 1 & 0 \\ 2 & 4 & 6 \end{bmatrix}$, $X = \begin{bmatrix} 1 & .5 & .5 \\ 3 & 4 & 5 \end{bmatrix}$, and $X = YD$, with $D = \begin{bmatrix} .5 & .5 & 0 \\ .5 & 0 & .5 \\ 0 & .5 & .5 \end{bmatrix}$ shows that Definition 4.6 cannot be simplified to simple transfers of type (A) and (B) that operate on only two columns at a time. This is one difference in going from vector majorization to matrix majorization.

The next, and last, example shows that with $X \prec^{\mathrm{ST}} Y$, it is possible for one column of X to be dominant even if there is no dominant column in Y.

$$X = \begin{bmatrix} 2.1 & 2.5 & 2.4 \\ 2.9 & 4 & 3.1 \end{bmatrix} \prec^{\mathrm{ST}} \begin{bmatrix} 2.1 & 3.0 & 1.9 \\ 2.9 & 4 & 3.1 \end{bmatrix} \prec^{\mathrm{ST}} \begin{bmatrix} 4 & 3 & 0 \\ 1 & 4 & 5 \end{bmatrix} = Y$$

This cannot happen if there is a nonnegative vector $\mathbf{a}$ such that $\mathbf{a}Y$ is a constant vector.

The following development suggests a simple operation other than (A) and (B) that also preserves the $\prec^{\mathrm{PC}}$ ordering, and which allows the transformation from Y to X in the first of Examples 4.8.

DEFINITIONS 4.9 Let $\mathbf{x}, \mathbf{y}$, and $\mathbf{z}$ be n–dimensional row vectors. Then $\mathbf{x}$ is $\mathbf{z}$–majorized by $\mathbf{y}$ (written $\mathbf{x} \prec_{\mathbf{z}} \mathbf{y}$) if $c\mathbf{z} + \mathbf{x} \prec c\mathbf{z} + \mathbf{y}$ for every $c > 0$.

In the case where $x \prec \mathbf{y}$, we define $S_{\mathbf{xy}} \equiv \{\mathbf{z} : \mathbf{x} \prec_{\mathbf{z}} \mathbf{y}\}$.

DEFINITION 4.10 *Type (C) (Directed reduction) operations.*

A type (C) operation is described as follows. Let Y have rows $\mathbf{y}_1, \ldots, \mathbf{y}_m$, and let $\mathbf{x} \prec \mathbf{y}_i$. Then $\mathbf{y}_i$ may be replaced by $\mathbf{x}$ if $S_{\mathbf{x}\mathbf{y}_i}$ is convex and $\mathbf{y}_j \in S_{\mathbf{x}\mathbf{y}_i}$ for each $j \neq i$.

To prove that a type C operation produces a PC-reduction, it is necessary first to develop properties of $\mathbf{z}$–majorization and the set $S_{\mathbf{xy}}$. Some immediate consequences of the definiton of $\mathbf{z}$–majorization are listed in the following lemma.

LEMMA 4.11
(a) If $\mathbf{x} \prec_{\mathbf{z}} \mathbf{y}$, then $\mathbf{x} \prec_{c\mathbf{z}} \mathbf{y}$ for every $c > 0$.
(b) $\mathbf{x} \prec_{\mathbf{z}} \mathbf{y}$ if and only if $\mathbf{z} + c\mathbf{x} \prec \mathbf{z} + c\mathbf{y}$ for every $c > 0$.
(c) If $\mathbf{x} \prec \mathbf{y}$, then $\mathbf{x} \prec_{\mathbf{e}} \mathbf{y}$, where $\mathbf{e}$ is the $1 \times n$ vector of ones.
(d) $\mathbf{x} \prec_{\mathbf{z}} \mathbf{y}$ if and only if

$$X = \begin{bmatrix} \mathbf{z} \\ \mathbf{x} \end{bmatrix} \prec^{\mathrm{PC}} \begin{bmatrix} \mathbf{z} \\ \mathbf{y} \end{bmatrix} = Y.$$

Part (a) of the lemma shows that the $\prec_{\mathbf{z}}$ ordering depends only on the direction, and not the actual value, of z. Part (b) is an alternate definition useful in proofs. Part (c) shows that the definition is not vacuous. Part (d) makes the key association with PC-majorization.

The role of $\mathbf{z}$ in $\mathbf{x} \prec_{\mathbf{z}} \mathbf{y}$ is to determine a set of permutation pairs $P_{\mathbf{xy}} \equiv \{(\pi, \nu)\}$ under which

$$\sum_{i=1}^{k} \mathbf{x}_{\pi(i)} \leq \sum_{i=1}^{k} \mathbf{y}_{\nu(i)}$$

for each $k = 1, \ldots, n$. Suppose that X and Y have the form given in Lemma 4.11d, with $X \prec^{\mathrm{PC}} Y$. Then

$$(1-\lambda)\mathbf{z} + \lambda\mathbf{x} \prec (1-\lambda)\mathbf{z} + \lambda\mathbf{y} \quad \text{for} \quad 0 \leq \lambda \leq 1.$$

If λ is sufficiently close to 0, Lemma 4.11d implies that $(\pi_{\mathbf{z}}, \pi_{\mathbf{z}})$ must be in $P_{\mathbf{xy}}$, where $\pi_{\mathbf{z}}$ is the permutation that puts the components of $\mathbf{z}$ in decreasing order. Similarly the pair $(\pi_{\mathbf{x}}, \pi_{\mathbf{y}})$ must be in $P_{\mathbf{xy}}$. As λ moves from 0 to 1, the ordering of $(1-\lambda)\mathbf{z} + \lambda\mathbf{x}$ changes from $\pi_{\mathbf{z}}$ to $\pi_{\mathbf{x}}$, and the ordering of $(1-\lambda)\mathbf{z} + \lambda\mathbf{y}$ changes from $\pi_{\mathbf{z}}$ to $\pi_{\mathbf{y}}$. In fact, the number of transpositions (Hamming distance) between either $(1-\lambda)\mathbf{z} + \lambda\mathbf{x}$ or $(1-\lambda)\mathbf{z} + \lambda\mathbf{y}$ and $\pi_{\mathbf{z}}$ is an increasing function of λ (cf. Theorem 5.2).

Lemma 4.11d describes, for $m = 2$, when one of the rows $\mathbf{y}$ of Y can be replaced by a new row $\mathbf{x}$ to achieve a PC-reduction; namely, when $\mathbf{x} \prec_{\mathbf{z}} \mathbf{y}$. We are thus interested in the question: Given $Y = \begin{bmatrix} \mathbf{z} \\ \mathbf{y} \end{bmatrix}$, what vectors $\mathbf{x}$ satisfy $\mathbf{x} \prec_{\mathbf{z}} \mathbf{y}$? It appears easiest to approach this problem indirectly, by first fixing $\mathbf{x}$ and $\mathbf{y}$ and then relating $P_{\mathbf{xy}}$ to the key construct of Definition 4.9, $S_{\mathbf{xy}}$.

LEMMA 4.12 *The set $S_{\mathbf{xy}}$ has the following properties:*

(a) If $\mathbf{z} \in S_{\mathbf{xy}}$*, then* $c\mathbf{z} \in S_{\mathbf{xy}}$*, for every* $c > 0$.
(b) If $\mathbf{z}$ *is similarly ordered to* $\mathbf{y}$*, then* $\mathbf{z} \in S_{\mathbf{xy}}$.
(c) If $\mathbf{z}$ *is oppositely ordered to* $\mathbf{x}$*, then* $\mathbf{z} \in S_{\mathbf{xy}}$.

PROOF Part (a) is an immediate consequence of Lemma 4.11a. Part (b) follows from

$$\sum_{i=1}^{k} (\mathbf{x} + c\mathbf{z})_{[i]} \leq \sum_{i=1}^{k} \mathbf{x}_{[i]} + c\mathbf{z}_{[i]} \leq \sum_{i=1}^{k} \mathbf{y}_{[i]} + c\mathbf{z}_{[i]} = \sum_{i=1}^{k} (\mathbf{y} + c\mathbf{z})_{[i]}.$$

Part (c) follows from the existence of some permutation π such that

$$\sum_{i=1}^{k} (\mathbf{y} + c\mathbf{z})_{[i]} = \sum_{i=1}^{k} \mathbf{y}_{[i]} + c\mathbf{z}_{\pi(i)} \geq \sum_{i=1}^{k} \mathbf{x}_{[i]} + c\mathbf{z}_{\pi(i)} \geq \sum_{i=1}^{k} (\mathbf{x} + c\mathbf{z})_{[i]}. \qquad \square$$

EXAMPLE 4.13 Let $\mathbf{y} = (1,2,3)$. We will show that, for each $\mathbf{x} = (x_1, x_2, x_3) \prec \mathbf{y}, S_{\mathbf{xy}}$ can be described as the intersection of half-spaces.

First note that for $\mathbf{x} = \mathbf{y}, S_{\mathbf{xy}}$ is all of $\mathbb{R}^n$. Otherwise,

a) If $x_1 \neq 1$ and $x_3 \neq 3$, then $S_{\mathbf{xy}}$ is constrained by $z_3 \geq z_1$.

b1) If $x_2 < 2$, then $S_{\mathbf{xy}}$ is constrained by $z_2 \geq z_1$.

b2) If $x_2 > 2$, then $S_{\mathbf{xy}}$ is constrained by $z_3 \geq z_2$.

Thus there are six possibilities for $S_{\mathbf{xy}}$. Either it is all of $\mathbb{R}^n$; is one of the three half spaces $\{z_2 \geq z_1\}, \{z_3 \geq z_1\}$ or $\{z_3 \geq z_2\}$; or is the intersection of two half spaces $\{z_3 \geq z_1\} \cap \{z_2 \geq z_1\}$ or $\{z_3 \geq z_1\} \cap \{z_3 \geq z_2\}$.

More generally, it appears that $S_{\mathbf{xy}}$ is the union of convex cones that are subsets of sets of the form $\{\mathbf{z} : z_{\pi(1)} \geq \cdots \geq z_{\pi(n)}\}$, where π ranges over a subset of permutations. As the following example illustrates, $S_{\mathbf{xy}}$ need not always be convex.

EXAMPLE 4.14 Let $\mathbf{x} = (3,5,3,1)$ and $\mathbf{y} = (0,2,4,6)$. Then $S_{\mathbf{xy}}$ is the union of 6 regions:

A. $z_4 \geq z_3 \geq z_2 \geq z_1$
B. $z_4 \geq z_2 \geq z_3 \geq z_1$
C. $z_4 \geq z_3 \geq z_1 \geq z_2$
D. $z_4 \geq z_1 \geq z_3 \geq z_2$
E. $z_3 \geq z_4 \geq z_1 \geq z_2$
F. $z_3 \geq z_4 \geq z_2 \geq z_1$.

The point $(0,1,.1,3)$ is in B, the point $(1,0,.1,3)$ is in D, but the average of these points, $(.5,.5,.1,3)$ is not in the union $S_{\mathbf{xy}}$ of the 6 regions.

Although the convexity of $S_{\mathbf{xy}}$ is not always guaranteed, the following theorem, which presumes the convexity of $S_{\mathbf{xy}}$ for a particular pair of vectors $(\mathbf{x}, \mathbf{y})$, can be useful for reducing a matrix Y when redistribution is restricted to a single resource.

THEOREM 4.15 *Let Y have row vectors $\mathbf{y}_1, \ldots, \mathbf{y}_m$. Suppose that X differs from Y only in the ith row, and that $S_{\mathbf{x}_i \mathbf{y}_i}$ is convex. Then $X \prec^{\text{PC}} Y$ if and only if $\mathbf{x}_i \prec_{\mathbf{y}_j} \mathbf{y}_i$ for each $j \neq i$.*

PROOF First suppose that $X \prec^{\text{PC}} Y$. For fixed $j \neq i$, and $k \in 1, 2, \ldots, m$, let

$$a_k = \begin{cases} c, & \text{if } k = i; \\ 1, & \text{if } k = j; \\ 0, & \text{otherwise.} \end{cases}$$

Then $\sum_{k=1}^m a_k \mathbf{x}_k \prec \sum_{k=1}^m a_k \mathbf{y}_k$ implies that $\mathbf{x}_j + c\mathbf{y}_i \prec \mathbf{x}_j + c\mathbf{x}_i$, and so $\mathbf{y}_i \prec_{\mathbf{x}_j} \mathbf{x}_i$.

We prove the converse by induction on m. By Lemma 4.11(d), the converse is true for $m = 2$. Suppose that it holds for some $m - 1$ where $m \geq 3$. Then there are at least two indices in $1, \ldots, m$ that are distinct from i. Without loss of generality, call these $m - 1$ and m. Then for any $\mathbf{a} \in \mathbb{R}_+^m$ define

$$\mathbf{z}_k = \begin{cases} \mathbf{y}_k, & \text{for } k = 1, \ldots, m-2; \\ a_{m-1}\mathbf{y}_{m-1} + a_m \mathbf{y}_m, & \text{for } k = m-1, \end{cases}$$

and

$$\mathbf{z}_k^* = \begin{cases} \mathbf{z}_k, & \text{for } k \neq i; \\ \mathbf{x}_i, & \text{for } k = i. \end{cases}$$

To invoke induction, we need to verify that

$$\mathbf{z}_i^* \prec_{\mathbf{z}_j} \mathbf{z}_i \quad \text{for each } j \in \{1, \ldots, m-1\} \backslash \{i\}. \tag{4.1}$$

For $j < m - 1$, (4.1) holds by assumption. For $j = m - 1$, (4.1) follows from the convexity and scale invariance (Lemma 4.12a) of $S_{\mathbf{z}_i^* \mathbf{z}_i}$. Now define

$$b_k = \begin{cases} a_k, & \text{for } k = 1, \ldots, m-2; \\ 1, & \text{for } k = m-1. \end{cases}$$

Then

$$\sum_{k=1}^{m} a_k \mathbf{x}_k = \sum_{k=1}^{m-1} b_k \mathbf{z}_k^* \prec \sum_{k=1}^{m-1} b_k \mathbf{z}_k = \sum_{k=1}^{m} a_k \mathbf{y}_k,$$

where the majorization follows from the induction assumption. Thus $X \prec^{\text{PC}} Y$. □

Theorem 4.15 proves that type (C) operations preserve the $\prec^{\text{PC}}$ ordering: It should be clear that a type (C) operation is distinct from types (A) and (B). Nevertheless, even with this additional operation, repeated application of operations (A), (B), and (C) do not generally produce all the matrices that are $\prec^{\text{PC}}$ smaller than an arbitrary initial matrix, as the following counterexample shows.

EXAMPLE 4.16 Let $X = \begin{bmatrix} 1.5 & 1.5 & 3 \\ 2.5 & 3.5 & 1 \end{bmatrix}$ and $Y = \begin{bmatrix} 1 & 2 & 3 \\ 4 & 1 & 2 \end{bmatrix}$. Using Theorem 3.2, it can be shown that $X \prec^{\mathrm{PC}} Y$. It is easy to argue that X and Y are not $\prec^{\mathrm{ST}}$ ordered. Also, it can be shown that X is not attainable from Y using type (C) operations. This example is a specific case of a more detailed study of 2×3 matrices in Theorem 5.6.

We do not know if there is yet another simple operation, which, in combination with (A), (B), and (C), will produce all matrices that are $\prec^{\mathrm{PC}}$ smaller than an arbitrary initial matrix.

5. Further Properties and PC-Reductions

In this section, we obtain additional properties for the $\prec^{\mathrm{PC}}$ and $\prec^{\mathrm{ST}}$ orderings. We show that the set $\mathcal{A}_Y = \{X : X \prec^{\mathrm{PC}} Y\}$ is convex and is contained in the convex set $\{X : X \prec^{\mathrm{MM}} Y\}$. The extreme points of $\{X : X \prec^{\mathrm{MM}} Y\}$ constitute the set, denoted by $\mathcal{C}_Y$, of matrices whose rows are permutations of the corresponding rows of Y. Methods for identifying the extreme points of $\mathcal{A}_Y$ are described and illustrated in the case $m = 2$, $n = 3$. One of the methods suggests a (non–simple) PC-reduction operation to be studied in further research.

In the course of this development, we show that, when restricted to $\mathcal{C}_Y$, the $\prec^{\mathrm{ST}}$ ordering is the same as the multivariate arrangement ordering of Boland and Proschan (1988).

THEOREM 5.1 *For fixed Y, the set $\mathcal{A}_Y = \{X : X \prec^{\mathrm{PC}} Y\}$ is a convex set.*

PROOF Suppose $X_1 \prec^{\mathrm{PC}} Y$ and $X_2 \prec^{\mathrm{PC}} Y$. Let λ be in the interval [0,1]. For $\mathbf{a} \in \mathbb{R}^m_+$, $\mathbf{a}(\lambda X_1 + (1-\lambda)X_2) = \lambda(\mathbf{a}X_1) + (1-\lambda)(\mathbf{a}X_2) \prec \mathbf{a}Y$ since $\mathbf{a}X_1 \prec \mathbf{a}Y$, $\mathbf{a}X_2 \prec \mathbf{a}Y$ and the set $\{\mathbf{x} : \mathbf{x} \prec \mathbf{a}Y\}$ is convex. Therefore $\lambda X_1 + (1-\lambda)X_2 \prec^{\mathrm{PC}} Y$. □

A natural question now concerns the extreme points of $\mathcal{A}_Y$ and relationships with convex hull results for vector majorization. First recall that $\mathcal{C}_Y$ is the set of the extreme points of $\{X : X \prec^{\mathrm{MM}} Y\}$, and that the extreme points of $\mathcal{A}_Y$ include some, but not necessary all, of the points in $\mathcal{C}_Y$.

Consider the $\prec^{\mathrm{ST}}$ and the $\prec^{\mathrm{PC}}$ orderings on $\mathcal{C}_Y$. The maximal matrices in $\mathcal{C}_Y$ with respect to $\prec^{\mathrm{ST}}$ or $\prec^{\mathrm{PC}}$ have similarly ordered rows. If Y has similarly ordered rows, then by Birkhoff's theorem, $\mathcal{A}_Y$ is the convex hull of the matrices in $\mathcal{C}_Y$. By Theorem 4.6, the $\prec^{\mathrm{ST}}$ and $\prec^{\mathrm{PC}}$ orderings on $\mathcal{C}_Y$ are the same if $n = 2$. For $m = 2$ and $n = 3$ or 4, the $\prec^{\mathrm{PC}}$ and $\prec^{\mathrm{ST}}$ orderings on $\mathcal{C}_Y$ can be shown to be the same by complete pairwise comparisons. Example 4.8 shows that the equivalence does not extend to $m = 2$, $n \geq 5$.

From results of Section 4, if two matrices in $\mathcal{C}_Y$ are $\prec^{ST}$ ordered, then they are $\prec^{PC}$ ordered.

We next note that the $\prec^{ST}$ ordering on $\mathcal{C}_Y$ is the same as the multivariate arrangement increasing ordering in Boland and Proschan (1988). This is the arrangement increasing ordering or decreasing in transposition ordering of Hollander, Proschan and Sethuraman (1977) in the case $m = 2$. Boland and Proschan's definition involve only transfers that operate on two columns at a time. The equivalence follows from the proposition below.

PROPOSITION 5.2 *Let Y be a $m \times n$ matrix with similarly ordered rows and let X be a matrix such that its ith row is a permutation of the ith row of Y. Then X can be obtained from Y using a sequence of type (B) operations that operate on two columns.*

PROOF Without loss of generality, suppose that each row of Y is increasing. X can be obtained from Y by operating in columns $(n-1, n)$ followed by $(n-2,n), \ldots, (1,n), (n-2, n-1), \ldots, (1, n-1), \ldots, (1,2)$. In the first stage, each row involves switches of the jth column with the nth column, $j = n-1, \ldots, 1$, until the nth column of X is obtained. In the next stage switches are made until the $(n-1)^{st}$ column is correct, etc. Note that the appropriate columns are ordered correctly after each transfer of type (B) on two columns in order that the later type (B) operations can be made. □

We now go to a further study of $\mathcal{A}_Y$. Consider the set $\mathcal{D}_Y = \{Z \in \mathcal{C}_Y : Z \prec^{ST} Y\}$ and let $\mathcal{B}_Y$ be the convex hull of the points in $\mathcal{D}_Y$. Then clearly $\mathcal{B}_Y \subset \mathcal{A}_Y$. By Theorem 4.6, $\mathcal{B}_Y = \mathcal{A}_Y$ if $n = 2$, and it can be shown that this is also valid sometimes when $n > 2$. One problem for future research is to deduce all conditions for which $\mathcal{B}_Y = \mathcal{A}_Y$.

From Theorem 3.2, the region $\mathcal{A}_Y$ can be specified precisely through a finite number of linear inequalities. The extreme points of $\mathcal{A}_Y$ can be enumerated by using some theory from linear programming and the simplex method. An alternative approach makes use of separating hyperplanes. Both approaches are not difficult to implement on a computer for a given Y but the combinatorial enumeration grows rapidly as m and n increase. However, both approaches led to the examples showing that the $\prec^{PC}$ and $\prec^{ST}$ orderings are not equivalent, and partly with the help of symbolic manipulation software, have been used to prove general results for $m = 2$, $n = 3$. We illustrate both approaches below. The separating hyperplane approach is shown first. Its advantage is that it suggests a more general reduction operation which is mentioned at the end of this section.

Since each matrix in $\mathcal{A}_Y$ has the same row sum vector as Y, $\mathcal{A}_Y$ is a set in $m(n-1)$ dimensional space. Hence $\mathcal{B}_Y$ can be partitioned into simplices, each with $m(n-1)+1$ vertices, where each vertex is in $\mathcal{D}_Y$. To see if $\mathcal{A}_Y$

extends beyond $\mathcal{B}_Y$, for each point Y^* in $\mathcal{C}_Y \backslash \mathcal{D}_Y$, simplices extending from $\mathcal{B}_Y$ to Y^* can be defined and then it can be checked whether any point in these simplices are in $\mathcal{A}_Y$. All such simplices can be found by enumerating sets of $m(n-1)$ matrices in $\mathcal{D}_Y$ and obtaining the hyperplane that these matrices lie on; the hyperplane can be found by solving a linear system or using a singular value decomposition depending on whether the matrix consisting of the column vectors formed from the first $n-1$ columns of each of the $m(n-1)$ matrices has full or less than full rank. The simplices involving Y^* that need to be checked are those resulting from hyperplanes that separate Y^* from $\mathcal{B}_Y$. The ideas will be clearer from the next result below for $m=2$, $n=3$ where Y has two oppositely ordered rows.

THEOREM 5.3 *Let $u_1 < u_2 < u_3$ and $v_1 < v_2 < v_3$. For $Y = \begin{bmatrix} u_1 & u_2 & u_3 \\ v_3 & v_2 & v_1 \end{bmatrix}$, $\mathcal{A}_Y = \mathcal{B}_Y$. Hence for this Y, if $X \prec^{\mathrm{PC}} Y$, then $X \prec^{\mathrm{ST}} Y$ and $X \prec^{\mathrm{UM}} Y$.*

PROOF Let $Y_1 = Y$, $Y_2 = \begin{bmatrix} u_1 & u_3 & u_2 \\ v_3 & v_1 & v_2 \end{bmatrix}$, $Y_3 = \begin{bmatrix} u_2 & u_1 & u_3 \\ v_2 & v_3 & v_1 \end{bmatrix}$, $Y_4 = \begin{bmatrix} u_2 & u_3 & u_1 \\ v_2 & v_1 & v_3 \end{bmatrix}$, $Y_5 = \begin{bmatrix} u_3 & u_1 & u_2 \\ v_1 & v_3 & v_2 \end{bmatrix}$, $Y_6 = \begin{bmatrix} u_3 & u_2 & u_1 \\ v_1 & v_2 & v_3 \end{bmatrix}$. These are the 6 points in $\mathcal{D}_Y$ and $\mathcal{B}_Y$ lies in a four-dimensional space determined by the row sums remaining constant. Let $X = \begin{bmatrix} x_{11} & x_{12} & x_{13} \\ x_{21} & x_{22} & x_{23} \end{bmatrix}$ be a generic point in this four-dimensional space.

Let $b_1 = (u_2 - u_1)/(v_3 - v_2)$, $b_2 = (u_3 - u_1)/(v_3 - v_1)$ and $b_3 = (u_3 - u_2)/(v_2 - v_1)$. These are the values of b such that $(1, b)Y$ is a vector with at least two components equal. Either b_1, b_2, b_3 are distinct or they are all the same. In the latter case, the conclusion follows from Proposition 5.4 below. Hence we now assume the former case. Note that b_2 is a convex combination of b_1 and b_3 so that it cannot be largest or smallest of the 3 b_j's. The 15 subsets of size 4 from $\mathcal{D}_Y$ and the equations of the corresponding hyperplanes for the each set of 4 are:

(a) Y_1, Y_2, Y_3, Y_4: $x_{11} + b_1 x_{21}$
(b) Y_1, Y_2, Y_3, Y_5: $x_{13} + b_3 x_{23}$
(c) Y_1, Y_2, Y_3, Y_6: $(v_3 - v_1)x_{11} + (u_3 - u_1)x_{21} + (v_2 - v_1)x_{12} + (u_3 - u_2)x_{22}$
(d) Y_1, Y_2, Y_4, Y_5: $(v_3 - v_2)x_{11} + (u_2 - u_1)x_{21} - (v_2 - v_1)x_{12} - (u_3 - u_2)x_{22}$
(e) Y_1, Y_2, Y_4, Y_6: $x_{12} + b_3 x_{22}$
(f) Y_1, Y_2, Y_5, Y_6: $x_{11} + b_2 x_{21}$
(g) Y_1, Y_3, Y_4, Y_5: $(v_2 - v_1)x_{11} + (u_3 - u_2)x_{21} + (v_3 - v_1)x_{12} + (u_3 - u_1)x_{22}$
(h) Y_1, Y_3, Y_4, Y_6: $x_{13} + b_2 x_{23}$
(i) Y_1, Y_3, Y_5, Y_6: $x_{12} + b_1 x_{22}$
(j) Y_1, Y_4, Y_5, Y_6: $(v_3 - v_1)x_{11} + (u_3 - u_1)x_{21} + (v_3 - v_2)x_{12} + (u_2 - u_1)x_{22}$
(k) Y_2, Y_3, Y_4, Y_5: $x_{12} + b_2 x_{22}$
(l) Y_2, Y_3, Y_4, Y_6: $(v_3 - v_2)x_{11} + (u_2 - u_1)x_{21} + (v_3 - v_1)x_{12} + (u_3 - u_1)x_{22}$

(m) Y_2, Y_3, Y_5, Y_6: $-(v_2-v_1)x_{11}-(u_3-u_2)x_{21}+(v_3-v_2)x_{12}+(u_2-u_1)x_{22}$
(n) Y_2, Y_4, Y_5, Y_6: $x_{13}+b_1x_{23}$
(o) Y_3, Y_4, Y_5, Y_6: $x_{11}+b_3x_{21}$.

For cases (c), (d), (g), (j), (l), (m), it is straightforward to check that the remaining 2 points in $\mathcal{D}_Y$ are on opposite sides of the hyperplane (by making use of the fact the b_2 is neither largest or smallest). The argument for the other cases is analogous to case (a) which is given next.

The equation $x_{11}+b_1x_{21}$ applied to Y_1, Y_2, Y_3, Y_4 leads to the constant $\alpha = u_1+b_1v_3 = u_2+b_2v_2$, and when applied to Y_5, Y_6 it leads to the constant $\beta = u_3+b_1v_1$. For a Y_7 in $\mathcal{C}_Y\backslash\mathcal{D}_Y$, suppose the equation leads to γ. The hyperplane $x_{11}+b_1x_{21}=\alpha$ separates Y_7 from Y_5, Y_6 only if $\beta<\alpha<\gamma$ or $\beta>\alpha>\gamma$. In either case, if X is any convex combination of Y_1, Y_2, Y_3, Y_4, Y_7 with a positive weight for Y_7, then $(1, b_1)X$ is not majorized by $(1, b_1)Y_1$. Hence it is not possible to extend $\mathcal{B}_Y$ towards Y_7.

The above argument works for the nine hyperplanes of the form (a) and any Y_7 not in $\mathcal{D}_Y$ so that the conclusion of the theorem follows. □

PROPOSITION 5.4 *Let Y be a $2\times n$ matrix. If, for a nonnegative nonzero vector (a_1, a_2), $(a_1, a_2)Y$ is a constant times a vector of ones, then $X \prec^{\mathrm{PC}} Y$ is equivalent to $X \prec^{\mathrm{UM}} Y$.*

PROOF Let $\mathbf{y}_1, \mathbf{y}_2$ be the first and second rows of Y. Then $\mathbf{y}_2 = c(1,\ldots,1)-a_1\mathbf{y}_1$ for a constant c. Suppose $X \prec^{\mathrm{PC}} Y$. Then $(a_1, a_2)X = c(1,\ldots,1)$ and $\mathbf{x}_2 = c(1,\ldots,1)-a_1\mathbf{x}_1$, where $\mathbf{x}_1, \mathbf{x}_2$ are the first and second rows of X. $X \prec^{\mathrm{PC}} Y$ implies $\mathbf{x}_1 \prec \mathbf{y}_1$ so that there is a doubly stochastic matrix D such that $\mathbf{x}_1 = \mathbf{y}_1 D$. The preceding equalities imply then that $\mathbf{x}_2 = \mathbf{y}_2 D$ and hence $X \prec^{\mathrm{UM}} Y$. □

We next illustrate the "linear inequalities" approach. Because of the second invariance property in Theorem 2.5, we can assume that the minimum component of each row of Y is 0 so that all components of Y are nonnegative. In the general case, one can subtract the ith row minimum c_i from the ith row, find the extreme points of $\mathcal{A}_{Y^*}$ for the resulting Y^*, and then add c_i to the ith row of each extreme point in $\mathcal{A}_{Y^*}$ to get $\mathcal{A}_Y$. Note that the ordering $\mathbf{a}X \prec \mathbf{a}Y = \mathbf{y}^*$ is equivalent to the following set of inequalities:

$$\sum_{i=1}^{m} a_i(x_{ij_1}+\cdots+x_{ij_k}) \le y^*_{[1]}+\cdots y^*_{[k]}, \quad j_1<\cdots<j_k, k=1,\ldots,n-1,$$

together with the sum constraint. The set $\mathcal{A}_Y$ can be represented by a finite number of inequalities of the above type together with the m row sum constraints. As in the simplex method for linear programming (see for example Gass (1985)), nonnegative slack variables can be introduced for each inequality. The total number of variables, N, is now mn plus the number of nonredundant inequalities, and the number of equations, M, is m plus the number of nonredundant inequalities. The extreme points can now be enumerated by setting in turn $N - M$ of the variables to zero and solving the resulting $M \times M$ linear system if the linear system is nonsingular; nonnegative solutions generated in this way correspond to the extreme points. However there are too many inequalities to prove a general theorem except for in a couple of "small" cases, given in the next two results.

These results involve one or two inversions of the second row relative to the first. For a $2 \times n$ matrix with a strictly increasing first row vector and with a second row vector $\mathbf{z} = (z_1, \dots, z_n)$, the number of inversions is the cardinality of the set $\{(j_1, j_2) : j_1 < j_2, z_{j_1} > z_{j_2}\}$.

THEOREM 5.5 *Let* $u_1 < u_2 < u_3$, $v_1 < v_2 < v_3$. *For* $Y = \begin{bmatrix} u_1 & u_2 & u_3 \\ v_1 & v_3 & v_2 \end{bmatrix}$, $\mathcal{A}_Y = \mathcal{B}_Y$.

PROOF Without loss of generality, we can assume that $u_1 = v_1 = 0$. From Theorem 3.2, $X \prec^{PC} Y$ if $\mathbf{a}X \prec \mathbf{a}Y$ for $\mathbf{a} = (1,0)$, $(0,1)$, and $(1, b)$ where $b = (u_3 - u_2)/(v_3 - v_2)$. In addition to nonnegativity constraints, the linear inequalities and equalities imposed by the three majorization orderings are:

$$x_{1j} \le u_3, \qquad x_{2j} \le v_3, \qquad j = 1, 2, 3,$$

$$0 = u_1 + bv_1 \le x_{1j} + bx_{2j} \le u_2 + bv_3 = u_3 + bv_2, \qquad j = 1, 2, 3,$$

$$x_{11} + x_{12} + x_{13} = u_2 + u_3, \qquad x_{21} + x_{22} + x_{23} = v_2 + v_3.$$

Note that the lower bound of the second set of inequalities are redundant given the nonnegativity constraints. By adding 9 nonnegative slack variables $s_1, \dots, s_9$, the 11 resulting equations are:

$$x_{1j} + s_j = u_3, \qquad x_{2j} + s_{3+j} = v_3, \qquad j = 1, 2, 3,$$

$$x_{1j} + bx_{2j} + s_{6+j} = u_2 + bv_3, \qquad j = 1, 2, 3,$$

$$x_{11} + x_{12} + x_{13} = u_2 + u_3, \qquad x_{21} + x_{22} + x_{23} = v_2 + v_3.$$

The extreme points can be found by setting in turn 4 of the 15 variables to zero and solving the 11×11 linear system. At most one of x_{1j} can be set to zero and at most one of x_{2j} can be set to zero and at most two of s_7, s_8, s_9 can be set to zero. If four zeros are chosen in this way, then either no nonnegative solution exists or YQ is the solution where Q is a permutation matrix. For other cases, one of $s_1, \dots, s_6$ is set to zero. By symmetry, assume that $s_1 = 0$ and $x_{11} = u_3$. Substitute $x_{13} = u_2 - x_{12}$, $x_{23} = v_2 + v_3 - x_{21} - x_{22}$. The linear system simplifies to:

$$x_{12} + t_1 = u_2, \quad x_{21} + t_2 = v_2, \quad x_{22} + t_3 = v_3, \quad x_{21} + x_{22} - t_4 \ge v_2,$$

where t_1, t_2, t_3, t_4 are nonnegative slack variables. Now set in turn 3 of the 7 variables to zero. 16 of the resulting systems are nonsingular, of which 4 lead to nonnegative solutions and the other 12 lead to solutions of the form YQ where Q is a permutation matrix. □

The last case to consider for $m = 2, n = 3$ is when there are two inversions of the second row relative to the first. For example, $Y = \begin{bmatrix} u_1 & u_2 & u_3 \\ v_3 & v_1 & v_2 \end{bmatrix}$, where $u_1 < u_2 < u_3$ and $v_1 < v_2 < v_3$. This is interesting because it is the simplest case where $\mathcal{A}_Y$ is strictly larger than $\mathcal{B}_Y$. To further illustrate the two approaches, a combination of them are used to find the extreme points of $\mathcal{A}_Y$ in this case. Not all details are provided.

THEOREM 5.6 *Let* $u_1 < u_2 < u_3$ *and* $v_1 < v_2 < v_3$. *Let* $Y = \begin{bmatrix} u_1 & u_2 & u_3 \\ v_3 & v_1 & v_2 \end{bmatrix}$. *Define* $u_{ij} = u_i - u_j$, $v_{ij} = v_i - v_j$, *for* $i > j$, *and* $d = u_{31}v_{31} - u_{21}v_{32}$. *In addition to the matrices in* $\mathcal{D}_Y$, *the extreme points of* $\mathcal{A}_Y$ *include*

$$(5.1) \quad Z = \begin{bmatrix} d^{-1}(u_{31}v_{21}u_1 + u_{32}v_{32}u_2) & d^{-1}(u_{32}v_{32}u_1 + u_{31}v_{21}u_2) & u_3 \\ d^{-1}(u_{32}v_{31}v_2 + u_{21}v_{21}v_3) & d^{-1}(u_{21}v_{21}v_2 + u_{32}v_{31}v_3) & v_1 \end{bmatrix},$$

and the matrices obtained by column permutations of this matrix.

OUTLINE OF DETAILS. Let $Y_1, \ldots, Y_6$ be as in the proof of Theorem 5.3. Let $Y_7 = Y$,

$$Y_8 = \begin{bmatrix} u_1 & u_3 & u_2 \\ v_3 & v_2 & v_1 \end{bmatrix}, Y_9 = \begin{bmatrix} u_2 & u_1 & u_3 \\ v_1 & v_3 & v_2 \end{bmatrix}, Y_{10} = \begin{bmatrix} u_2 & u_3 & u_1 \\ v_1 & v_2 & v_3 \end{bmatrix},$$

$$Y_{11} = \begin{bmatrix} u_3 & u_1 & u_2 \\ v_2 & v_3 & v_1 \end{bmatrix}, Y_{12} = \begin{bmatrix} u_3 & u_2 & u_1 \\ v_2 & v_1 & v_3 \end{bmatrix}.$$

These are the 12 points in $\mathcal{D}_Y$. Furthermore, let $Y_{13} = \begin{bmatrix} u_1 & u_2 & u_3 \\ v_2 & v_3 & v_1 \end{bmatrix}$.

Let $X = \begin{bmatrix} x_{11} & x_{12} & x_{13} \\ x_{21} & x_{22} & x_{23} \end{bmatrix}$ be a generic point in the four-dimensional space containing $\mathcal{B}_Y$. It is straightforward to show that the hyperplane $-v_{32}x_{11} + u_{21}x_{22} = u_2v_2 - u_1v_3$ contains Y_1, Y_3, Y_8, Y_9 and separates the other Y's in $\mathcal{D}_Y$ from Y_{13}.

Consider the simplex with vertices $Y_1, Y_3, Y_8, Y_9, Y_{13}$. Let X be the convex combination $\lambda_1 Y_1 + \lambda_3 Y_3 + \lambda_8 Y_8 + \lambda_9 Y_9 + \lambda_{13} Y_{13}$. Let $c_1 = u_{21}/v_{31}$ and $c_2 = u_{31}/v_{32}$. Since $X \prec^{\mathrm{MM}} Y$, by Theorem 3.2, $X \prec^{\mathrm{PC}} Y$ if $(1, c_j)X \prec (1, c_j)Y$, $j = 1, 2$. These two majorization orderings impose the constraints

$$\begin{aligned}(5.2)\ (u_1 + c_1v_3)\lambda_1 &+ (u_2 + c_1v_2)\lambda_3 + (u_1 + c_1v_3)\lambda_8 + (u_2 + c_1v_1)\lambda_9 \\ &+ (u_1 + c_1v_2)\lambda_{13} \geq u_1 + c_1v_3 = u_2 + c_1v_1,\end{aligned}$$

$$(5.3)\ (u_2 + c_1v_2)\lambda_1 + (u_1 + c_1v_3)\lambda_3 + (u_3 + c_1v_2)\lambda_8 + (u_1 + c_1v_3)\lambda_9 + (u_2 + c_1v_3)\lambda_{13} \le u_3 + c_1v_2,$$

$$(5.4)\ (u_1 + c_2v_3)\lambda_1 + (u_2 + c_2v_2)\lambda_3 + (u_1 + c_2v_3)\lambda_8 + (u_2 + c_2v_1)\lambda_9 + (u_1 + c_2v_2)\lambda_{13} \ge u_2 + c_2v_1,$$

$$(5.5)\ (u_2 + c_2v_2)\lambda_1 + (u_1 + c_2v_3)\lambda_3 + (u_3 + c_2v_2)\lambda_8 + (u_1 + c_2v_3)\lambda_9 + (u_2 + c_2v_3)\lambda_{13} \le u_1 + c_2v_3 = u_3 + c_2v_2$$

on $\lambda_1, \lambda_3, \lambda_8, \lambda_9, \lambda_{13}$. If the inequality (5.3) is multiplied by $(u_3+c_2v_2)/(u_3+c_1v_2)$, then $(u_i + c_1v_j)(u_3 + c_2v_2)/(u_3 + c_1v_2) \le u_i + c_2v_j$, for $(i,j) = (2,2)$, $(1,3)$, $(3,2)$, $(2,3)$ or $(3,3)$. Comparison with (5.5) then makes the inequality (5.3) redundant. Similarly if the inequality (5.4) is multiplied by $(u_2 + c_1v_1)/(u_2 + c_2v_1)$, then $(u_i + c_2v_j)(u_2 + c_1v_1)/(u_2 + c_2v_1) \ge u_i + c_1v_j$, for $(i,j) = (2,2)$, $(1,3)$, $(2,1)$, $(1,2)$ or $(1,1)$. Comparison with (5.2) then makes the inequality (5.4) redundant.

For the remaining two inequalities (5.2) and (5.5), substituting $\lambda_{13} = 1 - \lambda_1 - \lambda_3 - \lambda_8 - \lambda_9$ leads to

$$c_1v_{32}\lambda_1 + u_{21}\lambda_3 + c_1v_{32}\lambda_8 + c_1v_{32}\lambda_9 \ge c_1v_{32}$$

and

$$c_2v_{32}\lambda_1 + u_{21}\lambda_3 + u_{21}\lambda_8 + u_{21}\lambda_9 \ge u_{21}.$$

Other than the extreme points with $\lambda_1 = 1$, $\lambda_3 = 1$, $\lambda_8 = 1$, $\lambda_9 = 1$, the single nontrivial extreme point from these inequalities is when $\lambda_1 = u_{21}v_{21}/d$, $\lambda_3 = u_{32}v_{32}/d$, $\lambda_8 = \lambda_9 = 0$ (and $\lambda_{13} = u_{32}v_{21}/d$). This leads to the matrix given in (5.1).

Hence the separating hyperplane approach has extended $\mathcal{A}_Y$ beyond $\mathcal{B}_Y$. However in this case, it cannot show that ZQ, where Q is a permutation matrix, are the only other extreme points of $\mathcal{A}_Y$. The linear inequalities approach can be used to complete this last step. The details are more tedious than in Theorem 5.5.

Let $u_1 = v_1 = 0$ now without loss of generality. The linear inequalities and equalities imposed by $X \prec^{\mathrm{PC}} Y$ are:

$$x_{1j} \le u_3, \qquad x_{2j} \le v_3, \qquad j = 1,2,3,$$

$$u_2 \le x_{1j} + c_1x_{2j} \le u_3 + c_1v_2, \qquad j = 1,2,3,$$

$$u_2 \le x_{1j} + c_2x_{2j} \le c_2v_3, \qquad j = 1,2,3,$$

$$x_{11} + x_{12} + x_{13} = u_2 + u_3, \quad x_{21} + x_{22} + x_{23} = v_2 + v_3.$$

Slack variables $s_1, \ldots, s_{18}$ can be added (or subtracted) corresponding to the 18 inequalities. This leads to 20 linear equations in 24 variables. At most one of x_{1j} can be set to zero and at most one of x_{2j} can be set to zero, so that at least two of the slack variables must to set to zero to find the extreme points. By symmetries and constraints, there are 21 pairs $(s_{i_1}, s_{i_2}) = (0, 0)$ that have to be considered. Once a pair of slack variables is set to zero, the variables x_{ij} and the remaining s_k's can be expressed in terms of two slack variables, say s_{j_1}, s_{j_2}. Symbolic manipulation software helps for the algebra in this reduction. Extreme points from inequalities in two variables (in symbols) can be solved by hand. The outcome of all this is that the only extreme points are of the form YQ, Y_1Q, ZQ, where Q is a permutation matrix. □

REMARKS The two techniques for identifying the structure of $\mathcal{A}_Y$ can in general be implemented in computer programs. Numerical examples studied in this way may lead to more general theorems. Further research will attempt to determine when $\mathcal{A}_Y = \mathcal{B}_Y$ and when $\mathcal{A}_Y$ is strictly larger than $\mathcal{B}_Y$ for general (m, n) not covered by theorems in this paper. In addition, it would be useful to discover an approach that does not require enumeration.

The derivation in the first part of the proof of Theorem 5.6 suggest the following (non–simple) operation for extending to points in $\mathcal{A}_Y \backslash \mathcal{B}_Y$. Choose a matrix Y_0 in $\mathcal{C}_Y \backslash \mathcal{D}_Y$ satisfying some properties. Find a separating hyperplane that separates Y_0 from some points in $\mathcal{C}_Y$. Take a convex combination of Y_0 and matrices in $\mathcal{C}_Y$ that on the hyperplane. The linear inequalities imposed by the PC-ordering will put constraints on the possible convex combinations. A goal is to identify those Y_0 where the coefficient of Y_0 in convex combinations can be definitely positive.

Acknowledgements. The authors thank the referee for a careful reading and many detailed comments.

REFERENCES

ARNOLD, B. C. (1987). *Majorization and the Lorenz Order.* Springer-Verlag, New York.

BOLAND, P. J. AND PROSCHAN, F. (1988). Multivariate arrangement increasing functions with applications in probability and statistics. *J. Multivariate Anal.* **25** 286–298.

GASS, S. I. (1985). *Linear Programming – Methods and Applications.* Fifth edition. McGraw-Hill, New York.

HOLLANDER, M., PROSCHAN, F. AND SETHURAMAN, J. (1977). Functions decreasing in transposition and their applications in ranking problems. *Ann. Statist.* **5** 722–733.

JOE, H. (1985). An ordering of dependence for contingency tables. *Linear Alg. Appl.* **70** 89–103.

MARSHALL, A. W. AND OLKIN, I. (1979). *Inequalities: Theory of Majorization and its Applications.* Academic Press, New York.

TONG, Y. L. (1989). Probability inequalities for n–dimensional rectangles via multivariate majorization. In *Contributions to Probability and Statistics: Essays in Honor of Ingram Olkin.* L. J. Gleser, M. D. Perlman, S. J. Press, and A. R. Sampson, eds. Springer-Verlag, New York, 146-159.

DEPARTMENT OF STATISTICS
UNIVERSITY OF BRITISH COLUMBIA
VANCOUVER, B.C. CANADA V6T 1Z2

DEPARTMENT OF STATISTICS
OHIO STATE UNIVERSITY
COLUMBUS, OH 43210

Stochastic Inequalities
IMS Lecture Notes – Monograph Series
Volume 22 (1993)

COVARIANCE SPACES FOR MEASURES ON POLYHEDRAL SETS

By J. H. B. KEMPERMAN[1] and MORRIS SKIBINSKY

Rutgers University and University of Massachusetts

Let V be a given subset of $\mathbb{R}^n$. We are interested in determining the associated moment space $\mathcal{C}_r[V]$. The latter consists of all points $\mathbf{c} = (c(\mathbf{i}); |\mathbf{i}| \leq r)$ which can be realized as $c(\mathbf{i}) = \int x^{\mathbf{i}} \mu(dx)$, for all $|\mathbf{i}| \leq r$, by a measure μ on V. Here, $\mathbf{i} = (i_1, \ldots, i_n)$ with $i_j \in \mathbf{Z}_+$ and $|\mathbf{i}| = i_1 + \cdots + i_n$. Let $\mathcal{C}_r(V)$ be the analogous homogeneous moment space $\mathcal{C}_r(V)$, where one insists on $|\mathbf{i}| = r$. The calculation of $\mathcal{C}_r[V]$ is shown to be equivalent to that of $\mathcal{C}_r(W)$, with W as a suitable affine imbedding of V into $\mathbb{R}^{n+1}$. A central role is played by the dual $\mathcal{C}_r(V)^*$ of the convex cone $\mathcal{C}_r(V)$. One may interpret $\mathcal{C}_r(V)^*$ as the set of all homogeneous polynomials $f(x) = f(x_1, \ldots, x_n)$ on $\mathbb{R}^n$ of degree r that are nonnegative on V.

Detailed results are given only for the important case $r = 2$. Let $\mathcal{Q}_n$ be the linear space of all symmetric $n \times n$ matrices, supplied with the natural inner product $(A, B) = \mathrm{Tr}(AB)$. The pair $\mathcal{C}_2(V)$ and $\mathcal{C}_2(V)^*$ has a natural interpretation as a pair of dual convex cones in $\mathcal{Q}_n$. In fact, $\mathcal{C}_2(V)^*$ is the set of all $Q \in \mathcal{Q}_n$ such that $x^t Q x \geq 0$ for all $x \in V$. Special attention is given to the second order moment spaces $\mathcal{C}_2(K)$ and $\mathcal{C}_2[T]$ with

$$K = \{x \in \mathbb{R}^n : Ax \geq 0\}; \qquad T = \{x \in \mathbb{R}^n : Bx + e \geq 0\}.$$

Here A and B denote given $m \times n$ matrices. Our description of the latter moment spaces involves the crucial cone $\mathcal{P}_m = \{Q \in \mathcal{Q}_m : x^t Q x \geq 0 \text{ for all } x \in \mathbb{R}_+^m\}$.

These results are quite explicit when $m \leq 4$, as happens, for instance, when T is a triangle in $\mathbb{R}^2$ or a simplex in $\mathbb{R}^3$. This is largely due to the very simple structure of the cone $\mathcal{P}_m$ in the case $m \leq 4$, due to Diananda (1962). The remaining problem, of determining the second order moment spaces $\mathcal{C}_2(K)$ or $\mathcal{C}_2[T]$ for the case $m \geq 5$, is essentially equivalent to the long standing difficult open problem to determine the precise structure of the cone $\mathcal{P}_m$ when $m \geq 5$. Concrete applications will be given in subsequent papers.

---

[1]Research supported in part by National Science Foundation Grant DMS–9002856.
AMS 1991 ***subject classifications.*** 44A60, 60E15, 15A48, 49A55.
Key words and phrases. Finite multivariate moment problem, second order moment space, copositive matrices, nonnegative polynomials, dual cones, covariance spaces.

1. Moment Spaces

In the sequel, n and r are fixed positive integers. All sets considered are assumed to be measurable. All measures are finite–valued and nonnegative. For V as any subset of $\mathbb{R}^n$, let $\mathcal{M}_0(V)$ be the set of all measures μ on $\mathbb{R}^n$ that are supported by a finite subset of V. The larger set of all measures μ supported by V and possessing all moments of order $\leq r$ will be denoted as $\mathcal{M}(V)$. Thus $\mu \in \mathcal{M}(V)$ if μ is a measure on V such that the integral

$$c(\mathbf{i}) = c_\mu(\mathbf{i}) = \int x^{\mathbf{i}} d\mu = \int x_1^{i_1} \cdots x_n^{i_n} \mu(dx) \tag{1.1}$$

is absolutely convergent for all $\mathbf{i} = (i_1, \ldots, i_n) \in \mathbf{Z}_+^n$ with $|\mathbf{i}| = i_1 + \cdots + i_n \leq r$. Let

$$c[\mu] = (c_\mu(\mathbf{i}) : \mathbf{i} \in \mathbf{Z}_+^n; \ \ |\mathbf{i}| \leq r)$$

be the corresponding moment point (of order r); it has $\binom{n+r}{r}$ components. Let further

$$c(\mu) = (c_\mu(\mathbf{i}) : \mathbf{i} \in \mathbf{Z}_+^n; \ \ |\mathbf{i}| = r),$$

be the analogous "homogeneous" moment point (of order r); it has $\binom{n+r-1}{r}$ components.

We like to determine the set of all possible moment points $c[\mu]$ or $c(\mu)$, keeping V and r fixed. There is a considerable literature, see for instance Berg (1987), Cassier (1984) and Maserick (1977), on the analogous problem where $r = \infty$. Here, one is interested in characterizing the set of all *infinite* moment sequences $\{c_\mu(\mathbf{i}) : |\mathbf{i}| < \infty\}$ associated to the different measures μ on V. On the other hand, except for the very classical case $n = 1$, not much seems to be known about the *finite* case $2 \leq r < \infty$ we are considering.

Since only finitely many moments (1.1) are involved, we have for all $\mu \in \mathcal{M}(V)$ that there always exists a measure μ' on V of finite support ($\mu' \in \mathcal{M}_0(V)$) such that $c[\mu'] = c[\mu]$, thus also $c(\mu') = c(\mu)$. We will be interested in the two moment spaces

$$\begin{aligned} \mathcal{C}_r[V] &= \{c[\mu] : \mu \in \mathcal{M}(V)\} = \{c[\mu] : \mu \in \mathcal{M}_0(V)\} \quad \text{and} \\ \mathcal{C}_r(V) &= \{c(\mu) : \mu \in \mathcal{M}(V)\} = \{c(\mu) : \mu \in \mathcal{M}_0(V)\}. \end{aligned} \tag{1.2}$$

The right hand form (1.2) for $\mathcal{C}_r(V)$ essentially says that $\mathcal{C}_r(V)$ coincides with the convex cone spanned by the set of moments points $\{c(\delta_x) : x \in V\}$, one for each $x \in V$. Here, δ_x denotes the probability measure carried by $\{x\}$ thus

$$c(\delta_x) = (c(i) = x_1^{i_1} \cdots x_n^{i_n} : \mathbf{i} \in \mathbf{Z}_+^n; \ \ |\mathbf{i}| = r). \tag{1.3}$$

Similarly for $\mathcal{C}_r[V]$. By homogeneity, $c(\delta_{\lambda x}) = \lambda^r c(\delta_x)$ for all $x \in \mathbb{R}^n$; $\lambda \geq 0$. Consequently, $\mathcal{C}_r(V)$ will remain unchanged when the subset V of $\mathbb{R}^n$ is modified by replacing each $x \in V$ by any non-empty subset (such as a single point) of the corresponding half ray $\{\lambda x : \lambda > 0\}$, and also when one adds or deletes the element $x = 0$. The resulting subset W of $\mathbb{R}^n$ will be said to be equivalent to V, (since then $\mathcal{C}_r(W) = \mathcal{C}_r(V)$). In other words, V and W are *equivalent* when $\{\lambda x : \lambda > 0\} \cap V$ is non-empty if and only if $\{\lambda x : \lambda > 0\} \cap W$ is non-empty, this for each $x \in \mathbb{R}^n$ with $x \neq 0$. The convex cone W generated by V is equivalent to V, provided V is convex.

In many applications, the convex cone $K = \text{cone}(V)$ in $\mathbb{R}^n$ generated by V happens to be a pointed cone. Equivalently, there are numbers $\rho_1, \ldots, \rho_n$ such that $\rho_1 x_1 + \cdots + \rho_n x_n > 0$ for all $x \in V$ with $x \neq 0$. In that case, V is clearly equivalent to

$$(1.4)\ W = \{x \in \mathbb{R}^n : \rho_1 x_1 + \cdots + \rho_n x_n = 1;\ \ \lambda x \in V \text{ for some } \lambda > 0\},$$

in particular, $\mathcal{C}_r(W) = \mathcal{C}_r(V)$.

Let W be a subset of a hyperplane H in $\mathbb{R}^n$ of the form $\sum_j \rho_j x_j = 1$ and consider any measure $\nu \in \mathcal{M}(W)$. We assert that the "non-homogeneous" moment point $c[\nu]$ is then already determined by the corresponding "homogeneous" moment point $c(\nu)$. In fact, any moment $c_\nu(\mathbf{i})$ of order $|\mathbf{i}| \leq r$ as in (1.1) can be expressed as an explicit linear combinations of the moments $c_\nu(\mathbf{k})$ of order $|\mathbf{k}| = r$. After all, expanding

$$x^{\mathbf{i}} = x_1^{i_1} \cdots x_n^{i_n} = x_1^{i_1} \cdots x_n^{i_n} [\rho_1 x_1 + \cdots + \rho_n x_n]^{r-|\mathbf{i}|}, \quad (\text{for all } x \in W \subset H),$$

an integration relative to the measure ν yields that

$$(1.5) \qquad c_\nu(\mathbf{i}) = \sum_{|\mathbf{j}|=r-|\mathbf{i}|} \binom{r-|\mathbf{i}|}{\mathbf{j}} \rho^{\mathbf{j}} c_\nu(\mathbf{i}+\mathbf{j}), \quad \text{whenever } |\mathbf{i}| \leq r.$$

In particular, (1.5) sets up a 1:1 correspondence between $\mathcal{C}_r[W]$ and $\mathcal{C}_r(W)$, and it only remains to determine the "homogeneous" moment space $\mathcal{C}_r(W)$.

For T as an arbitrary subset of $\mathbb{R}^d$, one can reduce the study of the (non-homogeneous) moment space $\mathcal{C}_r[T]$ to a study of an associated homogeneous moment space $\mathcal{C}_r(W)$. Namely, take $W = \sigma T$ where $\sigma : \mathbb{R}^d \to \mathbb{R}^{d+1}$ is of the form

$$(1.6) \quad \sigma \mathbf{y} = (y_1, \ldots, y_d, x_n) \text{ where } x_n = \frac{1}{\rho_n}(1 - \rho_1 y_1 - \cdots - \rho_d y_d).$$

Here $n = d + 1$ and $\rho_n \neq 0$. Thus σ is a 1:1 affine map of $\mathbb{R}^{n-1}$ onto the hyperplane $H = \{x \in \mathbb{R}^n : \sum_j \rho_j x_j = 1\}$. For example, if $H = \{x \in \mathbb{R}^n : x_n = 1\}$ then $\sigma \mathbf{y} = (y_1, \ldots, y_{n-1}, 1)$. In all cases, $\pi = \sigma^{-1} : H \to \mathbb{R}^{n-1}$ is given by $\pi x = (x_1, \ldots, x_{n-1})$.

For μ as a measure on $\mathbb{R}^d$, one has $\mu \in \mathcal{M}(T)$ if and only if $\sigma\mu \in \mathcal{M}(W)$. Here $\nu = \sigma\mu$ is defined by $\nu(B) = \mu(\sigma^{-1}B) = \mu(\pi B)$, for all $B \subset H$.

Moreover, one has $c_\mu(i_1,\ldots,i_d) = c_\nu(i_1,\ldots,i_d,0)$ for all $\mathbf{i} = (i_1,\ldots,i_d) \in \mathbf{Z}_+^d$ with $|\mathbf{i}| \leq r$, showing that $\mathcal{C}_r[T]$ may be regarded as a simple image of $\mathcal{C}_r[W]$. This allows us to reduce the study of $\mathcal{C}_r[T]$ to that of $\mathcal{C}_r[W]$, and thus, by (1.5), to a study of $\mathcal{C}_r(W)$.

2. Duality: The Case $r = 2$

A typical "homogeneous" moment point $c = (c(\mathbf{i}) : |\mathbf{i}| = r)$ will be regarded as a point $c \in \mathbb{R}^{n_0}$, where $n_0 = \binom{n+r-1}{r}$. Here, $\mathbf{i}$ runs through all the n_0 tuples $\mathbf{i} = (i_1,\ldots,i_n) \in \mathbf{Z}_+^n$ with $|\mathbf{i}| = r$. The inner product (α, c) in $\mathbb{R}^{n_0}$ is given by $(\alpha, c) = \sum_{|\mathbf{i}|=r} \alpha(\mathbf{i})c(\mathbf{i})$.

Let $V \subset \mathbb{R}^n$. Recall that the homogeneous moment space $\mathcal{C}_r(V)$ is the convex cone generated by all points $c_x = c(\delta_x)$, one for each $x \in V$, (c_x having components $c_x(\mathbf{i}) = x^{\mathbf{i}}$). Thus, the dual of $\mathcal{C}_r(V)$ is the closed and convex cone given by

$$\begin{aligned} \mathcal{C}_r(V)^* &= \{\alpha \in \mathbb{R}^{n_0} : (\alpha, c) \geq 0 \text{ for all } c \in \mathcal{C}_r(V)\} \\ &= \{\alpha \in \mathbb{R}^{n_0} : (\alpha, c_x) \geq \text{ for all } x \in V\}. \end{aligned}$$

Equivalently,

$$\mathcal{C}_r(V)^* = \{\alpha \in \mathbb{R}^{n_0} : f_\alpha(x) \geq 0 \text{ for all } x \in V\}. \tag{2.1}$$

Here, $f_\alpha(x)$ denotes the homogeneous r^{th} degree polynomial

$$f_\alpha(x) = \sum_{|\mathbf{i}|=r} \alpha(\mathbf{i})x^{\mathbf{i}}, \qquad (x \in \mathbb{R}^n;\ \ \alpha \in \mathbb{R}^{n_0}). \tag{2.2}$$

As is well known, the second dual $(\mathcal{C}_r(V)^*)^*$ is precisely the closure of the original convex cone $\mathcal{C}_r(V)$. Thus

$$\mathrm{cl}(\mathcal{C}_r(V)) = \{c \in \mathbb{R}^{n_0} : (\alpha, c) \geq 0 \text{ for all } \alpha \in \mathcal{C}_r(V)^*\}. \tag{2.3}$$

In many applications, $\mathcal{C}_r(V)$ is already closed. This is true for instance when V is compact and, hence, also when V is equivalent (as defined in Section 1) to a compact subset W of $\mathbb{R}^n$. Formula (2.3) remains true when $\mathcal{C}_r(V)^*$ is replaced by a subset $\mathcal{E}$ of $\mathcal{C}_r(V)^*$, provided the convex cone generated by $\mathcal{E}$ is dense in $\mathcal{C}_r(V)^*$. Often the latter property holds for the set $\mathcal{E}_r(V)$ of all extreme members α of $\mathcal{C}_r(V)^*$ and then (2.3) implies that

$$\mathrm{cl}(\mathcal{C}_r(V)) = \{c \in \mathbb{R}^{n_0} : (\alpha, c) \geq 0 \text{ for all } \alpha \in \mathcal{E}_r(V)\}. \tag{2.4}$$

Formula (2.3) is our starting point. It reduces the problem of determining the moment space $\mathcal{C}_r(V)$ (or rather its closure) to the problem of determining all polynomials f_α as in (2.2) that are nonnegative on V. In this connection, see especially the interesting recent work of Micchelli and Pinkus (1989), which at least in spirit is close to the present paper.

Often (2.4) holds and then it suffices to determine all the extreme members of the collection of polynomials f_α as in (2.2) that are nonnegative on V. These problems tend to be very difficult. In a separate paper, we will apply the above ideas to the cubic case $r = 3$, in particular to the determination of $\mathcal{C}_3[T]$ when T is a planar triangle.

From now on in the present paper, we assume that $r = 2$. In that case $n_0 = n(n+1)/2$. Let $\mathcal{Q}_n$ denote the linear space of all real and symmetric $n \times n$ matrices $Q = (q_{ij})$. Note that $\mathcal{Q}_n$ has dimension n_0. A second degree homogeneous polynomial $f_\alpha(x)$, as in (2.2) with $r = 2$, can be written as $f_\alpha(x) = x^t Q x$ with $Q = (q_{ij}) \in \mathcal{Q}_n$. Namely, let $\alpha(\mathbf{i}) = q_{11}$ if $\mathbf{i} = (2, 0, \ldots, 0)$; $\alpha(\mathbf{i}) = 2q_{12} = 2q_{21}$ if $\mathbf{i} = (1, 1, 0, \ldots, 0)$, and so on.

As to the moment point $c = (c(\mathbf{i}) : |\mathbf{i}| = 2) \in \mathbb{R}^{n_0}$, we prefer to represent it in the form of a matrix $C = (c_{ij}) \in \mathcal{Q}_n$. Namely, let $c(\mathbf{i}) = c_{11}$ if $\mathbf{i} = (2, 0, \ldots, 0)$; $c(\mathbf{i}) = c_{12} = c_{21}$ if $\mathbf{i} = (1, 1, 0, \ldots, 0)$ and so on. The original inner product (α, c) in $\mathbb{R}^{n_0}$ now takes the form

$$(\alpha, c) = \sum_{|\mathbf{i}|=2} \alpha(\mathbf{i}) c(\mathbf{i}) = \sum_{i=1}^{n} q_{ii} c_{ii} + \sum_{1 \le i < j \le n} 2 q_{ij} c_{ij} = \sum_{i=1}^{n} \sum_{j=1}^{n} q_{ij} c_{ij} = \mathrm{Tr}(QC),$$

which indeed is the natural inner product in the linear space $\mathcal{Q}_n$. The dual of a subset $\mathcal{F}$ of $\mathcal{Q}_n$ is defined as the closed convex cone $\mathcal{F}^* = \{A \in \mathcal{Q}_n : \mathrm{Tr}(AB) \ge 0 \text{ for all } B \in \mathcal{F}\}$.

In the present case $r = 2$, the moment space $\mathcal{C}_2(V)$ can be identified with the convex cone in $\mathcal{Q}_n$ defined by

$$\text{(2.5)} \qquad \mathcal{C}_2(V) = \{C(\mu) : \mu \in \mathcal{M}(V)\} = \{C(\mu) : \mu \in \mathcal{M}_0(V)\}.$$

Here, $C(\mu) = C = (c_{ij}) \in \mathcal{Q}_n$ is given by

$$\text{(2.6)} \qquad c_{ij} = \int x_i x_j \mu(dx), \qquad (i, j = 1, \ldots, n).$$

Note that $C \in S$ where $S = S_n$ will denote the closed and convex cone of all $Q \in \mathcal{Q}_n$ that are nonnegative definite, (also written as $Q \gg 0$).

PROPOSITION 1 *Let $C \in \mathcal{Q}_n$. Then $C \in \mathcal{C}_2(V)$ if and only if C can be written as $C = U^t U$ with $U = (u_{ij})$ as an $s \times n$ matrix, such that*

$$\text{(2.7)} \qquad (u_{k1}, \ldots, u_{kn}) \in \{\lambda x : x \in V;\ \lambda \ge 0\}, \qquad (k = 1, \ldots, s).$$

Here, s may depend on C. In the special case $V = \mathbb{R}^n_+$ this requires precisely that $C = U^t U$ for some $U \ge 0$.

PROOF From (2.5), $C \in \mathcal{C}_2(V)$ if and only if $C = C(\mu)$ for some measure μ having a finite support $\{x^{(1)}, \ldots, x^{(s)}\} \subset V$. That is, $c_{ij} = \sum_k p_k x_i^{(k)} x_j^{(k)}$ for all $i, j = 1, \ldots, n$, where $p_k = \mu(\{x^{(k)}\}) > 0$ $(k = 1, \ldots, s)$. Letting $u_{kj} = \sqrt{p_k} x_j^{(k)}$ one obtains the stated assertion.

REMARK As is clear from the proof, the different ways of writing C as $C = U^t U$ with U satisfying (2.7) correspond precisely to the different measures μ on V of finite support such that $C = C(\mu)$. The stated property is equivalent to $x^t C x$ being representable as a finite sum $x^t C x = \sum_{k=1}^{s} L_k(x)^2$ with the $L_k(x)$ as linear forms $L_k(x) = u_{k1}x_1 + \cdots + u_{kn}x_n$ $(k = 1, \ldots, s)$ satisfying condition (2.7). In the special case $V = \mathbb{R}^n_+$, such matrices C are also said to be completely positive, see Hall (1986, p. 350).

Presently, (2.1) and (2.3) take the form

$$\text{(2.8)} \quad \mathrm{cl}(\mathcal{C}_2(V)) = \{C \in \mathcal{Q}_n : \mathrm{Tr}(QC) \geq 0 \text{ for all } Q \in \mathrm{cop}(V)\} = \mathrm{cop}(V)^*.$$

DEFINITION By $\mathrm{cop}(V)$ we mean the closed and convex cone defined by

$$\text{(2.9)} \qquad \mathrm{cop}(V) = \mathcal{C}_2(V)^* = \{Q \in \mathcal{Q}_n : x^t Q x \geq 0 \text{ for all } x \in V\}.$$

The matrices $Q \in \mathrm{cop}(V)$ are said to be V*-copositive*.

Note that (2.8) remains valid when $\mathrm{cop}(V)$ is replaced by a subset $\mathcal{E}$ of $\mathrm{cop}(V)$ such that the convex cone generated by $\mathcal{E}$ is dense in $\mathrm{cop}(V)$. Our main task would be the explicit determination of such a set $\mathcal{E}$ or if possible of the set $\mathrm{cop}(V)$ itself.

NOTATION Let $S = S_n$ denote the class of nonnegative definite $Q \in \mathcal{Q}_n$. Let further $N = N_n$ denote the closed and convex cone consisting of all nonnegative $Q \in \mathcal{Q}_n$, (written as $Q \geq 0$). Let further $N + S$ denote the set of all sums $Q_1 + Q_2$ with $Q_1 \in N$ and $Q_2 \in S$. Note that $N + S$ is precisely the convex cone generated by $N \cup S$.

It is evident that $S \subset \mathrm{cop}(V)$ for every $V \subset \mathbb{R}^n$. If $V \subset \mathbb{R}^n_+$, then also $N \subset \mathrm{cop}(V)$ and thus $N + S \subset \mathrm{cop}(V)$. The set $N + S$ can easily be much smaller than $\mathrm{cop}(V)$. For instance, let $1 \leq d < n$ and suppose each $x \in V$ satisfies $x_j = 0$ for $d < j \leq n$, while $x_j \geq 0$ otherwise, (so that $V \subset \mathbb{R}^n_+$). In this case, the condition $Q = (q_{ij}) \in \mathrm{cop}(V)$ clearly depends only on the elements q_{ij} with $1 \leq i, j \leq d$ and thus there is no need at all that $Q \in N + S$.

Of central importance is the case $V = \mathbb{R}^n_+$. We will write $\mathcal{P} = \mathcal{P}_n = \mathrm{cop}(\mathbb{R}^n_+)$. Thus $\mathcal{P}$ is the closed and convex cone consisting of all $Q \in \mathcal{Q}_n$ such that

$$\text{(2.10)} \qquad x^t Q x \geq 0 \text{ for all } x \geq 0.$$

Such matrices Q are said to be *copositive*. Note that $N + S \subset \mathcal{P}$.

Observe that $V = \mathbb{R}_+^n$ is equivalent to the *compact* simplex

$$S(n) = \{x = (x_1, \ldots, x_n) : x_i \geq 0 (i = 1, \ldots, n); x_1 + \cdots + x_n = 1\}. \tag{2.11}$$

This implies that $\mathcal{C}_2(\mathbb{R}_+^n) = \mathcal{C}_2(S(n))$ is closed, hence, (2.8) presently takes the form

$$\mathcal{C}_2(S(n)) = \mathcal{P}^* = \{C \in \mathcal{Q}_n : \mathrm{Tr}(QC) \geq 0 \text{ for all } Q \in \mathcal{P}\}. \tag{2.12}$$

Thus, *if* we would precisely know the class $\mathcal{P}$ then we would also know what $C \in \mathcal{Q}_n$ can be realized as $C = C(\mu)$ by a measure on $S(n)$ and, as a consequence, also what sets of first and second moments can be realized by a measure on the simplex $T(n-1)$ in $\mathbb{R}^{n-1}$. Here,

$$T(d) = \{y = (y_1, \ldots, y_d) \in \mathbb{R}^d : y_i \geq 0 \ \ (i = 1, \ldots, d); y_1 + \cdots + y_d \leq 1\} \tag{2.13}$$

For arbitrary n, one has that $N + S \subset \mathcal{P}$ thus

$$\mathcal{C}_2(S(n)) = \mathcal{P}^* \subset (N + S)^* = N \cap S. \tag{2.14}$$

After all $(N + S)^* = (N \cup S)^* = N^* \cap S^* = N \cap S$, since $N^* = N$ and $S^* = S$ as is easily seen, (see also Hall (1986, p. 353)). Here, $S^* = S$ says that $A \in \mathcal{Q}_n$ belongs to S if and only if $\mathrm{Tr}(AB) \geq 0$ for all $B \in S$.

Unfortunately, a precise description of the class $\mathcal{P} = \mathrm{cop}(\mathbb{R}_+^n)$ is only available when $n \leq 4$. In fact, Diananda (1962) showed that $\mathcal{P} = N + S$ when $n \leq 4$. Therefore,

$$\mathcal{C}_2(S(n)) = \mathcal{P}^* = (N + S)^* = N \cap S \text{ if } n \leq 4. \tag{2.15}$$

If $n \geq 5$ then $N+S$ is a proper subset of $\mathcal{P}$. This follows from an unpublished counterexample due to A. Horn, which is discussed in Hall (1986, p. 357). Specifically, Horn gave an explicit example with $n = 5$ of a member $H \in \mathcal{P}/(N + S)$. It is given below, see (4.13). It does not seem to be known whether or not $N + S$ is closed. Anyway, assuming $n \geq 5$, we can show (see the last remark of the paper) that even the closure of $N + S$ is a proper subset of $\mathcal{P}$, equivalently $\mathcal{P}^*$ is a proper subset of $N \cap S$. In this way, we arrive at the following result.

THEOREM 1 *In order that an $n \times n$ symmetric matrix $C = (c_{ij})$ admits a representation as in (2.6) with μ as a measure on the simplex $S(n)$ defined by (2.11), it is necessary that $C \in N \cap S$. Equivalently, C must be nonnegative definite and such that*

$$c_{ij} \geq 0 \textit{ for all } 1 \leq i < j \leq n. \tag{2.16}$$

This necessary condition $C \in N \cap S$ is also sufficient when $n \leq 4$, but not when $n \geq 5$.

3. The Polyhedral Case

Here, we assume that the subset V of $\mathbb{R}^n$ is a polyhedral cone of the form

$$K = \{x \in \mathbb{R}^n : Ax \geq 0\}. \tag{3.1}$$

Here, A is a given $m \times n$ real valued matrix. Recall that each $C = C(\mu)$ is necessarily nonnegative definite. A central result is as follows.

THEOREM 2 *Let $C \in \mathcal{Q}_n$ be nonnegative definite. Then in order that $C \in \text{cl}(\mathcal{C}_2(K))$ it is necessary and sufficient that*

$$\text{Tr}(A^t PAC) \geq 0 \text{ for all } P \in \mathcal{P}_m. \tag{3.2}$$

A necessary condition is that

$$ACA^t \geq 0 \tag{3.3}$$

Conversely, if $m \leq 4$ then (3.3) is also sufficient for $C \in \text{cl}(\mathcal{C}_2(K))$.

REMARK Let $P \in \mathcal{P}_m$, that is, $P \in \mathcal{Q}_m$ and $y^t Py \geq 0$ for all $y \in \mathbb{R}^n_+$. Hence, $x^t A^t PAx \geq 0$ for all $x \in K$. Integrating the latter inequality relative to a measure on μ on K such that $C(\mu) = C$, this confirms that (3.2) is a necessary condition for $C \in \mathcal{C}_2(K)$ and thus for $C \in \text{cl}(\mathcal{C}_2(K))$. In practice, we are yet unable to verify the necessary and sufficient condition (3.2) when $m \geq 5$, simply because then the class $\mathcal{P}_m$ is still largely unknown. The necessary condition (3.3) is equivalent to

$$\sum_{i=1}^{n}\sum_{j=1}^{n} a_{gi}a_{hj}c_{ij} \geq 0 \text{ for all } 1 \leq g < h \leq m, \tag{3.4}$$

altogether $\binom{m}{2}$ conditions. The necessity of (3.4) for $C \in \mathcal{C}_2(K)$ is obvious from an integration of the quadratic function $f(x) = (Ax)_g(Ax)_h$ relative to a measure μ on K such that $C(\mu) = C$. From (3.1), $f(x) \geq 0$ for all $x \in K$.

PROOF Suppose $P \in \mathcal{P}_m$ thus $y^t Py \geq 0$ when $y \in \mathbb{R}^m$, $y \geq 0$. Substituting $y = Ax$, we see that $x^t A^t PAx \geq 0$ for all $x \in K$, that is, $A^t PA \in \text{cop}(K)$ as defined by (2.9). It follows from (2.8) that (3.2) is a necessary condition.

Sufficiency. Suppose (3.2) holds. Let $\mathcal{G} = A^t\mathcal{P}_m A + S$ be the subset of $\text{cop}(K)$ which consists of all $Q \in \mathcal{Q}_n$ of the form $Q = A^t PA + B$ with $P \in \mathcal{P}_m$ and $B \in S$. We see from (3.2) and $C \in S$ that $\text{Tr}(QC) \geq 0$ for all $Q \in \mathcal{G}$. In view of (2.8), this would imply $C \in \text{cl}(\mathcal{C}_2(K))$, provided $\mathcal{G}$ can be shown to be dense in $\text{cop}(K)$.

In fact, let $Q \in \text{cop}(K)$ be given and consider $Q(\delta) = Q + \delta I$, with I as the $n \times n$ identity matrix and $\delta > 0$. It suffices to show that $Q(\delta) \in \mathcal{G}$ for all $\delta > 0$. In fact, merely using the fact that $x^t Q(\delta) x > 0$ whenever $Ax \geq 0$ and $x \neq 0$, this result $Q(\delta) \in \mathcal{G}$ is an immediate consequence of Theorem 4.2 due to Martin and Jacobson (1981).

By the way, Martin, Powell and Jacobson (1981, p. 53) showed by example that $\mathcal{G}$ can be a proper subset of $\text{cop}(K)$. See also Martin and Jacobson (1981, p. 246).

We already saw that condition (3.3) is necessary for $C \in \mathcal{C}_2(K)$ and thus for $C \in \text{cl}(\mathcal{C}_2(K))$. Also observe that (3.3) is equivalent to $\text{Tr}(A^t BAC) \geq 0$ for all $B \in N_m$, that is, for all nonnegative $B \in \mathcal{Q}_m$. As to the sufficiency of (3.3), assuming $m \leq 4$, it suffices to show that (3.3) implies (3.2). In fact, Diananda (1962) showed, for $m \leq 4$, that each $P \in \mathcal{P}_m$ is of the form $P = B + Q$ where $B \in N_m$ and $Q \in S_m$. Therefore

$$\text{Tr}(A^t PAC) = \text{Tr}(A^t BAC) + \text{Tr}(A^t QAC) \geq 0.$$

Here, it is also used that $A^t QA \in S_n$ and $C \in S_n$.

COMMENTS Suppose K is a direct product $K = K_1 \times \mathbb{R}^{n-d}$ with K_1 as a polyhedral cone in $\mathbb{R}^d$. Equivalently, $a_{ij} = 0$ when $j > d$. It is interesting to note that then the necessary and sufficient condition (3.2) depends only on the c_{ij} with $1 \leq i, j \leq d$. This feature also follows from (2.8) and the following Proposition 2, (which is related to Lemma 9 in Diananda (1962)). Here, $\text{span}(V)$ denotes the linear span of V.

PROPOSITION 2 *Let the subset V of $\mathbb{R}^n$ be a direct product $V = W \times \mathbb{R}^{n-d}$, with W as a subset of $\mathbb{R}^d$, $(1 \leq d < n)$. Assume that* $\text{span}(V) = \mathbb{R}^n$, *equivalently,* $\text{span}(W) = \mathbb{R}^d$. *Then a matrix $P \in \mathcal{Q}_n$ belongs to* $\text{cop}(V)$ *if and only if there exist matrices $Q \in \text{cop}(W)$ and $B \in S_n$ such that $x^t Px = y^t Qy + x^t Bx$, for all $x \in \mathbb{R}^n$. Here, we write $x \in \mathbb{R}^n$ as $x = (y, z)$ with $y \in \mathbb{R}^d$ and $z \in \mathbb{R}^{n-d}$. Thus $Q \in \mathcal{Q}_m$ and $y^t Qy \geq 0$ if $y \in \mathbb{R}^d_+$.*

PROOF Using an induction with respect to $k = n - d$, it suffices to consider the case $d = n - 1$. The stated condition is clearly sufficient. Conversely, assume that $P \in \text{cop}(V)$, that is, $x^t Px \geq 0$ whenever $x \in V$, that is, whenever $y \in W$. Let G be the upper principal $d \times d$ submatrix of P and let $L(y) = a_{n1} y_1 + \cdots + a_{n,n-1} y_{n-1}$. Then $y \in W$ implies that

$$x^t Px = a_{nn} z^2 + 2L(y)z + y^t Gy \geq 0 \text{ for all } z \in \mathbb{R}. \tag{3.5}$$

If $a_{nn} = 0$, it follows that $L(y) = 0$ for all $y \in W$, hence, $L(y) \equiv 0$ (since $\text{span}(W) = \mathbb{R}^d$) and the stated assertion holds with $Q = G$ and $B = 0$. In the case $a_{nn} > 0$ one may as well assume that $a_{nn} = 1$. Let $Q \in \mathcal{Q}_{n-1}$

be defined by $y^tQy \equiv y^tGy - L(y)^2$. It follows from (3.5) that $y^tQy \geq 0$, for all $y \in W$, hence, $Q \in \text{cop}(W)$. We further have for all $x \in \mathbb{R}^n$ that $x^tPx = y^tQy + x^tBx$, where $x^tBx \equiv (L(y)+z)^2 \geq 0$ thus $B \in S_n$.

4. Applications

Detailed applications of the preceding theory will be given in a separate paper. Here we will sketch just one application. Let T be a polyhedral convex subset of $\mathbb{R}^d$ of the form

$$T = \{y \in \mathbb{R}^d : By + e \geq 0\}. \tag{4.1}$$

Here, B is an $m \times d$ matrix and e an $m \times 1$ column vector. One may as well assume that $0 \in T$, hence, $e \geq 0$ and further that T has at least two points. For convenience, we will further assume that T is *compact* thus $m \geq d+1$; (the non–compact case can be handled equally well but is somewhat more delicate). Note that $e \neq 0$, (otherwise, T would be unbounded).

We like to determine the moment space $\mathcal{C}_2[T]$. Equivalently, we are interested in the necessary and sufficient conditions on the numbers c_0, ξ_i and c_{ij} $(i,j = 1,\ldots,d)$ in order that $\mu \in \mathcal{M}(T)$ can be found such that

$$\begin{aligned} c_0 &= \int \mu(dy); \quad \xi_i = \int y_i\mu(dy); \\ c_{ij} &= \int y_iy_j\mu(dy), \quad (i,j = 1,\ldots,d). \end{aligned} \tag{4.2}$$

If $c_0 = 1$ then $\xi_i = EY_i$ and $c_{ij} = EY_iY_j$ when $(Y_1,\ldots,Y_d)$ takes its values in T and has distribution μ. In terms of $X = (X_1,\ldots,X_n) = (Y_1,\ldots,Y_{n-1},1)$ one also has $c_0 = EX_n^2$ and $\xi_i = EX_iX_n$ $(1 \leq i < n)$.

Let $n = d+1$. Points $x \in \mathbb{R}^n$ will be written as $x = (y, x_n)$ where $y \in \mathbb{R}^{n-1}$ and $x_n \in \mathbb{R}$. It will be convenient to identify T with the compact set

$$\begin{aligned} V &= \{x = (y,x_n) \in \mathbb{R}^n : y \in T; \;\; x_n = 1\} \\ &= \{x \in \mathbb{R}^n : Ax \geq 0; \;\; x_n = 1\}. \end{aligned} \tag{4.3}$$

Here, A denotes the $m \times n$ matrix $A = (B, e)$. Further let

$$c_{nn} = \xi_n = c_0 \quad \text{and} \quad c_{in} = c_{ni} = \xi_i \quad (1 \leq i \leq n).$$

The above question now reduces to the problem of determining the *homogeneous* moment space $\mathcal{C}_2(V)$ of all $C = (c_{ij}) \in \mathcal{Q}_n$ which can be realized as $c_{ij} = \int x_ix_j\nu(dx)$ $(1 \leq i,j \leq n)$ by a finite measure ν on V. Since V is compact, the convex cone $\mathcal{C}_2(V)$ is closed.

Next consider the closed and convex cone

$$K = (x \in \mathbb{R}^n : Ax \geq 0). \tag{4.4}$$

We claim that K is precisely the convex cone generated by V. Since V is convex this implies that K and V are equivalent, thus $\mathcal{C}_2(K) = \mathcal{C}_2(V)$.

By $V \subset K$ one has cone$(V) \subset K$. As to the converse, assume that $x = (y, x_n) \neq 0$ belongs to K thus $Ax \geq 0$. It suffices to show that $\lambda x \in V$ for some $\lambda > 0$, equivalently, that $x_n > 0$. On the contrary, suppose that $x_n \leq 0$. Since $Ax = By + bx_n \geq 0$ one has that $By \geq (-x_n)e \geq 0$ (since $e \geq 0$). Hence, $y = 0$ since, otherwise, T would be unbounded, (in view of $T + \beta y \subset T$ for all $\beta \geq 0$). Next $x = (0, x_n) \neq 0$, thus, $x_n < 0$. But now $0 = By \geq (-x_n)e \geq 0$ implies that $e = 0$ and we have a contradiction.

We can now apply Theorem 2. The condition that the $n \times n$ matrix $C = (c_{ij})$ be nonnegative definite is easily checked. One may as well assume that $c_0 > 0$. It is natural to introduce the quantities

$$\sigma_{ij} = c_{ij} - \xi_i \xi_j / c_0, \qquad (i, j = 1, \ldots, n). \tag{4.5}$$

Especially note that $\sigma_{in} = \sigma_{ni} = 0$ for all i. If $c_0 = 1$ then the σ_{ij} can be regarded as covariances $\sigma_{ij} = \mathrm{Cov}(Y_i, Y_j)$ $(i, j = 1, \ldots, n)$, where $Y_n \equiv 1$. As is easily seen, the $n \times n$ matrix $C = (c_{ij})$ is nonnegative definite if and only if $\Sigma = (\sigma_{ij}; i, j = 1, \ldots, d)$ is nonnegative definite.

It remains to check the necessary and sufficient condition (3.2) of Theorem 2. Also in view of the remark following Theorem 2, that condition amounts to requiring that, for each $P = (p_{rs}) \in \mathcal{P}_m$, the left hand side of the obvious inequality

$$\sum_{r=1}^{m} \sum_{s=1}^{m} p_{rs} \left[\sum_{i=1}^{d} b_{ri} y_i + e_r \right] \left[\sum_{j=1}^{d} b_{sj} y_j + e_s \right] \geq 0 \quad \text{for all } y \in T, \tag{4.6}$$

must integrate to a nonnegative number relative to any measure μ on T that satisfies (4.2). In particular, for each choice of $1 \leq r < s \leq m$, the same must be true for the inequality

$$\left[\sum_{i=1}^{d} b_{ri} y_i + e_r \right] \left[\sum_{j=1}^{d} b_{sj} y_j + e_s \right] \geq 0 \qquad \text{for all } y \in T. \tag{4.7}$$

Naturally, (4.7) is merely a special case of (4.6) since $N \subset \mathcal{P}_m$.

For convenience we take $c_0 = 1$. Using (4.5), it is easily seen that (4.7) leads to the necessary condition, that for all $1 \leq r < s \leq m$,

$$\sum_{i=1}^{d} \sum_{j=1}^{d} b_{ri} b_{sj} \sigma_{ij} + \left[\sum_{i=1}^{d} b_{ri} \xi_i + e_r \right] \left[\sum_{j=1}^{d} b_{sj} \xi_j + e_s \right] \geq 0; \tag{4.8}$$

this also holds for $r = s$ because $\Sigma \gg 0$. Theorem 2 further implies that (4.8) is also sufficient (for μ to exist) provided $m \leq 4$. In particular, $\Sigma \gg 0$ together with (4.8) must imply that $\xi = (\xi_1, \ldots, \xi_d) \in T$, at least when $m \leq 4$.

The following Theorem 3 summarizes some of the above results and contains Theorem 1 as a special case. Here, the ξ_i and $\sigma_{ij} = \sigma_{ji}$ $(i, j = 1, \ldots, d)$ are given numbers.

THEOREM 3 *Let T be the form* (4.1). *In order that there exist random variables $Y_1, \ldots, Y_d$ satisfying* $\Pr((Y_1, \ldots, Y_d) \in T) = 1$; $EY_i = \xi_i$ *and* $\text{Cov}(Y_i, Y_j) = \sigma_{ij}$ $(i, j = 1, \ldots, d)$ *it is necessary that* $\Sigma = (\sigma_{ij})$ *be nonnegative definite and satisfies* (4.8). *If $m \leq 4$ then these necessary conditions are also sufficient.*

COUNTER EXAMPLE The following example shows that the above (rather obvious) necessary conditions on the ξ_j and c_{ij} are not sufficient anymore with $m \geq 5$. Choose $c_0 = 1$, $d = 4$ and T as the four-dimensional simplex $T = T(4)$, as in (2.13), consisting of all $y = (y_1, \ldots y_4) \in \mathbb{R}^4$ satisfying $y_i \geq 0$ $(i = 1, \ldots, 5)$. Here and below,

$$y_5 = 1 - (y_1 + y_2 + y_3 + y_4).$$

Thus, T is of the form (4.1) with $m = 5$. Namely, $b_{ij} = \delta_i^j$ if $i = 1, \ldots, 4$; $b_{5,j} = -1$ and $e_j = \delta_j^5$, $(j = 1, \ldots, 4)$. The special system (4.7) (of functions nonnegative on T) now consists of the 10 functions $y_r y_s$ $(1 \leq r, s \leq 5; r < s)$. Hence, the necessary conditions (4.8) take the form

$$(4.9)\; c_{rs} \geq 0; \quad \bar{c}_{r5} \geq 0 \quad (1 \leq r < s \leq 4), \quad \text{where } \bar{c}_{r5} = \bar{c}_{5r} = \xi_r - \sum_{j=1}^{4} c_{rj}.$$

Note that $\bar{c}_{r5}$ corresponds to the integral of $y_r y_5$. Let us choose

$$(4.10) \qquad c_0 = 1; \quad \xi_i = \frac{1}{5} \quad \text{and} \quad c_{ij} = d_{j-i} \quad (i, j = 1, \ldots, 4),$$

with $d_j = d_{-j} = d_{j-5}$. Put $d_0 = \alpha$; $d_{\pm 1} = \beta$; $d_{\pm 2} = \gamma$. For instance, $c_{14} = d_3 = d_{-2} = d_2 = \gamma$. We will further assume that

$$(4.11) \qquad \alpha > \beta > \gamma > 0 \quad \text{and} \quad \alpha + 2\beta + 2\gamma = \frac{1}{5},$$

(thus $\gamma < 1/25 < \alpha$). This implies that $\bar{c}_{r5} = d_{5-r} > 0$ $(1 \leq r \leq 4)$, such as $\bar{c}_{15} = d_4 = d_{-1} = \beta$ and $\bar{c}_{45} = \beta$. The necessary conditions (4.9) are now automatically satisfied. We will further assume that the 4×4 matrix $\Sigma = (\sigma_{ij})$ is strictly positive definite. It is easily seen that this is true if and only if

$$(4.12) \qquad \alpha(\alpha - \beta - \gamma) > \beta^2 - 3\beta\gamma + \gamma^2 \qquad \alpha(\alpha + \gamma) > 2\beta^2.$$

The necessary conditions of Theorem 3 are now all satisfied. Recall that these conditions essentially derive, as in (3.2), from the different members $P \in N+S \subset \mathcal{P}_5$. However, there is no guarantee that the moment problem on hand does have a solution, precisely because $\mathcal{P}_5$ happens to be strictly larger than $N+S$. One must also insist that the ξ_j and c_{ij} satisfy all the moment conditions associated, as in (3.2), to the different members $H \in \mathcal{P}_5/(N+S)$. One such matrix H is defined by the so-called Horn form

$$y^t H y = (y_1+y_2+y_3+y_4+y_5)^2 \qquad (4.13)$$
$$-4(y_1y_2+y_2y_3+y_3y_4+y_4y_5+y_5y_1),$$

(already mentioned at the end of Section 2). Since

$$\begin{aligned} y^tHy &= (y_1-y_2+y_3+y_4-y_5)^2+4y_2y_4+4y_3(y_5-y_4) \\ &= (y_1-y_2+y_3-y_4+y_5)^2+4y_2y_5+4y_1(y_4-y_5), \end{aligned}$$

from Hall (1986, p. 357), one has $y^tHy \geq 0$ for all $y \in \mathbb{R}^5_+$ thus $H \in \mathcal{P}_5$. Integrating y^tHy relative to a probability measure μ on $T(4)$ satisfying (4.2) (with $c_0 = 1$), this special matrix H leads to the new necessary condition

$$c_{12}+c_{23}+c_{34}+\bar{c}_{45}+\bar{c}_{51} \leq \frac{1}{4}; \qquad (4.14)$$

(permuting indices leads to a set of $120/5 = 24$ different necessary conditions of type (4.14)). In the present example, (4.14) is equivalent to $5\beta \leq 1/4$. Thus we have the desired counter example as soon as $\beta > 1/20$. In fact, choose $0 < \delta < \delta_0 = 2/(11+\sqrt{125}) = .09017$. Then the parameters

$$\gamma = 1/(80+10\delta); \qquad \alpha = 6\gamma; \qquad \beta = (4+\delta)\gamma \qquad (4.15)$$

do satisfy $\alpha > \beta > \gamma > 0$, further (4.11), (4.12) as well as $\beta > 1/20$.

Remark By the way, substituting (4.15), the first inequality (4.12) is equivalent to $1 - 11\delta - \delta^2 > 0$ and becomes an equality when $\delta = \delta_0$, showing that Σ is singular in this case. In the limiting case $\delta = 0$, there actually does exist a probability measure μ on $T(4)$ that satisfies (4.2) with the c_{ij} and ξ_i as above. Since the c_{ij} then satisfy (4.14) with the equality sign, that measure μ must be carried by $Z(H) = \{y \in T(4) : y^tHy = 0\}$. In fact, μ assigns mass 1/5 to each of five points $y^{(r)} \in Z(H)$ $(1 \leq r \leq 5)$. Here, $y = y^{(r)}$ has coordinates $y_r = 1/2$; $y_{r\pm1} = 1/4$; $y_{r\pm2} = 0$, where the indices are to be interpreted modulo 5.

Recall that $N+S$ is a proper subset of $\mathcal{P}_5$ since $H \in \mathcal{P}_5/(N+S)$, where H is defined by (4.13). The above construction implies that $N+S$ is not even dense in $\mathcal{P}_5$. Namely, consider the 5×5 matrix $C = C(\delta) = (c_{ij})$, defined as in (4.10) and (4.15) with $0 < \delta < \delta_0$ and $c_0 = 1$. The above proof essentially shows that $C(\delta) \in (N+S)^*/\mathcal{P}_5^*$. Hence, $\mathcal{P}_5^*$ is a proper subset of $(N+S)^* = N \cap S$, equivalently, $N+S$ cannot be dense in $\mathcal{P}_5$.

REFERENCES

BERG, C. (1987). The multidimensional moment problem and semigroups. *Proc. Symp. Appl. Math.* **37** American Mathematical Society, Providence, RI. 110–124.

CASSIER, G. (1984). Problème des moments sur un compact de $\mathbb{R}^n$ et décomposition de polynõmes a plusieurs variables. *J. Functional Anal.* **58** 254–266.

DIANANDA, P. H. (1962). On nonnegative forms in real variables some of all of which are nonnegative. *Proc. Cambridge Philo. Soc.* **58** 17–25.

HALL, JR., M. (1986). *Combinatorial Theory.* Wiley, New York.

MARTIN, D. H. AND JACOBSON, D. H. (1981). Copositive matrices and definiteness of quadratic forms subject to homogeneous linear inequality constraints. *Linear Alg. Appl.* **35** 227–258.

MARTIN, D. H., POWELL, M. J. D. AND JACOBSON, D. H. (1981). On a decomposition of conditionally positive–semidefinite matrices. *Linear Alg. Appl.* **39** 51–59.

MASERICK, P. H. (1977). Moments of measures on convex bodies. *Pacific J. Math.* **68** 135–152.

MICCHELLI, A. M. AND PINKUS, A. (1989). Nonnegative polynomials on polyhedra. In *Probability, Statistics and Mathematics, Papers in Honor of Samuel Karlin.* T. W. Anderson, K. B. Athreya and D. L. Iglehart, eds. Academic Press, San Diego, CA. 163–186.

DEPARTMENT OF STATISTICS
RUTGERS UNIVERSITY
NEW BRUNSWICK, NJ 08903

DEPARTMENT OF MATHEMATICS AND STATISTICS
UNIVERSITY OF MASSACHUSETTS
AMHERST, MA 01002

Stochastic Inequalities
IMS Lecture Notes – Monograph Series
Volume 22 (1993)

HYPERBOLIC–CONCAVE FUNCTIONS AND HARDY–LITTLEWOOD MAXIMAL FUNCTIONS

By ROBERT P. KERTZ[1] and UWE RÖSLER

Georgia Institute of Technology and Georg–August Universität Göttingen

A class of generalized convex functions, the hyperbolic–concave functions, is defined, and used to characterize the collection of Hardy–Littlewood maximal functions. These maximal functions and the probability measures associated with these maximal functions, the maximal probability measures, are used in representations and inequalities within martingale theory. A related collection of minimal probability measures is also characterized, through a class of hyperbolic–concave envelopes.

1. Introduction

In this paper, a new class of functions, the collection of hyperbolic–concave functions, is introduced to give natural characterizations of the collections of Hardy–Littlewood maximal probability measures (p.m.'s) and a related collection of minimal p.m.'s. These collections of probability measures play an important part in martingale theory and other areas of probability theory.

The Hardy–Littlewood maximal p.m.'s can be described as follows. Let μ be any p.m. on $\mathbb{R}$ with distribution function $F_\mu = F$ and left continuous inverse F_μ^{-1}, satisfying $\int_0^\infty x d\mu(x) < \infty$. The Hardy–Littlewood maximal function associated with μ is the function $H^{-1} = H_\mu^{-1}$ defined by

$$H^{-1}(u) := (1-u)^{-1} \int_u^1 F^{-1}(t)dt.$$

As a random variable on [0,1], with Borel sets and Lebesgue measure, H^{-1} has an associated p.m. μ^*, called the Hardy–Littlewood maximal p.m. associated with μ. These maximal p.m.'s appear in many areas of probability theory (Blackwell and Dubins (1963), Dubins and Gilat (1978), Hardy and Littlewood (1930), Kertz and Rösler (1990)).

In martingale theory the maximal p.m.'s appear in the following characterizations, see Blackwell and Dubins (1963), Dubins and Gilat (1978),

[1]Research supported in part by National Science Foundation Grant DMS-88-01818.
AMS 1991 *subject classifications.* Primary 60G44, 28A33.
Key words and phrases. Hardy–Littlewood maximal function, generalized convex functions, hyperbolic function, martingales, stochastic order, convex order.

Kertz and Rösler (1990): for any p.m. μ on $\mathbb{R}$ with $\int |x| d\mu(x) < \infty$,

$$\mu^* = \sup_{\prec_s} \{\nu : \text{ there is a martingale } (X_t)_{0 \le t \le 1} \text{ satisfying } X_1 \overset{\mathcal{D}}{=} \mu \text{ and } \sup_{0 \le t \le 1} X_t \overset{\mathcal{D}}{=} \nu\} \tag{1.1}$$

and

$$\{\nu : \text{ there is a martingale } (X_t)_{0 \le t \le 1} \text{ satisfying } X_1 \overset{\mathcal{D}}{=} \mu \text{ and } \sup_{0 \le t \le 1} X_t \overset{\mathcal{D}}{=} \nu\} = \{\nu \text{ is a p.m. on } \mathbb{R} : \mu \prec_s \nu \prec_s \mu^*\}.$$

Here $\prec_s$ denotes the stochastic order on p.m.'s, and $Y \overset{\mathcal{D}}{=} \mu$ denotes that Y has associated p.m. μ. Two collections of p.m.'s important for the characterizations in (1.1) are

$$\mathcal{P}^* := \{\nu \text{ is a p.m. on } \mathbb{R} : \nu \prec_s \mu^* \text{ for some p.m. } \mu \text{ with } \int_0^\infty x d\mu(x) < \infty\} \tag{1.2}$$

and

$$\mathcal{P}_0^* := \{\nu \text{ is a p.m. on } \mathbb{R} : \nu = \mu^* \text{ for some p.m. } \mu \text{ with } \int_0^\infty x d\mu(x) < \infty\}$$

In Kertz and Rösler (1991a), it was shown that $\mathcal{P}^*$, the set of p.m.'s dominated by maximal p.m.'s in the stochastic order, equals the set of p.m.'s ν on $\mathbb{R}$ satisfying $\limsup_{x \to \infty} x\nu[x, \infty) = 0$ (see also Kertz and Rösler (1991b)). The collection $\mathcal{P}_0^*$ is the set of Hardy–Littlewood maximal p.m.'s.

In the main section of this paper, Section 4, connections between maximal p.m.'s and hyperbolic–concave functions are given. In Theorem 4.2, it is shown that $\mathcal{P}_0^*$ is isomorphic to the set of hyperbolic–concave functions $\mu^*[., \infty)$ associated with p.m.'s in $\mathcal{P}^*$. Theorem 4.3 shows that maximal p.m.'s can be expressed in terms of their 'hyperbolic derivatives,' as defined in (2.7).

A related collection of 'minimal' p.m.'s is described as follows. For each p.m. $\nu \in \mathcal{P}^*$ (i.e., each p.m. ν on $\mathbb{R}$ with $\limsup_{x \to \infty} x\nu[x, \infty) = 0$), the minimal p.m. ν_Δ associated with ν is the p.m. on $\mathbb{R}$ satisfying

$$\nu_\Delta := \inf_{\prec_c} \{\mu \text{ is a p.m. on } \mathbb{R} : \int_0^\infty x d\mu(x) < \infty \text{ and } \nu \prec_s \mu^*\}.$$

Here $\prec_c$ denotes the convex order on right–tail–integrable p.m.'s. The existence of minimal p.m.'s ν_Δ was proved in Theorem 2.4 of Kertz and

Rösler (1991a). Also the importance of these minimal p.m.'s in martingale theory was made explicit there through the following characterizations: for any p.m. ν on $\mathbb{R}$ satisfying $\limsup_{x\to\infty} x\nu[x,\infty) = 0$ with finite $x_0 := \inf\{z : \nu(-\infty, z] > 0\}$,

$$\nu_\Delta = \inf_{\prec_c}\{\mu : \text{ there is a martingale } (X_t)_{0\le t\le 1} \text{ satisfying } X_1 \stackrel{\mathcal{D}}{=} \mu \text{ and } \sup_{0\le t\le 1} X_t \stackrel{\mathcal{D}}{=} \nu\}$$

and

$$\{\mu : \text{ there is a martingale } (X_t)_{0\le t\le 1} \text{ satisfying } X_1 \stackrel{\mathcal{D}}{=} \mu \text{ and } \sup_{0\le t\le 1} X_t \stackrel{\mathcal{D}}{=} \nu\}$$
$$= \{\mu \text{ is a p.m. on } \mathbb{R} : \int x d\mu(x) = x_0 \text{ and } \nu_\Delta \prec_c \mu \prec_s \nu\}.$$

In Theorem 4.4 it is shown that the minimal p.m.'s are characterized by the hyperbolic–concave envelopes of p.m.'s in $\mathcal{P}^*$.

The concept of hyperbolic–concave functions is defined in Section 2. Through a natural connection to convex functions, given by the map $x \to 1/x$, properties of convex functions carry over to give desirable properties of hyperbolic–concave functions. Some of these properties are listed in Sections 2 and 3. In particular, hyperbolic–concave envelopes are defined, and their properties are identified, in Section 3. The results in Sections 2 and 3 are applied to give characterizations and identifications in the central Section 4.

2. Hyperbolic–Concave Functions

In this Section, the concept of hyperbolic–concave functions is defined (Definition 2.2); and properties of these functions are given (Lemmas 2.5 and 2.7 and Proposition 2.6). Proofs are facilitated through a key equivalence between hyperbolic–concave functions and convex functions, given in Theorem 2.3. Standard definitions and properties associated with convex functions are used throughout this paper; for reference see Roberts and Varberg (1970) and Rockafellar (1970). Within this paper, intervals in the real numbers $\mathbb{R}$ may or may not contain their endpoints, are nonempty, but may be a singleton.

For any real numbers a and b with $a < b$, $k(\cdot, a, b)$ denotes the hyperbolic function from $\mathbb{R}$ into $(0, \infty]$ given by

$$k(x; a, b) = (b - a)/(x - a) \text{ if } x > a, \text{ and } = +\infty \text{ if } x \le a.$$

Let $\mathcal{H}_0$ denote the collection of such functions. The following two properties of these hyperbolic functions are easily verified.

LEMMA 2.1 *(i) Let* (x_1, y_1) *and* (x_2, y_2) *be any pairs of real numbers satisfying* $x_1 < x_2$ *and* $y_1 > y_2 > 0$. *Then there exists exactly one pair of real numbers* a *and* b *with* $a < b$ *for which the function* $k(\cdot; a, b)$ *satisfies* $k(x_1; a, b) = y_1$ *and* $k(x_2; a, b) = y_2$. *The numbers* a *and* b *are given by* $a = (x_1y_1 - x_2y_2)/(y_1 - y_2)$ *and* $b = (x_1y_1 - x_2y_2 + y_1y_2(x_2 - x_1))/(y_1 - y_2)$.

(ii) Let $k_1(\cdot) = k(\cdot; a_1, b_1)$ *and* $k_2(\cdot) = k(\cdot; a_2, b_2)$ *be two different functions in* $\mathcal{H}_0$. *Then there is some number* $\bar{x}$ *for which either (a)* $k_1(x) < k_2(x)$ *if* $a_1 < x < \bar{x}$, *and* $k_2(x) < k_1(x)$ *if* $\bar{x} < x < \infty$; *or (b)* $k_2(x) < k_1(x)$ *if* $a_2 < x < \bar{x}$, *and* $k_1(x) < k_2(x)$ *if* $\bar{x} < x < \infty$.

Function g from $\mathbb{R}$ into $[0, \infty]$ is said to be *nondegenerate* if g takes on at least one value in $(0, \infty)$. Let $\mathcal{G}$ denote the collection of nondegenerate, nonincreasing functions from $\mathbb{R}$ into $[0, \infty]$. For each function g in $\mathcal{G}$, let $w_0 := \sup\{x : g(x) = \lim_{y \to -\infty} g(y)\}$ if this set $\neq \emptyset$, and $= -\infty$ otherwise; and $x_0 := \inf\{x : g(x) = 0\}$ if this set $\neq \emptyset$, and $= +\infty$ otherwise. Observe that $w_0 \leq x_0$ if g is not identically constant; if $-\infty < w_0$, then $g(x) = \lim_{y \to -\infty} g(y)$ for all $x \in (-\infty, w_0)$; and if $x_0 < \infty$, then $g(x) = 0$ for all $x \in (x_0, \infty)$. For each g in $\mathcal{G}$, the function $1/g$ is defined on the interval $I_g := \mathrm{Dom}(1/g) = \{x : 0 < g(x) < \infty\}$. Observe that I_g is nonempty. Also, let w_1 be the extended real number

(2.1) $$w_1 := \sup\{x : g(x) = \text{limit of } g(y) \text{ as } y \downarrow \inf(I_g) \text{ over } y \text{ in } I_g\}$$
$$\text{if this set is nonempty, and } = w_0 \text{ otherwise.}$$

To aid in understanding these definitions, consider the following function: $g(x) = +\infty$ if $x \in (-\infty, 0]$, $= 1$ if $x \in (0, 2]$, $= -(1/2)(x - 3)$ if $x \in (2, 3]$, and $= 0$ if $x \in (3, \infty)$. For this function $w_0 = 0$, $w_1 = 2$, and $x_0 = 3$.

DEFINITION 2.2 Let f be any function in $\mathcal{G}$. We say that f is a *hyperbolic-concave function* if for any two pairs $(x, f(x))$ and $(y, f(y))$ with $x < y$ and $0 < f(y) < f(x) < \infty$ and any associated function $k(\cdot) = k(\cdot; a, b)$ from $\mathcal{H}_0$ satisfying $k(x) = f(x)$ and $k(y) = f(y)$, it follows that $k(z) \leq f(z)$ for all z in $[x, y]$.

Let $\mathcal{H}$ denote the collection of hyperbolic-concave functions. The approach taken here in defining the hyperbolic-concave functions is analogous to the approach taken to define $\mathcal{F}$-convex functions (see Roberts and Varberg (1970, Section 84)). The following result is very useful in the analysis of hyperbolic-concave functions.

THEOREM 2.3 *Let* $f \in \mathcal{G}$. *Then* f *is a hyperbolic-concave function if and only if* $1/f$ *is a convex function.*

PROOF This equivalence follows in a straightforward way, upon observing the following. Let x_1 and x_2 be in $\mathbb{R}$ with $x_1 < x_2$ and $\infty > f(x_1) > f(x_2) >$

0, and let $k(x) = k(x;a,b)$ be in $\mathcal{H}_0$ with $k(x_i) = f(x_i)$ for $i = 1,2$. Then for all z in $[x_1, x_2]$,

$$k(z) = \left(\left(\frac{x_2 - z}{x_2 - x_1} \right) \frac{1}{f(x_1)} + \left(\frac{z - x_1}{x_2 - x_1} \right) \frac{1}{f(x_2)} \right)^{-1}$$

(from substituting into $k(z;a,b)$ the expressions for a and b given in Lemma 2.1(i)). □

To obtain another useful characterization of hyperbolic–concave functions as a corollary of Theorem 2.3, we introduce the following definition. We say that function g in $\mathcal{G}$ has a *hyperbola of support* at x in I_g if there is a function k in $\mathcal{H}_0$ for which $k(x) = g(x)$ and $g \leq k$.

COROLLARY 2.4 *Let $f \in \mathcal{G}$. Then f is a hyperbolic–concave function if and only if, for each x in (w_1, x_0), there is a hyperbola of support for f at x, and f is right continuous at w_1, if $[w_1, w_1 + \epsilon) \subseteq I_f$ for some $\epsilon > 0$.*

PROOF To prove this equivalence, observe only that

f has hyperbola of support at x, for all x in (w_1, x_0)

$\Leftrightarrow$ for each $x \in (w_1, x_0)$ there is a function $\ell(y) = ((y-a)/(b-a))_+$, for some $a < b$, with $\ell(x) = 1/f(x)$ and $\ell \leq 1/f$ on I_f

$\Leftrightarrow$ for each $x \in (w_1, x_0)$, $1/f$ has a line of support at x. □

Using Theorem 2.3, one sees that the following are hyperbolic–concave functions: functions in $\mathcal{H}_0$; the functions $f(x) = x^{-\alpha}$ if $x > 0$, and $= \infty$ if $x \leq 0$, for $\alpha > 1$; the functions e^{-x} and $(1 + e^x)^{-1}$; the constant functions $f(x) \equiv a$, for $a > 0$; the function $f(x) = +\infty$ if $x < c$, $= a \in (0, \infty)$ if $x = c$, and $= 0$ if $x > c$; and the function $f(x) = 1$ if $x \leq 0$, $= 1 - x$ if $0 \leq x \leq 1$, and $= 0$ if $x \geq 0$. For comparison, some functions in $\mathcal{G} \backslash \mathcal{H}$ are the following: $g(x) = x^{-\alpha}$ if $x > 0$, and $= \infty$ if $x \leq 0$, for $0 < \alpha < 1$; the function $g(x) = (\log x)^{-1}$ if $x > 1$, and $= \infty$ if $x \leq 1$; and the function g given immediately after Expression (2.1). The following closure properties also follow from Theorem 2.3:

(2.2) (i) if $f \in \mathcal{H}$ and $0 < a < \infty$, then $af \in \mathcal{H}$;

(ii) for any index set Γ, if $f_\gamma \in \mathcal{H}$ for each $\gamma \in \Gamma$ and if $f(x) := \inf_{\gamma \in \Gamma} f_\gamma(x)$ is nondegenerate, then $f \in \mathcal{H}$.

The following lemma contains properties of hyperbolic–concave functions.

LEMMA 2.5 *Let $f \in \mathcal{H}$.*

(i) *The function f is continuous on the interior of I_f. If w_0 is the left endpoint of I_f, $w_0 \in I_f$, and $-\infty < w_0 < x_0 \leq \infty$, then f is right continuous at w_0.*

(ii) *For the number w_1 of (2.1), $f(x)$ is identically constant if $x < w_1$, $x \in I_f$, and $f(x)$ is strictly decreasing if $w_1 < x$, $x \in I_f$.*

(iii) *For $w_1 \leq x < y \leq x_0$, $(yf(y) - xf(x))/(f(y) - f(x))$ is nondecreasing in x and y; for $w_1 \leq x < x' < y < y' \leq x_0$,*

$$\frac{yf(y) - xf(x)}{f(y) - f(x)} \leq \frac{y'f(y') - x'f(x')}{f(y') - f(x')}.$$

PROOF The conclusions are straightforward from Theorem 2.3 and properties of convex functions. □

For each nonconstant function g in $\mathcal{G}$, define sets $\mathcal{S}_- = \mathcal{S}_-(g)$ and $\mathcal{S}_+ = \mathcal{S}_+(g)$ by

$$\text{(2.3)} \qquad \begin{aligned} \mathcal{S}_- &:= \{x : x > w_0 \text{ and } g(x) < g(y) \text{ for all } y < x\} \\ \mathcal{S}_+ &:= \{x : g(y) < g(x) < \infty \text{ for all } y > x\} \end{aligned}$$

and define functions $\Lambda^- g$ and $\Lambda^+ g$ on $\mathcal{S}_-$ and $\mathcal{S}_+$ respectively by

$$\text{(2.4)} \qquad \begin{aligned} \Lambda^- g(x) &:= \sup_{w \in (w_0, x)} (wg(w) - xg(x))/(g(w) - g(x)), \text{ and} \\ \Lambda^+ g(x) &:= \inf_{y \in (x, \infty)} (yg(y) - xg(x))/(g(y) - g(x)). \end{aligned}$$

Observe that sets $\mathcal{S}_-$ and $\mathcal{S}_+$ are contained in $[w_1, x_0]$. From Lemma 2.5(iii), it follows that if $f \in \mathcal{H}$, then $(w_1, x_0) \subseteq \mathcal{S}_- \cap \mathcal{S}_+$. Let $D^- g(x)$ and $D^+ g(x)$ denote respectively the left–hand derivative and the right–hand derivative of function g at x, and let $Dg(x)$ denote the derivative of g at x.

PROPOSITION 2.6 *Let f be any nonconstant function in $\mathcal{H}$.*

(i) *Functions $\Lambda^- f$ and $\Lambda^+ f$ have representations*

$$\text{(2.5)} \qquad \begin{aligned} \Lambda^- f(x) &= \lim_{y \to x-} (yf(y) - xf(x))/(f(y) - f(x)) \text{ and} \\ \Lambda^+ f(x) &= \lim_{y \to x+} (yf(y) - xf(x))/(f(y) - f(x)). \end{aligned}$$

(ii) $\Lambda^- f(x) \leq \Lambda^+ f(x)$.

(iii) *Functions $\Lambda^- f$ and $\Lambda^+ f$ are finite–valued and nondecreasing.*

(iv) *$\Lambda^+ f$ is right continuous on $\mathcal{S}_+ \backslash \{x_0\}$. $\Lambda^- f$ is left continuous on $\mathcal{S}_- \backslash \{x_0\}$. If f is left continuous at x_0, then so is $\Lambda^- f$.*

(v) $\Lambda^- f$ *and* $\Lambda^+ f$ *have representations*

$$(2.6)\quad \Lambda^- f(x) = x - \{f(x)D^-(1/f)(x)\}^{-1} = x + \{f(x)/D^- f(x)\};$$

and

$$\Lambda^+ f(x) = x - \{f(x)D^+(1/f)(x)\}^{-1} = x + \{f(x)/D^+ f(x)\}$$

if $x \in \mathcal{S}_+ \backslash \{x_0\}$, *and* $\Lambda^+ f(x_0) = x_0$ *if* $x_0 \in \mathcal{S}_+$.

(vi) $\Lambda^+ f(x) \le \Lambda^- f(y)$ *if* $x < y$; *for* $x \in (w_1, x_0)$, $\Lambda^- f(x+) = \Lambda^+ f(x)$ *and* $\Lambda^- f(x) = \Lambda^+ f(x-)$.

PROOF Conclusions (i), (ii) and (iii) follow directly from Lemma 2.5(iii). For conclusion (iv), use also Lemma 2.5(i); and for conclusion (v), use (2.5). For conclusion (vi), again use Lemma 2.5(iii), and use part (iv). □

For any nonconstant function f in $\mathcal{H}$, define the *hyperbolic derivative* Λf for x in $\mathcal{S}_- \cap \mathcal{S}_+$ by

$$(2.7)\quad \Lambda f(x) = \lim_{y \to x} (yf(y) - xf(x))/(f(y) - f(x)), \text{ if this limit exists.}$$

LEMMA 2.7 *Let* f *be any nonconstant function in* $\mathcal{H}$.

(i) *Within* $\mathcal{S}_- \cap \mathcal{S}_+$, *the set of discontinuity points of* $\Lambda^+ f$ *and of* $\Lambda^- f$ *coincide, and equals the set of discontinuity points within* $\mathcal{S}_- \cap \mathcal{S}_+$ *of* $D^+ f$ *and of* $D^- f$. *This set, denoted by* $\mathcal{D}$, *is countable.*

(ii) *Within* $\mathcal{S}_- \cap \mathcal{S}_+$,

$$\begin{aligned} x &\in (\mathcal{S}_- \cap \mathcal{S}_+) \backslash \mathcal{D} \\ &\Leftrightarrow \Lambda^- f(x) = \Lambda^+ f(x), \Lambda f(x) \text{ exists and equals this common value} \\ &\Leftrightarrow \text{there is a unique hyperbola of support for } f \text{ at } x. \end{aligned}$$

In this case,

$$\Lambda f(x) = x + \{f(x)/Df(x)\} = x - \{f(x)D(1/f)(x)\}^{-1};$$

and the unique hyperbola of support at x *is given by*

$$k(y; a, b) = f(x)(x - \Lambda f(x))/(y - \Lambda f(x)) \text{ if } y > \Lambda f(x), \text{ and } = +\infty \text{ otherwise}$$

(iii) *For all* x, y *with* $w_1 < x < y < x_0$,

$$(2.8)\quad f(y)/f(x) = \exp \int_x^y \{\Lambda^+ f(t) - t\}^{-1} dt,$$

and f *can be identified from its hyperbolic derivative.*

PROOF To obtain (2.8), use Theorem 2.3, Lemma 2.6(v) and standard results on convex functions found, e.g., in Freedman (1971, pp. 359–363). To verify the other conclusions, use Theorem 2.3 and Corollary 2.4, the representations in (2.6), and standard results for convex functions. □

3. Hyperbolic–Concave Envelopes

In this Section, hyperbolic–concave envelopes are defined (Definition 3.1), and properties of hyperbolic–concave envelopes are given. These hyperbolic–concave envelopes have a direct connection to convex envelopes of functions, given in (3.5). For this comparison, we recall the definition of convex envelopes; and for a class of functions of interest here, we recall some properties of these convex envelopes (for references to these results, see Rockafellar (1970, pp. 36, 51, 103, 157) and Roberts and Varberg (1970, p. 21). For any function f from an interval I to $\mathbb{R}$ which majorizes at least one affine function on I, the *convex envelope of* f, written env f, is the function defined on I by

(3.1) $$(\text{env } f)(x) = \sup\{A(x) : A \text{ is an affine function, } A \leq f\}.$$

The basic class of functions of interest in this paper is the class $\mathcal{G}^1$, the subset of $\mathcal{G}$ given by

(3.2) $$\mathcal{G}^1 := \{g(\cdot) : g \text{ is a left–continuous, nonincreasing function from } \mathbb{R} \text{ into } [0,1] \text{ with } \lim_{x\to-\infty} g(x) = 1 \text{ and } \limsup_{x\to\infty} xg(x) = 0\}.$$

Observe that if $g \in \mathcal{G}^1$, then g is nondegenerate and $\lim_{x\to\infty} g(x) = 0$; and if w_0 and x_0 are the numbers associated with g in Section 2, then $w_0 = \sup\{x : g(x) = 1\}$ if this set $\neq \emptyset$, and $= -\infty$ otherwise. For any function g in $\mathcal{G}^1$, the function $\text{env}(1/g)$, defined on $I_g := \text{Dom}(1/g)$ through (3.1), is a closed function. Thus it follows from the defining properties of $\mathcal{G}^1$ and properties of convex envelopes that for $g \in \mathcal{G}^1$,

(3.3) (i) $\text{env}(1/g)$ is the greatest convex function which is majorized by $1/g$ on I_g;

(ii) $\text{env}(1/g)$ is continuous on I_g;

(iii) $(\text{env}(1/g))(x) = \inf\{\mu : (x,\mu) \text{ is in the convex hull of the epigraph of } 1/g\} = \inf\{\lambda(1/g)(x_1) + (1-\lambda)(1/g)(x_2) : \lambda x_1 + (1-\lambda)x_2 = 1 \text{ for some } 0 \leq \lambda \leq 1,\ x_1 \leq x \leq x_2\}$; and

(iv) if $1/g$ is convex in I_g, then $1/g = \text{env}(1/g)$.

Let $\mathcal{H}^1 := \mathcal{G}^1 \cap \mathcal{H}$, the collection of hyperbolic–concave functions in $\mathcal{G}^1$. Numbers w_0, w_1 and x_0 associated with each function f in $\mathcal{H}^1$ satisfy $-\infty \le w_0 = w_1 \le x_0 \le +\infty$; it is the case that $w_0 = x_0$ if and only if $f(x) = 1$ if $x \le c$, and $= 0$ if $x > c$, for some $c \in \mathbb{R}$. If $f \in \mathcal{H}^1$ and $x_0 \in \mathbb{R}$, then $I_f = \mathrm{Dom}(1/f) = (-\infty, x_0)$ if $f(x_0) = 0$, and $= (-\infty, x_0]$ if $f(x_0) > 0$; $\mathcal{S}_- = (w_0, x_0]$, and if also $w_0 \in \mathbb{R}$, then $\mathcal{S}_+ = [w_0, x_0)$ if $f(x_0) = 0$, and $= [w_0, x_0]$ if $f(x_0) > 0$.

DEFINITION 3.1 For each g in $\mathcal{G}^1$, the *hyperbolic–concave envelope* of g is the function $\widehat{g}$ from $\mathbb{R}$ into $[0,1]$ defined by

$$\widehat{g}(x) := \inf\{k(x) : k \in \mathcal{H}_0, g \le k\}. \tag{3.4}$$

(Observe that the set in (3.4) is nonempty.) The collection of hyperbolic–concave envelopes of functions in $\mathcal{G}^1$ is denoted by $\widehat{\mathcal{G}^1}$. We show that $\widehat{\mathcal{G}^1} = \mathcal{H}_1$ in Theorem 3.4. Connections between hyperbolic–concave envelopes and convex envelopes, together with some other properties of hyperbolic–concave envelopes are given in the following.

LEMMA 3.2 *Let $g \in \mathcal{G}^1$ and let $I_g = \mathrm{Dom}(1/g)$. The hyperbolic–concave envelope $\widehat{g}$ has the following properties:*
(i) $\widehat{g}$ takes values in $[0,1]$; $\widehat{g}$ is nonincreasing; $\lim_{x\to-\infty}\widehat{g}(x) = 1$ and $\limsup_{x\to\infty} x\widehat{g}(x) = 0$; $w_0 := w_0(g) = w_0(\widehat{g})$, $x_0 := x_0(g) = x_0(\widehat{g})$, and $I_g = \mathrm{Dom}(1/g) = \mathrm{Dom}(1/\widehat{g}) = I_{\widehat{g}}$ with $\widehat{g}(x) = 1$ for all x in $(-\infty, w_0]$ and $\widehat{g}(x) = 0$ for all x in (x_0, ∞); and $\widehat{g}$ is continuous on $\mathbb{R}\backslash\{x_0\}$ and is left continuous at x_0;
(ii) $g \le \widehat{g}$; if $g(x) = 1$ for $x \le x_0$, and $= 0$ for $x > x_0$, for some $x_0 \in \mathbb{R}$, then $g = \widehat{g}$;
(iii) on I_g,

$$\mathrm{env}(1/g) = 1/\widehat{g}; \tag{3.5}$$

and
(iv) For each $x \in I_g$, one and only one of the following hold:

(a) $g(x) = \widehat{g}(x)$ and $1/g(x) = \mathrm{env}(1/g)(x)$; and for some $a < b$, $k(\cdot) = k(\cdot; a, b)$ is in $\mathcal{H}_0$ with $g \le k$ and $g(x) = k(x)$, and $\ell(y) = ((y-a)/(b-a))_+$ satisfies $\ell \le 1/g$ and $\ell(x) = 1/g(x)$; or

(b) $g(x) < \widehat{g}(x)$ and $1/g(x) > \mathrm{env}(1/g)(x)$; and for some $a < b$, and x_1, x_2, x_3, x_4 with $x_1 \le x_2 < x < x_3 \le x_4$, $k(\cdot) = k(\cdot; a, b)$ in $\mathcal{H}_0$ satisfies $\widehat{g} \le k$, $g(x_i) = \widehat{g}(x_i) = k(x_i)$ for $i = 1, \ldots, 4$, $\widehat{g}(y) = k(y)$ for $y \in [x_1, x_4]$, $\widehat{g}(y) < k(y)$ for $y \in (I_g\backslash[x_1, x_4]) \cup (x_0, \infty)$, and $g(y) < k(y)$ for $y \in (x_2, x_3)$; and $\ell(y) = ((y-a)/(b-a))_+$ satisfies $\ell \le \mathrm{env}(1/g)$, $1/g(x_i) = \mathrm{env}(1/g)(x_i) = \ell(x_i)$ for $i = 1, \ldots, 4$, $\mathrm{env}(1/g)(y) = \ell(y)$ for $y \in [x_1, x_4]$, $\mathrm{env}(1/g)(y) > \ell(y)$ for $y \in (I_g\backslash[x_1, x_4]) \cup (x_0, \infty)$, and $1/g(y) > \ell(y)$ for $y \in (x_2, x_3)$.

PROOF From the definition of $\widehat{g}$, it is immediate that $g \leq \widehat{g}$ and $\widehat{g}$ is nonincreasing; thus, also $0 \leq \widehat{g}$. The limit property $\limsup_{x\to\infty} x\widehat{g}(x) = 0$ follows easily from the analogous property of g.

We show $\widehat{g} \leq 1$. For any $0 < \epsilon < 1$, there is an $N > 1$ sufficiently large such that, for all $x \geq N$, $g(x) \leq \epsilon x^{-1} = k(x; 0, \epsilon)$. Let $\bar{x} < N$, and let δ be chosen sufficiently small so that $0 < \delta < 1$ and $\bar{x} + (1 + 2\delta)\delta^{-1} > N$. Let $k(\cdot) = k(\cdot; a, b)$ be the function in $\mathcal{H}_0$ passing through $(x_1, y_1) = (\bar{x} - \delta^{-1}, 1 + 2\delta)$ and $(x_2, y_2) = (\bar{x}, 1 + \delta)$. Then $g(x) \leq k(x)$ for all x (since if $x < N < \bar{x} + (1 + 2\delta)\delta^{-1}$, $g(x) \leq 1 \leq k(x)$; and if $x > N$, $g(x) \leq \epsilon x^{-1} < k(x)$ by Lemma 2.1(ii)); and thus $\widehat{g}(x) \leq k(x)$ for all x. Also, for $x \geq \bar{x}$, $\widehat{g}(x) \leq k(\bar{x}) = 1 + \delta$; since this can be done for each $0 < \delta < 1$ small, it follows that $\widehat{g} \leq 1$. It is immediate that $\widehat{g}(x) = 1$ for all x in $(-\infty, w_0]$, where $w_0 := w_0(g)$. From the definition of $\widehat{g}$ and appropriate choice of k's in $\mathcal{H}_0$ majorizing g, it similarly follows that $\widehat{g}(x) = 0$ for all x in (x_0, ∞), where $x_0 := x_0(g)$; and that if $g(x) = 1$ for $x \leq x_0$, and $= 0$ for $x > x_0$, for some $x_0 \in \mathbb{R}$, then $g = \widehat{g}$.

Next, to obtain (3.5) observe that for all $x \in I_g := \mathrm{Dom}(1/g)$,

$$
\begin{aligned}
(3.6)\quad \mathrm{env}(1/g)(x) &= \sup\{A(x) : A \text{ is an affine function, } A \leq 1/g\} \\
&= \sup\{A(x) : A(x) = ((x-a)/(b-a))_+ \\
&\qquad\qquad \text{for some and } a < b, A \leq 1/g\} \\
&= \{\inf\{k(x) : k \in \mathcal{H}_0, g \leq k\}\}^{-1} \\
&= 1/\widehat{g}(x).
\end{aligned}
$$

As immediate consequences of (3.3), (3.5), and the properties of g, one obtains that $w_0(g) = w_0(\widehat{g})$, $x_0(g) = x_0(\widehat{g})$, and $\mathrm{Dom}(1/g) = \mathrm{Dom}(1/\widehat{g})$; and that $\widehat{g}$ is continuous on $\mathbb{R}\backslash\{x_0\}$ and left continuous at x_0. By exploiting the correspondence between functions $k(\cdot; a, b)$ in $\mathcal{H}_0$ with $g \leq k$ and functions $\ell(y) = ((y-a)/(b-a))_+$ with $\ell \leq 1/g$, as in (3.6), and by using the nonnegativity and left continuity of g and the property $\limsup_{x\to\infty} xg(x) = 0$, one obtains that Lemma 3.2(iv) holds. □

PROPOSITION 3.3 *Let $g \in \mathcal{G}^1$. The hyperbolic–concave envelope $\widehat{g}$ is in $\mathcal{H}^1$ and is the smallest hyperbolic–concave function which majorizes g. The function $\widehat{g}$ has representation*

$$
\begin{aligned}
(3.7)\qquad \widehat{g}(x) = \sup\{k(x) : &\ k \text{ is in } \mathcal{H}_0 \text{ and passes through } (x_1, g(x_1)) \\
&\text{and } (x_2, g(x_2)) \text{ for some } x_1 < x < x_2\}.
\end{aligned}
$$

If g is a hyperbolic–concave function, then $g = \widehat{g}$.

PROOF Use Lemma 3.2(i) to obtain that $\widehat{g}$ is in $\mathcal{G}^1$. The results that $\widehat{g}$ is hyperbolic–concave, and is the smallest hyperbolic–concave function which majorizes g, follow from $g \le \widehat{g}$ and (3.5), (3.3)(i), and Theorem 2.3. The remaining conclusions follow from (3.5) and (3.3)(iii),(iv). □

THEOREM 3.4 *The collection of hyperbolic–concave functions in $\mathcal{G}^1$ and the collection of hyperbolic–concave envelopes of functions in $\mathcal{G}^1$ are equal; that is, $\widehat{\mathcal{G}^1} = \mathcal{H}^1$.*

PROOF If $\widehat{g} \in \widehat{\mathcal{G}^1}$ for some $g \in \mathcal{G}^1$, then $\widehat{g} \in \mathcal{H}^1$ from Proposition 3.3. If $h \in \mathcal{H}^1$, then $\widehat{h} = h$, from Proposition 3.3, and thus $h \in \widehat{\mathcal{G}^1}$. □

4. Characterizations of the Sets $\mathcal{P}^*$, $\mathcal{P}_0^*$, and the Set of Minimal p.m.'s

From the Introduction, recall the definitions of the collections of p.m.'s $\mathcal{P}^*$, the set of p.m.'s dominated by maximal p.m.'s (in the $\prec_s$ order); and $\mathcal{P}_0^*$, the set of Hardy–Littlewood maximal p.m.'s. Also recall, from Section 3, the collections of functions $\mathcal{G}^1$ of (3.2); and $\mathcal{H}^1$, the set of hyperbolic–concave functions in $\mathcal{G}^1$, which equals the set $\widehat{\mathcal{G}_1}$ of hyperbolic–concave envelopes of functions in $\mathcal{G}^1$ by Theorem 3.4. In this Section, the sets $\mathcal{P}^*$ and $\mathcal{P}_0^*$ are shown to be isomorphic to the sets $\mathcal{G}^1$ and $\mathcal{H}^1$. Moreover, an explicit identification between minimal p.m.'s ν_Δ, associated with p.m.'s ν in $\mathcal{P}^*$, and hyperbolic-concave envelopes $\widehat{f}$, associated with functions f in $\mathcal{G}^1$, is given in Theorem 4.4.

LEMMA 4.1 *There is an isomorphism between $\mathcal{G}^1$ and $\mathcal{P}^*$ identified by $g(x) = \nu[x,\infty)$ for $g \in \mathcal{G}^1, \nu \in \mathcal{P}^*$.*

PROOF As stated in the Introduction, Proposition 2.1 of Kertz and Rösler (1991a) gives that $\mathcal{P}^* = \{\nu \text{ is a p.m. on } \mathbb{R} : \limsup_{x\to\infty} x\nu[x,\infty) = 0\}$. From this representation and the usual identification of p.m.'s on $\mathbb{R}$ and distribution functions (see e.g., Section 10 of Loève (1963)), the conclusion is immediate. □

THEOREM 4.2 *There is an isomorphism between $\mathcal{H}^1$, the set of hyperbolic–concave functions in $\mathcal{G}^1$, and $\mathcal{P}_0^*$, the set of maximal p.m.'s on $\mathbb{R}$, identified by $f(x) = \mu^*[x,\infty)$ for $f \in \mathcal{H}^1, \mu^* \in \mathcal{P}_0^*$.*

PROOF In Lemma 2.6 of Kertz and Rösler (1991a), it was shown that for any p.m. ν on $\mathbb{R}$, $\nu \in \mathcal{P}_0^*$ if and only if the following holds

(4.1) (i) $\limsup_{w\nearrow 1}(1-w)F_\nu^{-1}(w) = 0$, and

(ii) $(1-w)F_\nu^{-1}(w)$ is a concave function.

Now, assume $\nu = \mu^* \in \mathcal{P}_0^*$; so, also F_ν^{-1} satisfies (4.1). First, observe that if $x_0 = x_0(\nu) < \infty$, then $\limsup_{x\to\infty} x\nu[x,\infty) = 0$; and if $x_0 = +\infty$, then

$$\begin{aligned}\limsup_{x\to\infty} x\nu[x,\infty) &= \limsup_{x\to\infty} F_\nu^{-1}(F_\nu(x))(1-F_\nu(x)) \\ &= \limsup_{w\nearrow 1}(1-w)F_\nu^{-1}(w) = 0.\end{aligned}$$

Second, observe that

$$(4.2)\qquad (1-w)F_\nu^{-1}(w) \text{ is concave iff } 1/(1-F_\nu(x)) \text{ is convex,}$$

for example, from a calculation based on the definitions of convexity and concavity for functions; and so $f(x) := \nu[x,\infty)$ is a hyperbolic–concave function, from Theorem 2.3. Thus f is a hyperbolic–concave function in $\mathcal{G}^1$, from Lemma 4.1, i.e., $f \in \mathcal{H}^1$.

On the other hand, assume $f \in \mathcal{H}^1$, i.e., f is a hyperbolic–concave function in $\mathcal{G}^1$. Let $\nu[x,\infty) := f(x)$; then ν is a p.m. in $\mathcal{P}^*$, from Lemma 4.1. From Lemma 2.5, f is continuous on $(-\infty, x_0)$ and strictly decreasing on $[w_0, x_0]$. It follows that $\limsup_{w\nearrow 1}(1-w)F_\nu^{-1}(w) = 0$ if $w_1 < 1$; and if $w_1 = 1$,

$$\begin{aligned}\limsup_{w\nearrow 1}(1-w)F_\nu^{-1}(w) &= \limsup_{x\to\infty}(1-F_\nu(x))F_\nu^{-1}(F_\nu(x)) \\ &= \limsup_{x\to\infty} x\nu[x,\infty) = 0.\end{aligned}$$

From the hyperbolic–concavity of $\nu[x,\infty)$, Theorem 2.3, and (4.2), it follows that $(1-w)F_\nu^{-1}(w)$ is concave. Thus F_ν^{-1} satisfies (4.1), and $\nu \in \mathcal{P}_0^*$. □

The following theorem shows that any Hardy–Littlewood maximal p.m. can be identified through its hyperbolic derivative.

THEOREM 4.3 *Let μ be any p.m. on $\mathbb{R}$ with $\int_0^\infty x\,d\mu(x) < \infty$, with associated Hardy–Littlewood maximal p.m. μ^*; and let f denote the function in $\mathcal{H}^1$ defined by $f(x) = \mu^*[x,\infty)$ for all $x \in \mathbb{R}$. Then μ^* can be identified through its hyperbolic derivative Λf by*

$$(4.3)\qquad \mu^*[x,\infty) = \exp\left(\int_{w_0}^{x}(\Lambda f(t) - t)^{-1}dt\right) \text{ for all } x \in [w_0, x_0].$$

PROOF Let $w_0 = w_0(\mu)$ and $x_0 = x_0(\mu)$. The representation (4.3) follows immediately from Lemma 2.7, since the function f is in $\mathcal{H}^1$. □

THEOREM 4.4 *(i) For each $\nu \in \mathcal{P}^*$, the minimal p.m. ν_Δ satisfies $(\nu_\Delta)^*[x,\infty) = \widehat{g}(x)$ for all $x \in \mathbb{R}$, where g is the function in $\mathcal{G}^1$ defined by $g(x) := \nu[x,\infty)$ for all $x \in \mathbb{R}$.*

(ii) For each $g \in \mathcal{G}^1$, the hyperbolic-concave envelope $\widehat{g}$ satisfies $\widehat{g}(x) = (\nu_\Delta)^*[x,\infty)$ for all $x \in \mathbb{R}$, where ν is the p.m. in $\mathcal{P}^*$ defined by $\nu[x,\infty) := g(x)$ for all $x \in \mathbb{R}$.

Thus, the following diagram commutes:

$$\begin{array}{ccc} \nu \text{ in } \mathcal{P}^* & \longrightarrow & g \text{ in } \mathcal{G}^1 \\ \downarrow & & \downarrow \\ \nu_\Delta & \longrightarrow & (\nu_\Delta)^*[\cdot,\infty) = \widehat{g}(\cdot) \end{array}$$

PROOF Let $\nu \in \mathcal{P}^*$, so that ν is a p.m. on $\mathbb{R}$ satisfying $\limsup_{x\to\infty} x\nu[x,\infty) = 0$, and consider the associated minimal p.m. ν_Δ, as defined in the Introduction, and p.m. $(\nu_\Delta)^*$; thus, ν_Δ is the unique p.m. on $\mathbb{R}$ satisfying

(4.4) (i) $\int_0^\infty x\,d\nu_\Delta(x) < \infty$;

(ii) $\nu \prec_s (\nu_\Delta)^*$; and

(iii) if $\bar{\mu}$ is any p.m. on $\mathbb{R}$ with $\int_0^\infty x\,d\bar{\mu}(x) < \infty$ and $\nu \prec_s \bar{\mu}^*$, then $(\nu_\Delta)^* \prec_s \bar{\mu}^*$.

(see Theorem 2.4 of Kertz and Rösler (1991a) for verification that such a p.m. ν_Δ exists). Define $g(x) := \nu[x,\infty)$ for all $x \in \mathbb{R}$; from Lemma 4.1, this function g is in $\mathcal{G}^1$. From Proposition 3.3, we know that $\widehat{g}$ is the unique function on $\mathbb{R}$ satisfying

(4.5) (i) $\widehat{g}$ is a hyperbolic–concave function in $\mathcal{G}^1$;

(ii) $g(x) \le \widehat{g}(x)$ for all $x \in \mathbb{R}$; and

(iii) if h is any hyperbolic–concave function in $\mathcal{G}^1$ satisfying $g(x) \le h(x)$ for all $x \in \mathbb{R}$, then $\widehat{g}(x) \le h(x)$ for all $x \in \mathbb{R}$.

Now, define $\bar{g}(x) := (\nu_\Delta)^*[x,\infty)$ for $x \in \mathbb{R}$. We claim that $\bar{g} = \widehat{g}$. From Theorem 4.2, $\bar{g}$ is a hyperbolic–concave function in $\mathcal{G}^1$; and from (4.4)(ii), it follows that $g(x) \le \bar{g}(x)$ for all $x \in \mathbb{R}$. To verify (4.5)(iii), we let h be any hyperbolic–concave function in $\mathcal{G}^1$ satisfying $g(x) \le h(x)$ for all $x \in \mathbb{R}$. From Theorem 4.2, there exists a p.m. $\bar{\mu}$ on $\mathbb{R}$ with $\int_0^\infty x\,d\bar{\mu}(x) < \infty$ for which $h(x) = (\bar{\mu})^*[x,\infty)$ for all $x \in \mathbb{R}$; and we have that $\nu[x,\infty) = g(x) \le h(x) = (\bar{\mu})^*[x,\infty)$ for all $x \in \mathbb{R}$. It follows from (4.4)(iii) that $\bar{g}(x) = (\nu_\Delta)^*[x,\infty) \le (\bar{\mu})^*[x,\infty) = h(x)$ for all $x \in \mathbb{R}$. Thus, $\widehat{g}(x) = (\nu_\Delta)^*[x,\infty)$ for all $x \in \mathbb{R}$.

For part (ii), let g be a function in $\mathcal{G}^1$; $\widehat{g}$ denotes the hyperbolic–concave envelope of g. From Lemma 4.1, $\nu[x,\infty) := g(x)$ defines a p.m. ν in $\mathcal{P}^*$; and from Theorem 4.2, there is a p.m. ρ on $\mathbb{R}$ with $\int_0^\infty x\,d\rho(x) < \infty$ and $\rho^*[x,\infty) = \widehat{g}(x)$ for all $\in \mathbb{R}$. We claim that $\rho = \nu_\Delta$; and thus $\widehat{g}(x) = (\nu_\Delta)^*[x,\infty)$ for all $x \in \mathbb{R}$. Now, $\nu[x,\infty) = g(x) \le \widehat{g}(x) = \rho^*[x,\infty)$ for all

$x \in \mathbb{R}$ from (4.5)(ii). Also, if $\bar{\mu}$ is any p.m. on $\mathbb{R}$ with $\int_0^\infty x d\bar{\mu}(x) < \infty$ and $\nu \prec_s \bar{\mu}^*$, then $h(x) := \bar{\mu}^*[x,\infty)$ is in $\mathcal{H}^1$ and satisfies $g(x) \leq h(x)$ for all $x \in \mathbb{R}$. It follows from (4.5)(iii) that $\rho^*[x,\infty) = \widehat{g}(x) \leq h(x) = (\bar{\mu})^*[x,\infty)$ for all $x \in \mathbb{R}$. Thus, ρ satisfies (4.4) and we have that $\rho = \nu_\Delta$. □

We remark that the second part of the proof of Theorem 4.4 gives another proof of the existence of the minimal p.m. ν_Δ associated with a p.m. $\nu \in \mathcal{P}^*$.

To illustrate these ideas, we include the following example. For $n \geq 1$, let $\nu = \sum_{i=0}^n p_i \epsilon_{y_i}$ where $y_0 < \ldots < y_n$ and $0 < p_i < 1$ for $i = 0, \ldots, n$ with $p_0 + \ldots + p_n = 1$, and ϵ_z = point mass at z. Then $\nu[x,\infty) = 0$ if $y_n < x < \infty$; $= \sum_{i=k+1}^n p_i$ if $y_k < x \leq y_{k+1}$ for $k = 0, \ldots, n-1$; and $= 1$ if $-\infty < x \leq y_0$. From the definition of ν^*, one obtains that

$$
\begin{aligned}
\nu^*[x,\infty) &= \left(\sum_{i=k+1}^n p_i(y_i - y_k)\right) /(x - y_k) \\
&\qquad \text{if } \sum_{i=k}^n p_i y_i \Big/ \sum_{i=k}^n p_i < x \leq \sum_{i=k+1}^n p_i y_i \Big/ \sum_{i=k+1}^n p_i \\
&\qquad \text{for } k = 0, \ldots, n-1 \\
&= 0 \text{ if } y_n < x < \infty, \text{ and } = 1 \text{ if } -\infty < x \leq \sum_{i=0}^n p_i y_i;
\end{aligned}
$$

and $f(x) = \nu^*[x,\infty)$ is a hyperbolic–concave function in $\mathcal{G}^1$. One can obtain for example from Kertz and Rösler (1991a) that $\nu_\Delta = \sum_{i=0}^k \pi_i \epsilon_{\Lambda(x_i)}$ and that

$$
\begin{aligned}
(\nu_\Delta)^*[x,\infty) &= \left(\sum_{m=\ell+1}^k \pi_m\right) (x_{\ell+1} - \Lambda(x_\ell))/(x - \Lambda(x_\ell)) \\
&\qquad \text{if } x_\ell \leq x \leq x_{\ell+1} \text{ for } \ell = 0, \ldots, k-1 \\
&= 0 \text{ if } x_k < x < \infty, \text{ and } = 1 \text{ if } -\infty < x \leq x_0,
\end{aligned}
$$

where

$$
\Lambda(x) = \inf_{y > x} \{(y\nu[y,\infty) - x\nu[x,\infty))/(\nu[y,\infty) - \nu[x,\infty))\};
$$

$x_0, \ldots, x_k$ are chosen as follows: $x_0 = y_0$, and having chosen $x_0 = y_0 < x_1 = y_{i_1} < \ldots < x_j = y_{i_j}$, the next number $x_{j+1} = y_{i_{j+1}}$ is the maximal number $y_\ell > x_j$ for which

$$
\Lambda(x_j) = (y_\ell \nu[y_\ell,\infty) - x_j \nu[x_j,\infty))/(\nu[y_\ell,\infty) - \nu[x_j,\infty)),
$$

and the last number $x_k = y_n$; and $\pi_0 = \nu(-\infty, x_1) = \sum_{i=0}^{i_1 - 1} p_i$, $\pi_j = \nu[x_j, x_{j+1}) = \sum_{i=i_j}^{i_{j+1}-1} p_i$ for $j = 1, \ldots, k-1$, and $\pi_k = \nu[x_k,\infty) = p_n$.

The function $h(x) := (\nu_\Delta)^*[x,\infty)$ is the hyperbolic–concave envelope of $g(x) := \nu[x,\infty)$.

In particular, let $\nu = \frac{1}{3}\epsilon_0 + \frac{1}{3}\epsilon_1 + \frac{1}{3}\epsilon_2$. Then $g(x) = \nu[x,\infty) = 0$ if $2 < x < \infty$, $= 1/3$ if $1 < x \le 2$, $= 2/3$ if $0 < x \le 1$, and $= 1$ if $x \le 0$; $h(x) = \nu^*[x,\infty) = 0$ if $2 < x < \infty$, $= (3(x-1))^{-1}$ if $3/2 \le x \le 2$, and $= x^{-1}$ if $1 \le x \le 3/2$, and $= 1$ if $x \le 1$. Also, $\nu_\Delta = \frac{1}{3}\epsilon_{-2} + \frac{1}{3}\epsilon_0 + \frac{1}{3}\epsilon_2$ and $\widehat{g}(x) = (\nu_\Delta)^*[x,\infty) = 0$ if $2 < x < \infty$, $= (2/3)x^{-1}$ if $1 \le x \le 2$, $= 2(x+2)^{-1}$ if $0 \le x \le 1$, and $= 1$ if $x \le 0$.

References

Blackwell, D. and Dubins, L. E. (1963). A converse to the dominated convergence theorem. *Illinois J. Math.* **7** 508–514.

Dubins, L. E. and Gilat, D. (1978). On the distribution of maxima of martingales. *Proc. Amer. Math. Soc.* **68** 337–338.

Freedman, D. (1971). *Markov Chains.* Holden–Day, San Francisco, CA.

Hardy, G. H. and Littlewood, J. E. (1930). A maximal theorem with function theoretic applications. *Acta Math.* **54** 81–116.

Kertz, R. P. and Rösler, U. (1990). Martingales with given maxima and terminal distributions. *Israel J. Math.* **68** 713–192.

Kertz, R. P. and Rösler, U. (1991a). Stochastic and convex orders and lattices of probability measures, with a martingale interpretation. *Israel J. Math.* To appear.

Kertz, R. P. and Rösler, U. (1991b). Complete lattices of probability measures, with applications to martingale theory, preprint.

Loève, M. (1963). *Probability Theory.* D. van Nostrand, New York.

Roberts, A. W. and Varberg, D. E. (1970). *Convex Functions.* Academic Press, New York.

Rockafellar, R. T. (1970). *Convex Analysis.* Princeton University Press, Princeton, NJ.

School of Mathematics
Georgia Institute of Technology
Atlanta, GA 30332–0160

Institut für Mathematische Stochastik
Georg–August Universität Göttingen
Lotzestrasse 13, 3400
Göttingen, Germany

Stochastic Inequalities
IMS Lecture Notes – Monograph Series
Volume 22 (1993)

LOWER BOUNDS ON MULTIVARIATE DISTRIBUTIONS WITH PREASSIGNED MARGINALS

By S. KOTZ and J. P. SEEGER

University of Maryland and BBN Communications

It is well known that the Fréchet lower bound on bivariate distributions with given marginals, F_1 and F_2, given by

$$\max\{F_1(x_1) + F_2(x_2) - 1, 0\},$$

cannot be extended for the case of three or more dimensions. To overcome this difficulty, in order to arrive at a sharp lower bound for multivariate distributions with preassigned marginals, we introduce the concept of the moment of inertia of a multivariate distribution about a given line in $\mathbb{R}^n$ and construct the distribution with the maximal moment of inertia about the line corresponding to the lower Fréchet bound. The multinormal case is discussed in some detail.

1. Introduction

In this paper we suggest an n-variate extension to the Fréchet lower bound for bivariate cumulative distribution functions (c.d.f.s). Recall that for $\Pi(F_1, F_2)$, the class of bivariate c.d.f.s with marginals F_1 and F_2, the Fréchet lower bound is defined as

$$H_*(x, y) = \max\{F_1(x) + F_2(y) - 1, 0\}$$

This does not lend itself to any straight-forward extension to the case of $\Pi(F_1, F_2, \ldots, F_n)$ when $n > 2$ where $\Pi(F_1, F_2, \ldots, F_n)$ is the class of all c.d.f.s whose univariate marginals are the c.d.f.s $F_1, F_2, \ldots, F_n$. However, by observing that the Fréchet upper bound for this class,

$$H^*(x_1, x_2, \ldots, x_n) = \min\{F_1(x_1), \ldots, F_n(x_n)\},$$

concentrates all the density on the curve

$$\{(x_1, \ldots, x_n) | F_1(x_1) = \cdots = F_n(x_n)\},$$

AMS 1991 *subject classifications.* 60E15, 62H99.
Key words and phrases. Stochastic dependence, dependence, multivariate dependence, multivariate normal distribution, fixed marginals, moment of inertia, Frechet bounds.

we were led to seek, as a lower bound, a c.d.f. which maximized the moment of inertia about this curve.

In the sections that follow we define the Fréchet bounds and prove that the lower bound is not extendable to classes of n-variate c.d.f.s for $n > 2$. We then define the moment of inertia as a measure of dependence of n-variate c.d.f.s and use it to present an alternative lower bound for $\Pi(F_1, \ldots, F_n)$. We conclude with specific applications to the multinormal distribution.

2. Assumptions and Notation

In this paper, when referring to $\Pi(F_1, F_2, \ldots, F_n)$ we shall assume the univariate marginals are continuous.

$$\Phi(x) = \frac{1}{\sqrt{2\pi}} \int_{-\infty}^{x} e^{-\frac{t^2}{2}} dt, \text{ the univariate standard normal c.d.f.}$$

$$\Phi_{\mu,\sigma}(x) = \frac{1}{\sqrt{2\pi}\sigma} \int_{-\infty}^{x} e^{-\frac{(t-\mu)^2}{2\sigma^2}} dt, \text{ the univariate } N(\mu, \sigma) \text{ c.d.f.}$$

3. Motivation

It was discovered by Hoeffding (1940) and later rediscovered by Fréchet (1951) that for any F and $G, \Pi(F, G)$ contains an upper bound and a lower bound. The upper bound is $\min\{F(x), G(y)\}$, denoted by $H^*_{F,G}(x, y)$ while the lower bound is $\max\{F(x) + G(y) - 1, 0\}$, denoted by $H_{*F,G}(x, y)$. That is, for any $H \in \Pi(F, G)$ and all $(x, y) \in \mathbb{R}^2$, $H_*(x, y) \leq H(x, y) \leq H^*(x, y)$.

For $n > 2, H^*(x_1, \ldots, x_n) = \min\{F_1(x_1), \ldots, F_n(x_n)\}$ is a valid extension of the bivariate Fréchet upper bound. That is, it is an element of $\Pi(F_1, \ldots, F_n)$ and an upper bound for this class. However, the corresponding n-dimensional extension of $H_*(x_1, \ldots, x_n) = \max\{1 - \sum_{i=1}^{n}(1 - F_i(x_i)), 0\}$ is not an element of $\Pi(F_1, \ldots, F_n)$ for $n > 2$. This was shown by Feron (1965) and Dall'Aglio (1960). Moreover, when H_* is not an element of $\Pi(F_1, \ldots, F_n)$, then it can be shown that this class contains no lower bound.

In fact, we have the well known

THEOREM 1 *Let $F_1, \ldots, F_n$ be continuous. Then $\Pi(F_1, \ldots, F_n)$ contains no lower bound.*

This result led us to seek alternative concepts for defining an extension of H_* for $n > 2$. We examined other ways in which H_* is extreme and sought to construct elements of $\Pi(F_1, \ldots, F_n)$ that share these qualities. The following lemmas are part of this pursuit.

4. Preliminary Results

While accepting that $\Pi(F_1, \ldots, F_n)$ contains no lower bound, insight about elements of this class with other extreme characteristics can be gained by observing properties such a bound would have were one to exist. We will then seek to construct distributions with such properties to observe whether they are in any sense extreme elements of $\Pi(F_1, F_2, \ldots, F_n)$.

LEMMA 1 *Let F be a univariate c.d.f. A lower bound for $\Pi(F, F, \ldots, F)$ must be (finitely) exchangeable, i.e. invariant under permutations of its arguments.*

PROOF Let $H_L(x)$ be a lower bound for $\Pi(F, \ldots, F)$. Let $\gamma(x)$ be a permutation of the components of $\mathbf{x}$, i.e. $\gamma : \mathbb{R}^n \xrightarrow[onto]{1-1} \mathbb{R}^n$ by $\gamma(x_1, \ldots, x_n) = (x_{i_1}, \ldots, x_{i_n})$ where $\{1, 2, \ldots, n\} = \{i_i, \ldots, i_n\}$. If $H_L(\mathbf{x})$ is not exchangeable, $\exists\ \gamma$ and $\mathbf{x}_1$ such that $H_L(\mathbf{x}_1) \neq H_L(\gamma(\mathbf{x}_1))$. Without loss of generality, let $H_L(\mathbf{x}_1) < H_L(\gamma(\mathbf{x}_1))$. Define $H_L'(\mathbf{x}) = H_L(\gamma^{-1}(\mathbf{x}))$. Since the marginals of H_L are all equal, $H_L' \in \Pi(F, F, \ldots, F)$. Also $H_L'(\gamma(\mathbf{x}_1)) = H_L\{\gamma^{-1}(\gamma(\mathbf{x}_1))\} = H_L(\mathbf{x}_1) < H_L(\gamma(\mathbf{x}_1))$ which contradicts the assumption that H_L is a lower bound for $\Pi(F, \ldots, F)$. □

5. The Moment of Inertia

We consider $H_{*F,G}$ as the distribution of extreme negative dependence. Intuitively, it is the distribution in $\Pi(F, G)$ giving the most probability to points, (x, y), for which x and y are far apart; i.e. points away from the diagonal $\{(x, y) | F(x) = G(y)\}$. Let $d((a, b); F, G)$ denote the distance of a point (a, b) from the curve $\{(x, y) | F(x) = G(y)\}$; i.e.

$$d((a, b); F, G) = \inf_{(z,w) \in \{F(x) = G(y)\}} \left\{ \sqrt{(a-z)^2 + (b-w)^2} \right\}.$$

Let (A, B) be a random vector distributed according to $H \in \Pi(F, G)$. Then, $d((A, B); F, G)$ is a random variable whose distribution is determined by H. The expected value, $E(d^2)$ which we shall label $\mu(H)$, can then be considered as a measure of (negative) dependence of H. Then, any $H_L \in \Pi(F, G)$ for which μ is maximized could be considered a distribution of extreme negative dependence.

EXAMPLE Let (X, Y) be uniformly distributed on $(0,1)^2$, i.e., $H_{x,y}(x,y)$ has p.d.f.

$$f_{x,y}(x,y) = \begin{cases} 1 \text{ for } (x,y) \in (0,1)^2 \\ 0 \text{ otherwise} \end{cases}$$

Since $F_x = F_y$, we may consider the moment of inertia about the line defined by $x = y$. Then, by definition

$$\mu(H) \equiv E_H\left\{d^2\left((s,t); F_x, F_y\right)\right\} = \int_0^1 \int_0^1 \frac{(s-t)^2}{2} ds dt = \frac{1}{12}.$$

More generally,

DEFINITION For any $H \in \Pi(F_1, \ldots, F_n)$, the moment of inertia of H about the curve $\{\mathbf{x} | F_1(x_1) = \ldots = F_n(x_n)\}$ is the expected value, according to H, of d^2, where d is the distance between points $\mathbf{x}$ in $\mathbb{R}^n$ and this curve.

As an example application of these concepts we shall deal with $\Pi(\Phi, \Phi, \ldots, \Phi)$, the class of standard multinormal distributions. A lower bound of $\Pi(\Phi, \ldots, \Phi)$ must be exhangeable, and its variance/covariance matrix must be of the form (a_{ij}) where $a_{ii} = 1$, $\forall\ i = 1, \ldots, n$ and $a_{ij} = a\ \forall\ i,j$ such that $i \neq j$. The upper bound for $\Pi(\Phi, \Phi, \ldots, \Phi)$ is $H^*(\mathbf{x}) = \min(\Phi(x_i), \ldots, \Phi(x_n))$ with density concentrated on the curve

$$\{\,\mathbf{x} | \Phi(x_1) = \ldots = \Phi(x_n)\,\} = \{\,\mathbf{x} | x_1 = \ldots = x_n\,\}$$

because Φ is continuous and strictly increasing. A further requirement for our lower bound should be that it maximize the moment of inertia about this line. Straightforward calculations show that this moment of inertia is:

$$\frac{1}{n} \sum_{i<j} \left[\sigma_{x_i}^2 + (E(x_i))^2 + \sigma_{x_j}^2 + (E(x_j))^2 - 2 \text{ cov } (x_i, x_j) - 2E(x_i)E(x_j)\right]$$

With the covariance matrix (a_{ij}), this moment becomes:

$$\frac{1}{n} \sum_{i<j} [1 + 0 + 1 + 0 - 2a - 2(0)] = \frac{1}{n} \frac{n(n-1)}{2} 2(1-a) = (n-1)(1-a)$$

To maximize it, we must thus minimize a.

REMARK 1 If $H_L(\mathbf{x})$ is a lower bound for $\Pi(\Phi, \ldots, \Phi)$, then it can be shown that its moment of inertia about the line $\{\,\mathbf{x} | x_1 = x_2 = \ldots = x_n\,\}$, denoted by $\mu(H_L)$, satisfies $\mu(H_L) \geq \mu(H)\ \forall\ H \in \Pi(\Phi, \ldots, \Phi)$. So if we find $H_M \in \Pi(\Phi, \ldots, \Phi)$ such that $\mu(H_M) \geq \mu(H)\ \forall\ H$ and there actually were an $H_L \in \Pi(\Phi, \ldots, \Phi)$ such that $H_L(\mathbf{x}) \leq H_M(\mathbf{x})\ \forall\ \mathbf{x}$, then $\mu(H_L) \geq \mu(H_M)$, and hence $\mu(H_L) = \mu(H_M)$. Namely, we may not have an actual lower bound of $\Pi(\Phi,\ldots,\Phi)$ using this procedure of maximization, but we will have a *correct upper bound* for $\mu(H)$.

LEMMA 2 *The determinant of the* $n \times n$ *matrix,* $A, (n \geq 2)$ *whose diagonal elements are* 1*'s and other elements are equal to some real number* a *is*

$$(n-1)(a+\frac{1}{n-1})(1-a)^{n-1}$$

PROOF See, for example, Graybill (1969). □

REMARK 2 Since the covariance matrix for a multinormal distribution must have a positive determinant, and since here $| a | \leq 1$ in order that A be a legitimate variance-covariance matrix, we must have $a > -\frac{1}{n-1}$. Hence, $-\frac{1}{n-1}$ is the lower bound for a which yields:

$$\sup_{H\in\Pi(\Phi,\ldots,\Phi)} \mu(H) = (n-1)\left(1-(-\frac{1}{n-1})\right) = n$$

The question arises, what multivariate distribution results in setting $a = -\frac{1}{n-1}$ in the covariance matrix of the form depicted above?

Since the determinant of:

$$\begin{pmatrix} 1 & -\frac{1}{n-1} & \cdots & -\frac{1}{n-1} \\ -\frac{1}{n-1} & 1 & & \\ \vdots & & \ddots & \vdots \\ -\frac{1}{n-1} & \cdots & & 1 \end{pmatrix}$$

is zero, the p.d.f. does not exist. We shall therefore write the p.d.f. in terms of a and observe the limiting distribution as $a \to -\frac{1}{n-1}$.

LEMMA 3 *Let* A *be the matrix described in* Lemma 2. *Then* $A^{-1} = (b_{ij})$ *where* $b_{ii} = -\frac{(n-2)a+1}{((n-1)a+1)(a-1)}$ $\forall\, i = 1, 2, \ldots, n$, *and* $b_{ij} = \frac{a}{((n-1)a+1)(a-1)}$ *for* $i \neq j$.

PROOF See, for example, Graybill (1969). □

Applying Lemmas 2 and 3 and straightforward calculations yield the following expression for the p.d.f. in terms of a:

$$\frac{\sqrt{c}}{(2\pi)^{\frac{n}{2}}\sqrt{(n-1)(1-a)^{n-1}}} \times$$

$$(1) \quad \exp\left\{-\frac{c}{2(n-1)(1-a)}\left[((n-2)a+1)\sum_{j=1}^{n} x_j^2 - 2a\sum_{i<j} x_i x_j\right]\right\}$$

where $c = \frac{n-1}{a(n-1)+1}$.

As $a \to -\frac{1}{n-1}$, the limit of the p.d.f. becomes

$$\lim_{c\to\infty} \frac{\sqrt{c}}{(2\pi)^{\frac{n}{2}}\sqrt{\frac{n^{n-1}}{(n-1)^{n-2}}}} \exp\left\{-\frac{c}{2n(n-1)}(\sum_{j=1}^{n} x_j)^2\right\}$$

$$= \begin{cases} 0 \text{ for all } \mathbf{x} \text{ such that } \sum_{j=i}^{n} x_j \neq 0 \\ \infty \text{ for all } \mathbf{x} \text{ such that } \sum_{i-i}^{n} x_i = 0 \end{cases}$$

This result should have been expected. The "density" is concentrated totally on the hyperplane $(\mathbf{x}| \sum_{i=1}^{n} x_i = 0)$ perpendicular to the line $(\mathbf{x}|x_i = \ldots = x_n)$ containing $(0,\ldots,0) = (E(x_1),\ldots,E(x_n))$.

Via direct computation, we can show that this concentration of density corresponds to a legitimate n-dimensional c.d.f. Labeling this c.d.f. $H_{-\frac{1}{n-1}}$, we see that $H_{-\frac{1}{n-1}}(\mathbf{t})$, is the value of the $(n-1)$-dimensional *mass* (density) contained in the $(n-1)$-dimensional simplex

$$\left\{\mathbf{x}\,|\,\sum_{i=1}^{n} x_i = 0\right\} \bigcap \{\mathbf{x}\,|\,x_1 \leq t_1,\ldots,x_n \leq t_n\}.$$

Integrating the p.d.f. (1), over this simplex, and taking the limit as $a \to -\frac{1}{n-1}$ yields

$$\frac{1}{(2\pi)^{\frac{n-1}{2}}\sqrt{|A|}} \exp\left\{-\frac{1}{2}(x_2,\ldots,x_n)A^{-1}(x_2\ldots x_n)'\right\} \tag{2}$$

where A is now the $(n-1)\times(n-1)$ matrix whose elements

$$a_{ij} = \begin{cases} 1 & i = j \\ -\frac{1}{n-1} & i \neq j \end{cases}$$

and whose determinant is (cf Lemma 2) $\frac{n^{n-2}}{(n-1)^{n-1}}$.

The result (2) was intuitively expected. The limiting density does exist and is obtained by placing an $(n-1)$-dimensional normal density in the $(n-1)$-dimensional hyperplane $\{\mathbf{x}|\sum_{=1}^{n} x_i = 0\}$.

An alternative approach to this analysis involves the application of multivariate characteristic functions and the Lévy-Cramér continuity theorem.

EXAMPLES

Case $n = 2$. In this case the hyperplane becomes $\{(x,y)\,|\,x + y = 0\}$ or the line $y = -x$. The value of this density at $(x,-x)$ by (2) is $\frac{1}{\sqrt{2\pi}}e^{-\frac{x^2}{2}}$, and the corresponding $H_L(x,y)$ is derived by noting that for all (x,y) such that $x+y \leq 0, H_L(x,y) = 0$, and for (x,y) such that $x+y > 0$, we must calculate the mass contained on the line segment between $(-y,y)$ and $(x,-x)$. In other words, here $H_L(x,y) = \frac{1}{\sqrt{2\pi}} \int_{-y}^{x} e^{-\frac{t^2}{2}} dt = \Phi(x) - \Phi(-y) = \Phi(x) + \Phi(y) - 1$. Thus, in this case $H_L(x,y) = \max\{\Phi(x) + \Phi(y) - 1, 0\}$ which is the lower Fréchet bound for $\Pi(\Phi, \Phi)$.

Case $n = 3$. Note that here $H_L(x,y,z)$ is concentrated on the plane $\{(x,y,z) \mid x+y+z=0\}$. Analogous but somewhat more involved calculations yield "the lower bound" of the form

$$H_L(u,v,w) = \max\left\{ \int_{-(u+w)}^{v} e^{-\frac{y^2}{2}} \left[\Phi\left(\frac{y+2w}{\sqrt{3}}\right) + \Phi\left(\frac{y+2u}{\sqrt{3}}\right) - 1 \right] dy, 0 \right\}$$

The calculation of $H_L(u,v,w)$ involves computation of the probability mass in the set

$$\{(x,y,z) \mid x \leq u, y \leq v, z \leq w\} \bigcap \{(x,y,z) \mid x+y+z=0\}.$$

(Details are available from the authors upon request.)

Finally, we note that if $H_L(\mathbf{x})$ is a lower bound for all n-dimensional standard multinormal c.d.f.s, then $H_L'(\mathbf{x}) \equiv H_L\left\{\frac{x_1-\mu_1}{\sigma_1}, \ldots, \frac{x_n-\mu_n}{\sigma_n}\right\}$ is a lower bound for all multinormal n-dimentional c.d.f.s in $\Pi(\Phi_{\mu_1,\sigma_1}, \ldots, \Phi_{\mu_n,\sigma_n})$.

It is also straightforward to calculate that the corresponding upper bound on moment of inertia $\mu(H)$ in this case, denoted by $M_{H_{L'}}$, is

$$M_{H_L'} = \frac{2n}{(n-1)\sum \sigma_i^2} \sum_{i<j} \sigma_i^2 \sigma_j^2$$

which coincides for $\sigma_i^2 = 1, i = 1, \ldots, n$ with

$$\sup_{H \in \Pi(\Phi,\ldots\Phi)} \mu(H) = n$$

CONCLUDING REMARK The approach suggested in this paper could be extended rather straightforwardly to other families of distributions. Of special interest may be the multivariate extensions of the Gumbel bivariate distribution (Gumbel (1960)) as well as other multivariate distributions with exponential or, more generally, Weibull marginals.

References

DALL'AGLIO, G. (1960). Les Fonctions Extrêmes de la Classes de Fréchet à Trois Dimensions. *Publ. Inst. Stat. Univ. Paris* **9** 175-188.

FERON, R. (1965). Sur les Tableaux de Corrélation dont les Marges sont Données, cas de l'espace à Trois Dimensions. *Publ. Inst. Stat. Univ., Paris* **5** 3-12.

FRÉCHET, M. (1951). Sur les Tableaux de Corrélation dont les Marges sont Données. *Ann. Univ. Lyon, Sect. A* **14** 53-77.

GRAYBILL, F. A. (1969). *Introduction to Matrices with Applications in Statistics.* 1st edition. Wadsworth, Belmont, CA.

GUMBEL, E. J. (1960). Bivariate Exponential Distributions. *J. Amer. Stat. Assoc.* **55** 698-707.

HOEFFDING, W. (1940). Masstabinvariante Korrelationstheorie. In *Schriften des mathematischen Instituts und des Instituts für Angewandte Mathematik der Universitat.* Berlin **5** 179-233.

MARDIA, K. V. (1970). *Families of Bivariate Distributions.* Hafner, Darien, CT.

NATAF, A. (1962). Détermination des Distributions de Probabilitiés dont les Marges sont Données. *Comptes Rendus de l'Académie des Sciences* **255** Paris, 42-43.

SKLAR, A. (1973). Random Variables, Joint Distribution Functions, and Copulas. *Kybernetica* **9** 449-460.

DEPARTMENT OF MANAGEMENT SCIENCE
AND STATISTICS
COLLEGE OF BUSINESS AND MANAGEMENT
UNIVERSITY OF MARYLAND
COLLEGE PARK, MD 20742

BBN COMMUNICATIONS
150 CAMBRIDGEPARK DRIVE
CAMBRIDGE, MA 02140

Stochastic Inequalities
IMS Lecture Notes – Monograph Series
Volume 22 (1993)

DEPENDENCE OF STABLE RANDOM VARIABLES

By MEI–LING TING LEE, SVETLOZAR T. RACHEV[1]
and GENNADY SAMORODNITSKY[2]

Harvard University, University of California, Santa Barbara, and Cornell University

The dependence structure of a multivariate normal distribution is characterized by its covariance matrix. However, in contrast to the normal case, discussion on dependence for α-stable random variables, $0 < \alpha < 2$, requires more care because variances do not exist. We review in this paper dependence concepts for α-stable random variables. A local measure of dependence is proposed. Also we illustrate how product–type stable laws arise naturally in applications.

1. Introduction

The study of dependence in random variables has yielded many useful results in statistical applications. For normal distributions, the dependence structure can be characterized by their covariance matrix. For example, Pitt (1982), Joag-Dev, Perlman and Pitt (1983) show that jointly normal random variables are associated if and only if their correlations are all nonnegative.

In contrast to normal vectors, a multivariate stable random vector cannot be specified in general by a finite number of numerical parameters. Moreover, when $0 < \alpha < 2$, no α-stable random variable has a finite second moment, and even the first moment does not exist when $\alpha \leq 1$. Therefore the investigation of dependence relationships among stable random variables is nontrivial. Using spectral measure as a tool, Lee, Rachev and Samorodnitsky (1990) derived necessary and sufficient conditions for association of stable random variables. In section 2, we will review some dependence results for stable random variables. Also we discuss the notion of geometric stable random variables.

In section 3 we focus on symmetric stable sub-Gaussian random variables. We show that except for the singular case, sub-Gaussian random vector

---

[1]Research supported in part by grants from National Science Foundation and Erasmus University, Rotterdam.

[2]Research supported by Office of Naval Research Grant N00014–90–J–1287.

AMS 1991 *subject classifications.* Primary 60E07, 62H20.

Key words and phrases. α–stable, dependence, association.

cannot be associated. It is therefore of interest to derive a measure of local strength of dependence based on Bjerve and Doksum (1990)'s correlation curve. In section 4 we discuss the relationship between product-type stable random vectors and subordinated processes.

2. Stable Random Vectors and Dependence

2.1 *Association of Stable Vectors*

Stable laws are very useful in statistical applications, they have been used to model the distribution of stock price changes (see e.g. Akgiray and Booth (1988), Du Mouchel (1983), Fama (1965), Mandelbrot (1963), and Mittnik and Rachev (1991)), and the distribution of the frailty factor in the context of biostatistics (see e.g. Hougaard (1986)).

A random vector $\mathbf{X}=(X_1, X_2, \ldots, X_n)$ is called α-stable, $0<\alpha\le 2$, if for any constants $A>0, B>0$, there is a $\mathbf{D}\in\mathbf{R}^n$ such that

$$A\mathbf{X}^{(1)} + B\mathbf{X}^{(2)} \stackrel{d}{=} (A^\alpha + B^\alpha)^{1/\alpha}\mathbf{X} + \mathbf{D},$$

where $\mathbf{X}^{(1)}, \mathbf{X}^{(2)}$ are independent copies of $\mathbf{X}$.

Normal distributions are special cases of stable distributions with index of stability $\alpha = 2$. An α-stable random vector is called strictly α-stable if $\mathbf{D}=\mathbf{0}$ for every A and B. An α-stable random vector $\mathbf{X}$ satisfying $\mathbf{X}\stackrel{d}{=}-\mathbf{X}$ is called symmetric α-stable ($S\alpha S$). Note that a $S2S$ vector is a zero-mean multivariate normal random vector.

Let $\phi_\alpha(\boldsymbol{\theta}) = \phi_\alpha(\theta_1, \ldots, \theta_n) = E\exp\{i(\boldsymbol{\theta}, \mathbf{X})\} = E\exp\{i\sum_{j=1}^n \theta_j X_j\}$ denote the characteristic function of an α-stable random vector in $\mathbf{R}^n$, where $(\boldsymbol{\theta}, \mathbf{X})$ denotes the inner product. When $n=1$, the characteristic function of an α-stable random variable, $0<\alpha\le 2$, has the form

$$(1)\quad \phi_\alpha(\theta) = \begin{cases} \exp[-\sigma^\alpha|\theta|^\alpha(1 - i\eta(\operatorname{sign}\theta)\tan\frac{\pi\alpha}{2}) + i\mu\theta] & \text{if } \alpha\ne 1 \\ \exp[-\sigma|\theta|(1 + i\eta\frac{2}{\pi}(\operatorname{sign}\theta)\ln|\theta|) + i\mu\theta] & \text{if } \alpha = 1. \end{cases}$$

where $\sigma\ge 0$ is typically referred to as scale parameter, $-1\le\eta\le 1$ is skewness parameter and $\mu\in R$ is location parameter (but do not rely too much on these names in the case $\alpha=1$). Conversely, a random variable X with characteristic function given by (1) is α-stable, and we say that X has a stable distribution $S_\alpha(\sigma,\eta,\mu)$.

When $\eta=1$, the random variable X is said to be *totally right skewed.* If also $0<\alpha<1$, and $\mu=0$, then X has the positive real line as its support, in which case it has the Laplace transform $E[\exp(-\theta X)] = \exp(-c\sigma^\alpha\theta^\alpha)$, where $c = (\cos\frac{\pi\alpha}{2})^{-1}$.

In the case $n \geq 2$, there is a similar representation for the characteristic function (ch.f) of α-stable vectors. Namely, a random vector $\mathbf{X} = (X_1, \ldots, X_n)$ is α-stable, $0 < \alpha \leq 2$, if ane only if there is a finite Borel measure m on the unit sphere S_n of $\mathbf{R}^n$ and a vector $\boldsymbol{\mu}^0 = (\mu_1^0, \ldots, \mu_n^0)$ in $\mathbf{R}^n$ such that:
(a) If $\alpha \neq 1$

$$(2)\quad \phi_\alpha(\boldsymbol{\theta}) = \exp\{-\int_{S_n} |(\boldsymbol{\theta}, \mathbf{s})|^\alpha (1 - i \operatorname{sign}((\boldsymbol{\theta}, \mathbf{s})) \tan \frac{\pi\alpha}{2}) m(ds) + i(\boldsymbol{\theta}, \boldsymbol{\mu}^0)\}$$

(b) If $\alpha = 1$
(2a)

$$\phi_\alpha(\boldsymbol{\theta}) = \exp\{-\int_{S_n} |(\boldsymbol{\theta}, \mathbf{s})|(1 + i\frac{2}{\pi} \operatorname{sign}((\boldsymbol{\theta}, \mathbf{s})) \ln |(\boldsymbol{\theta}, \mathbf{s})|) m(ds) + i(\boldsymbol{\theta}, \boldsymbol{\mu}^0)\},$$

where $\mathbf{s} = (s_1, \ldots, s_n) \in S_n$. The pair $(m, \boldsymbol{\mu})$ is unique when $0 < \alpha < 2$, the measure m is then called the *spectral measure* of the α-stable random vector $\mathbf{X}$.

Specifically, if $\mathbf{X}$ is symmetric α-stable, then it has characteristic function of the form

$$(3)\qquad \phi_\alpha(\boldsymbol{\theta}) = \exp\{-\int_{S_n} |\theta_1 s_1 + \ldots + \theta_n s_n|^\alpha m(ds)\}$$

where Γ is a finite symmetric measure on the Borel subsets of the unit sphere S_n.

Note that an α-stable random vector $\mathbf{X}$ has independent components if and only if its spectral measure m is discrete and concentrated on the intersection of the axes with the unit sphere S_n. See Samorodnitsky and Taqqu (1991) for a review on properties of multivariate stable random vectors.

Random variables $X_1, \ldots, X_n$ are called *associated* if for any functions $f, g\colon \mathbb{R}^n \to \mathbb{R}$, nondecreasing in each argument, we have $\operatorname{cov}(f(\mathbf{X}), g(\mathbf{X})) \geq 0$ whenever the covariance exists. The concept of *association* was introduced by Esary, Proschan, and Walkup (1967) to obtain bounds related to coherent functions(co-ordinatewise increasing) occurring in the theory of reliability. In a completely different context, Fortuin, Kasteleyn, and Ginibre (1971), considered the association concept for the Ising model of statistical physics. Association represents a strong form of positive dependence.

Pitt (1982), Joag-Dev, Perlman and Pitt (1983) show that nonnegatively correlated normal variables are associated. Inspired by their results, Lee, Rachev and Samorodnitsky (1990) derive the following theorem.

THEOREM 1 *Let $X_1, \ldots, X_n$ be jointly α-stable random variables, $0 < \alpha < 2$, with characteristic function given by* (2). *Then $X_1, \ldots, X_n$ are associated if and only if the spectral measure m satisfies the condition*

$$m(S_n^-) = 0$$

where $S_n^- = \{(s_1, \ldots, s_n) \in S_n$: *for some* $i, j \in \{1, \ldots, n\}, s_i > 0$ *and* $s_j < 0\}$.

Note that a result related to the sufficiency part of the above theorem was obtained by Resnick (1988) in terms of Poisson representation of an infinitely divisible random vector.

We list here some notions of dependence. A bivariate density function $f(x, y)$ of two arguments is said to be *totally positive of order 2* (abbreviated TP_2) if for all $x_1 < x_2$, and $y_1 < y_2$,

$$\begin{vmatrix} f(x_1, y_1) & f(x_1, y_2) \\ f(x_2, y_1) & f(x_2, y_2) \end{vmatrix} \geq 0.$$

A joint density function $f(x_1, \ldots, x_n)$ of n arguments is said to be TP_2 *in pairs* if $f(x_1, \ldots, x_i, \ldots, x_j, \ldots, x_n)$ is TP_2 in (x_i, x_j) for all $i \neq j$ and all fixed values of the remaining arguments. If a random vector has a TP_2-in-pairs density then it is associated. See Karlin (1968), Barlow and Proschan (1981), and Tong (1990) for a review. Random variables $X_1, \ldots, X_n$ are called *positive upper orthant dependent* (PUOD) if

$$P(X_1 > x_1, \ldots, X_n > x_n) \geq P(X_1 > x_1) \ldots P(X_n > x_n)$$

for any $x_1, \ldots, x_n$, and they are called *positive lower orthant dependent* (PLOD) if

$$P(X_1 \leq x_1, \ldots, X_n \leq x_n) \geq P(X_1 \leq x_1) \ldots P(X_n \leq x_n)$$

for any $x_1, \ldots, x_n$. That is, if $X_1, \ldots, X_n$ are PUOD or PLOD, then they are more likely to take on larger values together or smaller values together. Lehmann (1966) shows that for the bivariate case, X_1, X_2 are PUOD if and only if they are PLOD; however, for higher dimensional cases, the equivalence no longer holds. It is also well known that association implies both PUOD and PLOD, but in general these implications cannot be reversed. For stable random variables, Lee, Samorodnitsky and Rachev (1990) show, as a result of theorem 1, that PLOD or PUOD implies association.

COROLLARY 1 *Let* $X_1, \ldots, X_n$ *be jointly* α*-stable. Then the notion of association is equivalent to PLOD or PUOD.*

Following Alam and Saxena (1981), we call random variables $X_1, \ldots, X_n$ *negatively associated* if for any $1 \leq k < n$, and $f: \mathbb{R}^k \to \mathbb{R}, g: \mathbb{R}^{n-k} \to \mathbb{R}$, nondecreasing in each argument, $\operatorname{cov}(f(\mathbf{Y}), g(\mathbf{Z})) \leq 0$ whenever the covariance exists, where $\mathbf{Y}$ and $\mathbf{Z}$ are any k and $(n-k)$-dimensional random vectors correspondingly, representing a partition of the set $(X_1, \ldots, X_n)$ into two subset of sizes k and $n-k$ accordingly. In the normal case, negative association has been characterized by Joag-dev and Proschan (1983). Lee, Samorodnitsky and Rachev (1990) derive the following theorem.

THEOREM 2 *Let $X_1,\ldots,X_n$ be jointly α-stable random variables $0<\alpha<2$, with characteristic function given by (2). Then $X_1,\ldots,X_n$ are negatively associated if and only if the spectral measure m satisfies the condition*

$$m(S_n^+)=0$$

where $S_n^+=\{(s_1,\ldots,s_n)\in S_n\colon$ for some $i\neq j,\ s_i\cdot s_j>0\}$.

Recently, some other properties related to dependence concepts of stable random vectors have also been discussed. Multiple regressions on stable random vectors have been considered by Wu and Cambanis (1991) and Samorodnitsky and Taqqu (1991). A version of Slepian-type inequalities, due to Fernique (1975), was extended to stable random vectors in terms of Levy measures by Samorodnitsky and Taqqu (1990).

2.2. *Geometric Stable Vectors*

A random vector $\mathbf{Y}=(Y_1,\ldots,Y_n)$ is called *geometric stable* if there exist (1) a sequence of i.i.d. random vectors $\mathbf{X}^{(1)},\mathbf{X}^{(2)},\ldots$, (2) independent of the $\mathbf{X}^{(i)}$'s a geometric r.v. $T(p)$ with mean $1/p$ $(0<p<1)$, and (3) constants $\mathcal{A}(p)>0$ and $\mathcal{C}(p)\in\mathbb{R}^n$ such that

$$\mathcal{A}(p)\sum_{i=1}^{T(p)}(\mathbf{X}^{(i)}+\mathcal{C}(p))\xrightarrow{d}\mathbf{Y},\ \text{as } p\to 0 \tag{4}$$

Geometric stable random vectors (GSRV's) are used in reliability queuing theory, financial modelling and its study goes back to the works of A. Renyi, H. Robbins and B. V. Gnedenko, see the surveys in Rachev (1991), and Rachev and Sengupta (1991).

We now characterize the class of GSRV's. The characterization is in terms of "dual" representation (see (b) below) of the ch.f's of GSRV, and α-stable random vectors, which, as it follows (see (a)-(c)), share one and the same domain of attraction.

LEMMA 1 *For a random vector $\mathbf{Y}$ the following are equivalent:*
(a) $\mathbf{Y}$ is GSRV.
(b) The characteristic function $f_{\mathbf{Y}}$ of $\mathbf{Y}$ admits the representation

$$f_{\mathbf{Y}}(\boldsymbol{\theta})=\frac{1}{1-\ln\phi_\alpha(\boldsymbol{\theta})},\quad \boldsymbol{\theta}\in\mathbb{R}^n, \tag{5}$$

where $\phi_\alpha(\boldsymbol{\theta})$ is the ch.f of α-stable vector (see (2), (2a)*).*
(c) $\mathbf{X}^{(1)}$ *in* (4) *belongs to the domain of attraction of α-stable random vector with ch.f. ϕ_α defined by* (5).
(d) $\mathbf{X}^{(1)}$ *in* (4) *has polar coordinates ρ and Θ such that, as $R \to \infty$,*

$$P(\rho > R, \Theta \in A)/P(\rho > kR, \Theta \in B)$$

$$\to k^\alpha \tilde{m}(A)/\tilde{m}(B) \tag{6}$$

for any $k > 0$ and any Borel sets A and B of S_n with $\tilde{m}(B) \neq 0$, $\tilde{m}$ stands for the spectral measure of ϕ_α in (5) *rewritten in polar coordinates.*

PROOF The proof is essentially given in Mittnik and Rachev (1991). The limit relation (6) follows from the characterization of the domain of attraction of multivariate α-stable law (see Rvačeva (1962), Resnick and Greenwood (1979)).

Taking into account the characterization (5) of GSRV $\mathbf{Y}$ we say that $\mathbf{Y}$ and ϕ_α share one and the same spectral measure Γ and index of stability α given in (2), (2a).

Necessary conditions for association of GSRV's follow from lemma 1 and the condition for association of stable vectors in the previous section. Sufficient conditions, however, are not obvious, and this problem is still open.

3. Product-type Stable Random Vectors

3.1. *Stable Laws Derived from Products*

Let $\mathbf{Z} = (Z_1, \ldots, Z_n)$ be an arbitrary symmetric α'-stable random vector. Let T be a positive α/α'-stable random variable, $0 < \alpha < \alpha'$, independent of $\mathbf{Z}$ and having Laplace transform $E\exp(-\theta T) = \exp(-\theta^{\alpha/\alpha'})$, $\theta \geq 0$. Then the random vector defined by

$$\mathbf{X} = T^{1/\alpha'}\mathbf{Z} \tag{7}$$

is symmetric α-stable. They are sometimes referred to as product-type stable random vectors. In section 4, we will see that these product-type stable random vectors can be obtained naturally from stable processes directed by an operational time stable process.

It is clear that components of the product-type stable vector $\mathbf{X}$ are conditionally independent. If one further assumes that $Z_1, \ldots, Z_n$ are i.i.d., then components of $\mathbf{X}$ are positively dependent by mixture as considered by Shaked (1977), Tong (1977, 1980), and Shaked and Tong (1985). We note that components of the derived vectors $\mathbf{X}$ can be strongly dependent. This fact can be demonstrated by the following example.

EXAMPLE 1 For any fixed positive integers $k_1 > 1$ and $k_2 > 1$, assume that $\{Z_i,\ i = 1,\dots,n\}$ are n i.i.d. totally right skewed strictly $1/k_2$-stable random variables. Let T be a totally right skewed strictly $1/k_1$-stable variable, independent of $\{Z_i, i = 1,\dots,n\}$. Then the derived vector $(X_1,\dots,X_n) = T^{1/\alpha'}\mathbf{Z}$, with $\alpha' = 1/k_1$, is $1/k_1k_2$ stable and its components are TP_2 in pairs. See Theorem 6 in section 4 for general results. We show in the following section that there are many cases where components of a product-type stable random vector are neither associated nor positive orthant dependent.

On the other hand, note that if $\mathbf{X} = E^{1/\alpha'}\mathbf{Z}$ where $\mathbf{Z}$ is a symmetric α'-stable random vector and if (1) E is independent of $\mathbf{Z}$, (2) E is an exponential random variable then $\mathbf{X}$ is GSRU, see equation (4). If E has arbitrary distribution on $\mathbf{R}_+$ then $\mathbf{X} = E^{1/\alpha'}\mathbf{Z}$ is called Robbins mixture: for applications of Robbins mixtures to reliability theory, queueing and finance modelling we refer to Szasz (1972), Szynal (1976), Karolev (1988), Melamed (1988), Rachev and Ruschendorf (1991), Rachev and Samorodnitsky (1991).

3.2. *Sub-Gaussian Random Vectors*

When $\mathbf{Z} \equiv \mathbf{G} = (G_1,\dots,G_n)$ is a zero mean Gaussian vector (i.e. S_2S) in $\mathbf{R}^n$, independent of the positive α/α'-stable random vector T in (7), then the derived vector $\mathbf{X} = T^{1/2}\mathbf{G}$ is called a *sub-Gaussian $S\alpha S$* random vector, with governing Gaussian vector $\mathbf{G}$. It is shown in section 4 that sub-Gaussian vectors arise naturally in stable processes which are subordinated to Gaussian processes. Sub-Gaussian vectors form a special class of stable vectors which, unlike general stable vectors, can be characterized by finitely many parameters. Specifically, the sub-Gaussian vector $\mathbf{X}$ derived above has a characteristic function of the form

$$\phi_\alpha(\boldsymbol{\theta}) = \exp\{-|\frac{1}{2}\sum_{i=1}^{n}\sum_{j=1}^{n}\theta_i\theta_j R_{ij}|^{\alpha/2}\}, \tag{8}$$

where $R_{ij} = E(G_iG_j),\ i,j = 1,\dots,n$ are the covariances of the underlying Gaussian vector $(G_1,\dots,G_n)$.

Let m_0 be a uniform (i.e. rotationally invariant) finite Borel measure on S_n. Then for some $c \geq 0$

$$\int_{S_n} |\sum_{j=1}^{n}\theta_j s_j|^\alpha m_0(ds) = c\ [\frac{1}{2}\sum_{j=1}^{n}\theta_j^2]^{\alpha/2}.$$

Therefore, a $S\alpha S$ random vector , $\alpha < 2$, is sub-Gaussian with a governing Gaussian vector having i.i.d. components if and only if its spectral measure is uniform on the unit sphere S_n. A uniform spectral measure does not satisfy the required condition for association as was stated in Theorem 1. As a consequence, we have

COROLLARY 2 *If a $S\alpha S$ random vector $\mathbf{X}$ is sub-Gaussian with a governing Gaussian vector having i.i.d. $N(0,\sigma^2)$ components, then components of $\mathbf{X}$ are positively dependent by mixture but they are neither associated nor positively orthant dependent.*

In general, the spectral measure of a sub-Gaussian vector is a transform of the uniform measure m_0 on S_n. Hence we have

THEOREM 3 *A non-degenerate (i.e. having non-zero components) sub-Gaussian vector $(T^{1/2}G_1, \ldots, T^{1/2}G_n)$ with governing Gaussian vector $\mathbf{G}$, as defined in equation (7), is associated if and only if $G_1 = c_2G_2 = \ldots = c_nG_n$ a.e. for some $c_2 > 0, \ldots, c_n > 0$.*

PROOF Suppose $P(G_1 = c_2G_2) = 0$ for any $c_2 \in R$. Assume that the stable random vector $(T^{1/2}G_1, \ldots, T^{1/2}G_n)$ is associated. Then the sub-vector $(T^{1/2}G_1, T^{1/2}G_2)$ is also associated. Hence $(T^{1/2}G_1, T^{1/2}G_2)$ has a spectral measure which is concentrated on the parts of the unit circle specified in Theorem 1, namely, in the first and third quadrant of the unit circle. On the other hand, any Gaussian vector can be written as a linear combination of i.i.d. $N(0,1)$'s. The condition $P(G_1 = c_2G_2) = 0$ for any $c_2 \in R$ implies that this linear transformation for the vector (G_1, G_2) is of full rank. Therefore the spectral measure of $(T^{1/2}G_1, T^{1/2}G_2)$ is a rigid transformation of the uniform measure on the unit sphere, and it maps the unit sphere onto the entire unit sphere. This leads to a contradiction. Similarly, the case $G_1 = c_2G_2$ for some $c_2 < 0$ can be ruled out.

3.3. *Conditional Moments for Sub-Gaussians Random Vectors*

Samorodnitsky and Taqqu (1991) derived regression equations for general stable random vectors. In particular, they show that when $(X_1, X_2) = (T^{1/2}G_1, T^{1/2}G_2)$ in equation (7) is a non-degenerate sub-Gaussian $S\alpha S$, (even when $\alpha \leq 1$), random vector with governing Gaussian vector $\mathbf{G}$,

$$E(X_2|X_1 = x) = \frac{\operatorname{Cov}(G_1, G_2)}{\operatorname{Var}G_1}x = R_{12}{R_{11}}^{-1}x \qquad a.e. \tag{9}$$

and $E(X_2^2|X_1 = x) < \infty$ a.e. if $1 < \alpha < 2$ (note that the unconditional second moment is infinite.).

We can calculate the conditional variance as follows.

$$\begin{aligned} E(G_2|G_1) &= R_{12}{R_{11}}^{-1}G_1, \\ \mathrm{Var}(G_2|G_1) &= R_{22} - R_{12}^2 R_{11}^{-1}, \\ E(X_2^2|X_1 = x) &= E(E(TG_2^2|T, G_1)|T^{1/2}G_1 = x) \\ &= E(T|X_1 = x)\mathrm{Var}\,(G_2|G_1) + R_{12}^2 {R_{11}^2}^{-1} x^2. \end{aligned}$$

Hence

$$\mathrm{Var}\,(X_2|X_1 = x) = E(T|X_1 = x)(R_{22} - R_{12}^2 R_{11}^{-1}). \tag{10}$$

Note that from equation (9) we see that the conditional mean of X_2 given X_1 is completely determined by their governing Gaussian vector $\mathbf{G}$. Equation (10), however, demonstrates how the random variable T influences the variance of the conditional law of X_2 given X_1.

Wu and Cambanis (1991) show that

$$\mathrm{Var}(X_2|X_1 = x) = (1/2)(R_{22} - R_{12}^2 R_{11}^{-1}) f(x;\alpha)^{-1} \int_{|x|}^{\infty} u f(u;\alpha)du, \tag{10a}$$

where $f(x;\alpha)$ is the density function of a $S_\alpha(1,0,0)$ random variable (i.e. having the characteristic function $\exp[-|t|^\alpha]$).

Moreover, if T in $\mathbf{X} = \sqrt{T}\mathbf{G}$ is standard exponentially distributed rather than positive stable distributed, then $\mathbf{X}$ has a multivariate Laplace distribution, that is, its ch.f has the form

$$\phi_\alpha(\boldsymbol{\theta}) = \frac{1}{1 + \frac{1}{2}\sum_{i=1}^n \sum_{j=1}^n \theta_i \theta_j R_{ij}}.$$

See Feldman and Rachev (1991) and Rachev and Sengupta (1991) for application of Laplace distributions in random fields, U-statistics and modelling commodity prices.

3.4. *Local Correlation Functions*

In an effort to characterize the dependence for sub-Gaussian vectors, we derive in this section a notion of local correlation for sub-Gaussian random vectors.

Bjerve and Doksum (1990) introduced a measure of local strength of dependence by combining ideas from nonparametric regression and of Galton (1888). They define the function

$$\rho(x) = \frac{\sigma\beta(x)}{[\{\sigma\beta(x)\}^2 + \sigma^2_{X_2|X_1=x}]^{1/2}} \tag{11}$$

where $\sigma^2 = \mathrm{Var}(X_1)$, $\sigma^2_{X_2|X_1=x} = \mathrm{Var}(X_2|X_1 = x)$, and $\beta(x) = \frac{d}{dx}E(X_2|X_1 = x)$ is the slope of the nonparametric regression. More generally, they note that the conditional mean $E(X_2|X_1 = x)$ can be replaced by a location function. The variances $\sigma^2 = \mathrm{Var}(X_1)$ and $\sigma^2_{X_2|X_1=x} = \mathrm{Var}(X_2|X_1 = x)$ can be replaced by squares of corresponding scale functions.

For an $S\alpha S$ random vector (X_1, X_2), Samorodnitsky and Taqqu (1991) show that in many cases the first, and when $1 < \alpha \leq 2$, second conditional moments exist if a certain integrability condition holds. When $1 < \alpha < 2$, the *covariation* is designed to replace the covariance. It is a useful quantity and it appears naturally in the context of regression for stable random variables.

DEFINITION Let X_1 and X_2 be jointly $S\alpha S$ with $\alpha > 1$ and let m be the spectral measure of the random vector (X_1, X_2). The covariation of X_1 and X_2 is defined to be the real number

$$[X_1, X_2]_\alpha = \int_{S_2} s_1 |s_2|^{\alpha-1} \mathrm{sign}(s_2) m(d\mathbf{s}).$$

As a result, we have

$$[X_1, X_1]_\alpha = \int_{S_2} |s_1|^\alpha m(d\mathbf{s}) = \sigma_1^\alpha,$$

where σ_1 is the scale parameter of the $S\alpha S$ random variable X_1. Let $\mathcal{F}_\alpha$ be a linear space of jointly $S\alpha S$ random variables. Then when $\alpha > 1$, the covariation induces a norm on $\mathcal{F}_\alpha$ such that

$$\|X\|_\alpha = ([X, X]_\alpha)^{1/\alpha}.$$

Convergence in $\|.\|_\alpha$ is equivalent to convergence in probability and convergence in L^p for any $p < \alpha$. Moreover, for sub-Gaussian vector $(X_1, \ldots, X_n)$ with ch.f defined in (8), we have

$$\begin{aligned} [X_i, X_j]_\alpha &= 2^{-\alpha/2} R_{ij} R_{jj}^{(\alpha-2)/2}, \quad i, j = 1, \ldots, n, \quad \text{and} \\ \|X_i\|_\alpha &= 2^{-1/2} R_{ii}^{1/2}. \end{aligned}$$

Note that when $\alpha = 2$, we have $[X_1, X_2]_2 = \frac{1}{2}\,\mathrm{Cov}(X_1, X_2)$, and $[X_1, X_1]_2 = \frac{1}{2}\,\mathrm{Var}(X_1) = \sigma_1^2$. For an $S\alpha S$ random variable X, $1 < \alpha < 2$, we use $k_\alpha \|X\|_\alpha$, with

$$k_\alpha = [\frac{\alpha\Gamma(1 - 1/\alpha)}{\Gamma(1/\alpha)}]^{1/2},$$

to replace the notion of $\{\mathrm{Var}(X)\}^{1/2}$ in the calculation of a notion of local correlation function. Note that $k_2 = \sqrt{2}$ is consistent with the above computation.

Hence we can define, for an $S\alpha S$ random vector (X_1, X_2), the local correlation curve as

$$\rho(x) = \frac{\beta(x) k_\alpha \|X_1\|_\alpha}{[\{\beta(x) k_\alpha \|X_1\|_\alpha\}^2 + \sigma^2_{X_2|X_1=x}]^{\frac{1}{2}}} \tag{12}$$

whenever the conditional variance is finite *a.e.* Note that, for the special case when $\alpha = 2$, $\rho(x) \equiv$ the correlation coefficient of (X_1, X_2).

For $S\alpha S$ sub-Gaussian vector $(X_1, X_2) = (T^{1/2}G_1,\ T^{1/2}G_2)$ with governing Gaussian vector (G_1, G_2) we have by (9),(10) and (10a)

$$\rho(x) = \frac{k_\alpha R_{12}}{[k_\alpha^2 R_{12}^2 + (R_{11}R_{22} - R_{12}^2) f(x;\alpha)^{-1} \int_{|x|}^{\infty} u f(u;\alpha) du]^{1/2}}. \tag{13}$$

It follows that $\rho(x)$ is an even function, $\rho(0) = R_{12}/(R_{11}R_{22})^{1/2}$ = the correlation coefficient of G_1 and G_2. Observe that

$$\rho(x) \sim \frac{k_\alpha}{\alpha(\alpha-1)^{1/2}} R_{12}(R_{11}R_{22} - R_{12}^2)^{-1/2}|x|^{-1} \text{ as } |x| \to \infty. \tag{14}$$

Furthermore, $\rho(x) \equiv 0$ when G_1 and G_2 are independent; $\rho(x) \equiv 1$, when $G_1 = G_2$.

4. Subordination

In this section we will show that product-type stable random vectors considered in section 3 can appear naturally in applications. Let $\{\mathbf{X}(t)\}$ be a Markov process with continuous transition probabilities and $\{\mathbf{T}(t)\}$ a process with nonnegative independent increments, then $\{\mathbf{X}(\mathbf{T}(t))\}$ is again a Markovian process. The process $\{\mathbf{X}(\mathbf{T}(t))\}$ is said to be *subordinate to the parent process* $\{\mathbf{X}(t)\}$ *using the operational time* $\mathbf{T}(t)$. The process $\{\mathbf{T}(t)\}$ is called the *directing process.* The role of the directing process is to inject some additional randomness into the parent process through its time parameter t. In equipment usage, for example, $\{\mathbf{X}(t)\}$ may represent cumulative wear on a machine component after t hours of operation and $\{\mathbf{T}(t)\}$ may represent the number of hours that the machine has operated after t hours of calendar time have passed. The process $\{\mathbf{T}(t)\}$ thereby captures the random delays and accelerations of operational use of the machine over calendar time. The term *subordination* was first introduced by Bochner (1955). Subordinated processes have also been referred to as *derived processes* by Cohen (1962). Various properties of derived processes were investigated by Stam (1965). See Mandelbrot and Taylor (1967), Clark (1973), Rachev and Ruschendorf (1991), Rachev and Samorodnitsky (1991), for modelling stock returns and option pricing via subordinated processes. Whitmore and Lee (1991) considered statistical inferences for subordinated processes and applications.

DEFINITION A stochastic process $\{\mathbf{X}(t)\}$ is called an *α-stable Lévy motion* with skewness parameter η and scale parameter σ if
(1) $\mathbf{X}(0) = 0$.
(2) $\{\mathbf{X}(t)\}$ has stationary independent increments.
(3) $\mathbf{X}(t) - \mathbf{X}(s)$ has the distribution $S_\alpha(\sigma(t-s)^{1/\alpha}, \eta, 0)$ for any $\sigma > 0$, $0 \le s < t < \infty$, and for some $0 < \alpha \le 2$, and $-1 \le \eta \le 1$.
When $\sigma = 1$, $\{\mathbf{X}(t)\}$ is said to be a *standard α-stable Lévy motion.* An α-stable Lévy motion is $1/\alpha$-self similar unless both $\alpha = 1$ and $\eta \neq 0$. The role that stable Lévy motion plays among stable processes is similar to the role that Brownian motion plays among Gaussian processes. See Samorodnitsky and Taqqu (1992) for properties of Lévy motions.

The following results was given in Lee and Whitmore (1991). Related results can also be found in Stam (1966, p. 137-138), and Samorodnitsky and Taqqu (1992).

THEOREM 4 *Assume that $\{\mathbf{X}(t)\}$ is a standard α-stable Lévy motion with $\alpha \neq 1$ and skewness parameter η, and that $\{\mathbf{T}(t)\}$ is a standard β-stable Lévy motion with $0 < \beta < 1$ and the skewness parameter* 1. *Assume also that processes $\{\mathbf{X}(t)\}$ and $\{\mathbf{T}(t)\}$ are independent. Then*
(a) *If $\alpha\beta \neq 1$, then the process $\{\mathbf{X}(\mathbf{T}(t))\}$ is an $\alpha\beta$-stable Lévy motion such that*

$$\mathbf{X}(\mathbf{T}(t)) - \mathbf{X}(\mathbf{T}(s)) \sim S_{\alpha\beta}((\kappa(t-s))^{1/\alpha\beta}, \xi, 0),$$

with $\xi = \tan(\zeta\beta)/(\tan\frac{\alpha\beta\pi}{2})$, $\kappa = (\cos\zeta\beta)(1+\eta^2\tan^2\frac{\pi\alpha}{2})^{\beta/2}(\cos\frac{\pi\alpha}{2})^{-1}$, *and* $\zeta = \arctan(\eta\tan\frac{\pi\alpha}{2})$.
(b) *If $\alpha\beta = 1$ then the process $\{\mathbf{X}(\mathbf{T}(t))\}$ is of the form $\kappa(L(t) + t\tan\beta\zeta)$, where L is a standard symmetric* 1*-stable (Cauchy) motion, and κ and ζ are as in* (a).
(c) *The process $\{\mathbf{X}(\mathbf{T}(t))\}$ has the same one-dimensional distributions as that of the process $\{(\mathbf{T}(t))^{1/\alpha}\mathbf{X}(1)\}$.*

As a special case of Theorem 4, consider k independent Brownian motions $\{\mathbf{X}_j(t)\}, j = 1, 2, \ldots, k$, such that $\mathbf{X}_j(0) = 0$ and $\mathbf{X}_j(t) \sim N(0, t/c)$ for $j = 1, 2, \ldots, k$, where $c = (\cos\frac{\pi\alpha}{4})^{-1}$ and $\alpha < 2$. Assume that $\{\mathbf{T}(t)\}$ is a standard totally right skewed $\alpha/2$ stable Lévy motion, independent of processes $\{\mathbf{X}_j(t)\}, j = 1, 2, \ldots, k$. Then for any $t > 0$ fixed, the random vector $\langle \mathbf{X}_1(\mathbf{T}(t)), \ldots, \mathbf{X}_k(\mathbf{T}(t))\rangle$ equals in distribution the symmetric α-stable random vector $\langle \mathbf{T}(t)^{1/2}\mathbf{X}_1(1), \ldots, \mathbf{T}(t)^{1/2}\mathbf{X}_k(1)\rangle$. The latter vector is a sub-Gaussian vector with a governing Gaussian vector having i.i.d. $N(0, t/c)$ components as was discussed in section 3.

Markov processes with TP_2 transition densities are useful in shock models. For example, inverse Gaussian processes, gamma processes and some right skewed stable processes have TP_2 transition densities. Lee and Whitmore (1991) show that if the transition densities of both the parent process and the operational time process are TP_2, then transition density of the derived subordinated process is also TP_2. They have the following results.

THEOREM 5 *Assume that the process* $\{\mathbf{X}_j(t)\}$ *has a transition density function* $f_{j,t}(x)$ *which is* TP_2 *in* t *and* x, *for* $j = 1, \ldots, k$, *and that* $\{\mathbf{T}(t)\}$ *has a transition density function* $u_t(s)$ *which is* TP_2 *in* t *and* s. *If the process* $\{\mathbf{T}(t)\}$ *is independent of processes* $\{\mathbf{X}_j(t)\}, j = 1, 2, \ldots, k$, *then*
(a) *The transition density function* $h_{j,t}(x)$ *of the subordinated process* $\{\mathbf{X}_j(\mathbf{T}(t))\}$ *is* TP_2 *in* t *and* x, *for* $j = 1, \ldots, k$.
(b) *For any* $t > 0$ *fixed, the random vector* $\langle \mathbf{X}_1(\mathbf{T}(t)), \ldots, \mathbf{X}_k(\mathbf{T}(t)) \rangle$ *is* TP_2 *in pairs.*

THEOREM 6 *For any two positive integers* $k_1 > 1, k_2 > 1$ *fixed, assume that* $\{\mathbf{X}_i(t)\}$ *are* n *i.i.d. totally right skewed* $1/k_2$*-stable Lévy motions,* $i = 1, \ldots, n$. *Let* $\{\mathbf{T}(t)\}$ *be a totally right skewed* $1/k_1$*-stable Lévy motion independent of processes* $\{\mathbf{X}_i(t)\}$, $i = 1, \ldots, n$. *Then, for any* $t > 0$ *fixed, the vector* $\langle \mathbf{X}_1(\mathbf{T}(t)), \ldots, \mathbf{X}_n(\mathbf{T}(t)) \rangle$ *is* $1/(k_1 k_2)$*-stable and is* TP_2 *in pairs.*

Finally, as a special case of subordination, consider an exponential time-change of α-stable Lévy motion $\{\mathbf{X}(t)\}$. Let $\mathbf{Y}(t) = \mathbf{X}(Et)$, $t \geq 0$ where E is standard exponential random variable independent of $\mathbf{X}$. Then by the self-similarity of Lévy motion, $\mathbf{Y} \stackrel{d}{=} E^{1/\alpha}\mathbf{X}$, unless $\alpha = 1$ and $\eta \neq 0$. The finite dimensional distributions of the process $\{\mathbf{Y}(t)\}$ are geometric stable. In fact, readily from Lemma 1 (c) we get

$$E \exp(i\{(Y(t_1), \ldots, Y(t_k)), \boldsymbol{\theta}\}) = \frac{1}{1 + \int_{S_n} |(\boldsymbol{\theta}, \mathbf{s})|^{\alpha} \Gamma_{t_1, \ldots, t_k}(ds)},$$

where $m_{t_1, \ldots, t_k}$ is the spectral measure of $(X(t_1), \ldots, X(t_k))$. See Rachev and Resnick (1991) for similar results on max-stable random processes and exponential time change leading to geometric max-stable processes.

REFERENCES

AKGIRAY, V. AND BOOTH, G.G. (1988). The stable–law model of stock returns. *J. Bus. Econ. Stat.* **6** 51–57.

ALAM, K. AND SAXENA, K.M.L. (1981). Positive dependence in multivariate distributions. *Comm. Statist. - Theor. Meth.* **A10** 1183–1186.

BARLOW, R.E. AND PROSCHAN, F. (1981). *Statistical Theory of Reliability and Life Testing.* Holt, Reinhart and Winston, New York.

BJERVE, S. AND DOKSUM, K. (1990). Correlation Curves: Measures of association as functions of covariate values, preprint.

BOCHNER, S. (1955). *Harmonic Analysis and the Theory of Probability.* University of California Press, Berkeley, CA.

BRINDLEY, J.E.C. AND THOMPSON, J.W.A. (1972). Dependence and aging aspects of multivariate survival. *J. Amer. Statist. Assoc.* **67** 821–829.

CLARK, P.K. (1973). A subordinated stochastic process model with finite variance for speculative prices. *Econometrica* **41** 135–155.

COHEN, J.W. (1962). Derived Markov chains. *Proc. Kon. Ned. Ak. vanWetensch.* **A65** 55–92.

DU MOUCHEL, W. H. (1983). Estimating the stable index α in order to measure tail thickness: A critique. *Ann. Statist.* **11** 1019–1031.

ESARY, J.D., PROSCHAN, F. AND WALKUP, D.W. (1967). Association of random variables, with applications. *Ann. Math. Statist.* **38** 1466–1474.

FAMA, E. (1965). The behavior of stock market prices. *J. Business* **38** 34–105.

FELDMAN, R. AND RACHEV, S.T. (1991). U–statistics of random–size samples and limit theorems for systems of Markovian particles with non–Poisson initial distributions, preprint.

FELLER, W. (1971). *An Introduction to Probability Theory and its Applications.* Volume 2, 3rd edition, Wiley, New York.

FERNIQUE, X. (1975). Regularité des trajectoires des fonctions aléatoires gaussiennes. In *Lecture Notes in Mathematics* **V.480** 1–96, Springer–Verlag, New York,

FORTUIN, C., KASTELYN, P. AND GINIBRE, J. (1971). Correlation inequalities on some partially ordered sets. *Comm. Math. Phys.* **22** 89–103.

GALTON, F. (1888). Co–relations and their measurement, chiefly from anthropometric data. *Proc. Roy. Soc. London* **45** 135–145.

HOUGAARD, P. (1986). Survival models for heterogeneous populations derived from stable distributions. *Biometrika* **73** 387–396.

JOAG–DEV, K. AND PROSCHAN, F. (1983). Negative association of random variables, with applications. *Ann. Statist.* **11** 286–295.

KARLIN, S. (1968). *Total Positivity.* Vol. 1, Stanford University Press, Stanford, CA.

KOROLEV, V. YU. (1988). The asymptotic distributions of random sums. *Lecture Notes Math.* **1412** Springer–Verlag, New York, 110–123.

LEE, M.–L.T., RACHEV, S. AND SAMORODNITSKY, G. (1990). Association of stable random variables. *Ann. Probab.* **18** 1759–1764.

LEE, M.–L.T. AND WHITMORE, G.A. (1993). Stochastics processes directed by randomized operational time. *J. Appl. Prob.* (to appear).

LEHMANN, E.L. (1966). Some concepts of dependence. *Ann. Math. Statist.* **37** 1137–1153.

MANDELBROT, B.B. (1963). New methods in statistical economics *J. Polit. Econ.* **71** 621–660.

MANDELBROT, B.B. AND TAYLOR, H.M. (1967). On the distribution of stock price differences. *Oper. Res.* **15** 1057–1062.

MELAMED, J.A. (1988). Limit theorems in the set up of summation of a random number of independent identically distributed random variables. *Lecture Notes in Math.* **1412** Springer–Verlag, New York, 194–228.

MITTNIK, S. AND RACHEV, S.T. (1991). Alternative multivariate stable distributions. In *Stable Processes and Related Topics*, S. Cambanis, G. Samorodnitsky, and M. S. Taqqu, eds. A selection of papers from MSI workshop 1990, Birkhauser, Boston, MA. 107–120.

PITT, L.D. (1982). Positively correlated normal variables are associated. *Ann. Probab.* **10** 496–499.

RACHEV, S.T. AND RÜSELENDORF, L. (1991). On the Cox, Ross and Rubinstein model for option prices, preprint.

RACHEV, S.T. AND SAMORODNITSKY, G. (1991). Option pricing formulae for speculative prices modelled by subordinated stochastic process, preprint.

RACHEV, S.T. (1991). Geometric infinitely divisible and geometric stable distributions on Banach spaces, preprint.

RACHEV, S.T. AND SENGUPTA, A. (1991). Geometric stable distributions and Laplace–Weibull mixtures. *Statist. Decision* **9** 327–373.

RACHEV, S.T. AND RESNICK, S. (1991). Max–geometric infinite divisibility and stability. *Stochastic Models* **2** 191–218.

RESNICK, S.I. (1988). Association and multivariate extreme value distributions. *Gani Festschrift: Studies in Statistical Modeling and Statistical Science*, C.C. Heyde, ed. Statistical Society of Australia, 261–271.

RESNICK, S. AND GREENWOOD, P. (1979). A bivariate stable characterization and domains of attraction. *J. Multivariate Anal.* **10** 206–221.

RVAČEVA, E.L. (1962). On domains of attraction of multidimensional distributions. Selected Translations in *Math. Statist. Prob. Theory* **2** 183–205.

SAMORODNITSKY, G. AND TAQQU, M. (1990). Stochastic monotonicity and Slepian–type inequalities for infinitly divisible and stable random vectors, preprint.

SAMORODNITSKY, G. AND TAQQU, M. (1991). Conditional moments for stable random variables. *Stoch. Proc. Appl.* (to appear).

SAMORODNITSKY, G. AND TAQQU, M. (1992). Non–Gaussian stable processes, preprint.

SHAKED, M. (1977). A concept of positive dependence for exchangeable variables. *Ann. Statist.* **5** 505–515.

SHAKED, M. AND TONG, Y.L. (1985). Some partial orderings of exchangeable random variables by positive dependence. *J. Multivariate Anal.* **17** 339–349.

STAM, A.J. (1965). Derived stochastic processes. *Compositio Math.* **17** 102–140.

SZASZ, D. (1972). Limit theorems for the distributions of the sums of a random number of random variables. *Ann. Math. Statist.* **43** 1902–1913.

SZYNAL, D. (1976). On limit distribution theorems for sums of a random number of random variables appearing in the study of rarefaction of a recurrent process. *Zasfosowania matematyki* **15** 277–288.

TONG, Y.L. (1977). An ordering theorem for conditionally independent and identically distributed random variables. *Ann. Statist.* **5** 505–515.

TONG, Y.L. (1980). *Probability Inequalities in Multivariate Distributions.* Academic Press, New York.

TONG, Y.L. (1990). *The Multivariate Normal Distribution.* Springer–Verlag, New York.

WHITMORE, G.A. AND LEE, M.-L.T. (1991). Inferences for Poisson-Hougaard processes, preprint.

WU, W. AND CAMBANIS, S. (1991). Conditional Variance of Symmetric Stable Variables. In *Stable Processes and Related Topics*, S. Cambanis, G. Samorodnitsky, and M. S. Taqqu, eds. A selection of papers from MSI workshop, 1990, Birkhauser, Boston, MA. 85-100.

DEPARTMENT OF STATISTICS
HARVARD UNIVERSITY
CAMBRIDGE, MA 02138

DEPT. OF STATISTICS & APPLIED PROBABILITY
UNIVERSITY OF CALIFORNIA, SANTA BARBARA
SANTA BARBARA, CA 93106

SCHOOL OF OPERATIONS RESEARCH
AND INDUSTRIAL ENGINEERING
CORNELL UNIVERSITY
ITHACA, NY 14853

Stochastic Inequalities
IMS Lecture Notes – Monograph Series
Volume 22 (1993)

A MULTIVARIATE STOCHASTIC ORDERING BY THE MIXED DESCENDING FACTORIAL MOMENTS WITH APPLICATIONS[1]

By CLAUDE LEFÈVRE and PHILIPPE PICARD

Université Libre de Bruxelles and Université de Lyon 1

A stochastic order relation for discrete random vectors is introduced that relies on the mixed descending factorial moments. Connection with more usual orderings is pointed out through a hierarchical classification. The order relation is then used for comparing the state of a population which is subjected to certain damage processes by death, sampling or infection. In particular, for the multipopulation collective epidemic model, it allows us to establish in which sense the ultimate numbers of susceptibles do decrease with the infectivity level of the infectives. This paper extends to the multivariate case a recent work by the authors.

1. Introduction

In a previous paper (Lefèvre and Picard (1991)), we introduced an order relation for $\mathbb{N}$-valued random variables, unusual in the literature, that relies on the descending factorial moments; for this reason, we called it *the factorial ordering.* Our original motivation came from the epidemic context, namely to make precise in which probabilistic terms the total damage caused by the disease in a collective Reed-Frost epidemic model can indeed be viewed as an increasing function of the infection intensity exerted by the infectives. Further applications occur when comparing certain sampling procedures through the number of unsampled individuals. In particular, we used the ordering to obtain qualitative results for a reinforcement-depletion urn model and for a non-linear death process.

Our purpose here is to construct a multivariate version of this ordering based on the mixed descending factorial moments, and then to illustrate its relevance with some applications in the same fields. The ordering is derived in Section 2 through a hierarchical classification of various potential order

[1]Research supported in part by the Institut National de la Santé et de la Recherche Médicale under contrat n°898014.

AMS 1991 *subject classifications.* 60C05, 60E15, 60K99, 92A15.

Key words and phrases. Stochastic order relations, mixed factorial moments, compartmental urn model, death process, collective epidemic model, family of polynomials.

relations for discrete random vectors. Connection with more classical orderings follows easily. In Section 3, we use it to compare the size of a population subjected to certain damage schemes. This allows us to generalize the qualitative analysis for the urn model and the death process mentioned above. Section 4 is concerned with the collective epidemic model, this time for an heterogeneous population. Thanks to the ordering, we are in a position to establish a monotonicity property of the ultimate numbers of susceptibles with respect to the infectivity level of the infectives. To this end, we adopt the approach developed recently in Picard and Lefèvre (1990) and which has recourse to a special family of polynomials with several variables defined in Lefèvre and Picard (1990). The method is direct, though rather technical, and has the merit to emphasize the interest and the flexibility of these polynomials.

2. Ordering Random Vectors by the Mixed Descending Factorial Moments

A number of stochastic order relations have been proposed to compare random vectors (see, e.g., Stoyan (1983)). We are going to derive a hierarchical classification of various potential multivariate stochastic orderings for discrete vectors. As a consequence, the ordering of interest by the mixed descending factorial moments will then emerge in a simple and natural way. For simplicity, but without loss of generality, we only consider bidimensional random vectors. We mention that the presentation below extends the one followed in Lefèvre and Picard (1991) for the univariate case; a letter I will be added to the numbering when referring to the associated formula in that paper.

2.1. *A Sequence of Remarkable Cones of Functions*

Let us consider the cone $\mathcal{F}_2$ of the functions $f(x_1, x_2)$ from $\mathbb{N}^2$ to $\mathbb{R}^+$. We start by constructing in $\mathcal{F}_2$ a sequence of remarkable cones $\mathcal{F}_2^{(1)}, \ldots, \mathcal{F}_2^{(6)}$. Put $1(A)$ as the indicator function of A, and for i, $j \in \mathbb{N}$, let $j_{[i]} = j(j-1)\ldots(j-i+1)$ and $j^{[i]} = j(j+1)\ldots(j+i-1)$, with $j_{[0]} = j^{[0]} = 1$.

DEFINITION 2.1 For $j = 1, \ldots, 5$ and $(i_1, i_2) \in \mathbb{N}^2$, let

$$e_{i_1,i_2}^{(j)}(x_1, x_2) = e_{i_1}^{(j)}(x_1) e_{i_2}^{(j)}(x_2), \tag{2.1}$$

with $e_i^{(j)}(x)$ given respectively by

$$\begin{aligned} &e_i^{(1)}(x) = 1(x \geq i), \\ (2.2) \qquad &e_0^{(2)}(x) = 1 \text{ and } e_i^{(2)}(x) = (x - i + 1)^+, i = 1, 2, \ldots \\ &e_i^{(3)}(x) = x_{[i]} \;, \; e_i^{(4)}(x) = x^i \;, \; e_i^{(5)}(x) = x^{[i]}. \end{aligned}$$

Then, $\mathcal{F}_2^{(j)}, j = 1, \ldots, 5$, is defined as the cone of the functions $f(x_1, x_2)$ in $\mathcal{F}_2$ that can be expressed as a linear combination (finite or not) with positive coefficients of the functions of the family $\{e_{i_1,i_2}^{(j)}(x_1, x_2), (i_1, i_2) \in \mathbb{N}^2\}$; in short, $\mathcal{F}_2^{(j)}$ is said to be generated by the $e_{i_1,i_2}^{(j)}(x_1, x_2)$. Similarly, $\mathcal{F}_2^{(6)}$ is the cone generated by the elements of the family $\{e_{a_1,a_2}^{(6)}(x_1, x_2)$, for any reals $a_1, a_2 > 1\}$, where

$$e_{a_1,a_2}^{(6)}(x_1, x_2) = e_{a_1}^{(6)}(x_1) e_{a_2}^{(6)}(x_2),$$

with

$$(2.3) \qquad e_a^{(6)}(x) = a^x.$$

The first three cones can be characterized equivalently as follows. We denote by $\Delta^{i_1,i_2}(\nabla^{i_1,i_2})f(j_1, j_2)$, (i_1, i_2) and $(j_1, j_2) \in \mathbb{N}^2$, the forward (backward) difference of $f(x_1, x_2)$ of orders i_1 in x_1 and i_2 in x_2 evaluated at $(x_1, x_2) = (j_1, j_2)$. For $i, j \in \mathbb{N}$, we put $i \wedge j = \min(i, j)$.

PROPERTY 2.2 *$\mathcal{F}_2^{(j)}$, $j = 1, 2, 3$, is the cone of the functions $f(x_1, x_2)$ in $\mathcal{F}_2$ such that, for $(i_1, i_2) \in \mathbb{N}^2$, $\nabla^{1\wedge i_1, 1\wedge i_2} f(i_1, i_2) \geq 0$ when $j = 1$, $\nabla^{2\wedge i_1, 2\wedge i_2} f(i_1, i_2) \geq 0$ when $j = 2$, and $\Delta^{i_1,i_2} f(0, 0) \geq 0$ when $j = 3$.*

PROOF Fix $j = 1, 2$ or 3. We first observe that any function $f(x_1, x_2)$ in $\mathcal{F}_2$ can be expanded in terms of the $e_{i_1,i_2}^{(j)}(x_1, x_2)$ as

$$(2.4) \qquad f(x_1, x_2) = \sum_{i_1=0}^{\infty} \sum_{i_2=0}^{\infty} \alpha_{i_1,i_2}^{(j)} e_{i_1,i_2}^{(j)}(x_1, x_2)$$

for some appropriate coefficients $\alpha_{i_1,i_2}^{(j)}$. Indeed, by (2.1) and (2.2), the summation in (2.4) is, for any given $(x_1, x_2) \in \mathbb{N}^2$, a finite sum, so that the $\alpha_{i_1,i_2}^{(j)}$ may be determined recursively. Now, we proved in (I, 2.2) and (I, 2.5) that

$$(2.5) \qquad \nabla^{1\wedge i} e_k^{(1)}(i) = \nabla^{2\wedge i} e_k^{(2)}(i) = \Delta^i e_k^{(3)}(0)/i! = 1(k = i) \;, \quad i, k \in \mathbb{N}.$$

Combining (2.5) with (2.1), (2.2) and (2.4), we obtain successively that

$$\begin{aligned} &\nabla^{1\wedge i_1, 1\wedge i_2} f(i_1, i_2) = \alpha_{i_1,i_2}^{(1)}, \\ (2.6) \qquad &\nabla^{2\wedge i_1, 2\wedge i_2} f(i_1, i_2) = \alpha_{i_1,i_2}^{(2)}, \\ &\Delta^{i_1,i_2} f(0, 0)/i_1! i_2! = \alpha_{i_1,i_2}^{(3)}, \quad (i_1, i_2) \in \mathbb{N}^2. \end{aligned}$$

By definition, $f(x_1, x_2)$ is in $\mathcal{F}_2^{(j)}$ iff the $\alpha_{i_1,i_2}^{(j)}$ are in $\mathbb{R}^+$. From (2.6), the above characterizations of $\mathcal{F}_2^{(j)}$, $j = 1,2,3$, are then straightforward. □

REMARK 2.3 From Property 2.2, we easily deduce that $\mathcal{F}_2^{(j)}$, $j = 1,2,3$, contains any function $f(x_1, x_2)$ in $\mathcal{F}_2$ that can be factorized as

$$f(x_1, x_2) = f_1(x_1) f_2(x_2), \tag{2.7}$$

where $f_1(x)$ and $f_2(x)$ are functions from $\mathbb{N}$ to $\mathbb{R}^+$ which are increasing for $j = 1$, increasing and convex for $j = 2$, and such that $\Delta^i f_1(0)$ and $\Delta^i f_2(0) \geq 0$, $i \in \mathbb{N}$, for $j = 3$.

We now show that the six cones decrease in the inclusion sense.

PROPERTY 2.4 $\mathcal{F}_2^{(j)} \supset \mathcal{F}_2^{(j+1)}$, $j = 1, \ldots, 5$.

PROOF Using Property 2.2, we observe that $\mathcal{F}_2^{(1)} \supset \mathcal{F}_2^{(2)}$ obviously, and $\mathcal{F}_2^{(2)} \supset \mathcal{F}_2^{(3)} \supset \mathcal{F}_2^{(4)}$ because by (2.1) and (2.2),

$$\nabla^{2\wedge i_1, 2\wedge i_2} e_{k_1,k_2}^{(3)}(i_1, i_2) = \nabla^{2\wedge i_1} i_{1,[k_1]} \nabla^{2\wedge i_2} i_{2,[k_2]} \geq 0,$$

$$\nabla^{i_1,i_2} e_{k_1,k_2}^{(4)}(0,0) = \Delta^{i_1} 0^{k_1} \Delta^{i_2} 0^{k_2} \geq 0,$$

for (i_1, i_2) and $(k_1, k_2) \in \mathbb{N}^2$, respectively. Moreover, $\mathcal{F}_2^{(4)} \supset \mathcal{F}_2^{(5)} \supset \mathcal{F}_2^{(6)}$ since $x^{[i]}$, $i \in \mathbb{N}$, can be generated by the x^j, $j \in \mathbb{N}$, and a^x, $a > 1$, by the $x^{[j]}$, $j \in \mathbb{N}$ (see (I, 2.7)). □

2.2. *The Induced Stochastic Order Relations*

Let us denote by $\mathcal{D}_2$ the space of the $\mathbb{N}^2$-valued random vectors. To each of the cones of Definition 2.1, we can associate an order relation on $\mathcal{D}_2$ as follows. Let $\mathbf{X} = (X_1, X_2)$ and $\hat{\mathbf{X}} = (\hat{X}_1, \hat{X}_2)$ be r.v.s in $\mathcal{D}_2$.

DEFINITION 2.5 $\mathbf{X}$ is smaller than $\hat{\mathbf{X}}$ in the $\leq_j$ sense (written $\mathbf{X} \leq_j \hat{\mathbf{X}}$), $j = 1, \ldots, 6$, when

$$E[f(X_1, X_2)] \leq E[f(\hat{X}_1, \hat{X}_2)] \text{ for any function in } \mathcal{F}^{(j)}, \tag{2.8}$$

that is, equivalently,

$$\begin{aligned} E[e_{i_1,i_2}^{(j)}(X_1, X_2)] &\leq E[e_{i_1,i_2}^{(j)}(\hat{X}_1, \hat{X}_2)] \text{ for } (i_1, i_2) \in \mathbb{N}^2, \\ &\quad \text{when } j = 1, \ldots, 5, \\ E[e_{a_1,a_2}^{(6)}(X_1, X_2)] &\leq E[e_{a_1,a_2}^{(6)}(\hat{X}_1, \hat{X}_2)] \text{ for } a_1, a_2 > 1, \\ &\quad \text{when } j = 6. \end{aligned} \tag{2.9}$$

REMARK 2.6 From (2.8), (2.9) and Remark 2.3, we obtain directly the following characterization of the first three orderings. For $j = 1, 2, 3$, $\mathbf{X} \leq_j \hat{\mathbf{X}}$ iff

$$E[f_1(X_1)f_2(X_2)] \leq E[f_1(\hat{X}_1)f_2(\hat{X}_2)] \tag{2.10}$$

for any functions f_1 and f_2 from $\mathbb{N}$ to $\mathbb{R}^+$ that are increasing when $j = 1$, increasing and convex when $j = 2$, and such that $\Delta^i f_1(0)$ and $\Delta^i f_2(0) \geq 0$, $i \in \mathbb{N}$, when $j = 3$.

The orderings $\leq_1$ and $\leq_2$ correspond to those introduced by Bergmann (1978), for discrete or not random vectors. At our knowledge, the four others have not been investigated so far in the literature. In fact, from (2.1), (2.2) and (2.9), we see that $\leq_3$ compares the mixed descending factorial moments of $\mathbf{X}$ and $\hat{\mathbf{X}}$, $\leq_4$ their moments about zero, $\leq_5$ their ascending factorial moments and $\leq_6$ the expected value of increasing exponentials of their components.

By Property 2.4, the six orderings in Definition 2.5 decrease in the strength sense.

PROPERTY 2.7 *$\mathbf{X} \leq_j \hat{\mathbf{X}}$ implies $\mathbf{X} \leq_{j+1} \hat{\mathbf{X}}$, $j = 1, \ldots, 5$.*

These order relations generalize those defined in (I, Section 2) for $\mathbb{N}$-valued random variables. Furthermore, the following connection is immediate from (2.9).

PROPERTY 2.8 *For $j = 1, \ldots, 6$,*

$$\mathbf{X} \leq_j \hat{\mathbf{X}} \text{ implies } X_1 \leq_j \hat{X}_1 \text{ and } X_2 \leq_j \hat{X}_2. \tag{2.11}$$

When X_1, X_2, as well as $\hat{X}_1, \hat{X}_2$, are independent, then the converse of (2.11) is true.

2.3. *The So-Called Factorial Ordering*

For the sequel, we will mainly use the order relation $\leq_3$. As it compares random vectors through their mixed descending factorial moments, we keep the name given in I of *factorial ordering*, with the notation $\leq_F$. In addition, we will limit our attention to the subspace $\mathcal{D}_{n_1,n_2}$ in $\mathcal{D}_2$ of the random vectors $\mathbf{X} = (X_1, X_2)$ with X_1 and X_2 valued in the sets $\{0, 1, \ldots, n_1\}$ and $\{0, 1, \ldots, n_2\}$, respectively. Thus, for $\mathbf{X}, \hat{\mathbf{X}} \in \mathcal{D}_{n_1,n_2}$, $\mathbf{X} \leq_F \hat{\mathbf{X}}$ when

$$E[X_{1,[i_1]}X_{2,[i_2]}] \leq E[\hat{X}_{1,[i_1]}\hat{X}_{2,[i_2]}] \quad , \quad 0 \leq i_1 \leq n_1, 0 \leq i_2 \leq n_2. \tag{2.12}$$

We note that within $\mathcal{D}_{n_1,n_2}$, all the orderings of Definition 2.5 satisfy the axioms of partial order relation. Moreover, $\leq_6$ is now closely tied with the probability generating function ordering ($\leq_g$ in Stoyan (1983)). For $\mathbf{X}, \hat{\mathbf{X}} \in \mathcal{D}_2$, $\mathbf{X} \leq_g \hat{\mathbf{X}}$ when

$$E\left(z_1^{X_1} z_2^{X_2}\right) \geq E\left(z_1^{\hat{X}_1} z_2^{\hat{X}_2}\right) \quad , \quad 0 \leq z_1, z_2 \leq 1. \tag{2.13}$$

From (2.3) and (2.9), we thus deduce that for $\mathbf{X}, \hat{\mathbf{X}} \in \mathcal{D}_{n_1,n_2}$,

$$\mathbf{X} \leq_6 \hat{\mathbf{X}} \text{ iff } \mathbf{n} - \mathbf{X} \geq_g \mathbf{n} - \hat{\mathbf{X}}, \tag{2.14}$$

where $\mathbf{n} - \mathbf{X}$ denotes the vector $(n_1 - X_1, n_2 - X_2)$. As a consequence, when comparing $\leq_F$ with more usual orderings, we have the following implications: for $\mathbf{X}, \hat{\mathbf{X}} \in \mathcal{D}_{n_1,n_2}$,

$$\mathbf{X} \leq_2 \hat{\mathbf{X}} \Rightarrow \mathbf{X} \leq_F \hat{\mathbf{X}} \Rightarrow \mathbf{n} - \mathbf{X} \geq_g \mathbf{n} - \hat{\mathbf{X}}. \tag{2.15}$$

3. Comparison of the Outcome of Certain Damage Procedures

We are going to show that the factorial ordering is a well-adapted notion when comparing certain damage procedures through the number of unhurt individuals. As main applications, we will use it for two particular situations, namely a non-linear death process and a reinforcement-depletion urn model. The results extend in several ways those obtained in (I, Section 3) - and earlier ones.

3.1. *A Single Population Subjected to a Death Risk*

Consider a population of initial size n which shares a death risk. We denote by $T_i, i = 1, \ldots, n$, the lifetime of individual i. The T_i are assumed to be exchangeable; this hypothesis, however, could be removed without difficulty. We are interested in the number X_t of individuals surviving at time $t, t \in \mathbb{R}^+$ (or $\mathbb{N}$).

Let t_1, t_2 be any two instants with $t_1 < t_2$. Fix then k_1 and k_2 in $[1, n]$. We can write that

$$\binom{X_{t_1}}{k_1} = \sum 1(T_{\alpha_1} > t_1, \ldots, T_{\alpha_{k_1}} > t_1), \tag{3.1}$$

where the sum is over the $\binom{n}{k_1}$ groups of k_1 distinct individuals $\alpha_1, \ldots, \alpha_{k_1}$. An analogous formula is valid for $\binom{X_{t_2}}{k_2}$. From (3.1), we then obtain that

$$\binom{X_{t_1}}{k_1}\binom{X_{t_2}}{k_2} = \tag{3.2}$$

$$\sum 1(T_{\alpha_1} > t_1, \ldots, T_{\alpha_{k_1}} > t_1; T_{\beta_1} > t_2, \ldots, T_{\beta_{k_2}} > t_2),$$

where the sum is over the $\binom{n}{k_1}\binom{n}{k_2}$ groups of k_1 distinct individuals $\alpha_1, \ldots, \alpha_{k_1}$ and k_2 distinct individuals $\beta_1, \ldots, \beta_{k_2}$. Certain of the α_i and β_j individuals may be identical, of course; in fact, there can exist $k = 0, \ldots, \min(k_1, k_2)$ individuals in common. For a given value of k, since $t_1 < t_2$, the sum contains $\binom{n}{k_1 - k, k_2}$ indicator functions of the type $1(T_1 > t_1, \ldots, T_{k_1-k} > t_1$ - provided $k_1 - k \geq 1$; $T_{k_1+1} > t_2, \ldots, T_{k_1+k_2} > t_2)$. Therefore, taking the expectation in (3.2) yields for the mixed descending factorial moments

$$(3.3) \qquad E\left[X_{t_1,[k_1]} X_{t_2,[k_2]}\right] = \sum_{k=0}^{\min(k_1,k_2)} k_{1,[k]} n_{[k_1+k_2-k]} P(T_1 > t_1, \ldots, T_{k_1-k} > t_1; T_{k_1+1} > t_2, \ldots, T_{k_1+k_2} > t_2).$$

We note that when k_2 (e.g.)$= 0$, (3.3) is easily adapted and becomes

$$(3.4) \qquad E\left[X_{t_1,[k_1]}\right] = n_{[k_1]} P(T_1 > t_1, \ldots, T_{k_1} > t_1).$$

Consider now a similar model characterized by the lifetimes $\hat{T}_i, i = 1, \ldots, n$. Let $\hat{X}_t, t \in \mathbb{R}^+$, be the state of the population at time t. Using the factorial ordering (2.12), we deduce from (3.3) and (3.4) the following comparison.

PROPOSITION 3.1 *Let $t_1 \neq t_2$. If for any $\tau_1, \ldots, \tau_i$, $i \in [1, n]$, taken in $\{t_1, t_2\}$,*

$$(3.5) \qquad P(T_1 > \tau_1, \ldots, T_i > \tau_i) \leq P(\hat{T}_1 > \tau_1, \ldots, \hat{T}_i > \tau_i),$$

then

$$(3.6) \qquad (X_{t_1}, X_{t_2}) \leq_F (\hat{X}_{t_1}, \hat{X}_{t_2}).$$

3.1.1. A non-linear death process

A special case of the model arises when $X_t, t \in \mathbb{R}^+$, is governed by a non-linear Markovian death process. Here, given $X_t = x$, $x = 1, \ldots, n$, each of the x individuals still alive at t can die, during $(t, t + dt)$, independently of the others and with the probability $\xi(x)dt$, where $\xi(x)$ is some positive function of the current state x.

Now, the (unconditional) lifetimes T_i are clearly exchangeable and interdependent. Ball and Donnelly (1987) investigated the nature of that dependence; see also Lefèvre and Michaletzky (1990) and an amendment by Donnelly (1991). They proved that if

(3.7) $\quad \xi(x), x = 1, \ldots, n,$ forms an increasing (decreasing) sequence,

then for any $\tau_1, \ldots, \tau_i, i \in [1, n]$,

$$(3.8) \qquad P(T_1 > \tau_1, \ldots, T_i > \tau_i) \leq (\geq) P(T_1 > \tau_1) \ldots P(T_i > \tau_i).$$

We can use this result to compare the model with an approximated one where the individuals would behave independently. Specifically, consider a population of n individuals whose lifetimes $\hat{T}_i, i = 1, \ldots, n$, are i.i.d., with the same marginal distribution as the original T_i. Thus, for any $\tau_1, \ldots, \tau_i$, $i \in [1, n]$,

$$(3.9) \qquad P(\hat{T}_1 > \tau_1, \ldots, \hat{T}_i > \tau_i) = P(T_1 > \tau_1) \ldots P(T_i > \tau_1).$$

Let $\hat{X}_t, t \in \mathbb{R}^+$, be the new population state at time t. From (3.7), (3.8), (3.9) and Proposition 3.1, we then deduce that

(3.10) the condition (3.7) implies $(X_{t_1}, X_{t_2}) \leq_F (\geq_F)(\hat{X}_{t_1}, \hat{X}_{t_2}), \quad t_1 \neq t_2.$

3.1.2. A reinforcement-depletion urn model

We turn now to an urn model, developed by Shenton (1981), with successive reinforcement-depletions of random size. The urn contains initially n white balls and m black balls. At stage $t, t = 1, 2, \ldots$, the black balls are reinforced by the addition of a random number R_t of extra black balls. All the balls are then uniformly mixed, and depletion occurs as a sample of balls, of the same size R_t, is drawn without replacement from the urn. Attention centers on the number X_t of white balls that remain in the urn just after stage t.

This model can be viewed as a particular case of the model above by simply assimilating the sampling of a white ball from the urn to its death. Thus, $T_i, i = 1, \ldots, n$, represents here the time period white ball i will spend in the urn. These T_i are exchangeable. Moreover, let $\tau_1 \leq \tau_2 \leq \ldots \leq \tau_i$, $i \in [1, n]$, be the termination time of various stages. By conditioning on the event $A = [R_t = r_t, t = 1, \ldots, \tau_i]$, we obtain

$$(3.11) \qquad P(T_1 > \tau_1, \ldots, T_i > \tau_i | A) =$$
$$\prod_{j=1}^{i} \prod_{u_j = \tau_{j-1}+1}^{\tau_j} \binom{n + m + r_{u_j} - (i - j + 1)}{r_{u_j}} \Big/ \binom{n + m + r_{u_j}}{r_{u_j}},$$

where we put $\tau_0 = 0$ and $\prod_{u_j} = 1$ when $\tau_{j-1} = \tau_j$. From (3.11), we then deduce that

$$(3.12)\quad P(T_1 > \tau_1, \ldots, T_i > \tau_i) = \left\{ \prod_{j=1}^{i} \left[(n+m)_{[i-j+1]} \right]^{\tau_j - \tau_{j-1}} \right\} \times E \left\{ 1 / \prod_{j=1}^{i} \prod_{u_j = \tau_{j-1}+1}^{\tau_j} (n + m + R_{u_j})_{[i-j+1]} \right\}.$$

Recently, Donnelly and Whitt (1989;Section 4) examined for the model some effects of more variable reinforcement-depletion sizes. Their results can be strengthened, as shown below for their Corollary 4.1; for brevity, that corollary is not recalled. We begin by introducing a further stochastic ordering for random vectors which was proposed before by Bergmann (1978). In the notations of Section 2, $\mathbf{X}$ is smaller that $\hat{\mathbf{X}}$ in the $\leq_{2d}$ sense when the inequality (2.10) holds for any functions f_1 and f_2 from $\mathbb{N}$ to $\mathbb{R}^+$ that are decreasing and convex. It is easily seen that an analogous inequality is then valid for the cone of the functions $f(x_1, x_2)$ in $\mathcal{F}_2$ that are generated by the elements of the family $\{(i_1 - x_1)^+(i_2 - x_2)^+, (i_1, i_2) \in \mathbb{N}^2\}$. Now, consider another urn model with random sizes $\hat{R}_t, t = 1, 2, \ldots$, and let $\hat{T}_i, i = 1, \ldots, n$, be the lifetime of white ball i. We observe that (3.12) is the expectation of the product of τ_i functions with arguments $R_t, t = 1, \ldots, \tau_i$, respectively, each of these functions being decreasing and convex. Therefore, applying $\leq_{2d}$, we deduce from (3.12) and Proposition 3.1 that

$$(3.13)\quad (R_t, t = 1, \ldots, t_2) \leq_{2d} (\hat{R}_t, t = 1, \ldots, t_2) \text{ implies } (X_{t_1}, X_{t_2}) \leq_F (\hat{X}_{t_1}, \hat{X}_{t_2}), t_1 < t_2.$$

3.2. *A Multipopulation Subjected to a Sampling*

Consider a bipopulation of n_1 individuals of type 1 and n_2 individuals of type 2, subjected jointly to a sampling procedure. Let $B_{1,i}(B_{2,j})$ be the event that individual i in population 1 (j in population 2) is not drawn. The $B_{1,i}$, $i = 1, \ldots, n_1$, are supposed to be exchangeable, as well as the $B_{2,j}$, $j = 1, \ldots, n_2$; this hypothesis, however, is not essential. We are concerned with the vector (X_1, X_2) of the numbers of unsampled individuals from populations 1 and 2, respectively.

Fix k_1 in $[1, n_1]$ and k_2 in $[1, n_2]$. Arguing as for (3.2), we obtain that

$$(3.14)\quad \binom{X_1}{k_1}\binom{X_2}{k_2} = \sum 1 \left(B_{1,\alpha_1} \cap \ldots \cap B_{1,\alpha_{k_1}} \cap B_{2,\beta_1} \cap \ldots \cap B_{2,\beta_{k_2}} \right),$$

where the sum is over the $\binom{n_1}{k_1}\binom{n_2}{k_2}$ groups of k_1 distinct individuals $\alpha_1,\ldots,\alpha_{k_1}$ in population 1 and k_2 distinct individuals $\beta_1,\ldots,\beta_{k_2}$ in population 2. Let $q(k_1,k_2)$ denote the probability that any given group of individuals of that kind is not drawn from the population; thus

$$q(k_1,k_2) = P(B_{1,1}\cap\ldots\cap B_{1,k_1}\cap B_{2,1}\cap\ldots\cap B_{2,k_2}). \tag{3.15}$$

From (3.14) and (3.15), we then deduce that

$$E\left[X_{1,[k_1]}X_{2,[k_2]}\right] = n_{1,[k_1]}n_{2,[k_2]}q(k_1,k_2). \tag{3.16}$$

We note that when k_2 (e.g.)$=0$, (3.16) is still true provided we put $q(k_1,0) = P(B_{1,1}\cap\ldots\cap B_{1,k_1})$, with $q(0,0)=1$.

Suppose now that another sampling is based on the parameters $\hat{q}(k_1,k_2)$, $0\le k_1\le n_1$, $0\le k_2\le n_2$, and let $(\hat{X}_1,\hat{X}_2)$ be the resulting size of the unsampled populations. The characterization of $\underset{F}{\le}$ below follows then directly from (3.16).

PROPOSITION 3.2

(3.17) $(X_1,X_2)\le_F(\hat{X}_1,\hat{X}_2)$ *iff*

$$q(k_1,k_2)\le\hat{q}(k_1,k_2)\ ,\ 0\le k_1\le n_1, 0\le k_2\le n_2.$$

3.2.1. A sampling with random size

Let us examine the special sampling that consists in taking, with or without replacement, random numbers of individuals R_1 and R_2, possibly dependent, from populations 1 and 2, respectively. Such a situation can arise, for instance, when modelling the infection process in epidemic models (see 4.3(i) below).

For a sampling with replacement, we obtain, for $0\le k_1\le n_1$, $0\le k_2\le n_2$,

$$q(k_1,k_2) = E\left\{[(n_1-k_1)/n_1]^{R_1}[(n_2-k_2)/n_2]^{R_2}\right\}. \tag{3.18}$$

Consider a similar sampling with random sizes $\hat{R}_1$ and $\hat{R}_2$. Using the definition (2.13) of the $\le_g$ ordering, we then deduce from (3.17) and (3.18) that

$$(R_1,R_2)\ge_g(\hat{R}_1,\hat{R}_2) \text{ implies } (X_1,X_2)\le_F(\hat{X}_1,\hat{X}_2). \tag{3.19}$$

When the sampling is done without replacement, we have, for $0\le k_1\le n_1$, $0\le k_2\le n_2$,

$$q(k_1,k_2) = E\left[(n_1-R_1)_{[k_1]}(n_2-R_2)_{[k_2]}\right]/n_{1,[k_1]}n_{2,[k_2]}, \tag{3.20}$$

so that from (3.17) and (3.20),

(3.21) $(X_1, X_2) \leq_F (\hat{X}_1, \hat{X}_2)$ iff $(n_1 - R_1, n_2 - R_2) \leq_F (n_1 - \hat{R}_1, n_2 - \hat{R}_2)$.

3.2.2. An extended urn model

We generalize the reinforcement-depletion urn model described in 3.1.2 by putting in the urn balls of three different colours, white, red and black, in initial numbers n_1, n_2 and m, respectively. As before, at stage $t, t = 1, 2, \ldots$, the black colour is reinforced with a random number R_t of balls, and just after, R_t balls are drawn without replacement from the urn. Interest centers on the vector (X_{1,t_1}, X_{2,t_2}) where $X_{1,t_1}(X_{2,t_2})$ represents the number of white (red) balls that remain in the urn immediately after stage $t_1(t_2)$.

From (3.15) and (3.16), we have, for $k_1 \in [1, n_1]$, $k_2 \in [1, n_2]$,

$$E\left[X_{1,t_1,[k_1]} X_{2,t_2,[k_2]}\right] = n_{1,[k_1]} n_{2,[k_2]} q(k_1, k_2; t_1, t_2), \tag{3.22}$$

where $q(k_1, k_2; t_1, t_2)$ denotes the probability that any given group of k_1 white balls and k_2 red balls is still in the urn just after stages t_1 and t_2, respectively. Choose $t_1 \leq t_2$, for example. By first conditioning on $[R_t, t = 1, \ldots, t_2]$, we then obtain

$$\begin{aligned} q(k_1, k_2; t_1, t_2) &= E\left\{ \prod_{j=1}^{2} \prod_{u_j = t_{j-1}+1}^{t_j} \frac{\binom{n_1 + n_2 + m + R_{u_j} - k_j - k_{j+1}}{R_{u_j}}}{\binom{n_1 + n_2 + m + R_{u_j}}{R_{u_j}}} \right\} \\ &= \left\{ \prod_{j=1}^{2} \left[(n_1 + n_2 + m)_{[k_j + k_{j+1}]} \right]^{t_j - t_{j-1}} \right\} \\ &\quad \times E\left\{ 1 \Big/ \prod_{j=1}^{2} \prod_{u_j = t_{j-1}+1}^{t_j} (n_1 + n_2 + m + R_{u_j})_{[k_j + k_{j+1}]} \right\}, \end{aligned} \tag{3.23}$$

where we put $t_0 = 0$, $k_3 = 0$ and $\prod_{u_2} = 1$ if $t_1 = t_2$. We note that the formulae (3.22) and (3.23) are easily adapted when k_2 (e.g.)$= 0$.

Consider now another urn model with random sizes $\hat{R}_t, t = 1, 2, \ldots$, and let $\hat{X}_{1,t_1}(\hat{X}_{2,t_2})$ be the resulting number of white (red) balls just after stage $t_1(t_2)$. Using, as for (3.13), the $\leq_{2d}$ ordering, we deduce from (3.17), (3.22) and (3.23) that

$$\begin{aligned} &(R_t, t = 1, \ldots, t_2) \leq_{2d} (\hat{R}_t, t = 1, \ldots, t_2) \text{ implies} \\ &\qquad (X_{1,t_1}, X_{2,t_2}) \leq_F (\hat{X}_{1,t_1}, \hat{X}_{2,t_2}), t_1 \leq t_2. \end{aligned} \tag{3.24}$$

4. Comparison of the Final Outcome of Collective Epidemics

In (I, Section 4), we showed the effect of increased infection intensity on the total damage caused by a collective epidemic process. We are going to generalize our analysis by examining this time the case of an heterogeneous population. For clarity, that population is supposed to contain only two different groups of individuals. The approach relies on results derived in Lefèvre and Picard (1990) and Picard and Lefèvre (1990); these results will be referred with a supplementary letter II or III, respectively. We begin by establishing a comparison property that involves a family of polynomials with several variables introduced in II. We then apply it to the epidemic model formulae obtained in III.

4.1. *A Property of the Family of Polynomials*

Let us recall the definition (II, 4.1) of these polynomials, given here for two variables. For $j = 1, 2$, let $U^{(j)} = \{u^{(j)}_{i_1,i_2}, (i_1, i_2) \in \mathbb{N}^2\}$ be a fixed family of real numbers. To $U^{(1)}, U^{(2)}$ is attached a unique family of polynomials $G_{k_1,k_2}(x_1, x_2|U^{(1)}, U^{(2)})$ of degrees k_1 in x_1, k_2 in x_2, $(k_1, k_2) \in \mathbb{N}^2$, defined recursively by

$$G_{0,0}(x_1, x_2|U^{(1)}, U^{(2)}) = 1,$$

and when $k_1 + k_2 \geq 1$,

$$G_{k_1,k_2}(x_1, x_2|U^{(1)}, U^{(2)}) = \frac{x_1^{k_1}}{k_1!}\frac{x_2^{k_2}}{k_2!} - \tag{4.1}$$

$$\sum_{D(k_1,k_2)} \left[\frac{(u^{(1)}_{i_1,i_2})^{k_1-i_1}}{(k_1-i_1)!} \frac{(u^{(2)}_{i_1,i_2})^{k_2-i_2}}{(k_2-i_2)!} \right] G_{i_1,i_2}(x_1, x_2|U^{(1)}, U^{(2)}),$$

where $D(k_1, k_2)$ denotes the set of indexes $\{(i_1, i_2)$, with $0 \leq i_1 \leq k_1$, $0 \leq i_2 \leq k_2$ and $i_1 + i_2 < k_1 + k_2\}$. Observe that $G_{k_1,k_2}(\ )$, $k_1 + k_2 \geq 1$, depends only on the $u^{(1)}_{i_1,i_2}$ and $u^{(2)}_{i_1,i_2}$ with $(i_1, i_2) \in D_{k_1,k_2}$.

For our purpose, we need to establish a monotonicity property of certain expansions constructed from these polynomials with respect to the parameters in $U^{(1)}$ and $U^{(2)}$. Let $f(x_1, x_2)$ be a function with derivatives $f^{(i_1,i_2)}(x_1, x_2)$, $(i_1, i_2) \in \mathbb{N}^2$, and let $\hat{A}, A, \hat{U}^{(1)}, \hat{U}^{(2)}, U^{(1)}, U^{(2)}$ be six families of real numbers $\hat{a}_{i_1,i_2}, a_{i_1,i_2}, \hat{u}^{(1)}_{i_1,i_2}, \hat{u}^{(2)}_{i_1,i_2}, u^{(1)}_{i_1,i_2}, u^{(2)}_{i_1,i_2}$, respectively, $(i_1, i_2) \in \mathbb{N}^2$. Given these elements, fix $(k_1, k_2) \in \mathbb{N}^2$ and consider the polynomial

$$h_{k_1,k_2}(x_1,x_2|\hat{A},\hat{U}^{(1)},\hat{U}^{(2)},f,A,U^{(1)},U^{(2)})$$

$$(4.2) \qquad = \sum_{i_1=0}^{k_1}\sum_{i_2=0}^{k_2} \hat{a}_{i_1,i_2} f^{(i_1,i_2)}(\hat{u}^{(1)}_{i_1,i_2},\hat{u}^{(2)}_{i_1,i_2}) G_{i_1,i_2}(x_1,x_2|\hat{U}^{(1)},\hat{U}^{(2)})$$

$$- \sum_{i_1=0}^{k_1}\sum_{i_2=0}^{k_2} a_{i_1,i_2} f^{(i_1,i_2)}(u^{(1)}_{i_1,i_2},u^{(2)}_{i_1,i_2}) G_{i_1,i_2}(x_1,x_2|U^{(1)},U^{(2)}).$$

Note that $h_{k_1,k_2}(\ )$, $(k_1,k_2)\in \mathbb{N}^2$, depends only on the $\hat{a}_{i_1,i_2}$, $\hat{u}^{(1)}_{i_1,i_2}$, $\hat{u}^{(2)}_{i_1,i_2}$, a_{i_1,i_2}, $u^{(1)}_{i_1,i_2}$, $u^{(2)}_{i_1,i_2}$ with $0\le i_1\le k_1$, $0\le i_2\le k_2$. We remark that the two sums in (4.2) correspond to finite Abel expansions of the function $f(x_1,x_2)$ of the type given in (II, 4.5).

PROPERTY 4.1 *Let* $(k_1,k_2)\in \mathbb{N}^2$. *If for* $j=1,2$ *and* $0\le i_1\le k_1$, $0\le i_2\le k_2$, *the following conditions hold*

(4.3)

$u^{(j)}_{i_1,i_2}$ *and* $\hat{u}^{(j)}_{i_1,i_2}$ *are decreasing sequences in* i_1(i_2 *fixed*) *and* i_2(i_1 *fixed*),

$\hat{u}^{(j)}_{i_1,i_2}\ge u^{(j)}_{i_1,i_2}$,

$\hat{a}_{i_1,i_2}\ge \max\{0,a_{\ell_1,\ell_2},$ *for* $i_1\le \ell_1\le k_1, i_2\le \ell_2\le k_2\}$,

$f^{(i_1,i_2)}(u^{(1)}_{i_1,i_2},u^{(2)}_{i_1,i_2})\ge 0$, *and*

$f^{(k_1+1,k_2)}(x_1,x_2)$ *and* $f^{(k_1,k_2+1)}(x_1,x_2)\ge 0$ *for* $x_1\ge u^{(1)}_{k_1,k_2}, x_2\ge u^{(2)}_{k_1,k_2}$,

then

$$(4.4) \qquad h_{k_1,k_2}(x_1,x_2|\hat{A},\hat{U}^{(1)},\hat{U}^{(2)},f,A,U^{(1)},U^{(2)})\ge 0$$

$$\text{for } x_1\ge \hat{u}^{(1)}_{0,0}, x_2\ge \hat{u}^{(2)}_{0,0}.$$

PROOF This can be shown by extending to the multivariate case the argument by induction followed for Property (I, 4.2). The proof is then direct, though rather technical, and uses properties of the polynomials given in (II, Section 4); it is omitted. □

4.2. *Varying Infectivity in Collective Epidemics*

The multipopulation collective epidemic model introduced in (III, Section 4) describes the spread of an infectious disease in a closed population subdivided in several (here two) groups (men and women, for example). Each group j, $j=1,2$, is partitioned in three classes of individuals, the susceptibles, the infectives and the removed cases. Initially, these are in numbers n_j, m_j and 0, respectively, and infection is then propagated as follows. Any infective remains infected during a random period of time. All the infectious periods are independent and, for j given, identically distributed, the common distribution being that of a variable D_j, say. These D_j are in general $\mathbb{R}$-valued, but can be discrete and possibly constant. While infected, the individual behaves independently of the others and can contact susceptibles of the two groups. Specifically, he will fail to transmit the infectious agents within any given set of k_1 susceptibles in group 1 and k_2 susceptibles in group 2, k_1 in $[1, n_1]$ and k_2 in $[1, n_2]$, with a random probability that depends on his infectious period. We make for these random variables, k_1, k_2 fixed, the same hypotheses as for the infectious periods, the common distribution for j given being that of a variable $Q_j(k_1, k_2)$, say. After that, the infective becomes a removed case and plays no further role in the infection process.

Let T denote the end of the epidemic, when there are no more infectious present in the population. We are interested by the vector $(S_{1,T}, S_{2,T})$ that represents the ultimate numbers of susceptibles surviving the disease in groups 1 and 2, respectively. Using the polynomials (4.1), we obtained, inter alia, in (III, 4.15) the formula (4.5) below for the mixed descending factorial moments of that vector. For $j = 1,2$ and $k_1 \in [1, n_1]$, $k_2 \in [1, n_2]$, let

$$q_j(k_1, k_2) = E[Q_j(k_1, k_2)] \tag{4.5}$$

be the expected value of the different probabilities of non-infection; when k_2 (e.g)$= 0$, put $q_j(k_1, 0) = q_j(k_1, -)$, with $q_j(0,0) = 1$. Then, for $0 \leq k_1 \leq n_1$, $0 \leq k_2 \leq n_2$,

$$E\left[S_{1,T,[k_1]} S_{2,T,[k_2]}\right] = \sum_{i_1=k_1}^{n_1} \sum_{i_2=k_2}^{n_2} \left\{ n_{1,[i_1]} n_{2,[i_2]} [q_1(i_1, i_2)]^{n_1+m_1-i_1} \right. \\ \left. \times [q_2(i_1, i_2)]^{n_2+m_2-i_2} \right\} G_{i_1-k_1, i_2-k_2}\left[1, 1 | \mathcal{E}^{k_1,k_2} U^{(1)}, \mathcal{E}^{k_1,k_2} U^{(2)}\right], \tag{4.6}$$

where for $j = 1,2$, $\mathcal{E}^{k_1,k_2} U^{(j)}$ is the family $\{u^{(j)}_{k_1+i_1, k_2+i_2}, (i_1, i_2) \in \mathbb{N}^2\}$, with

$$u^{(j)}_{i_1,i_2} = q_j(i_1, i_2) \quad , \quad 0 \leq i_1 \leq n_1, 0 \leq i_2 \leq n_2, \tag{4.7}$$

the $u^{(j)}_{i_1,i_2}$ for other indexes being superfluous and omitted.

Intuitively, one expects that lower infectivity levels translated by smaller $q_j(k_1, k_2)$ should generate larger ultimate numbers of susceptibles. Hereafter, we prove that this is indeed true provided comparison on $(S_{1,T}, S_{2,T})$ is made through the factorial ordering. Thus, consider a similar bipopulation collective epidemic characterized now by the variables $\hat{D}_j$ and $\hat{Q}_j(k_1, k_2)$, $j = 1,2$ and $0 \leq k_1 \leq n_1$, $0 \leq k_2 \leq n_2$. Let $\hat{q}_j(k_1, k_2)$ be the associated expectations (4.5), and denote by $(\hat{S}_{1,T}, \hat{S}_{2,T})$ the resulting final numbers of susceptibles.

PROPOSITION 4.2 *If for* $j = 1, 2$ *and* $0 \leq i_1 \leq n_1$, $0 \leq i_2 \leq n_2$,

$$q_j(i_1, i_2) \leq \hat{q}_j(i_1, i_2), \tag{4.8}$$

then

$$(S_{1,T}, S_{2,T}) \leq_F (\hat{S}_{1,T}, \hat{S}_{2,T}). \tag{4.9}$$

PROOF We have to show that for $0 \leq k_1 \leq n_1$, $0 \leq k_2 \leq n_2$,

$$E\left[\hat{S}_{1,T,[k_1]}\hat{S}_{2,T,[k_2]}\right] - E\left[S_{1,T,[k_1]}S_{2,T,[k_2]}\right] \geq 0. \tag{4.10}$$

Define $\mathcal{E}^{k_1,k_2}A = \{a_{k_1+i_1,k_2+i_2}, (i_1, i_2) \in \mathbb{N}^2\}$, with

$$a_{i_1,i_2} = \left(u^{(1)}_{i_1,i_2}\right)^{m_1}\left(u^{(2)}_{i_1,i_2}\right)^{m_2}, \quad 0 \leq i_1 \leq n_1, 0 \leq i_2 \leq n_2, \tag{4.11}$$

the other a_{i_1,i_2} being superfluous for the discussion. Putting

$$f(x_1, x_2) = x_1^{n_1-k_1}x_2^{n_2-k_2}, \tag{4.12}$$

(4.6) can be expressed as

$$E\left[S_{1,T,[k_1]}S_{2,T,[k_2]}\right] = n_{1,[k_1]}n_{2,[k_2]}\sum_{i_1=0}^{n_1-k_1}\sum_{i_2=0}^{n_2-k_2} a_{k_1+i_1,k_2+i_2}f^{(i_1,i_2)}\left(u^{(1)}_{k_1+i_1,k_2+i_2}, u^{(2)}_{k_1+i_1,k_2+i_2}\right) \times G_{i_1,i_2}\left[1, 1|\mathcal{E}^{k_1,k_2}U^{(1)}, \mathcal{E}^{k_1,k_2}U^{(2)}\right]. \tag{4.13}$$

Write then the formula (4.13) associated with the alternative model. Using the definition (4.2), we see that the difference in (4.10) just corresponds to

$$n_{1,[k_1]}n_{2,[k_2]}h_{n_1-k_1,n_2-k_2}\left[1, 1|\mathcal{E}^{k_1,k_2}\hat{A}, \mathcal{E}^{k_1,k_2}\hat{U}^{(1)}, \mathcal{E}^{k_1,k_2}\hat{U}^{(2)}, f, \mathcal{E}^{k_1,k_2}A, \mathcal{E}^{k_1,k_2}U^{(1)}, \mathcal{E}^{k_1,k_2}U^{(2)}\right]. \tag{4.14}$$

Now, let us examine $h_{n_1-k_1,n_2-k_2}(\ )$ in (4.14). It is easily verified that thanks to the hypothesis (4.8), the conditions (4.3) in Property 4.1 are well satisfied. Since $1 \geq \hat{u}^{(1)}_{k_1,k_2}$ and $\hat{u}^{(2)}_{k_1,k_2}$, we deduce (4.10) from (4.4) and (4.14). □

The factorial ordering between final epidemic outcomes has been obtained under rather weak conditions on the model parameters $q_j(k_1, k_2)$. We will show in a forthcoming paper how it can be strengthened under stronger conditions on these parameters.

4.3. *Some Specific Applications*

For illustration, let us discuss some applications of this comparison result for two particular cases of the model.

(i) Suppose that the infectious periods are of length 1 ($D_1 = D_2 = 1$), and each infective can contact, independently of the others, a random number of individuals per time unit. All these numbers of contacts are independent, and for infectives of group $j, j = 1, 2$, identically distributed, the common distribution for contacts within groups 1 and 2 being that of the vector $(R_{j,1}, R_{j,2})$, say. The contacts then occur with or without replacement amongst the $N_1 = n_1 + m_1$ and $N_2 = n_2 + m_2$ individuals of these groups.

Clearly, the model here can be viewed as the iterative version of a sampling scheme such as described in 3.2.1. For the case with replacement, we have, for $j = 1,2$ and $0 \leq k_1 \leq n_1$, $0 \leq k_2 \leq n_2$,

$$q_j(k_1, k_2) = E\left\{[(N_1 - k_1)/N_1]^{R_{j,1}}[(N_2 - k_2)/N_2]^{R_{j,2}}\right\}. \tag{4.15}$$

From (4.15) and Proposition 4.2, we then deduce that (in obvious notations)

$$(R_{j,1}, R_{j,2}) \geq_g (\hat{R}_{j,1}, \hat{R}_{j,2}), j = 1, 2, \text{ implies (4.9)}. \tag{4.16}$$

In a similar way, for the sampling without replacement,

$$q_j(k_1, k_2) = E\left[(N_1 - R_{j,1})_{[k_1]}(N_2 - R_{j,2})_{[k_2]}\right] / N_{1,[k_1]}N_{2,[k_2]}, \tag{4.17}$$

and we deduce that

$$(N_1 - R_{j,1}, N_2 - R_{j,2}) \leq_F (N_1 - \hat{R}_{j,1}, N_2 - \hat{R}_{j,2}), \quad j = 1, 2, \text{ implies (4.9)}. \tag{4.18}$$

(ii) Consider the situation above where the infectious periods are r.v.s which are independent and, for infectives of group $j, j = 1, 2$, distributed as D_j. Suppose now that while infected, each infective can contact, independently of the others, any susceptible present at the points of a Poisson process. All these processes are independent, and for infectives of group $j, j = 1, 2$, the associated contact rates within groups 1 and 2 are equal to $\beta_{j,1}$ and $\beta_{j,2}$, respectively. Here thus, for $j = 1, 2$ and $0 \leq k_1 \leq n_1$, $0 \leq k_2 \leq n_2$,

$$Q_j(k_1, k_2) = \exp\left[-(k_1\beta_{j,1} + k_2\beta_{j,2})D_j\right]. \tag{4.19}$$

The so-called general epidemic model corresponds to the case where the D_j are exponentially distributed. When $D_1 = D_2 = 1$, then $Q_j(k_1, k_2) = (q_{j,1})^{k_1}(q_{j,2})^{k_2}$ with $q_{j,1} = \exp(-\beta_{j,1})$, $q_{j,2} = \exp(-\beta_{j,2})$, and the model reduces to the Reed-Frost process (see, e.g., Bailey (1975)).

We start by assessing the effect of varying the infectious periods. For this, we are going to use a weak stochastic ordering based on the Laplace-Stieltjes transform ($\leq_L$ in Stoyan (1983)): for random variables $X_1, \hat{X}_1$ valued in $\mathbb{R}^+$, $X_1 \leq_L \hat{X}_1$ when

$$E[\exp(-\theta X_1)] \geq E[\exp(-\theta \hat{X}_1)] \ , \ \theta \in \mathbb{R}^+. \tag{4.20}$$

From (4.19), (4.20) and Proposition 4.2, we thus deduce that

$$D_j \geq_L \hat{D}_j, j = 1, 2, \text{ implies (4.9)} \ . \tag{4.21}$$

This result can be exploited to construct bounds for the ultimate numbers of suceptibles when only partial information on the D_j is available. For example, suppose that D_j, $j = 1, 2$, are known to belong to the $\mathcal{L}(\bar{\mathcal{L}})$ class introduced by Klefsjö (1983); in other words, we have

$$D_j \geq_L (\leq_L) Exp(\mu_j), \tag{4.22}$$

where $\mu_j = E(D_j)$ and $Exp(\mu_j)$ denotes an exponential variable with the same mean μ_j. From (4.21) and (4.22), we then obtain that the general epidemic with exponential infectious periods with the same means provides an upper (lower) $\leq_F$ bound for the statistic $(S_{1,T}, S_{2,T})$.

Another comparison of interest is between the original epidemic model with random infectious periods and an approximated Reed-Frost process in which, by definition, infectious periods are all equal to 1. Take first the expectation in (4.19). Using the fact that the set of a single random variable is associated (Esary, Proschan and Walkup (1967)), we can write that for $j = 1, 2$ and $0 \leq k_1 \leq n_1$, $0 \leq k_2 \leq n_2$,

$$q_j(k_1, k_2) \geq \{E\left[\exp\left(-k_1 \beta_{j,1} D_j\right)\right]\} \{E\left[\exp\left(-k_2 \beta_{j,2} D_j\right)\right]\} . \tag{4.23}$$

Then, applying twice Jensen's inequality in (4.23), we obtain that

$$q_j(k_1, k_2) \geq (\hat{q}_{j,1})^{k_1} (\hat{q}_{j,2})^{k_2} \geq (q^*_{j,1})^{k_1} (q^*_{j,2})^{k_2}, \tag{4.24}$$

where for $i = 1, 2$,

$$\begin{cases} \hat{q}_{j,i} = E[\exp(-\beta_{j,i} D_j)], \\ q^*_{j,i} = \exp[-\beta_{j,i} E(D_j)]. \end{cases} \tag{4.25}$$

Therefore, from (4.24), (4.25) and Proposition 4.2, we deduce that the Reed-Frost model with the $\hat{q}_{j,i}$ as probabilities of non-infection yields to a lower $\leq_F$ bound for $(S_{1,T}, S_{2,T})$. Moreover, replacing the infectious periods by their mean leads to a further Reed-Frost model that predicts an even $\leq_F$ smaller number of susceptibles. We mention that for the latter comparison, the factorial ordering could be strengthened.

Acknowledgments

The authors would like to thank the referees for helpful comments and suggestions on an earlier version.

References

Bailey, N.T.J. (1975). *The Mathematical Theory of Infectious Diseases and its Applications.* Griffin, London.

Ball, F. and Donnelly, P. (1987). Interparticle correlation in death processes with application to variability in compartmental models. *Adv. Appl. Prob.* **19** 755-766.

Bergmann, R. (1978). Some classes of semi-ordering relations for random vectors and their use for comparing covariances. *Math. Nachr.* **82** 103-114.

Donnelly, P. (1991). Personal communication.

Donnelly, P. and Whitt, W. (1989). On reinforcement-depletion compartmental urn models. *J. Appl. Prob.* **26** 477-489.

Esary, J.D., Proschan, F. and Walkup, D.W. (1967). Association of random variables, with applications. *Ann. Math. Statist.* **38** 1466-1474.

Klefsjö, B. (1983). A useful ageing property based on the Laplace transform. *J. Appl. Prob.* **20** 615-626.

Lefevre, Cl. and Michaletzky, G. (1990). Interparticle dependence in a linear death process subjected to a random environment. *J. Appl. Prob.* **27** 491-498.

Lefevre, Cl. and Picard, Ph. (1990). A non-standard family of polynomials and the final size distribution of Reed-Frost epidemic processes. *Adv. Appl. Prob.* **22** 25-48.

Lefevre, Cl. and Picard, Ph. (1991). An unusual stochastic order relation with some applications in sampling and epidemic theory. To appear in *Adv. Appl. Prob.* **25**.

Picard, Ph. and Lefevre, Cl. (1990). A unified analysis of the final size and severity distribution in collective Reed-Frost epidemic processes. *Adv. Appl. Prob.* **22** 269-294.

Shenton, L.R. (1981). A reinforcement-depletion urn problem - I. Basic theory. *Bull. Math. Biol.* **43** 327-340.

Stoyan, D. (1983). *Comparison Methods for Queues and other Stochastic Models.* Wiley, New York.

Institut de Statistique, C.P. 210
Boulevard du Triomphe
Universite Libre de Bruxelles
B-1050 Bruxelles
Belgique

Mathematiques Appliquees
43 Boulevard du
11 Novembre 1918
Universite de Lyon 1
F-69622 Villeurbanne
France

Stochastic Inequalities
IMS Lecture Notes – Monograph Series
Volume 22 (1993)

ALLOCATION THROUGH STOCHASTIC SCHUR CONVEXITY AND STOCHASTIC TRANSPOSITION INCREASINGNESS[1]

By LIWAN LIYANAGE and J. GEORGE SHANTHIKUMAR

University of Western Sidney and University of California, Berkeley

Consider a stochastic allocation problem where a total resource of R units are to be allocated among m competing facilities in a system. An allocation of r_i units to facility i results in a random response $X_i(r_i), i = 1, \ldots, m$. The system response is then defined by the random variable $Y(\mathbf{r}) = h(X_1(r_1), \ldots, X_m(r_m))$ where $h : \mathbb{R}^m \to \mathbb{R}$ is the system performance function. Let $\mathcal{S} \subset \mathbb{R}_+^m$ be the set of all feasible allocations. We are then interested in the stochastic allocation problem $\min\{Eg(Y(\mathbf{r})) : \sum_{i=1}^m r_i = R, \mathbf{r} \in \mathcal{S}\}$ for some utility function g. The aim of the paper is to obtain a partial or a full characterization of the optimal solution to this problem with minimal restriction on g. For this we introduce notions of *stochastic Schur convexity* and *stochastic transposition increasingness* and identify sufficient conditions on $X_i(r_i), i = 1, \ldots, m$ and h under which $Y(\mathbf{r})$ will be either stochastically Schur convex or transposition increasing with respect to $\mathbf{r}$. Then under appropriate condition on g it can be shown that the stochastic Schur convexity of $Y(\mathbf{r})$ will imply the optimality of balanced resource allocation and the transposition increasingness will imply a partial characterization of the optimal solution thus reducing the computational effort needed to find the optimal solution. Several examples in the telecommunication, manufacturing and reliability/performability systems are presented to illustrate the main results of this paper.

1. Introduction

Consider a system consisting of m facilities that compete for a limited resource with a capacity of R units. An allocation of r_i units to facility i results in a random response $X_i(r_i), i = 1, \ldots, m$. The overall system response

[1]Research supported in part by National Science Foundation Grant ECS–8811234.
AMS 1991 *subject classification.* 60E15, 62N05.
Key words and phrases. Stochastic allocation, stochastic convexity, stochastic Schur convexity, stochastic transposition increasingness, resequencing queue, flexible manufacturing systems, minimal repair, reliability/performability.

is then defined by the random variable $Y(\mathbf{r}) = h(X_1(r_1), \ldots, X_m(r_m))$ where $h : \mathbb{R}^m \to \mathbb{R}$ is the system performance function. Let $\mathcal{S} \subset \mathbb{R}^m_+$ be the set of all feasible allocations. In this paper we are interested in the stochastic allocation problem

$$\min\{Eg(Y(\mathbf{r})) : \sum_{i=1}^{m} r_i = R, \mathbf{r} \in \mathcal{S}\}, \tag{1.1}$$

where g is an appropriate utility function chosen by the decision maker. For example consider a flexible manufacturing system consisting of m machine cells. A total of R flexible machines are available that needs to be allocated among the m machine cells. Let $X_i(r_i)$ be the stationary number of parts in cell i if r_i flexible machines are allocated to cell $i, i = 1, \ldots, m$. Suppose the performance of the system is measured by the total number of parts in it: i.e. $h(\mathbf{x}) = \sum_{i=1}^m x_i, \mathbf{x} \in \mathcal{Z}^m_+$. A stochastic allocation problem for this scenario is then to obtain, if possible, an optimal allocation that minimizes $\sum_{i=1}^m X_i(r_i)$ in the usual stochastic sense: i.e. we restrict g to be an increasing but an arbitrary function.

The purpose of this paper is to obtain a partial or a full characterization of the optimal solution to the stochastic allocation problem with minimal restriction on g. For this we define notions of *stochastic Schur convexity* (Section 2) and *stochastic transposition increasingness* (Section 3) and find sufficient conditions on $X_i(r_i), i = 1, \ldots, m$ and h under which $Y(\mathbf{r})$ will be either stochastically Schur convex or transposition increasing with respect to $\mathbf{r}$. Then under appropriate condition on g it will be shown that the Schur convexity of $Y(\mathbf{r})$ will imply the optimality of balanced resource allocation (Section 2) and the transposition increasingness will imply a partial characterization of the optimal solution thus reducing the computational effort needed to obtain the optimal solution (Section 3). Several examples in the telecommunication, manufacturing and reliability/performability systems are presented in Section 4 to illustrate the main results of this paper.

2. Stochastic Schur Convexity

In this section we will define stochastic Schur convexity and give sufficient conditions on $X_i(r_i), i = 1, \ldots, m$ and h under which $Y(\mathbf{r})$ is stochastically schur convex with respect to $\mathbf{r}$. For this we will need the following definitions of majorization and Schur functions (e.g. see Marshall and Olkin (1979) for more details). For any $\mathbf{x} \in \mathbb{R}^m, x_{[1]} \geq x_{[2]} \geq \cdots \geq x_{[m]}$ denotes the decreasing rearrangement of the coordinates of $\mathbf{x}$ and for any $\mathbf{x}, \mathbf{y} \in \mathbb{R}^m, \mathbf{x} \geq \mathbf{y}$ denotes the usual coordinatewise ordering. Throughout this paper the terms 'increasing' and 'decreasing' are not used in the strict sense.

DEFINITION 2.1 Let $\mathbf{x}, \mathbf{y} \in \mathbb{R}^m$. Then $\mathbf{x}$ *majorizes* $\mathbf{y}$ if

$$(2.1) \qquad \sum_{i=1}^{k} x_{[i]} \geq \sum_{i=1}^{k} y_{[i]}, k = 1, \ldots, m, \text{ and } \sum_{i=1}^{m} x_{[i]} = \sum_{i=1}^{m} y_{[i]},$$

or equivalently

$$(2.2) \qquad \sum_{i=k}^{m} x_{[i]} \leq \sum_{i=k}^{m} y_{[i]}, k = 1, \ldots, m, \text{ and } \sum_{i=1}^{m} x_{[i]} = \sum_{i=1}^{m} y_{[i]}.$$

We denote this $\mathbf{x} \geq_m \mathbf{y}$. When the requirement of the equality $\sum_{i=1}^m x_{[i]} = \sum_{i=1}^m y_{[i]}$ is dropped from (2.1) [(2.2)] we say that $\mathbf{x}$ *weakly sub-majorizes* [*sup-majorizes*] $\mathbf{y}$ and is denoted $\mathbf{x} \geq_{wm} [\geq^{wm}]\mathbf{y}$.

The following lemma (e.g. see Marshall and Olkin (1979)) allows one to simplify the analysis of majorization, often making it sufficient to prove the desired result just for the two dimensional case.

LEMMA 2.2

(i) $\mathbf{x} \geq_m \mathbf{y} \Leftrightarrow$ *there exist a finite number (say k) of vectors* $\mathbf{x}^{(i)}, i = 1, \ldots, k$ *such that* $\mathbf{x} = \mathbf{x}^{(1)} \geq_m \cdots \geq_m \mathbf{x}^{(k)} = \mathbf{y}$ *and such that* $\mathbf{x}^{(i)}$ *and* $\mathbf{x}^{(i+1)}$ *differ in two coordinates only,* $i = 1, \ldots, k-1$.

(ii) $\mathbf{x} \geq_{wm} \mathbf{y} \Leftrightarrow$ *there exists a vector* $\mathbf{z}$ *such that* $\mathbf{x} \geq \mathbf{z}$ *and* $\mathbf{z} \geq_m \mathbf{y}$.

(iii) $\mathbf{x} \geq^{wm} \mathbf{y} \Leftrightarrow$ *there exists a vector* $\mathbf{z}$ *such that* $\mathbf{x} \leq \mathbf{z}$ *and* $\mathbf{z} \geq_m \mathbf{y}$.

DEFINITION 2.3 A function $\phi : \mathbb{R}^m \to \mathbb{R}$ is *Schur convex* [*concave*] if $\mathbf{x} \geq_m \mathbf{y}$ implies $\phi(\mathbf{x}) \geq [\leq]\phi(\mathbf{y})$. It is *increasing* Schur convex [concave] if it is increasing and Schur convex [concave]; i.e. (see Lemma 2.2 (ii) [(iii)]), if $\mathbf{x} \geq_{wm} [\geq^{wm}]\mathbf{y}$ implies $\phi(\mathbf{x}) \geq [\leq]\phi(\mathbf{y})$.

Note that all Schur convex and Schur concave functions are symmetric: i.e. for any permutation π of $\{1, \ldots, m\}, \mathbf{x} \in \mathbb{R}^m$ and $\mathbf{x}_\pi = (x_{\pi(i)}, i = 1, \ldots, m)$ one has $\phi(\mathbf{x}_\pi) = \phi(\mathbf{x})$. Recall that a random variable V is said to be larger than a random variable W in the sense of the usual stochastic [increasing convex, increasing concave] ordering if $E\psi(V) \geq E\psi(W)$ for all increasing [increasing and convex, increasing and concave] functions ψ. We denote this $V \geq_{st} [\geq_{icx}, \geq_{icv}]W$ (e.g. see Ross (1983)).

DEFINITION 2.4 *(Stochastic Schur Convexity)*: A real valued random variable $Z(\mathbf{x})$ parametrized by $\mathbf{x} \in \mathcal{X} \subset \mathbb{R}^m$ is *stochastically Schur convex* in the sense of the usual stochastic [increasing convex, increasing concave] ordering if for any $\mathbf{x}, \mathbf{y} \in \mathcal{X}, \mathbf{x} \geq_m \mathbf{y}$ implies $Z(\mathbf{x}) \geq_{st} [\geq_{icx}, \geq_{icv}] Z(\mathbf{y})$. We denote this $\{Z(\mathbf{x}), \mathbf{x} \in \mathcal{X}\} \in S - SchurCX(st)[S - SchurCX(icx), S - SchurCX(icv)]$.

If in addition $Z(\mathbf{x})$ is stochastically increasing [decreasing] then $Z(\mathbf{x})$ is stochastically increasing [decreasing] and Schur convex. We denote this $\{Z(\mathbf{x}), \mathbf{x} \in \mathcal{X}\} \in SI - SchurCX(st)$ $[\{Z(\mathbf{x}), \mathbf{x} \in \mathcal{X}\} \in SD - SchurCX(st)]$ etc. That is (see Lemma 2.2 (ii) [(iii)]), for any $\mathbf{x}, \mathbf{y} \in \mathcal{X}, \mathbf{x} \geq_{wm} [\geq^{wm}] \mathbf{y}$ implies $Z(\mathbf{x}) \geq_{st} [\geq_{st}] Z(\mathbf{y})$ etc.

If $-Z(\mathbf{x})$ is stochastically Schur convex then we say that $Z(\mathbf{x})$ is stochastically Schur concave.

As an immediate consequence of the definition of stochastic Schur convexity one has the following characterization of the optimal solution to the allocation problem (1.1).

THEOREM 2.5 *Let* $Y(\mathbf{r}) = h(X_1(r_1), \ldots, X_m(r_m))$. *Suppose* $\{Y(\mathbf{r}), \mathbf{r} \in \mathcal{S}\} \in S - SchurCX(st)[S - SchurCX(icx), S - SchurCX(icv)]$. *Then for any increasing [increasing and convex, increasing and concave] function* g *and* $\mathbf{r}, \mathbf{s} \in \mathcal{S}$ *one has,*

$$\mathbf{r} \geq_m \mathbf{s} \Rightarrow Eg \circ h(X_1(r_1), \ldots, X_m(r_m)) \geq Eg \circ h(X_1(s_1), \ldots, X_m(s_m)).$$

Suppose $\{Y(\mathbf{r}), \mathbf{r} \in \mathcal{S}\} \in SI - SchurCX(st)[SI - SchurCX(icx), SI - SchurCX(icv)]$. *Then for any increasing [increasing and convex, increasing and concave] function* g *and* $\mathbf{r}, \mathbf{s} \in \mathcal{S}$ *one has,*

$$\mathbf{r} \geq_{wm} \mathbf{s} \Rightarrow Eg \circ h(X_1(r_1), \ldots, X_m(r_m)) \geq Eg \circ h(X_1(s_1), \ldots, X_m(s_m)).$$

In either case if $\mathbf{r}^* := (\frac{R}{m}, \ldots, \frac{R}{m})$ *is in* $\mathcal{S}$ *then*

$$Eg \circ h(X_1(r_1), \ldots, X_m(r_m)) \geq Eg \circ h(X_1(\frac{R}{m}), \ldots, X_m(\frac{R}{m})),$$

i.e. $\mathbf{r}^*$ *is an optimal solution to problem* (1.1).

From the above theorem it is clear that it is worthwhile to search for sufficient conditions on $X_i(r_i), i = 1, \ldots, m$ and h for $Y(\mathbf{r})$ to be stochastically Schur convex. We shall do this next in this section (and examples where such conditions are naturally satisfied are given in Section 4.) Before that we note the following easily verified lemma.

LEMMA 2.6 *Let* $F_Z(t; \mathbf{x}) = P\{Z(\mathbf{x}) \leq t\}$ *and* $\bar{F}_Z(t; \mathbf{x}) = P\{Z(\mathbf{x}) > t\}$ *be respectively, the cumulative and survival functions of* $Z(\mathbf{x})$. *Then* $\{Z(\mathbf{x}), \mathbf{x} \in \mathcal{X}\} \in S - SchurCX(st)[S - SchurCX(icx), S - SchurCX(icv)] \Leftrightarrow \bar{F}_Z(t; \mathbf{x})[\int_t^\infty \bar{F}_Z(s; \mathbf{x}) ds, -\int_{-\infty}^t F_Z(s; \mathbf{x}) ds]$ *is Schur convex in* $\mathbf{x}$.

In the remainder of this section we will assume that $\{X_i(\theta), \theta \in \Theta\}, i = 1, \ldots, m$ are m probabilistically identical and mutually independent collections of random variables. We will now present a result on stochastic majorization that extends a result of Proschan and Sethuraman (1976) for random variables with proportional hazard rates. We will need the following definition for this result (see Example 4.3 of Shaked and Shanthikumar (1990a)).

DEFINITION 2.7 (*Stochastic Convexity in the Hazard Rate*): Let $\{Z(\theta), \theta \in \Theta\}$ be a collection of absolutely continuous positive random variables with hazard rate functions $\{\gamma(\cdot;\theta), \theta \in \Theta\}$. Then $\{Z(\theta), \theta \in \Theta\}$ is said to be stochastically increasing and convex in the sense of hazard rate ordering if $\gamma(t;\theta)$ is pointwise decreasing and concave in θ for each fixed $t \in \mathbb{R}_+$. We denote this $\{Z(\theta), \theta \in \Theta\} \in SICX(hr)$.

REMARK 2.8 Observe that since $\bar{F}(t;\theta) = P\{Z(\theta) > t\} = exp\{-\int_{u=0}^{t} \gamma(u;\theta)du\}, t \in \mathbb{R}_+$ from Theorem 3.16 of Shaked and Shanthikumar (1990a) it follows that $\{Z(\theta), \theta \in \Theta\} \in SICX(hr) \Rightarrow \{Z(\theta), \theta \in \Theta\} \in SICX(st) \Rightarrow \{Z(\theta), \theta \in \Theta\} \in SICX(sp) \Rightarrow \{Z(\theta), \theta \in \Theta\} \in SICX$.

THEOREM 2.9 *Suppose* $\{X_i(\theta), \theta \in \Theta\} \in SICX(hr)$. *Then for any increasing and symmetric function* h *and* $Y(\mathbf{x}) = h(X_1(x_1), \ldots, X_m(x_m))$ *one has* $\{Y(\mathbf{x}), \mathbf{x} \in \Theta^m\} \in SI - SchurCX(st)$. *That is for any* $\mathbf{x}, \mathbf{y} \in \Theta^m$, *increasing symmetric function* h *and increasing function* g,

$$(2.3) \qquad \mathbf{x} \geq_{wm} \mathbf{y} \quad \rightarrow \quad \begin{aligned} &Eg \circ h(X_1(x_1), \ldots, X_m(x_m)) \\ \geq\ &Eg \circ h(X_1(y_1), \ldots, X_m(y_m)). \end{aligned}$$

Since a Schur convex function is a symmetric function, one also has

$$(2.4) \qquad \mathbf{x} \geq_{wm} \mathbf{y} \quad \rightarrow \quad \begin{aligned} &(X_1(x_1), \ldots, X_m(x_m)) \\ \geq_{wm:st}\ &(X_1(y_1), \ldots, X_m(y_m)). \end{aligned}$$

Here $\mathbf{V} \geq_{wm:st} \mathbf{W}$ *stands for the weak stochastic majorization order of two random vectors* $\mathbf{V}$ *and* $\mathbf{W}$. *That is* $Ef(\mathbf{V}) \geq Ef(\mathbf{W})$ *for all increasing Schur convex function* f *(see* Marshall and Olkin (1979, Chapter 11G)*).*

PROOF We will first establish the theorem for the case $m = 2$ and $\mathbf{x} \geq_m \mathbf{y}$. Then the result for the case $\mathbf{x} \geq_{wm} \mathbf{y}$ will follow from Lemma 2.2 (ii) and the stochastic monotonicity of $X_i(\theta)$ in θ. Suppose, without a loss of generality, $x_1 \leq y_1 \leq y_2 \leq x_2$ and $x_1 + x_2 = y_1 + y_2$. Define

$$(2.5) \qquad \hat{\gamma}(t;y_1) = \gamma(t;x_1) + \frac{y_1 - x_1}{x_2 - x_1}\{\gamma(t;x_2) - \gamma(t;x_1)\}, t \in \mathbb{R}_+,$$

and

(2.6) $$\hat{\gamma}(t;y_2) = \gamma(t;x_1) + \frac{y_2 - x_1}{x_2 - x_1}\{\gamma(t;x_2) - \gamma(t;x_1)\}, t \in \mathbb{R}_+.$$

From the concavity of the hazard rate function $\gamma(\cdot;\theta)$ in θ it is immediate that

(2.7) $$\hat{\gamma}(t;y_i) \leq \gamma(t;y_i), t \in \mathbb{R}_+, i = 1,2.$$

Let $Z_i, \hat{Z}_i, i = 1,2$ be four mutually independent random variables such that $Z_i =^{st} X_i(x_i), i = 1,2$ and $\hat{Z}_i$ has a hazard rate function $\hat{\gamma}(\cdot;y_i), i = 1,2$. Then from (2.7) one sees that $\hat{Z}_i$ is larger than $X_i(y_i)$ in the hazard rate ordering, $i = 1,2$ (e.g. see Ross (1983)). That is

(2.8) $$\hat{Z}_i \geq_{hr} X_i(y_i), i = 1,2.$$

Observe that the hazard rate function of $\min\{\hat{Z}_1, \hat{Z}_2\}$ is $\sum_{i=1}^2 \hat{\gamma}(\cdot;y_i) = \sum_{i=1}^2 \gamma(\cdot;x_i)$ which is the same as that for $\min\{Z_1, Z_2\}$. Therefore

(2.9) $$\min\{\hat{Z}_1, \hat{Z}_2\} =^{st} \min\{Z_1, Z_2\}.$$

For any $s, t \in \mathbb{R}_+$ such that $t \geq s$, consider

(2.10) $$\begin{aligned} &P\{\max\{Z_1, Z_2\} > t | \min\{Z_1, Z_2\} = s\} \\ &= \{\frac{f(s;x_1)\bar{F}(s;x_2)}{f(s;x_1)\bar{F}(s;x_2) + f(s;x_2)\bar{F}(s;x_1)}\}\frac{\bar{F}(t;x_2)}{\bar{F}(s;x_2)} \\ &+\{\frac{f(s;x_2)\bar{F}(s;x_1)}{f(s;x_1)\bar{F}(s;x_2) + f(s;x_2)\bar{F}(s;x_1)}\}\frac{\bar{F}(t;x_1)}{\bar{F}(s;x_1)} \\ &= \frac{\gamma(s;x_1)exp\{-\int_s^t \gamma(u;x_2)du\} + \gamma(s;x_2)exp\{-\int_s^t \gamma(u;x_1)du\}}{\gamma(s;x_1) + \gamma(s;x_2)}. \end{aligned}$$

Similarly one sees that

(2.11) $$\begin{aligned} &(P\{\max\{\hat{Z}_1, \hat{Z}_2\} > t | \min\{\hat{Z}_1, \hat{Z}_2\} = s\} \\ &= \frac{\hat{\gamma}(s;y_1)exp\{-\int_s^t \hat{\gamma}(u;y_2)du\} + \hat{\gamma}(s;y_2)exp\{-\int_s^t \hat{\gamma}(u;y_1)du\}}{\hat{\gamma}(s;y_1) + \hat{\gamma}(s;y_2)}. \end{aligned}$$

Since $\gamma(u;x_1) \geq \hat{\gamma}(u;y_1) \geq \hat{\gamma}(u;y_2) \geq \gamma(u;x_2)$ and $\gamma(u;x_1) + \gamma(u;x_2) = \hat{\gamma}(u;y_1) + \hat{\gamma}(u;y_2)$ one has

$$\begin{gathered} \gamma(s;x_1)exp\{-\int_s^t \gamma(u;x_2)du\} + \gamma(s;x_2)exp\{-\int_s^t \gamma(u;x_1)du\} \geq \\ \hat{\gamma}(s;y_1)exp\{-\int_s^t \hat{\gamma}(u;y_2)du\} + \hat{\gamma}(s;y_2)exp\{-\int_s^t \hat{\gamma}(u;y_1)du\} \\ \text{for all } u \in \mathbb{R}_+. \end{gathered}$$

Therefore, from (2.10) and (2.11) one sees that

$$(2.12) \qquad P\{\max\{Z_1, Z_2\} > t | \min\{Z_1, Z_2\} = s\} \geq P\{\max\{\hat{Z}_1, \hat{Z}_2\} > t | \min\{\hat{Z}_1, \hat{Z}_2\} = s\}, t \in \mathbb{R}_+.$$

Now combining (2.9) and (2.12) one concludes that

$$(2.13) \quad (\min\{Z_1, Z_2\}, \max\{Z_1, Z_2\}) \geq_{st} (\min\{\hat{Z}_1, \hat{Z}_2\}, \max\{\hat{Z}_1, \hat{Z}_2\}).$$

Then from (2.8) one has

$$(\min\{\hat{Z}_1, \hat{Z}_2\}, \max\{\hat{Z}_1, \hat{Z}_2\}) \geq_{st} (\min\{X_1(y_1), X_2(y_2)\}, \max\{X_1(y_1), X_2(y_2)\})$$

and hence for any increasing symmetric function h

$$h(X_1(x_1), X_2(x_2)) \geq_{st} h(X_1(y_1), X_2(y_2)).$$

That is we have established the theorem for $m = 2$. Extension to the general case can be routinely carried out using Lemma 2.2 (i) (e.g. see Marshall and Olkin (1979), Shanthikumar (1987)).

Proschan and Sethuraman (1976) showed that if the random variables $\{X_i(\theta), \theta \in \Theta\}$ have proportional hazard rates, i.e. if $\gamma(u;\theta) = \theta\alpha(u), \theta \in \Theta$ for each fixed u, then (2.4) holds. Here we have established a stronger conclusion (2.3) with a condition weaker than the proportionality of the hazard rates.

For the above result we need the strong condition of stochastic convexity in the hazard rate on $\{X_i(\theta), \theta \in \Theta\}$. We will next present a result that is weaker than Theorem 2.9, but requires only a weaker stochastic convexity condition on $\{X_i(\theta), \theta \in \Theta\}$. For this we will need the following definitions.

DEFINITION 2.10 (*Stochastic Convexity in the Usual Stochastic Ordering*): Let $\{Z(\theta), \theta \in \Theta\}$ be a collection of random variables with survival function $\bar{F}(\cdot;\theta)$. Then $\{Z(\theta), \theta \in \Theta\}$ is said to be stochastically increasing and linear [convex, concave] in the sense of usual stochastic ordering (see Shaked and Shanthikumar 1990a) if $\bar{F}(t;\theta)$ is pointwise increasing and linear [convex, concave] in θ for each fixed t. We denote this $\{Z(\theta), \theta \in \Theta\} \in SIL(st)[SICX(st), SICV(st)]$.

DEFINITION 2.11 A function $\phi : \mathbb{R}^m \to \mathbb{R}$ is submodular [supermodular] if for any $\mathbf{x}, \mathbf{y} \in \mathbb{R}^m$ we have

$$\phi(\mathbf{x}) + \phi(\mathbf{y}) \geq [\leq] \phi(\mathbf{x} \vee \mathbf{y}) + \phi(\mathbf{x} \wedge \mathbf{y}).$$

Here $\mathbf{x} \vee \mathbf{y} = (\max\{x_1, y_1\}, \ldots, \max\{x_m, y_m\})$ and $\mathbf{x} \wedge \mathbf{y} = (\min\{x_1, y_1\}, \ldots, \min\{x_m, y_m\})$. Now we can present the next theorem.

THEOREM 2.12 *Suppose* $\{X_i(\theta), \theta \in \Theta\} \in SIL(st)$. *Then for any increasing symmetric submodular [supermodular] function* h *and* $Y(\mathbf{x}) = h(X_1(x_1), \ldots, X_m(x_m))$ *one has* $\{Y(\mathbf{x}), \mathbf{x} \in \Theta^m\} \in SI{-}SchurCX(icv)[SI{-}SchurCV(icx)]$. *That is for any* $\mathbf{x}, \mathbf{y} \in \Theta^m$, *increasing and symmetric submodular [supermodular] function* h *and increasing and concave [convex] function* g,

$$\mathbf{x} \geq_{wm} [\geq^{wm}] \ \mathbf{y} \to Eg \circ h(X_1(x_1), \ldots, X_m(x_m)) \geq [\leq] \ Eg \circ h(X_1(y_1), \ldots, X_m(y_m)). \tag{2.14}$$

PROOF We will first establish the theorem for the case $m = 2$ and $\mathbf{x} \geq_m \mathbf{y}$. Then the result for the case $\mathbf{x} \geq_{wm} [\geq^{wm}]\mathbf{y}$ will follow from Lemma 2.2 (ii) [(iii)] and the stochastic monotonicity of $X_i(\theta)$ in θ. Suppose, without a loss of generality, $x_1 \leq y_1 \leq y_2 \leq x_2$ and $x_1 + x_2 = y_1 + y_2$. Then from Theorem 3.9 of Shaked and Shanthikumar (1990a) for a collection of $SIL(st)$ random variables, it is known that there exist four random variables $\{\hat{X}_i; \hat{Y}_i, i = 1, 2\}$ defined on a common probability space such that

$$\begin{aligned} \hat{X}_i &=^{st} X(x_i); \hat{Y}_i =^{st} X(y_i), i = 1, 2 \\ \hat{X}_1 &= \min\{\hat{Y}_1, \hat{Y}_2\} \\ \hat{X}_2 &= \max\{\hat{Y}_1, \hat{Y}_2\}. \end{aligned} \tag{2.15}$$

Observe that $\hat{X}_1 \leq \hat{X}_2$. Therefore if $\{\hat{X}_i^{(j)}; \hat{Y}_i^{(j)}, i = 1, 2\}, j = 1, 2$ are two independent samples of $\{\hat{X}_i; \hat{Y}_i, i = 1, 2\}$ one has,

$$\hat{X}_1^{(1)} \leq \hat{X}_2^{(1)} \text{ and } \hat{X}_1^{(2)} \leq \hat{X}_2^{(2)}. \tag{2.16}$$

Consider specific realizations $x_i^{(j)}$ of $\hat{X}_i^{(j)}$ and $y_i^{(j)}$ of $\hat{Y}_i^{(j)}, i = 1, 2; j = 1, 2$. Then by (2.16) one has $x_1^{(1)} \leq x_2^{(1)}$ and $x_1^{(2)} \leq x_2^{(2)}$. There are only the following four cases one may encounter:

$$\begin{aligned} &(a) & x_1^{(1)} &= y_1^{(1)} (\Leftrightarrow x_2^{(1)} = y_2^{(1)}), & x_1^{(2)} &= y_1^{(2)} (\Leftrightarrow x_2^{(2)} = y_2^{(2)}) \\ &(b) & x_1^{(1)} &= y_1^{(1)} (\Leftrightarrow x_2^{(1)} = y_2^{(1)}), & x_1^{(2)} &= y_2^{(2)} (\Leftrightarrow x_2^{(2)} = y_1^{(2)}) \\ &(c) & x_1^{(1)} &= y_2^{(1)} (\Leftrightarrow x_2^{(1)} = y_1^{(1)}), & x_1^{(2)} &= y_1^{(2)} (\Leftrightarrow x_2^{(2)} = y_2^{(2)}) \\ &(d) & x_1^{(1)} &= y_2^{(1)} (\Leftrightarrow x_2^{(1)} = y_1^{(1)}), & x_1^{(2)} &= y_2^{(2)} (\Leftrightarrow x_2^{(2)} = y_1^{(2)}). \end{aligned}$$

It is easily verified that for cases (a) and (d) and any symmetric function ϕ,

$$\phi(x_1^{(1)}, x_2^{(2)}) + \phi(x_1^{(2)}, x_2^{(1)}) = \phi(y_1^{(1)}, y_2^{(2)}) + \phi(y_1^{(2)}, y_2^{(1)}). \tag{2.17}$$

So consider case (b). If ϕ is symmetric and submodular [supermodular] one sees that

$$\begin{aligned} &\phi(x_1^{(1)}, x_2^{(2)}) + \phi(x_1^{(2)}, x_2^{(1)}) = \phi(x_1^{(1)}, x_2^{(2)}) + \phi(x_2^{(1)}, x_1^{(2)}) \\ &\geq [\leq] \phi(x_1^{(1)}, x_1^{(2)}) + \phi(x_2^{(1)}, x_2^{(2)}) = \phi(y_1^{(1)}, y_2^{(2)}) + \phi(y_2^{(1)}, y_1^{(2)}) \\ &= \phi(y_1^{(1)}, y_2^{(2)}) + \phi(y_1^{(2)}, y_2^{(1)}). \end{aligned} \tag{2.18}$$

The first equality follows by the symmetry of ϕ, the second inequality follows from the submodularity [supermodularity] of ϕ and $x_1^{(j)} \leq x_2^{(j)}, j = 1, 2$, the third equality follows because of the conditions of case (b) and the final equality follows by the symmetry of ϕ. Similarly it can be shown that under case (c),

$$(2.19) \quad \phi(x_1^{(1)}, x_2^{(2)}) + \phi(x_1^{(2)}, x_2^{(1)}) \geq [\leq] \phi(y_1^{(1)}, y_2^{(2)}) + \phi(y_1^{(2)}, y_2^{(1)}).$$

Therefore

$$\phi(\hat{X}_1^{(1)}, \hat{X}_2^{(2)}) + \phi(\hat{X}_1^{(2)}, \hat{X}_2^{(1)}) \geq [\leq] \phi(\hat{Y}_1^{(1)}, \hat{Y}_2^{(2)}) + \phi(\hat{Y}_1^{(2)}, \hat{Y}_2^{(1)}).$$

Hence

$$(2.20) \qquad E\phi(X_1(x_1), X_2(x_2)) \geq [\leq] E\phi(X_1(y_1), X_2(y_2)).$$

Observing that if g is an increasing concave [convex] function and h is an increasing submodular [supermodular] function then $\phi = g \circ h$ is a submodular [supermodular] function the proof of the theorem for the case $m = 2$ is complete. Extension to the general case can be routinely carried out using Lemma 2.2 (i).

The following result then follows easily from Theorem 2.12.

THEOREM 2.13 *Suppose $\{X_i(\theta), \theta \in \Theta\} \in SICX(st)[SICV(st)]$. Then for any increasing symmetric submodular [supermodular] function h and $Y(\mathbf{x}) = h(X_1(x_1), \ldots, X_m(x_m))$ one has $\{Y(\mathbf{x}), \mathbf{x} \in \Theta^m\} \in SI{-}SchurCX(icv)[SI{-}SchurCV(icx)]$. That is for any $\mathbf{x}, \mathbf{y} \in \Theta^m$, increasing and symmetric submodular [supermodular] function h and increasing and concave [convex] function g,*

$$(2.21) \qquad \mathbf{x} \geq_{wm} [\geq^{wm}] \ \mathbf{y} \rightarrow Eg \circ h(X_1(x_1), \ldots, X_m(x_m)) \geq [\leq] \ Eg \circ h(X_1(y_1), \ldots, X_m(y_m)).$$

Consider the case where the system performance is measured by the maximum [minimum] of the individual responses (e.g. parallel [series] reliability system.) That is $h(\mathbf{x}) = \max\{x_i, i = 1, \ldots, m\}$ [$= \min\{x_i, i = 1, \ldots, m\}$]. It is easily verified that h in this case is a submodular [supermodular] function. If we then apply the above result for this case we will have to restrict g to be increasing and concave [convex]. But the following easily verified lemma will allow us to strengthen this result.

LEMMA 2.14 *For any increasing function $\psi : \mathbb{R} \rightarrow \mathbb{R}$, the function $\phi : \mathbb{R}^m \rightarrow \mathbb{R}$ defined by $\phi(\mathbf{x}) = \psi(\max\{x_1, \ldots, x_m\})$ [$= \psi(\min\{x_1, \ldots, x_m\})$] for $\mathbf{x} \in \mathbb{R}^m$ is symmetric, increasing and submodular [supermodular].*

Combining Lemma 2.14 with Theorem 2.13 (Equation 2.20) one obtains

THEOREM 2.15 *Suppose* $\{X_i(\theta), \theta \in \Theta\} \in SICX(st)[SICV(st)]$. *Then if we define* $Y(\mathbf{x}) = \max\{X_1(x_1), \ldots, X_m(x_m)\}$ $[= \min\{X_1(x_1), \ldots, X_m(x_m)\}]$ *one has* $\{Y(\mathbf{x}), \mathbf{x} \in \Theta^m\} \in SI - SchurCX(st)[SI - SchurCV(st)]$. *That is for any* $\mathbf{x}, \mathbf{y} \in \Theta^m$ *and increasing function* g,

$$\begin{aligned} \mathbf{x} \geq_{wm} \mathbf{y} \quad \rightarrow \quad & Eg(\max\{X_1(x_1), \ldots, X_m(x_m)\}) \\ \geq \quad & Eg(\max\{X_1(y_1), \ldots, X_m(y_m)\}) \end{aligned}$$

$$\begin{aligned} [\mathbf{x} \geq^{wm} \mathbf{y} \quad \rightarrow \quad & Eg(\min\{X_1(x_1), \ldots, X_m(x_m)\}) \\ \leq \quad & Eg(\min\{X_1(y_1), \ldots, X_m(y_m)\})]. \end{aligned}$$

REMARK 2.16 Since $P\{\min\{X_1(x_1), \ldots, X_m(x_m)\} > t\} = \Pi_{i=1}^m \bar{F}(t; x_i)$ it is immediate that $\bar{F}(t; \theta)$ is increasing and logconcave in θ will imply that $\{Y(\mathbf{x}), \mathbf{x} \in \Theta^m\} \in SI - SchurCV(st)$. This is a stronger conclusion than that in theorem 2.15.

3. Stochastic Transposition Increasingness

In this section we will define stochastic transposition increasingness and give sufficient conditions on $X_i(r_i), i = 1, \ldots, m$ and h under which $Y(\mathbf{r}) := h(X_1(r_1), \ldots, X_m(r_m))$ is stochastically transposition increasing with respect to $\mathbf{r}$. For this we will need the following definition of transposition increasing functions (which is slightly different from that given in Hollander, Proschan and Sethuraman (1981), and Marshall and Olkin (1979)).

DEFINITION 3.1 Let $\mathbf{y} \in \mathbb{R}^m$ and $\mathbf{x}$ be a permutation of $\mathbf{y}$. Then $\mathbf{x}$ is more *arranged* than $\mathbf{y}$ if $\mathbf{x}$ can be obtained from $\mathbf{y}$ by a finite number of successive pairwise interchanges of two coordinates at a time such that each interchange results in a decreasing order for the interchanged elements. We denote this $\mathbf{x} \geq_a \mathbf{y}$. (e.g. $(1,5,4,3) \geq_a (1,5,3,4)$; $(1,5,3,4) \geq_a (1,4,3,5)$ and $(1,5,4,3) \geq_a (1,4,3,5)$.)

Note that the above definition of the arrangement ordering ($\geq_a$) allows one to simplify the analysis of transposition increasingness, often making it sufficient to prove the desired result just for the two dimensional case.

DEFINITION 3.2 A function $\phi : \mathbb{R}^m \rightarrow \mathbb{R}$ is *transposition increasing* [*decreasing*] if $\mathbf{x} \geq_a \mathbf{y}$ implies $\phi(\mathbf{x}) \geq [\leq] \phi(\mathbf{y})$.

DEFINITION 3.3 *(Stochastic Transposition Increasingness)*: A real valued random variable $Z(\mathbf{x})$ parametrized by $\mathbf{x} \in \mathcal{X} \subset \mathbb{R}^m$ is *stochastically transposition increasing* in the sense of the usual stochastic [increasing convex, increasing concave] ordering if for any $\mathbf{x}, \mathbf{y} \in \mathcal{X}, \mathbf{x} \geq_a \mathbf{y}$ implies $Z(\mathbf{x}) \geq_{st} [\geq_{icx}, \geq_{icv}] Z(\mathbf{y})$. We denote this $\{Z(\mathbf{x}), \mathbf{x} \in \mathcal{X}\} \in S - TI(st)[S - TI(icx), S - TI(icv)]$.

If $-Z(\mathbf{x})$ is stochastically transposition increasing then we say that $Z(\mathbf{x})$ is stochastically transposition decreasing (and denote $S - TD(st)$ etc.)

As an immediate consequence of the definition of stochastic transposition increasingness one has the following partial characterization of the optimal solution to the allocation problem (1.1).

THEOREM 3.4 *Let $Y(\mathbf{r}) = h(X_1(r_1), \ldots, X_m(r_m))$. Suppose $\{Y(\mathbf{r}), \mathbf{r} \in \mathcal{S}\} \in S - TI(st)[S - TI(icx), S - TI(icv)]$. Then*
(i) for any increasing [increasing and convex, increasing and concave] function g and $\mathbf{r}, \mathbf{s} \in \mathcal{S}$ one has,

$$\mathbf{r} \geq_a \mathbf{s} \Rightarrow Eg \circ h(X_1(r_1), \ldots, X_m(r_m)) \geq Eg \circ h(X_1(s_1), \ldots, X_m(s_m))$$

(ii) If $\mathcal{S} = \mathcal{S}_0 := \{\mathbf{r} : \sum_{i=1}^m r_i = R, r_i \geq 0\}$, then an optimal solution $\mathbf{r}^*$ to (1.1) *will satisfy $r_1^* \leq \cdots \leq r_m^*$.*

Note that when we have a discrete resource of R units there can exist a large number of feasible solutions for (1.1). For example with $R = 10$ and $m = 10$ there are total of 92,378 different possible solutions (i.e. $|\mathcal{S}_0|$ = 92,378). The number of solutions that satisfy the characterization given above is only 42 (see Table 1 of Shanthikumar and Yao (1988)). From the above discussion it is clear that it is worthwhile to search for sufficient conditions on $X_i(r_i), i = 1, \ldots, m$ and h for $Y(\mathbf{r})$ to be stochastically transposition increasing. We shall do this next in this section (and examples where such conditions are naturally satisfied are given in Section 4.) Before that we note the following easily verified lemma.

LEMMA 3.5 *Let $F_Z(t; \mathbf{x}) = P\{Z(\mathbf{x}) \leq t\}$ and $\bar{F}_Z(t; \mathbf{x}) = P\{Z(\mathbf{x}) > t\}$ be respectively, the cumulative and survival functions of $Z(\mathbf{x})$. Then $\{Z(\mathbf{x}), \mathbf{x} \in \mathcal{X}\} \in S - TI(st)[S - TI(icx), S - TI(icv)] \Leftrightarrow \bar{F}_Z(t; \mathbf{x})[\int_t^\infty \bar{F}_Z(s; \mathbf{x}) ds, -\int_{-\infty}^t F_Z(s; \mathbf{x}) ds]$ is transposition increasing in $\mathbf{x}$.*

In the remainder of this section we will assume that $\{X_i(\theta), \theta \in \Theta\}, i = 1, \ldots, m$ are m mutually independent collections of random variables.

THEOREM 3.6 *Let $\gamma_i(\cdot; \theta)$ be the hazard rate function of the absolutely continuous positive random variable $X_i(\theta), \theta \in \Theta, i = 1, \ldots, m$. Suppose for each $t \in \mathbb{R}_+$, we have $\gamma_i(t; \theta)$ componentwise monotone and submodular in $(i, \theta) \in \{1, \ldots, m\} \times \Theta$. Then for any increasing and symmetric function h and $Y(\mathbf{x}) = h(X_1(x_1), \ldots, X_m(x_m))$ one has $\{Y(\mathbf{x}), \mathbf{x} \in \Theta^m\} \in S - TD(st)$. That is for any $\mathbf{x}, \mathbf{y} \in \Theta^m$, increasing symmetric function h and increasing* *function g,*

$$\mathbf{y} \geq_a \mathbf{x} \rightarrow Eg \circ h(X_1(y_1), \ldots, X_m(y_m)) \leq Eg \circ h(X_1(x_1), \ldots, X_m(x_m)). \tag{3.1}$$

PROOF We will first establish the theorem for the case $m = 2$ and $\gamma_i(\cdot;\theta)$ decreasing in i and θ. The other case where $\gamma_i(\cdot;\theta)$ is increasing in i and θ can be similarly proved. Suppose, without a loss of generality, $x_1 = y_2 \leq y_1 = x_2$. Since for each $t \in \mathbb{R}_+$, $\gamma_i(t;\theta)$ is componentwise decreasing and submodular in $(i,\theta) \in \{1,\ldots,m\} \times \Theta$ one has

$$\gamma_1(t;x_1) \geq \gamma_1(t;y_1) \geq \gamma_2(t;x_2); \quad \gamma_1(t;x_1) \geq \gamma_2(t;y_2) \geq \gamma_2(t;x_2) \tag{3.2}$$

and

$$\gamma_1(t;x_1) + \gamma_2(t;x_2) \leq \gamma_1(t;y_1) + \gamma_2(t;y_2). \tag{3.3}$$

Therefore there exist $\hat{\gamma}_1(t;y_1)$ and $\hat{\gamma}_2(t;y_2)$ such that

$$\gamma_2(t;x_2) \leq \hat{\gamma}_1(t;y_1) \leq \gamma_1(t;y_1); \quad \gamma_2(t;x_2) \leq \hat{\gamma}_2(t;y_2) \leq \gamma_2(t;y_2), \tag{3.4}$$

and

$$\gamma_1(t;x_1) + \gamma_2(t;x_2) = \hat{\gamma}_1(t;y_1) + \hat{\gamma}_2(t;y_2). \tag{3.5}$$

Let $Z_i, \hat{Z}_i, i = 1,2$ be four mutually independent random variables such that $Z_i =^{st} X_i(x_i), i = 1,2$ and $\hat{Z}_i$ has a hazard rate function $\hat{\gamma}_i(\cdot;y_i), i = 1,2$. Then from (3.4) one sees that $\hat{Z}_i$ is larger than $X_i(y_i)$ in the hazard rate ordering, $i = 1,2$. That is

$$\hat{Z}_i \geq_{hr} X_i(y_i), i = 1,2. \tag{3.6}$$

Now using a derivation same as that employed in the proof of Theorem 2.9 it can be shown that

$$(\min\{Z_1,Z_2\}, \max\{Z_1,Z_2\}) \geq_{st} (\min\{\hat{Z}_1,\hat{Z}_2\}, \max\{\hat{Z}_1,\hat{Z}_2\}). \tag{3.7}$$

Then from (3.6) one has

$$(\min\{\hat{Z}_1,\hat{Z}_2\}, \max\{\hat{Z}_1,\hat{Z}_2\}) \geq_{st} (\min\{X_1(y_1),X_2(y_2)\}, \max\{X_1(y_1),X_2(y_2)\})$$

and hence for any increasing symmetric function h

$$h(X_1(x_1),X_2(x_2)) \geq_{st} h(X_1(y_1),X_2(y_2)).$$

That is we have established the theorem for $m = 2$. Extension to the general case can be routinely carried out using the property of arrangement ordering $\geq_a$ as given in Definition 3.1 (e.g. see Marshall and Olkin (1979), Shanthikumar (1987)).

For the above result we need the strong condition of submodularity of the hazard rate of $X_i(\theta)$. We will next present a result that is weaker than Theorem 3.6, but requires only the supermodularity [submodularity] of the survival function of $X_i(\theta)$.

THEOREM 3.7 *Let $\bar{F}_i(\cdot;\theta)$ be the survival function of $X_i(\theta), \theta \in \Theta, i = 1, \ldots, m$. Suppose for each fixed $t \in \mathbb{R}, \bar{F}_i(t;\theta)$ is componentwise increasing and supermodular [submodular] in $(i,\theta) \in \{1,\ldots,m\} \times \Theta$. Then for any increasing symmetric submodular [supermodular] function h and $Y(\mathbf{x}) = h(X_1(x_1),\ldots,X_m(x_m))$ one has $\{Y(\mathbf{x}), \mathbf{x} \in \Theta^m\} \in S{-}TD(icv)[S{-}TI(icx)]$. That is for any $\mathbf{x}, \mathbf{y} \in \Theta^m$, increasing and symmetric submodular [supermodular] function h and increasing and concave [convex] function g,*

$$(3.8) \qquad \mathbf{y} \geq_a \mathbf{x} \to Eg \circ h(X_1(y_1),\ldots,X_m(y_m)) \leq [\geq]\ Eg \circ h(X_1(x_1),\ldots,X_m(x_m)).$$

PROOF We will first establish the theorem for the case $m = 2$. Suppose, without a loss of generality, $x_1 = y_2 \leq y_1 = x_2$. Since for each $t \in \mathbb{R}_+$, $\bar{F}_i(t;\theta)$ is componentwise increasing and supermodular [submodular] in $(i,\theta) \in \{1,\ldots,m\} \times \Theta$ one has

$$(3.9) \qquad \bar{F}_1(t;x_1) \leq [\bar{F}_1(t;y_1), \bar{F}_2(t;y_2)] \leq \bar{F}_2(t;x_2)$$

and

$$(3.10) \qquad \bar{F}_1(t;x_1) + \bar{F}_2(t;x_2) \geq [\leq]\bar{F}_1(t;y_1)] + \bar{F}_2(t;y_2).$$

Therefore there exist $\bar{F}_1^*(t;y_1)$ and $\bar{F}_2^*(t;y_2)$ such that

$$(3.11) \qquad \bar{F}_1^*(t;y_1) \geq [\leq]\bar{F}_1(t;y_1); \quad \bar{F}_2^*(t;y_2) \geq [\leq]\bar{F}_2(t;y_2),$$

and

$$(3.12) \qquad \bar{F}_1(t;x_1) + \bar{F}_2(t;x_2) = \bar{F}_1^*(t;y_1) + \bar{F}_2^*(t;y_2).$$

Let $Z_i, \hat{Z}_i, i = 1,2$ be four mutually independent random variables such that $Z_i =^{st} X_i(x_i), i = 1,2$ and $\hat{Z}_i$ has the survival function $\bar{F}_i^*(\cdot;y_i), i = 1,2$. Then from (3.11) one sees that $\hat{Z}_i$ is larger [smaller] than $X_i(y_i)$ in the usual stochastic ordering, $i = 1,2$. That is

$$(3.13) \qquad \hat{Z}_i \geq_{st} [\leq_{st}]X_i(y_i), i = 1,2.$$

Now using a derivation similar to that employed in the proof of Theorem 3.4 of Shaked and Shanthikumar (1990a), it can be shown that there exist four random variables $\{\hat{X}_i; \hat{Y}_i, i = 1,2\}$ defined on a common probability space such that

$$
\begin{aligned}
\hat{X}_i &=^{st} X(x_i); \hat{Y}_i =^{st} \hat{Z}_i, i = 1,2 \\
\hat{X}_1 &= \min\{\hat{Y}_1, \hat{Y}_2\} \\
\hat{X}_2 &= \max\{\hat{Y}_1, \hat{Y}_2\}.
\end{aligned} \tag{3.14}
$$

Now using a derivation same as that employed in the proof of Theorem 2.12 it can be shown that for any increasing symmetric submodular [supermodular] function ϕ,

$$\phi(\hat{X}_1^{(1)}, \hat{X}_2^{(2)}) + \phi(\hat{X}_1^{(2)}, \hat{X}_2^{(1)}) \geq [\leq] \phi(\hat{Y}_1^{(1)}, \hat{Y}_2^{(2)}) + \phi(\hat{Y}_1^{(2)}, \hat{Y}_2^{(1)}).$$

Therefore from (3.13) one sees that

$$E\phi(X_1(x_1), X_2(x_2)) \geq [\leq] E\phi(X_1(y_1), X_2(y_2)). \tag{3.15}$$

Observing that if g is an increasing concave [convex] function and h is an increasing submodular [supermodular] function then $\phi = g \circ h$ is a submodular [supermodular] function the proof of the theorem for the case $m = 2$ is complete. Extension to the general case can be routinely carried out using the property of arrangement ordering presented in Definition 3.1.

Combining Lemma 2.14 with the above Theorem 3.7 (Equation 3.15) one obtains

THEOREM 3.8 *Let $\bar{F}_i(\cdot;\theta)$ be the survival function of $X_i(\theta), \theta \in \Theta, i = 1,\ldots,m$. Suppose for each fixed $t \in \mathbb{R}$, $\bar{F}_i(t;\theta)$ is componentwise increasing and supermodular [submodular] in $(i,\theta) \in \{1,\ldots,m\} \times \Theta$. Then if we define $Y(\mathbf{x}) = \max\{X_1(x_1),\ldots,X_m(x_m)\}$ [$= \min\{X_1(x_1),\ldots,X_m(x_m)\}$] one has $\{Y(\mathbf{x}), \mathbf{x} \in \Theta^m\} \in S-TD(st)[S-TI(st)]$. That is for any $\mathbf{x},\mathbf{y} \in \Theta^m$ and increasing function g,*

$$
\begin{aligned}
\mathbf{y} \geq_a \mathbf{x} \;\rightarrow\; & Eg(\max\{X_1(y_1),\ldots,X_m(y_m)\}) \\
\leq\; & Eg(\max\{X_1(x_1),\ldots,X_m(x_m)\})
\end{aligned}
$$

$$
\begin{aligned}
[\mathbf{y} \geq_a \mathbf{x} \;\rightarrow\; & Eg(\min\{X_1(y_1),\ldots,X_m(y_m)\}) \\
\geq\; & Eg(\min\{X_1(x_1),\ldots,X_m(x_m)\})].
\end{aligned}
$$

4. Applications

Allocation problems of the kind (1.1) described in Section 1 arise in many different areas (e.g. see Jean-Marie and Gun (1990), Shanthikumar (1988), Shanthikumar and Stecke (1986), Shanthikumar and Yao (1988),

Yao and Shanthikumar (1987) and the papers referenced in there). In these papers each problem is analyzed and solved in its specific context. The results presented in Sections 2 and 3 now provide a unified way to solve many of these allocation problems. In this section we will present several applications of the results derived in Sections 2 and 3 to problems arising in telecommunication, manufacturing and reliability/performability systems.

4.1 *Parallel Queues with Resequencing*

Consider a single stage queueing system consisting of m parallel servers. The n-th customer arrives at time A_n and requires a service of length $B_n^{(i)}$ if serviced by the i- th server, $n = 1, 2, \ldots$. Customers on its arrival are assigned to one of the m parallel servers according to some assignment rule. Suppose the n-th customer is assigned the server $U_n, n = 1, 2, \ldots$. Each of the m servers is assumed to have a buffer with an unlimited capacity to store the waiting customers. Customers leaving this single stage are stored in a resequencing area where each customer is allowed to leave as soon as only after all the customers arrived to the system before it are released from the resequencing area. Queueing systems of this kind serve as models of telecommunication systems (e.g. see Baccelli, Makowski and Shwartz (1989), Gun (1989), Harrus and Plateau (1982), Jean-Marie (1987), Jean-Marie and Gun (1990)), of distributed database systems (e.g. see Kamoun, Kleinrock and Muntz (1981)) and of flexible assembly systems (e.g. see Buzacott (1990), Buzacott and Shanthikumar (1992)).

Let $V_i(t)$ be the workload at server i at time $t, t \in \mathbb{R}_+; i = 1, \ldots, m$. Then

$$(4.1) \qquad \begin{aligned} V_{U_n}(A_n) &= V_{U_n}(A_n-) + B_n^{(U_n)}, n = 1, 2, \ldots, \\ \frac{d}{dt} V_i(t) &= I\{V_i(t) > 0\}, t \in \mathbb{R}_+. \end{aligned}$$

Define the maximum workload at time t by

$$(4.2) \qquad \hat{V}(t) = \max\{V_1(t), \ldots, V_m(t)\}, t \in \mathbb{R}_+.$$

Then it is very easy to see that the sojourn time S_n of the n-th customer through the single stage and the sojourn time T_n of the n-th customer through the system (including the time spent, if any, at the resequencing area) are given by

$$(4.3) \qquad S_n = V_{U_n}(A_n) \text{ and } T_n = \hat{V}(A_n), n = 1, 2, \ldots$$

Define $W_n = \hat{V}(A_n-), n = 1, 2, \ldots$ A typical portion of a sample path of $\{\hat{V}(t), t \in \mathbb{R}_+\}$ is shown in Figure 1.

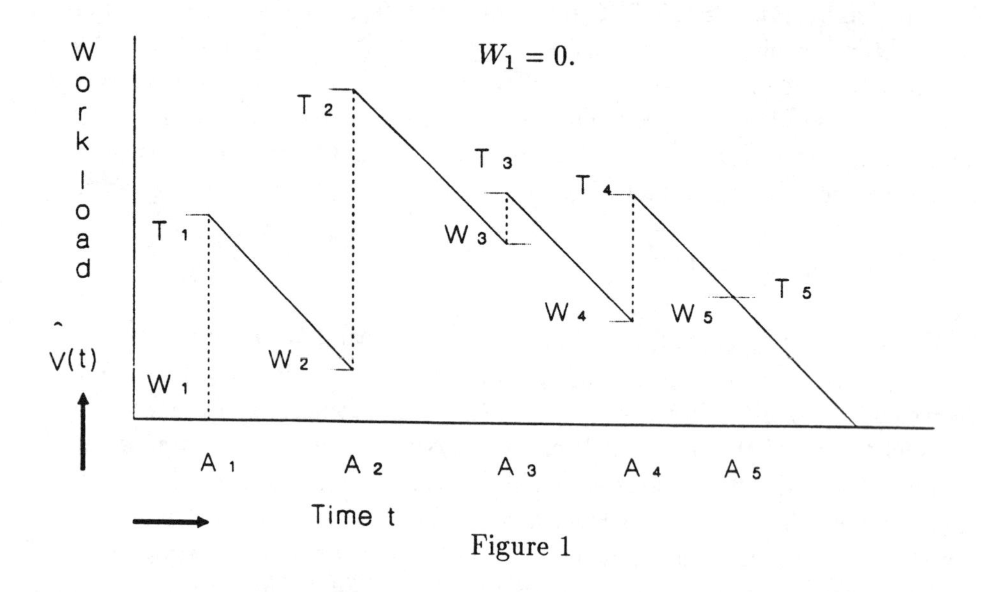

Figure 1

Assuming their existence let F_T, F_W and $F_{\hat{V}}$ be the stationary distribution of T_n, W_n and $\hat{V}(t)$. If the limit $\lambda = \lim_{n\to\infty}\{n/A_n\}$, and the density $f_{\hat{V}}$ of $F_{\hat{V}}$ exist, then equating the rates of up- and down-crossings over level x (e.g. see Cohen (1977), Shanthikumar (1980)) one gets

$$\lambda(\bar{F}_T(x) - \bar{F}_W(x)) = f_{\hat{V}}(x), x > 0. \tag{4.4}$$

Therefore

$$\bar{F}_T(x) = \bar{F}_W(x) + \frac{1}{\lambda} f_{\hat{V}}(x), x > 0. \tag{4.5}$$

Note that a similar relationship for a special case is derived in Jean-Marie and Gun (1990). Particularly they assume that (i) $\{A_n - A_{n-1}, n = 1,2,\ldots\}$, $\{B_n^{(i)}, n = 1,2,\ldots\}, i = 1,\ldots,m$ and $\{U_n, n = 1,2,\ldots\}$ are all mutually independent sequences of *i.i.d.* random variables, that (ii) the arrival process is Poisson and that (iii) $\{U_n, n = 1,2,\ldots\}$ is a sequence of multinomial trials with $P\{U_n = i\} = p_i, i = 1,\ldots,m$. Now we will make the same set of assumptions and look at the allocation of the probabilities $(p_1,\ldots,p_m)$ that will stochastically minimize the stationary system sojourn time $T(\mathbf{p})$. Since Poisson arrivals see time averages (see Wolff (1982)) it is clear that when these stationary distributions exist (i.e. when $0 \le \lambda p_i < \mu_i = 1/EB_n^{(i)}, i = 1,\ldots,m$), $F_W(x) = F_{\hat{V}}(x), x \in \mathbb{R}_+$. Hence (4.5) reduces (after integrating) to

$$\int_t^\infty \bar{F}_T(x)dx = \int_t^\infty \bar{F}_W(x)dx + \frac{1}{\lambda}\bar{F}_W(t), t > 0. \tag{4.6}$$

Let $W(\mathbf{p})$ be a generic random variable with distribution function F_W. Then from (4.6) and Lemma 2.6 one sees that if $\{W(\mathbf{p}), \mathbf{p} \in \mathcal{X}\} \in S-$

$SchurCX(st)$ then $\{T(\mathbf{p}), \mathbf{p} \in \mathcal{X}\} \in S - SchurCX(icx)$, where $\mathcal{X}$ is the set of values of $\mathbf{p}$ for which the stationary distributions exist. Since the stationary workloads at the m servers are independent because of Poisson arrival process and multinomial splitting one sees that

$$W(\mathbf{p}) = \max\{X_1(p_1), \ldots, X_m(p_m)\}, \tag{4.7}$$

where $X_i(p_i)$ is the stationary workload in an M/G/1 queueing system with arrival rate λp_i and service times $\{B_n^{(i)}, n = 1, 2, \ldots\}$ and $\{X_i(p_i)\}, i = 1, \ldots, m$ are independent sequences. From Theorem 4.17 of Shaked and Shanthikumar (1989) (also see Gun (1989)) it is known that $\{X_i(p_i), 0 \leq p_i < \mu_i/\lambda\} \in SICX(st)$. Then from (4.6) and Theorem 2.15 one sees that if $\{B_n^{(i)}, i = 1, \ldots, m\}$ are identical, then $\{T(\mathbf{p}), \mathbf{p} \in \mathcal{X}\} \in S - SchurCX(icx)$. Therefore balancing the allocation probabilities will minimize T in the increasing convex ordering. This conclusion was first derived by Jean-Marie and Gun (1990). The stochastic Schur convexity result for T is, however, new.

4.2 *Flexible Assembly Line*

Consider an m-stage serial assembly line. Parts arrive at this assembly line according to a Poisson process with rate λ. The nominal processing times (i.e. if only one standard worker is used) at stage i are *i.i.d.* exponential random variables with mean $1/\mu_i, i = 1, \ldots, m$. Suppose we have a total of R workers available and that we can assign them among the m stages of the assembly line. Problem of this kind arise naturally in many settings in the manufacturing systems (e.g. see Buzacott and Shanthikumar (1992), Shanthikumar and Yao (1988)). We will consider two typical problems that arise in this context (see Chapter 6 of Buzacott and Shanthikumar (1992)).

PROBLEM 1 Suppose the average workload assigned per stage is the same for all stages (i.e. $\mu_i = \mu, i = 1, \ldots, m$). The number of workers available is more than the number of stages (i.e. $R > m$). If we allocate r workers to the same stage the collective processing rate is $c(r), r = 1, \ldots R$. In an ideal case we would expect $c(n) = n$, but for the sake of generality, we will assume that $c(n)$ is increasing and concave in n. Let $T(\mathbf{r})$ be the stationary sojourn time of an arbitrary part through the assembly line if we allocate r_i workers to stage $i, i = 1, \ldots, m$. Let $X_i(n)$ be an exponential random variable with constant hazard rate $\gamma_i(n) = c(n)\mu - \lambda > 0$. Also assume that $\{X_i(n)\}, i = 1, \ldots, m$ are mutually independent. Then (e.g. see Jackson

(1963))

$$T(\mathbf{r}) = \sum_{i=1}^{m} X_i(r_i), \quad r_i : c(r_i) > \lambda/\mu. \tag{4.8}$$

Since $\{X_i(n), n : c(n) > \lambda/\mu\} \in SDCX(hr)$ from Theorem 2.9 one sees that $\{T(\mathbf{r}), r_i : c(r_i) > \lambda/\mu, i = 1, \ldots, m\} \in SD - SchurCX(st)$. Therefore a balanced worker allocation will stochastically minimize the total sojourn time T in the usual stochastic ordering.

PROBLEM 2 Suppose the stages are numbered such that the average workload assigned to stage i is larger than that assigned to stage $i+1$ (i.e. $1/\mu_i \geq 1/\mu_{i+1}$) ,$i = 1, \ldots, m$. The number of workers available is more than the number of stages (i.e. $R > m$). If we allocate r workers to the stage i the collective processing rate is $c_i(r), r = 1, \ldots R$. Note that $c_i(1) = \mu_i$. Let $T(\mathbf{r})$ be the stationary sojourn time of an arbitrary part through the assembly line if we allocate r_i workers to stage $i, i = 1, \ldots, m$. Let $X_i(n)$ be an exponential random variable with constant hazard rate $\gamma_i(n) = c_i(n) - \lambda > 0$. Also assume that $\{X_i(n)\}, i = 1, \ldots, m$ are mutually independent. Then (e.g. see Jackson (1963))

$$T(\mathbf{r}) = \sum_{i=1}^{m} X_i(r_i), \quad r_i : c_i(r_i) > \lambda, i = 1, \ldots, m. \tag{4.9}$$

Suppose $c_i(n)$ is componentwise increasing and submodular in (i, n). Then one finds that $\gamma_i(n)$ is componentwise increasing and submodular. Then from Theorem 3.6 one sees that $\{T(\mathbf{r}), r_i : c_i(r_i) > \lambda, i = 1, \ldots, m\} \in S - TD(st)$. Therefore allocating more workers to the stages with smaller indices (i.e., stages with more workload) will stochastically reduce the total sojourn time T in the usual stochastic ordering.

4.3 *Reliability/Performability System*

Consider a reliability/performability system (e.g., see Shaked and Shanthikumar (1990b)) consisting of m components. Suppose a total budget of R dollars is to be allocated among the m components. Suppose $X_i(r_i)$ is the lifetime of component i if r_i dollars is allocated for it. Suppose the performability function is $c(n)$ (i.e. when n components are alive the rate at which performance is accumulated is $c(n)$). For example $c(n)$ could be the production rate of a manufacturing system when n machines are working. We assume that $c(0) = 0$. The total performance as a function of the component lifetimes is

$$Y(\mathbf{r}) = h(X_1(r_1), \ldots, X_m(r_m)), \tag{4.10}$$

where

$$h(\mathbf{x}) = \int_0^\infty c(\sum_{i=1}^m I\{x_i > t\})dt, \mathbf{x} \in \mathbb{R}_+^m. \tag{4.11}$$

We will need the following characterization of h.

LEMMA 4.1 *Let h be defined as in (4.11). If c is an increasing and concave [convex] function, then h is an increasing submodular [supermodular] function.*

PROOF Observe that

$$\frac{d}{dx_j}h(\mathbf{x}) = c(1 + \sum_{i=1}^m I\{x_i > x_j\}) - c(\sum_{i=1}^m I\{x_i > x_j\}) \tag{4.12}$$

Observing that $c(1+n) - c(n)$ is decreasing [increasing] in n if c is concave [convex] and that $\sum_{i=1}^m I\{x_i > x_j\}$ is increasing in x_i it can be concluded that h is submodular [supermodular].

For example if $c(0) = 0, c(n) = 1, n = 1, 2, \ldots$ [$c(n) = 0, n = 0, 1, \ldots m-1; c(n) = n + 1 - m, n = m, m+1, \ldots$], then $Y(\mathbf{r})$ will be the lifetime of a parallel [series] reliability system (e.g. see Barlow and Proschan (1975)). As expected since $c(n)$ is concave [convex] in n, the lifetime of a parallel [series] reliability system is submodular [supermodular] in the component lifetimes. Combining Theorem 2.13 with Lemma 4.1 one obtains

THEOREM 4.2 *Suppose the families $\{X_i(r_i)\}$ of the lifetimes of the components of the reliability/performability system are independent and identical. Then if $\{X_i(r_i)\} \in SICX(st)[SICV(st)]$ and the performability function is concave [convex], then $\{Y(\mathbf{r})\} \in SI - SchurCX(icv)$ [$SI - SchurCV(icx)$].*

Acknowledgement

The authors would like to thank the referee and Moshe Shaked for their valuable comments on an earlier version of this paper.

REFERENCES

BACCELLI, F., MAKOWSKI, A. M. AND SHWARTZ, A. (1989). The fork-join queue and related systems with synchronization constraints: Stochastic ordering and computable bounds. *Adv. Appl. Prob.* **21** 629-660.

BARLOW, R. E. AND PROSCHAN, F. (1975). *Statistical Theory of Reliability and Life Testing*, Holt, Rinehart and Winston, New York.

BUZACOTT, J. A. (1990). Abandoning the moving assembly line: Models of human operators and job sequencing. *Int. J. Prod. Res.* **28** 821-835.

BUZACOTT, J. A. AND SHANTHIKUMAR, J. G. (1992). *Stochastic Models of Manufacturing Systems.* Prentice Hall, Englewood Cliffs, NJ.

COHEN, J. W. (1977). On up-and-down crossings. *J. Appl. Prob.* **14** 405-410.

GUN, L. (1989). Performance Evaluation and Optimization of Parallel Systems with Synchronization. Ph.D. Thesis, Department of Electrical Engineering, University of Maryland, College Park, MD.

HARRUS, G. AND PLATEAU, B. (1982). Queueing analysis of a reordering issue. *IEEE Trans. Software Eng.* **SE-8** 113-123.

HOLLANDER, M., PROSCHAN, F. AND SETHURAMAN, J. (1981). Decreasing in transposition property of overlapping sums, and applications. *J. Multivariate Anal.* **11** 50-57.

JACKSON, J. R. (1963). Jobshop-like queueing systems. *Mgt. Sci.* **10** 131-142.

JEAN-MARIE, A. (1987). Load balancing in a system of two queues with resequencing. *Proc. Performance '87.* Brussels, Belgium. 75-88

JEAN-MARIE, A. AND GUN, L. (1990). Parallel queues with resequencing. INRIA, Centre de Sophia Antipolis, Valbonne Cedex, France.

KAMOUN, F., KLEINROCK, L. AND MUNTZ, R. (1981). Queueing analysis of ordering issue in a distributed database concurrency control mechanism. *Proc. 2nd Intl. Conf. on Distributed Computing Systems.* Versailles, France. 13-23.

MARSHALL, A. W. AND OLKIN, I. (1979). *Inequalities: Theory of Majorizations and its Applications.* Academic Press, New York.

PROSCHAN, F. AND SETHURAMAN, J. (1976). Stochastic comparison of order statistics from heterogeneous populations, with applications in reliability. *J. Multivariate Anal.* **6** 608-616.

ROSS, S. M. (1983). *Stochastic Processes.* Wiley, New York.

SHAKED, M. AND SHANTHIKUMAR, J. G. (1988). Stochastic convexity and its applications. *Adv. Appl. Prob.* **20** 427-446.

SHAKED, M. AND SHANTHIKUMAR, J. G. (1990a). Convexity of a set of stochastically ordered random variables. *Adv. Appl. Prob.* **22** 160-177.

SHAKED, M. AND SHANTHIKUMAR, J. G. (1990b). Reliability and Maintainability. Chapter 13 in *Handbook in Operations Research and Management Science*, Vol. 2, D. P. Heyman and M. J. Sobel, eds. North–Holland, Amsterdam. 653-713.

SHAKED, M. AND SHANTHIKUMAR, J. G. (1992). Regular, sample path and strong stochastic convexity: A review. In *Stochastic Orders and Decision Under Risk.* K. Mosler and M. Scarcini, eds., Institute of Mathematical Statistics, Hayward, CA. 320–333.

SHANTHIKUMAR, J. G. (1980). Some analyses on the control of queues using level crossings of regenerative processes. *J. Appl. Prob.* **17** 814-821.

SHANTHIKUMAR, J. G. (1987). Stochastic majorization of random variables with proportional equilibrium rates. *Adv. Appl. Prob.* **19** 854-872.

SHANTHIKUMAR, J. G. AND STECKE, K. (1986). Reducing work-in-process inventory in certain classes of flexible manufacturing systems. *Eur. J. Opnl. Res.* **26** 266-271.

SHANTHIKUMAR, J. G. AND YAO, D. (1988). On server allocation in multiple center manufacturing systems. *Oper. Res.* **36** 333-342.

WOLFF, R. W. (1982), Poisson arrivals see time average. *Oper. Res.* **30** 223-231.

YAO, D. AND SHANTHIKUMAR, J. G. (1987). The optimal input rates to a system of manufacturing cells. *INFOR, Canad. J. Opnl. Res. Inf. Proc.* **25** 57-65.

SCHOOL OF BUSINESS AND TECHNOLOGY
UNIVERSITY OF WESTERN SIDNEY
CAMBELLTOWN, NSW 2560
AUSTRALIA

WALTER A. HAAS SCHOOL OF BUSINESS
UNIVERSITY OF CALIFORNIA, BERKELEY
BERKELEY, CA 94720

Stochastic Inequalities
IMS Lecture Notes – Monograph Series
Volume 22 (1993)

EXTREMAL PROBLEMS FOR PROBABILITY DISTRIBUTIONS: A GENERAL METHOD AND SOME EXAMPLES

By L. MATTNER

Universität Hamburg

A general method for treating extremal problems for probability distributions is presented. It is based on a Lagrange multiplier rule for constrained extremal problems in cones of Banach spaces. Some concrete problems are discussed.

1. Introduction

The purpose of this article is to report on a general method for solving extremal problems for probability distributions, as well as to present some examples developed in detail in the author's thesis Mattner (1990a) of which Mattner (1990b) is the relevant part in this context.

Additionally, a new and, hopefully, illuminating example (number 2 below) is treated.

The idea underlying the method to be presented is quite simple, namely: Just apply the existing Lagrange multiplier theory for extremal problems in Banach Spaces and modify it slightly, in such a way that the essential side condition of positivity is taken care of. This will lead to a necessary condition to be satisfied by any solution of a given extremal problem, provided that the functional to be extremized as well as functionals representing side conditions are sufficiently well-behaved, e.g. continuously Fréchet-differentiable.

Before stating a general theorem, let us look at a specific example which in fact motivated my study.

EXAMPLE 1 Let X and Y denote independent and identically distributed real random variables with

$$E[X] = 0, \quad Var(X) = 1. \tag{1}$$

AMS 1991 *subject classifications.* Primary 60E15; Secondary 49K27, 62H05.

Key words and phrases. Characterizations of probability distributions, expectation inequalities, extremal problems, Lagrange multiplier rule.

The problem is to maximize the expected distance of X and Y, under the above constraints:

$$E[|X - Y|] \stackrel{!}{=} \max. \tag{2}$$

This problem has received its solution a long time ago: the maximum is attained if and only if X and Y are uniformly distributed over the interval $[-\sqrt{3}, \sqrt{3}]$. In fact, Plackett (1947) considered a more general problem. He gave an argument which, for the present case, essentially reduces to writing the expected distance as

$$E[|X - Y|] = -2 \int_{-\infty}^{\infty} F(1 - F)dx, \tag{3}$$

where F denotes the distribution function of X, and performing a variation with respect to F. But he did this only formally, neither worrying about the existence of a solution nor making sure in an adequate way that his variations where still distribution functions. Nevertheless, he somehow arrived at the solution stated above. A proof of its correctness was first indicated by Moriguti (1951, footnote 5, p. 534) and given explicitly by Hartley and David (1954, p. 88), roughly speaking by applying the Cauchy-Schwarz inequality to the right-hand side of (3) after first manipulating that integral in such a way as to make sure that equality will hold for the presumed solution of Plackett.

2. A General Method

Thus it appears that the method used to solve Plackett's problem is unsystematic and also intrinsically univariate, the latter since it relies on manipulations involving the distribution function. These remarks apply as well to a more recent proof based on Terrell (1983) and given in Baringhaus and Henze (1990). Hence before trying to solve similar and perhaps more complicated problems, one should look for a general method yielding Plackett's result. To this end we observe that the problem may be viewed as an extremal problem in a Banach space.

Namely, let

$$M := \{\mu : \mu \text{ signed Borel measure with } ||\mu|| < \infty\},$$

where

$$||\mu|| := \int (1 + x^2)\, |d\mu(x)| \tag{4}$$

and $|d\mu(x)|$ denotes integration with respect to the total variation of μ. Plackett's problem, which I prefer to write as a minimization problem, may

then be written as follows:
"Minimize

$$\varphi_0(\mu) := -\int\int |x-y|\, d\mu(x)\, d\mu(y)$$

subject to the constraints

$$\begin{array}{rcll} \psi_1(\mu) & := & \int x\, d\mu(x) & = 0, \\ \psi_2(\mu) & := & \int x^2 d\mu(x) - 1 & = 0, \\ \psi_3(\mu) & := & \mu(\mathbb{R}) - 1 & = 0, \\ \mu & \geq & 0." & \end{array}$$

If the last condition were absent, we would just have an extremal problem in a Banach space with finitely many one-dimensional side conditions. What makes things slightly complicated is that $\mu \geq 0$ is an inequality constraint of infinite dimensional character. However, it may be written as $\mu \in C$, where C is the cone of the positive measures in M. And in fact, there is a Lagrange multiplier rule applicable in such cases:

THEOREM 1 *Let Z be a Banach space,*

$$\begin{array}{rcll} \varphi_0 & : & Z & \to \mathbb{R}, \\ \varphi_i & : & Z & \to \mathbb{R}, \quad i = 1, \ldots, m, \\ \psi_j & : & Z & \to \mathbb{R}, \quad j = 1, \ldots, n, \end{array}$$

continuously Fréchet-differentiable, and

(5) C *a convex cone in* Z.

Define

$$\mathcal{L}(z) := \lambda_0 \varphi_0(z) + \sum_{i=1}^{m} \lambda_i \varphi_i(z) + \sum_{j=1}^{n} \alpha_j \psi_j(z).$$

If $z \in Z$ minimizes φ_0 subject to

$$\begin{array}{rcll} \varphi_i(z) & \leq & 0, & i = 1, \ldots, m, \\ \psi_j(z) & = & 0, & j = 1, \ldots, n, \\ z & \in & C, & \end{array}$$

then there exist $\lambda_0, \lambda_1, \ldots, \lambda_m, \alpha_1, \ldots, \alpha_n \in \mathbb{R}$ with

(*i*) *not all λ_i and α_j vanish,*
(*ii*) $\lambda_i \geq 0, \quad i = 0, \ldots, m,$
(*iii*) $\langle \mathcal{L}'(z), w \rangle \geq 0, \quad w \in C,$
(*iv*) $\langle \mathcal{L}'(z), z \rangle = 0.$

This is proved in Mattner (1990a,b) by an application of an abstract multiplier rule given in Tikhomirov (1986). Here we try to explain the theorem by looking at two special cases, which should make it easy to memorize it.

In case one we assume that C is the whole space. Then condition (iii) may be applied to any w as well als to $-w$. So it just says that $\mathcal{L}'(z)$ is the zero functional. This contains condition (iv), and the theorem reduces to the ordinary Lagrange multiplier rule for Banach spaces.

In case two we assume that the ordinary constraints given by the φ_i and ψ_j are absent. Consider a typical point $z \in C$. It will usually lie on the boundary. In fact, it is easy to see, that the cone of positive measures in Plackett's example has empty interior. If now z minimizes φ_0 in C, then those one-sided directional derivatives $\langle \varphi_0'(z),\ w\rangle$ of φ_0 for which $w \in C$ have to be nonnegative. This yields condition (iii) with $\mathcal{L} = \varphi_0$. In the direction of z we may even perform a two-sided derivative, which accordingly has to vanish. This is condition (iv). So the theorem is seen to be true in both cases, and it is at least plausible that it is true in general.

3. Examples

We will now look at several examples, beginning with Plackett's problem.

EXAMPLE 1 (CONT.) A standard argument involving tightness, Fatou's lemma and integration to the limit of uniformly integrable sequences (see Mattner (1990a), pp. 16-17, for details) shows that a solution of the extremal problem exists. The functionals involved here are either quadratic or linear, and continuous by choice of the norm (4). Hence the derivative of the Lagrange functional $\mathcal{L}$ is given by

$$\begin{aligned} \langle \mathcal{L}'(\mu), \nu\rangle &= -2\lambda_0 \textstyle\int\int |x-y|\,d\mu(y)\,d\nu(x) + \alpha_1 \int x d\nu(x) \\ &\quad + \alpha_2 \textstyle\int x^2 d\nu(x) + \alpha_3 \nu(\mathbb{R}) \\ &= \textstyle\int l(x)\,d\nu(x), \end{aligned}$$

where

$$l(x) = -2\lambda_0 \int |x-y| d\mu(y) + \alpha_1 x + \alpha_2 x^2 + \alpha_3.$$

Assuming that μ is any solution, we may apply (iii) to any Dirac measure $\nu = \delta_x$, in order to get

$$l(x) \begin{cases} \geq 0, & x \in \mathbb{R} \\ = 0, & x \in \text{supp}\ \mu, \end{cases}$$

where the equality in supp μ, the support of μ, follows from the nonnegativity of l and (iv).

This is the typical preliminary result obtained when working with the above Lagrange multiplier rule: We have an integral relation for the extremal measure μ consisting of an integral inequality in the whole space and an integral equality in the support of μ. A difficulty is that we don't know the support since we don't know μ. From this point on, the arguments to follow will have to make use of the specific properties of the extremal problem at hand.

Assume for a moment that λ_0 vanishes. Then l is a nonnegative polynomial of degree at most 2, not vanishing identically and hence having at most one zero. This implies that the support of μ is a singleton, which is impossible since the variance of μ equals one. This contradiction shows that we may assume without loss of generality that $2\,\lambda_0 = 1$.

It is now convenient to rewrite l as

$$l(x) = \int_x^\infty (1 - F(x))\,dy + \int_{-\infty}^x F(y)\,dy + \alpha_1 x + \alpha_2 x^2 + \alpha_3,$$

where F denotes the distribution function of μ and the familiar area formula for the expectation of a random variable has been applied. Clearly, l possesses at least one-sided derivatives given by

$$l'(x\pm) = 2F(x\pm) - 1 + \alpha_1 + 2\alpha_2 x.$$

We have

$$0 \le 2\,(F(x+) - F(x-)) = l'(x-) - l'(x+) \le 0, \quad x \in \text{supp}\ \mu,$$

the latter inequality holding because l assumes its minimum value at every $x \in$ supp μ. This shows that F is continuous and that l' exists and vanishes in the support of μ. Hence

$$F(x) = a + bx, \quad x \in \text{supp}\ \mu.$$

Using the continuity of F, it readily follows that μ is uniform over some interval, which is determined by the mean and variance.

The purpose of the above example was to illustrate the multiplier method in one of the simplest nontrivial cases, rather than to match other proofs of Plackett's result in brevity. The remaining examples were not previously treated by simpler methods and illustrate various aspects of the present method.

EXAMPLE 2 This example will show that the problem of the support of the extremal measure is not a trivial matter. Bentkus (1991) raised the question of what happens if the constraint

$$E[|X|^3] \le \beta$$

is added in Example 1.

It makes things slightly easier to omit the condition on the mean, which turns out to be satisfied anyway if we replace (1) by

$$E[X^2] = 1, \; E[|X|^3] \le \beta. \tag{6}$$

The argument given in Example 1 applies virtually without change, leading to the existence of a solution μ for every $\beta > 1$ and to the condition

$$F(x) = \frac{1}{2} + bx + cx|x| =: \varphi(x), \quad x \in \operatorname{supp} \mu, \tag{7}$$

for the corresponding continuous distribution function F, where c is known to be nonnegative.

In case $b \ge 0, \varphi$ is strictly increasing, which forces F to agree with φ on an interval symmetrical with respect to the origin, i.e. F has a density f given by

$$f(x) = b + 2c|x|, \quad |x| \le A. \tag{8}$$

However, in case $b < 0, \varphi$ increases in $(-\infty, \frac{b}{c}]$, decreases in $[\frac{b}{c}, -\frac{b}{c}]$, and increases again in $[-\frac{b}{c}, \infty)$, strictly in each case. Now (7) and continuity of F allow for several possibilities. In each case F has to coincide with φ in one or two compact intervals and be constant in the complementary intervals. Instead of calculating $E[|X - Y|]$ for each of these possible F, it is more convenient to observe that the extremal distribution has to be symmetrical with respect to the origin, leaving a density given by

$$f(x) = b + 2c|x|, \quad -\frac{b}{c} \le |x| \le A \tag{9}$$

as the only possibility. In fact, if the distribution of X is extremal satisfying (6), so is that of $-X$. The representation (3) shows that the functional to be minimized is strictly convex on the set of the probability measures satisfying (6). Hence, the solution is unique, i.e. X is distributed as $-X$.

Taking $\int f dx = \int x^2 f(x) dx = 1$ into account, we get from (8) and (9) by trite calculations

$$\begin{aligned} f(x) &= (1-\alpha)\tfrac{1}{2}\sqrt{\tfrac{2+\alpha}{6}} + \alpha\tfrac{2+\alpha}{6}|x|, \quad |x| \le \sqrt{\tfrac{6}{2+\alpha}}, \\ \beta(f) &:= \int |x|^3 f(x) dx = \tfrac{3\sqrt{6}}{10}\, \tfrac{5+3\alpha}{(2+\alpha)^{\frac{3}{2}}}, \\ E_f[|X-Y|] &= \sqrt{\tfrac{6}{2+\alpha}}\left(\tfrac{2}{3} + \tfrac{\alpha}{6} - \tfrac{\alpha^2}{30}\right) \end{aligned}$$

for $b \geq 0$ and

$$\begin{aligned}
f(x) &= \left(\tfrac{1}{\alpha}\sqrt{1-\alpha+\tfrac{2}{3}\alpha^2-\tfrac{\alpha^3}{6}}|x| - \tfrac{1-\alpha}{\alpha}\right)\sqrt{1-\alpha+\tfrac{2}{3}\alpha^2-\tfrac{\alpha^3}{6}}, \\
& (1-\alpha)\left(1-\alpha+\tfrac{2}{3}\alpha^2-\tfrac{\alpha^3}{6}\right)^{-\frac{1}{2}} \leq |x| \\
&\leq \left(1-\alpha+\tfrac{2}{3}\alpha^2-\tfrac{\alpha^3}{6}\right)^{-\frac{1}{2}}, \\
\beta(f) &:= \int |x|^3 f(x)dx = \frac{1-\frac{3}{2}\alpha+\frac{3}{2}\alpha^2-\frac{3}{4}\alpha^3+\frac{3}{20}\alpha^4}{\left(1-\alpha+\frac{2}{3}\alpha^2-\frac{\alpha^3}{6}\right)^{\frac{3}{2}}}, \\
E_f[|X-Y|] &= \frac{1-\frac{\alpha}{3}+\frac{2}{15}\alpha^2}{\sqrt{1-\alpha+\frac{2}{3}\alpha^2-\frac{\alpha^3}{6}}}
\end{aligned}$$

for $b \leq 0$, where in each case α is allowed to vary between zero and one. We observe that, for the above densities f, $E_f[|X-Y|]$ is strictly increasing in $\beta(f)$, by considering both as functions of α. The largest value of $\beta(f)$ is $\frac{3\sqrt{3}}{4}$, corresponding to the uniform density. Hence we may conclude that the f with $\beta(f) = \min\left(\beta, \frac{3\sqrt{3}}{4}\right)$ is the solution of our problem. The support of these solutions is disconnected for $1 < \beta < \frac{4\sqrt{2}}{5}$ and connected for $\frac{4\sqrt{2}}{5} \leq \beta < \infty$, which was hardly obvious at the outset.

Incidentally, it follows from the above that we get an extremal problem without solution if we replace (6) by

$$E[X^2] = 1, \quad E[|X|^3] = \beta$$

for some $\beta > \frac{4\sqrt{2}}{5}$.

The next two examples concern multivariate extremal problems.

EXAMPLE 3 The expected distance makes sense also for multivariate random variables. How large can it be given the second moment of the euclidean norm? The answer is given by the following theorem.

THEOREM 2 *If X and Y are independent and identically distributed random vectors in d-space, then*

$$E[|X-Y|] \leq \sqrt{2}\sqrt{E[|X|^2]} \cdot \begin{cases} \sqrt{\frac{2}{3}}, & d=1, \\ \frac{\pi}{\sqrt{6}}, & d=2, \\ \frac{\Gamma^2(\frac{d}{2})}{\Gamma(\frac{d}{2}-\frac{1}{4})\Gamma(\frac{d}{2}+\frac{1}{4})}, & d \geq 3, \end{cases}$$

where equality and $E|X|^2 = 1$ occurs if and only if

$$\begin{aligned}
& X \sim U([-\sqrt{3}, \sqrt{3}]), & d=1 \\
& f_X(x) = \tfrac{1}{2\pi}\frac{\frac{2}{3}}{\sqrt{1-\frac{2}{3}|x|^2}}\, 1\left(|x| \leq \sqrt{\tfrac{3}{2}}\right), & d=2 \\
& X \sim U(\{x \in \mathbb{R}^d : |x| = 1\}), & d \geq 3.
\end{aligned}$$

Here $|\cdot|$ denotes the euclidean norm and U stands for uniform distribution.

This is proved in Mattner (1990a,b) and, independently and in a different way, in Buja, Logan, Reeds and Shepp (1990). Of course, for $d = 1$ we just get a reformulation of Plackett's result.

The case $d = 2$ is particularly interesting. A heuristic for it runs as follows. The multiplier rule leads as before to an integral relation for a measure μ maximizing the expected distance given $E[|X|^2] = 1$:

$$(10) \qquad - \int_{\mathbb{R}^2} |x-y| \, d\mu(y) + \alpha_1 + \alpha_2 \, |x|^2 \begin{cases} \geq 0, & x \in \mathbb{R}^2 \\ = 0, & x \in \operatorname{supp} \mu. \end{cases}$$

Let us formally apply the Laplacian to the above equality. Because of $\Delta|\cdot| = \frac{1}{|\cdot|}$, valid in $\mathbb{R}^2$, this should lead to something like

$$\int_{\mathbb{R}^2} \frac{d\mu(y)}{|x-y|} = 4\,\alpha_2 \quad (x \in \operatorname{supp} \mu \subset \mathbb{R}^2).$$

Now read this relation "three-dimensionally": under the plausible assumption that the support of μ is a circular disk, it says that the spatial potential of μ is constant on it and hence μ has to be the electrostatic equilibrium distribution of unit charge on that disk. The latter is known and has, taking the condition $E[|x|^2] = 1$ into account, the density given in the theorem.

For a rigorous proof note that the euclidean norm is a so-called negative-definite function, which almost by definition implies that the functional

$$\mu \mapsto - \int \int |x-y| \, d\mu(x) \, d\mu(y)$$

is convex on the set of those probability measures with finite second moments. The convexity is in fact strict (This is true for arbitrary dimensions. We encountered the one-dimensional and elementary case in the previous example.). An application of the Kuhn-Tucker theorem shows that (10) is also sufficient for μ to be extremal. A not completely trivial calculation shows that the density of the theorem fulfills (10). See Mattner (1990a,b) for details.

So far we have solved given extremal problems. Now we are reversing the question and ask for an extremal problem having a given solution.

EXAMPLE 4 Can we characterize the uniform distribution over a ball in d-space as the solution of an extremal problem similar to Plackett's? Replacing $|x-y|$ by some unspecified function $K_d(x,y)$ and proceeding formally as above, we get

$$\int K_d(x,y) \, d\mu(y) = \alpha_1 + \alpha_2 \, |x|^2 \quad (x \in \operatorname{supp} \mu)$$

as a necessary condition for any extremal μ. Applying again the Laplacian to this equation, we get a constant on the right-hand side and want to get $f(x)$, a density of μ, on the left-hand side. This will be the case if $\Delta_x K_d(x,y)$ (Laplacian with respect to x) is the Dirac measure located at y. This suggests to take $K_d(x,y) = u(x-y)$ with u a constant multiple of a fundamental solution of the Laplacian. Thus we are led to guess the following theorem, proved in Mattner (1990a,b).

THEOREM 3 *If*

$$K_d(x,y) = \begin{cases} -|x-y|, & d=1, \\ \log \frac{1}{|x-y|}, & d=2, \\ \frac{1}{|x-y|^{d-2}}, & d \geq 3, \end{cases}$$

and if X and Y are independent and identically distributed random vectors in d-space with $E[|x|^2] = 1$, then $E[K_d(X,Y)]$ is minimal if and only if $X \sim U(\{x \in \mathbb{R}^d : |x| \leq r_d\})$ for some suitable r_d.

EXAMPLE 5 Can we characterize any given probability distribution as in Example 4? It is proved in Mattner (1990a,b) that the answer is "yes" under some regularity conditions, as well as that it is often "no" if we try $K(x,y) = u(x-y)$ for some function u:

THEOREM 4 *If $|u(x)| \leq A(1+x^2)$ for some finite A and if $E[u(X-Y)]$ is extremal for $X \sim N(0,1)$ under the constraints* (1), *then u is a polynomial of degree at most* 2 *and there are other extremal distributions.*

Acknowledgements

It is a pleasure to thank Professors Moshe Shaked and Y. L. Tong for organizing this conference and inviting me to present this paper. Financial support by the National Science Foundation and the Deutsche Forschungsgemeinschaft is gratefully acknowledged. Also I wish to thank a referee for his careful reading of this paper.

REFERENCES

BARINGHAUS, L., AND HENZE, N. (1990). A consistent test for uniformity with unknown limits based on D'Agostinos D. *Statist. Prob. Lett.* **9** 299-304.

BENTKUS, V. (1991). Personal communication.

BUJA, A., LOGAN, B. F., REEDS, J. R. AND SHEPP, L. A. (1990). Inequalities and positive-definite functions arising from a problem in multidimensional scaling. Preprint.

HARTLEY, H. O. AND DAVID, H. A. (1954). Universal bounds for mean range and extreme observation. *Ann. Math. Statist.* **25** 85-99.

MATTNER, L. (1990a). Behandlung einiger Extremalprobleme für Wahrscheinlichkeitsverteilungen. Dissertation, Universität Hannover.

MATTNER, L. (1990b). Extremal problems for probability distributions. Preprint.

MORIGUTI, S. (1951). Extremal properties of extreme value distributions. *Ann. Math. Statist.* **22** 523-536.

PLACKETT, R. L. (1947). Limits of the ratio of mean range to standard deviation. *Biometrika* **34** 120-122.

TERRELL, G. R. (1983). A characterization of rectangular distributions. *Ann. Probab.* **11** 823-826.

TIKHOMIROV, V. M. (1986). *Theory of Extremal Problems.* Wiley, New York.

INSTITUT FÜR MATHEMATISCHE STOCHASTIK
UNIVERSITÄT HAMBURG
BUNDESSTR. 55
D-2000 HAMBURG 13
GERMANY

Stochastic Inequalities
IMS Lecture Notes – Monograph Series
Volume 22 (1993)

CONCENTRATION INEQUALITIES FOR MULTIVARIATE DISTRIBUTIONS: II. ELLIPTICALLY CONTOURED DISTRIBUTIONS[1]

By MICHAEL D. PERLMAN

University of Washington

In part I of this study it was shown that $\Sigma_1 \leq \Sigma_2 \Rightarrow P_{\Sigma_1}(C) \geq P_{\Sigma_2}(C)$ under various convexity and symmetry assumptions on the set $C \subset \mathbb{R}^p$, where P_Σ denoted the p-variate normal distribution with mean vector 0 and positive definite covariance matrix Σ. In Part II extensions of these results to the family of elliptically contoured distributions are considered. The proof of the concentration inequality of Fefferman, Jodeit, and Perlman (1972) for convex centrally symmetric sets C is examined to determine whether it can be extended to sets C with other convexity and/or symmetry properties. Whereas it does not appear that this proof remains applicable, in the bivariate case ($p = 2$) an alternate geometric argument not only extends the concentration inequalities for convex G-invariant sets C and for G-decreasing sets C in Part I to elliptically contoured distributions, but also enlarges the class of groups G for which the concentration inequality for G-decreasing sets is valid. Also, sharpened forms of these concentration inequalities are presented for elliptically contoured distributions that are not absolutely continuous with respect to Lebesgue measure.

5. A Concentration Inequality for Convex Centrally Symmetric Sets

In Part I of this study[2] it was shown that

$$\Sigma_1 \leq \Sigma_2 \Rightarrow P_{\Sigma_1}(C) \geq P_{\Sigma_2}(C) \tag{5.0}$$

under various convexity and symmetry assumptions on the set $C \in \mathbb{R}^p$, where P_Σ denoted the p-variate normal distribution with mean vector 0 and positive definite covariance matrix Σ. It is evident that such concentration

[1]Research supported in part by National Science Foundation Grant No. DMS-89-02211.
AMS 1991 *subject classifications.* Primary 60E15; Secondary 52A40.
Key words and phrases. Multivariate concentration inequalities, elliptically contoured distributions, convex set, group invariance, orthogonal group, cyclic group, dihedral group.

[2]Eaton and Perlman (1991). Part I comprised Sections 1-4; Part II comprises Sections 5-7.

inequalities for multivariate normal distributions in Theorems 3.1, 3.2, and 3.3 of Part I remain valid when P_Σ is taken to be a scale mixture over $\lambda > 0$ of normal distributions on $\mathbb{R}^p$ with mean 0 and covariance matrix $\lambda\Sigma$, e.g., a multivariate Student-t distribution. Like the normal distribution itself, such a scale mixture is both unimodal and elliptically contoured. It is somewhat surprising that the first of these theorems, and possibly the other two as well, remain valid for *all* elliptically contoured distributions *without* assuming unimodality.

Fefferman, Jodeit, and Perlman (1972) substantially strengthened the concentration inequality in Theorem 3.1 for convex centrally symmetric sets $C \in \mathbb{R}^p$ by extending it from normal to elliptically contoured distributions (see also Das Gupta *et al* (1972), Theorem 3.3). Surprisingly, their proof is also based on Anderson's convolution theorem, Theorem 2.1, as was the proof of Theorem 3.1 in the normal case, although Anderson's theorem is now applied in a quite different way. In this section we review their proof in detail to determine whether or not it can be extended to sets C with other convexity and/or symmetry properties. Whereas it does not appear that their method of proof remains applicable, in the bivariate case ($p = 2$) an alternate geometric argument not only extends Theorem 3.2 (for convex G-invariant sets) and Theorem 3.3 (for G-decreasing sets) to elliptically contoured distributions but also enlarges the class of groups G to which Theorem 3.3 applies. These bivariate results are given in Theorems 6.1 and 6.2 of Section 6. In Section 7, sharpened forms of the concentration inequalities in Sections 5 and 6 are presented for elliptically contoured distributions that are not absolutely continuous with respect to Lebesgue measure on $\mathbb{R}^p$ and which therefore may assign nonzero probability to the boundary of C.

DEFINITION 5.1 The random vector $X \in \mathbb{R}^p$ has an *elliptically contoured distribution*, denoted by $X \sim EC_p(\Sigma)$, if its characteristic function $\varphi(t) \equiv E\{\exp(it'X)\}$, $t \in \mathbb{R}^p$, has the form $\varphi(t) = \gamma(t'\Sigma t)$ for some function γ, where Σ is a $p \times p$ positive definite matrix. Equivalently,

$$X \sim EC_p(\Sigma) \Leftrightarrow X \stackrel{d}{=} \Sigma^{1/2} Z, \tag{5.1}$$

where $\Sigma^{1/2}$ is the $p \times p$ positive definite matrix such that $(\Sigma^{1/2})^2 = \Sigma$ and where Z is an orthogonally invariant random vector in $\mathbb{R}^p$. If X has a probability density function f on $\mathbb{R}^p$ then $X \sim EC_p(\Sigma)$ iff $f(x) = |\Sigma|^{-1/2} g(x'\Sigma^{-1}x)$ for some function g; in particular, the multivariate normal distribution $N_p(0, \Sigma)$ is $EC_p(\Sigma)$.

The following notation is used: B and S denote the unit ball and unit sphere in $\mathbb{R}^p$, ν denotes the uniform probability measure on S, and $D \equiv \text{Diag}(d_1, \ldots, d_p)$ denotes a $p \times p$ diagonal matrix with $0 < d_i \leq 1$ for $i =$

$1,\dots,p$, so D is a *contraction*. The class of all convex centrally symmetric sets in $\mathbb{R}^p$ is denoted by $\mathcal{C}_1$.

THEOREM 5.1 (Fefferman, Jodeit, and Perlman (1972)). *Suppose that* $X \sim EC_p(\Sigma)$. *If* $C \in \mathcal{C}_1$ *and* C *is closed, then* $\Sigma_1 \le \Sigma_2 \Rightarrow P_{\Sigma_1}(C) \ge P_{\Sigma_2}(C)$.

PROOF By (5.1),

$$X \sim EC_p(\Sigma) \Rightarrow P_\Sigma(C) \equiv P_\Sigma(X \in C) = P(Z \in \Sigma^{-1/2}C), \tag{5.2}$$

where $\Sigma^{-1/2} = (\Sigma^{1/2})^{-1} = (\Sigma^{-1})^{1/2}$. Since Z is orthogonally invariant, $Z \stackrel{d}{=} R \cdot U$, where R and U are independent, U is uniformly distributed on the sphere $S \equiv \{x \in \mathbb{R}^p : \|x\| = 1\}$, and $0 \le R < \infty$. Therefore

$$P_\Sigma(C) = E\{P[U \in R^{-1}\Sigma^{-1/2}C|R]\} \equiv E\{\nu(R^{-1}\Sigma^{-1/2}C)\}. \tag{5.3}$$

Since $C \in \mathcal{C}_1 \Leftrightarrow R^{-1}C \in \mathcal{C}_1$ (provided $R > 0$) it therefore suffices to compare $\nu(\Sigma_1^{-1/2}C)$ and $\nu(\Sigma_2^{-1/2}C)$ for $C \in \mathcal{C}_1$.

By the Singular Value Decomposition

$$\Sigma_2^{-1/2}\Sigma_1^{1/2} = \psi' D\Gamma, \tag{5.4}$$

where ψ and Γ are $p \times p$ orthogonal matrices, $D = \mathrm{Diag}(d_1,\dots,d_p)$, and $d_1,\dots,d_p$ are the singular values of $\Sigma_2^{-1/2}\Sigma_1^{1/2}$. Since $0 < \Sigma_1 \le \Sigma_2$, $0 < d_i \le 1$ for $i = 1,\dots,p$, so D is a contraction. Because ν is orthogonally invariant,

$$\begin{aligned} \nu(\Sigma_1^{-1/2}C) &= \nu(\Gamma\Sigma_1^{-1/2}C) = \nu(K) \\ \nu(\Sigma_2^{-1/2}C) &= \nu(\psi' D\Gamma\Sigma_1^{-1/2}C) = \nu(DK), \end{aligned} \tag{5.5}$$

where

$$K \equiv K(C;\Sigma_1,\Gamma) = \Gamma\Sigma_1^{-1/2}C \in \mathcal{C}_1. \tag{5.6}$$

Thus the desired result is equivalent to the following assertion: for every closed $K \in \mathcal{C}_1$ and every diagonal contraction mapping D,

$$\nu(K) \ge \nu(DK). \tag{5.7}$$

This inequality is nontrivial since DK need not be contained in K. By means of the Divergence Theorem, however, it can be shown that[3]

$$\begin{aligned} \frac{\partial}{\partial d_i}[\nu(DK)] &\equiv \frac{\partial}{\partial d_i}\int_S I_{DK}(x)d\nu(x) \\ &\doteq -d_i^{-1}\frac{\partial^2}{\partial\beta^2}[\int_B I_{DK}(x-\beta\theta_i)dx]_{\beta=0}, \end{aligned} \tag{5.8}$$

[3]The equality $\doteq$ in (5.8) may hold only for almost every d_i, so a more careful argument is needed which makes use of the assumption that C, and hence K, is closed. First, if K is not bounded, consider the bounded set $K^* \equiv K \cap (m^{-1}B)$, where $m \equiv min(d_1,\dots,d_p) < 1$. Then $K^* \in \mathcal{C}_1, \nu(K^*) = \nu(K)$, and $\nu(DK^*) = \nu(DK)$, so it would suffice to establish

where I_E denotes the indicator function of the set E and θ_i is the unit vector with i-th component 1. Since both B and $DK \in \mathcal{C}_1$, Anderson's Theorem 2.1 implies that

$$(5.9) \qquad \psi(\theta) \equiv \int_B I_{DK}(x-\theta)dx$$

is centrally symmetric and ray-decreasing in θ, hence has a local (in fact, global) maximum at 0, so the second derivative in (5.8) is nonpositive. Therefore $\nu(DK)$ is nondecreasing in each $d_i, i = 1, \ldots, p$, which establishes (5.7). □

In Section 3 of Part I we saw that for multivariate normal distributions, the method of proof of Theorem 3.1 could be used to establish Theorems 3.2 and 3.3 simply by replacing Theorem 2.1 by Theorems 2.2 and 2.4 respectively. Unfortunately this is not so for elliptically contoured distributions. In the proof of Theorem 5.1, Anderson's Theorem 2.1 was applied to show that ψ in (5.9) has a maximum at $\theta = 0$ when C (*and therefore* DK!!) $\in \mathcal{C}_1$. In order to extend Theorem 3.2 to elliptically contoured distributions by this method, it would be necessary to apply Theorem 2.2 to show that ψ has a maximum at $\theta = 0$ when $C \in \mathcal{C}_G$, where G is a compact subgroup of the orthogonal group $\mathcal{O}_p$ that acts effectively on $\mathbb{R}^p$, $\mathcal{C}_G$ is the class of all convex G-invariant subsets of $\mathbb{R}^p$, and Σ_1 is G-invariant (for detailed definitions, see Section 3; recall from (5.6) that K, and hence ψ, depends on Σ_1). Now it can be shown[4] that

$$(5.10) \qquad \begin{aligned} \Sigma_1 \text{ is } G-\text{invariant} &\Rightarrow \Sigma_1^{1/2} \text{ is } G-\text{invariant} \\ &\Rightarrow \Sigma_1^{-1/2} \text{ is } G-\text{invariant}, \end{aligned}$$

so

$$(5.11) \qquad C \in \mathcal{C}_G \Rightarrow \Sigma_1^{-1/2} C \in \mathcal{C}_G \Rightarrow K \in \mathcal{C}_{\tilde{G}},$$

where K is defined in (5.6) and

$$(5.12) \qquad \tilde{G} = \Gamma G \Gamma' = \{\Gamma g \Gamma' | g \in G\}$$

(5.7) with K replaced by K^*. Thus we may assume that K is in fact compact. Since K is compact, convex, and centrally symmetric, by considering its supporting hyperplanes we see that it is the decreasing limit of a sequence of compact convex centrally symmetric polyhedra in $\mathbb{R}^p$, so we may assume that K is such a polyhedron. Then we may construct a sequence of smooth centrally symmetric unimodal functions u_ϵ which converges to I_K everywhere in $\mathbb{R}^p$ except possibly on ∂K as $\epsilon \to 0$, but $\nu(\partial K) = 0$ since K is a polyhedron. If we now replace $I_{DK}(x) \equiv I_K(D^{-1}x)$ by $u_\epsilon(D^{-1}x)$ in (5.8) then $\doteq$ becomes $=$ for every d_i, so $\int_S u_\epsilon(D^{-1}x)d\nu(x)$ is nondecreasing in each d_i, hence so is $\nu(DK)$. (See Fefferman *et al* (1972) for further details.)

[4]If Σ_1 is G-invariant then so is $f(\Sigma_1)$, where f is any polynomial with real coefficients. But $(\Sigma_1)^{1/2} = f(\Sigma_1)$ where f is any real polynomial such that $f(\lambda_i) = (\lambda_i)^{1/2}$ for $i = i, \ldots, p$, where $\lambda_1, \ldots, \lambda_p$ are the eigenvalues of Σ_1. (The coefficients of f may depend on $\lambda_1, \ldots, \lambda_p$ and hence on Σ_1.) Similarly, $(\Sigma_1)^{-1}$ and $(\Sigma_1)^{-1/2}$ are G-invariant. (We thank Steen A. Andersson for this observation.)

is also a compact effective subgroup of $\mathcal{O}_p$. Unfortunately, although the linear transformation $K \to DK$ preserves convexity *it need not preserve* $\widetilde{G}$*-invariance* (unlike Theorem 5.1 where $G = \widetilde{G} = \{\pm I\}$ with I the $p \times p$ identity matrix), so we cannot conclude that $DK \in \mathcal{C}_{\widetilde{G}}$ and thus are unable to apply Theorem 2.2 to ψ in (5.9). Similarly, when $C \in \mathcal{M}_G$ (the class of all G-decreasing subsets of $\mathbb{R}^p$) with G a compact effective reflection group, then $\widetilde{G}$ is also a compact effective reflection group but we cannot conclude that $DK \in \mathcal{M}_{\widetilde{G}}$, hence cannot apply Theorem 2.4 to extend Theorem 3.3 to elliptically contoured distributions.

Despite these difficulties, we conjecture that the concentration inequalities for the classes $\mathcal{C}_G$ and $\mathcal{M}_G$ in Theorems 3.2 and 3.3 remain valid for elliptically contoured distributions. To support this conjecture, in Section 6 we present an alternate geometric argument, similar to that in Section 1 of Fefferman, Jodeit, and Perlman (1972), which establishes these results in the bivariate case, i.e., when $p = 2$. In fact Theorem 6.2, the extension of Theorem 3.3 thus obtained, is *strictly stronger* than Theorem 3.3 in the bivariate case in that it applies to a *larger* class of groups G (acting on $\mathbb{R}^2$) than the class of effective reflection groups.

REMARK 5.1 If $-I \in G$ then $\mathcal{C}_G \subseteq \mathcal{C}_1$, so in this case the extension of Theorem 3.2 to elliptically contoured distributions is implied by Theorem 5.1 *without* the assumption that Σ_1 is G-invariant (also see Remark 3.1 of Part I). □

REMARK 5.2 Because it suffices to show only that ψ in (5.9) has a *local* maximum at $\theta = 0$, the method of proof in this section may succeed in extending Theorems 3.2 and 3.3 to elliptically contoured distributions provided that suitable *local* versions of Theorems 2.2 and 2.4 can be found. Note too that one of the sets in (5.9), namely B, is a ball, so the full generality of these latter theorems would not be needed. Furthermore, even the existence of a local maximum at $\theta = 0$ is not necessary; it would suffice to show that C is locally concave at $\theta = 0$. □

6. Bivariate Concentration Inequalities for Elliptically Contoured Distributions

In this section unless otherwise noted, $p = 2$, B and S denote the closed unit disk and unit circle in $\mathbb{R}^2$, respectively, ν denotes the uniform measure on S with $\nu(S) = 1$, and $D \equiv \text{Diag}(d_1, d_2)$ is a contraction ($0 < d_1, d_2 \leq 1$). Lemma 6.1 presents the basic geometric construction by means of which we shall extend Theorems 3.2 and 3.3 to elliptically contoured distributions in the bivariate case. This argument, based on that on pp. 114-5 in Fefferman *et al* (1972), is an alternative to that used to derive (5.7) in the proof of Theorem 5.1 above (but see Remark 6.3).

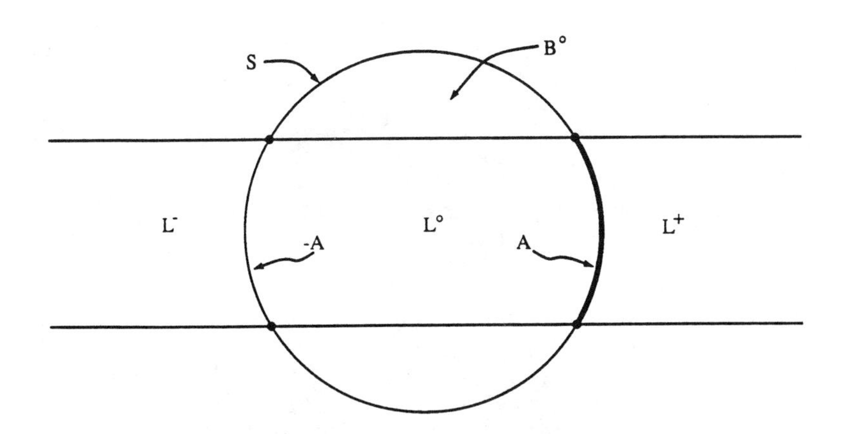

Figure 6.1. The arc A, the strip $L \equiv L(A)$, and the ball $B \equiv B^\circ \cup S$.

DEFINITION 6.1 For any closed arc $A \subset S$ with $0 \leq \text{arclength}(A) < \pi$, define $L \equiv L(A)$ to be the closed centrally symmetric strip such that $L \cap S = (-A) \cup A$ (see Figure 6.1.) Then L can be expressed as the disjoint union

$$L = L^- \cup L^\circ \cup L^+, \tag{6.1}$$

where $L^\circ \equiv L^\circ(A) = L(A) \cap B^\circ$ with B° the *open* unit disk, $L^- \equiv L^-(A) \supset (-A)$, and $L^+ \equiv L^+(A) \supset A$; note that L^- and L^+ are both *closed* sets. Note too that if $A_1, \ldots, A_m$ are disjoint then $L^+(A_1), \ldots, L^+(A_m)$ (hence also $D[L^+(A_1)], \ldots, D[L^+(A_m)]$) are disjoint.

LEMMA 6.1 *Let K be a closed subset of $\mathbb{R}^2$ with $K \cap S = \cup\{A_j | j = 1, \ldots, m\}$, a disjoint union of closed arcs such that $0 \leq \text{arclength}(A_j) < \pi$. Define $L_j = L(A_j)$, $L_j^- = L^-(A_j)$, and $L_j^+ = L^+(A_j)$. If*

$$K \backslash B^\circ \subseteq \cup L_j^+ \tag{6.2}$$

then

$$\nu(K) \geq \nu(DK). \tag{6.3}$$

PROOF If we define

$$K(j) = (K \backslash B^\circ) \cap L_j^+, \tag{6.4}$$

(see Figure 6.2) then $K(1), \ldots, K(m)$ are disjoint and (6.2) implies that

$$K \backslash B^\circ = \cup K(j). \tag{6.5}$$

Express K as the disjoint union $K = (K \backslash B^\circ) \cup (K \cap B^\circ)$. Since D is a contraction, $DB^\circ \cap S = \emptyset$, hence $DK \cap S = D(K \backslash B^\circ) \cap S$, so $\nu(DK) =$

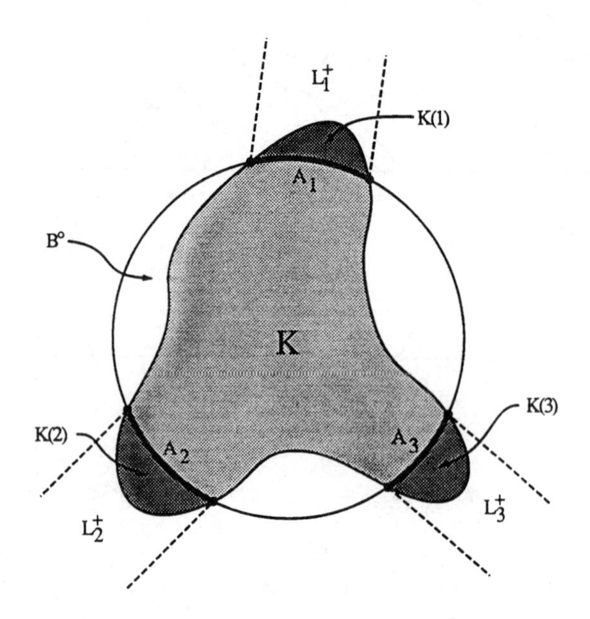

Figure 6.2. A set K (shaded) that satisfies (6.2).

$\nu[D(K\backslash B^\circ)] = \nu\{\cup D[K(j)]\}$ by (6.5). By (6.4), however, $D[K(j)] \subseteq D(L_j^+)$, hence

$$
\begin{aligned}
\nu(DK) &\le \nu\{\cup[D(L_j^+)]\} &&= \textstyle\sum \nu[D(L_j^+)] \\
&= \tfrac{1}{2}\textstyle\sum \nu(DL_j) &&\le \tfrac{1}{2}\textstyle\sum \nu(L_j) \\
&= \textstyle\sum \nu(L_j^+) &&= \textstyle\sum \nu(A_j) \\
&= \nu(K).
\end{aligned}
\tag{6.6}
$$

The second equality in (6.6) follows from (6.1), the inclusion $D(L_j^\circ) \subset B^\circ$, and the relation $L_j^- = -L_j^+$ (implied by the central symmetry of L_j):

$$\nu(DL_j) = \nu[D(L_j^-)] + \nu[D(L_j^+)] = 2\nu[D(L_j^+)]. \tag{6.7}$$

The second inequality in (6.6) follows since the width of the strip DL_j cannot exceed that of L_j as D is a contraction. Thus (6.3) is established. □

REMARK 6.1 In Lemma 6.1 suppose in addition that K is star-shaped with respect to the origin. Then for each $x \in K\backslash B^\circ$ the closed line segment $[0, x]$ intersects the unit circle S at a unique point $y(x) \in K \cap S \equiv \cup A_j$. Thus if we define

$$K_j = \{x \in K\backslash B^\circ \mid y(x) \in A_j\} \tag{6.8}$$

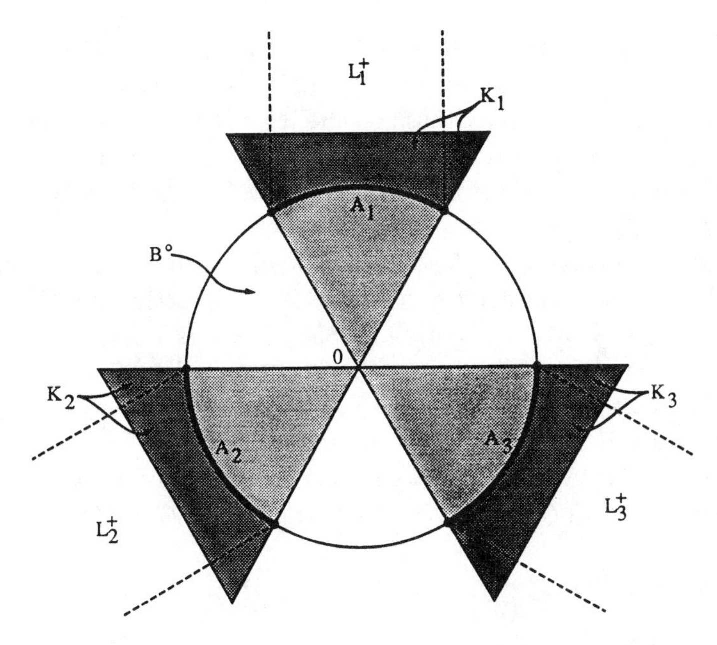

Figure 6.3. A star-shaped set K (shaded) that does not satisfy (6.9).

(see Figure 6.3), then $K_1, \ldots, K_m$ are disjoint and $\cup K_j = K \backslash B^\circ$. It is readily verified that $K(j) \subseteq K_j$ and that (6.2) is equivalent to the condition that for each $j = 1, \ldots, m$

$$K_j \subseteq L_j^+. \tag{6.9}$$

□

REMARK 6.2 For the validity of Lemma 6.1 it is not necessary that the strips $L_j \equiv L(A_j)$ be centrally symmetric, only that $L_j \cap S = (A_j^-) \cup A_j$ where A_j^- is any closed arc such that the relative interiors of A_j^- and A_j do not intersect. Note that this condition still implies that $\nu(A_j^-) = \nu(A_j)$ since L_j is a strip. Again L_j can be decomposed as in (6.1), where now $L_j^- \equiv L^-(A_j) \supset A_j^-$ and $L_j^+ \equiv L^+(A_j) \supset A_j$. Similarly, decompose DL_j as $(DL_j)^- \cup (DL_j)^\circ \cup (DL_j)^+$ where $(DL)^\circ = DL \cap B^\circ$. Then the proof of Lemma 6.1 remains valid with the following three modifications: (i) the first equality in (6.6) must be replaced by the inequality $\leq$, for now $L_1^+, \ldots, L_m^+$ (hence $D(L_1^+), \ldots, D(L_m^+)$) need not be disjoint; (ii) although $L_j^- \neq -L_j^+$ if L_j is not centrally symmetric, it follows from the fact that D is a contraction that $D(L_j^-) \cap S = (DL_j)^- \cap S$ and $D(L_j^+) \cap S = (DL_j)^+ \cap S$, hence $\nu[D(L_j^-)] = \nu[(DL_j)^-] = \nu[(DL_j)^+] = \nu[D(L_j^+)]$; (iii) $\nu(DL_j) \leq \nu(L_j)$ since the strip DL_j is both narrower and closer to the origin than L_j . □

Theorems 6.1 and 6.2 below extend Theorems 3.2 and 3.3 from normal distributions to elliptically contoured distributions in the bivariate case. To prove these extensions we shall apply Lemma 6.1 and Remark 6.1 to the set $K \equiv K(C; \Sigma_1, \Gamma)$ defined in (5.6). If G is a compact subgroup of the orthogonal group $\mathcal{O}_2$ that acts effectively on $\mathbb{R}^2$ and $C \in \mathcal{C}_G$ (or $\mathcal{M}_G$) then $K \in \mathcal{C}_{\widetilde{G}}$ (or $\mathcal{M}_{\widetilde{G}}$) (see (5.11) or (6.14)) so K is star-shaped (apply Lemma 3.1 of Part I), hence to apply these results it must be verified that K satisfies (6.9). This will be accomplished in the proofs of Theorems 6.1 and 6.2 by means of the convexity (or monotonicity) and $\widetilde{G}$-invariance of K. ($\widetilde{G}$ is defined in (5.12).)

Before proceeding with the statements and proofs of Theorems 6.1 and 6.2 we describe the compact subgroups $G \subseteq \mathcal{O}_2$ acting on $\mathbb{R}^2$. It is well known (e.g., see Grove and Benson (1985), Theorem 2.2.1) that if G is finite then either G is the cyclic group $\mathcal{C}_2^n$ of order n generated by the rotation through angle $2\pi/n$ or else G is the dihedral group $\mathcal{H}_2^n$ of order $2n$ generated by $\mathcal{C}_2^n$ and a single reflection in $\mathbb{R}^2$, where $n = 1, 2, \ldots$. The group $\mathcal{C}_2^n(\mathcal{H}_2^n)$ is the group of all rotations (all rotations and reflections) that leave a regular n-gon invariant, and G is a finite reflection group iff $G = \mathcal{H}_2^n$ for some $n \geq 1$. Thus

$$\mathcal{C}_2^1 = \{I\}, \qquad \mathcal{C}_2^2 = \{\pm I\},$$

$$\mathcal{H}_2^1 \cong \left\{ \begin{pmatrix} \pm 1 & 0 \\ 0 & 1 \end{pmatrix} \right\}, \quad \mathcal{H}_2^2 \cong \left\{ \begin{pmatrix} \pm 1 & 0 \\ 0 & \pm 1 \end{pmatrix} \right\} = \mathcal{D}_2,$$

where $\mathcal{D}_2$ is the group of sign changes of coordinates in $\mathbb{R}^2$ (recall Section 2 of Part I). Thus $\mathcal{C}_2^1$ and $\mathcal{H}_2^1$ do not act effectively on $\mathbb{R}^2$, $\mathcal{C}_2^2$ and $\mathcal{H}_2^2$ act effectively but not irreducibly, while $\mathcal{C}_2^n$ and $\mathcal{H}_2^n$ act effectively and irreducibly for $n \geq 3$ (see Section 3 of Part I for definitions). Finally, the only infinite compact subgroups of $\mathcal{O}_2$ are $\mathcal{O}_2$ itself and $\mathcal{SO}_2$, the subgroup of all proper rotations of $\mathbb{R}^2$, both of which act effectively and irreducibly.

THEOREM 6.1 *Suppose that* $X \sim EC_2(\Sigma)$. *If* $C \in \mathcal{C}_G$ *and* C *is closed, then* $\Sigma_1 \leq \Sigma_2 \Rightarrow P_{\Sigma_1}(C) \geq P_{\Sigma_2}(C)$ *provided that* Σ_1 *is* G-*invariant and* G *acts effectively on* $\mathbb{R}^2$ *(i.e.,* $G \neq \mathcal{C}_2^1$ *or* $\mathcal{H}_2^1$*).*

PROOF As in the proof of Theorem 5.1, the desired result is equivalent to the assertion that (5.7) holds for every closed $K \in \mathcal{C}_{\widetilde{G}}$ and every contraction D, where $\widetilde{G}$ is defined in (5.12).

If $G = \mathcal{O}_2$ or $\mathcal{SO}_2$ then $\widetilde{G} = G$ and $\mathcal{C}_{\widetilde{G}}$ is simply the class of all open or closed disks centered at 0, so $DK \subseteq K$ and (5.7) is trivially valid. Thus we may assume that $G = \mathcal{C}_2^n$ or $\mathcal{H}_2^n$, $n \geq 2$. If n is *even*, however, then $-I \in G$ and the desired result is already a consequence of Theorem 5.1 (see Remark 5.1). Since $\mathcal{H}_2^n \supset \mathcal{C}_2^n$ it therefore suffices to establish (5.7) when $G = \mathcal{C}_2^n$ for $n \geq 3$ and n *odd* (for, $G \supset G' \Rightarrow \mathcal{C}_G \subset \mathcal{C}_{G'}$).

In fact, the following argument establishes (5.7) when G is the rotation group $\mathcal{C}_2^n$ for *any* $n \geq 2$. First, note that $\widetilde{G} = G$ and that we may assume that $K \in \mathcal{C}_G$ is compact, convex and G-invariant, hence is the limit of a

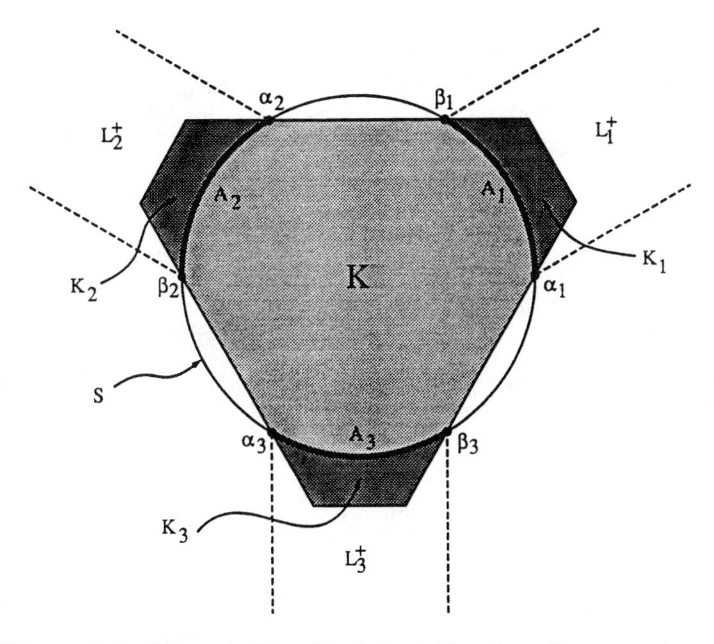

Figure 6.4. The set $K \in C_G$ (shaded); $G = C_2^n$, $n = m = 3$.

decreasing sequence of closed convex G-invariant polygons (recall Footnote 3). Thus it suffices to establish (5.7) when K is such a polygon.

In this case either $K \cap S = S$ and (5.7) is trivial, or $K \cap S = \emptyset$ so $DK \cap S = \emptyset$ and (5.7) is also trivial, or else $K \cap S = \cup A_j$, the union of $m \geq 2$ disjoint closed arcs $A_1, \ldots, A_m$, some possibly degenerate at single points (see Figure 6.4; note that $-A_j$ does not necessarily appear in $\{A_1, \ldots, A_m\}$ if n is odd). For $j = 1, \ldots, m$, let $\alpha_j \equiv \exp(i\theta_j)$ and $\beta_j \equiv \exp(i\varphi_j)(i = \sqrt{-1})$ denote the endpoints of arc A_j in counterclockwise order. Without loss of generality assume that

$$0 \leq \theta_1 \leq \varphi_1 < \theta_2 \leq \varphi_2 < \cdots < \theta_m \leq \varphi_m < 2\pi, \tag{6.10}$$

i.e., $A_1, \ldots, A_m$ are arranged in consecutive counterclockwise order on the unit circle S. Since K is C_2^n-invariant so is $\cup A_j$, hence m is a multiple of n and for each $j = 1, \ldots, m$,

$$\begin{gathered} 0 < \theta_{j+1} - \theta_j \leq 2\pi/n \leq \pi \\ 0 < \varphi_{j+1} - \varphi_j \leq 2\pi/n \leq \pi, \end{gathered} \tag{6.11}$$

where $\theta_{m+1} \equiv \theta_1 + 2\pi$, $\varphi_{m+1} \equiv \varphi_1 + 2\pi$. By (6.10) this implies that for each $j = 1, \ldots, m$,

$$\begin{gathered} 0 \leq \varphi_j - \theta_j < 2\pi/n \leq \pi \\ 0 < \theta_{j+1} - \varphi_j \leq 2\pi/n \leq \pi. \end{gathered} \tag{6.12}$$

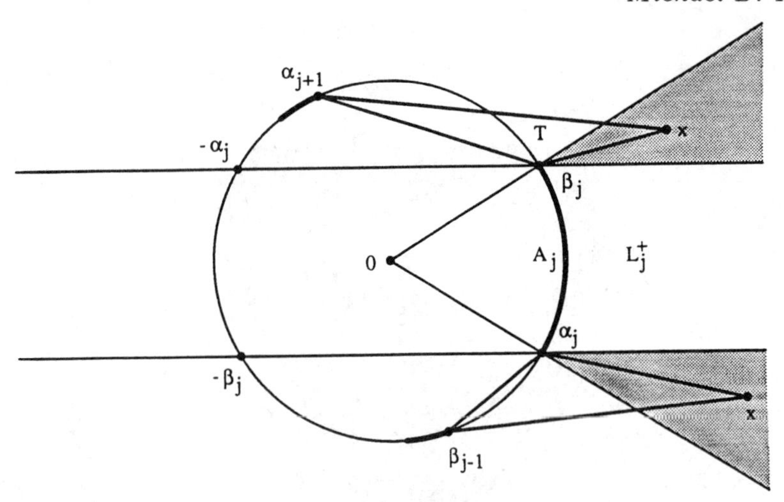

Figure 6.5.

In particular, $0 \leq \text{arclength}(A_j) < \pi$ for $j = 1, \ldots, m$. Also, because $K \in \mathcal{C}_G \subset \mathcal{M}_G$ and G acts effectively on $\mathbb{R}^2$, the line segment $[0, x] \subset K$ whenever $x \in K$ (apply Lemma 3.1 of Part I), hence K is star-shaped with respect to the origin. By Lemma 6.1 and Remark 6.1, therefore, in order to establish (5.7) it suffices to show that K satisfies (6.9), where $L_j^+ = L^+(A_j)$ and K_j is defined in (6.8).

The inclusion (6.9) is trivial if $\alpha_j = \beta_j$ so assume that $\alpha_j \neq \beta_j$, i.e., that $\theta_j < \varphi_j$. If (6.9) fails, consider $x \in K_j \backslash L_j^+$. By definition of K_j, x must lie in one of the two shaded wedge-shaped regions indicated in Figure 6.5. If x lies in the upper shaded region, consider the triangle T with vertices x, β_j, and α_{j+1}.[5] Clearly $T \subset K$, since K is convex and $x, \beta_j, \alpha_{j+1} \in K$. But T must intersect the open arc (β_j, α_{j+1}), hence this arc has a nonempty intersection with K, which contradicts the fact that this arc is contained in $S \backslash K$. If it is assumed that x lies in the lower shaded region in Figure 6.5, replace β_j, α_{j+1} by α_j, β_{j-1} to get a similar contradiction. Thus (6.9) is confirmed. □

REMARK 6.3 This method of proof does not extend in any obvious way to the multivariate case $p \geq 3$. To see this, suppose that K is a closed convex G-invariant polyhedron in $\mathbb{R}^3$; let B (or S) denote the closed unit ball (or sphere) in $\mathbb{R}^3$ and ν denote the uniform measure on S. As in the above proof it suffices to consider the case where $K \cap S = \cup A_j$, the union of disjoint closed subsets $A_1, \ldots, A_m$ of S. Since K is star-shaped, again the sets K_j are well-defined by (6.8) and $\cup K_j = K \backslash B^\circ$. Unlike the case $p = 2$, however, the

[5]It is essential to verify that Figure 6.5 accurately depicts the location of α_{j+1}, i.e., that α_{j+1} lies in the half-open arc $(\beta_j, -\alpha_j]$. But this is equivalent to the condition $\varphi_j < \theta_{j+1} \leq \theta_j + \pi$, which follows from (6.12). (Note that we define $\alpha_{m+1} = \alpha_1, \theta_{m+1} = \theta_1 + 2\pi$). Similarly, β_{j-1} lies in the half-open arc $(\alpha_j, -\beta_j]$. (Define $\beta_0 = \beta_m, \varphi_0 = \varphi_m - 2\pi$.)

sets A_j no longer need have a simple form, so it is not apparent how to define sets $L_1^+, \ldots, L_m^+$ such that $L_j^+ \cap B^\circ = \emptyset$, $L_j^+ \cap S = A_j$, $\nu[D(L_j^+)] \leq \nu(L_j^+)$, and such that (6.9) holds. (If it is possible to find such sets L_j^+ then (5.7) would follow as in (6.6).) Nonetheless we conjecture that Theorem 6.1, like Theorem 5.1, is valid for $p \geq 3$. (Also see Remark 5.1.) □

REMARK 6.4 If Σ_1 is not assumed G-invariant in Theorem 6.1 then $K \equiv K(C; \Sigma_1, \Gamma)$, although still star-shaped, need not be $\tilde{G}$-invariant. In this case it is easy to find examples where the sets K_j constructed from the arcs A_j as in (6.8) do not satisfy (6.9) – for example, take $G = \mathcal{C}_2^3$ and C an equilateral triangle centered at 0, then choose Σ_1 such that K is an isosceles triangle with altitude(K) >> 1 >> base(K). Nonetheless, we conjecture that Theorem 6.1 remains valid (when $p = 2$) even if Σ_1 is not G-invariant. By Remark 5.1, this is true (in fact, true for all $p \geq 2$) if $-I \in G$. However, Example 3.1 of Part I shows that Theorem 6.1 may fail when $p \geq 3$ if Σ_1 is not G-invariant and $-I \notin G$, even if the probability distribution is normal. □

For the bivariate case ($p = 2$), Theorem 6.2 below not only extends Theorem 3.3 from normal distributions to elliptically contoured distributions but also applies to almost every effective subgroup G of $\mathcal{O}_2$, including the rotation groups $\mathcal{C}_2^n$, $n \geq 4$, whereas Theorem 3.3 applies only to the reflection groups $\mathcal{H}_2^n$, $n \geq 2$ (as well as to $\mathcal{O}_2$ itself).

Recall (Section 3, Part I) that the class $\mathcal{M}_G$ of all G-decreasing subsets of $\mathbb{R}^p$ is closed under unions, whereas $\mathcal{C}_G$ is not, although both are closed under intersections. In fact,

$$C \in \mathcal{M}_G \Leftrightarrow C = \cup\{C_G(x) | x \in C\}, \tag{6.13}$$

where $C_G(x)$ denotes the convex hull of the G-orbit of x; note that $C_G(x) \in \mathcal{C}_G$ is a closed convex G-invariant polygon for every $x \in \mathbb{R}^p$. Recall also that every $C \in \mathcal{M}_G$ is G-invariant. It is readily verified from (6.13) (recall (5.6) and (5.10)-(5.12)) that if Σ_1 is G-invariant, then

$$C \in \mathcal{M}_G \Rightarrow \Sigma_1^{-1/2} C \in \mathcal{M}_G \Rightarrow K \in \mathcal{M}_{\tilde{G}}. \tag{6.14}$$

THEOREM 6.2 *Suppose that $X \sim EC_2(\Sigma)$. If $C \in \mathcal{M}_G$ and C is closed, then $\Sigma_1 \leq \Sigma_2 \Rightarrow P_{\Sigma_1}(C) \geq P_{\Sigma_2}(C)$ provided that Σ_1 is G-invariant and G acts effectively on $\mathbb{R}^2$ (i.e., $G \neq \mathcal{C}_2^1$ or $\mathcal{H}_2^1$), but also $G \neq \mathcal{C}_2^2$ or $\mathcal{C}_2^3$.*

PROOF If $G = \mathcal{O}_2$ or $S\mathcal{O}_2$ then $\mathcal{M}_G = \mathcal{C}_G$ and the result is trivial. Two cases remain.

(i) $G = \mathcal{C}_2^n$, $n \geq 4$. Again $\tilde{G} = G$. As in the proof of Theorem 6.1, the desired result is equivalent to the assertion that (5.7) holds for every compact $K \in \mathcal{M}_G$ and every contraction $D = \text{Diag}(d_1, d_2)$ $(0 < d_1, d_2 \leq 1)$. Such a set K is the limit of a sequence $\{K_\lambda\}$ of finite unions of closed convex

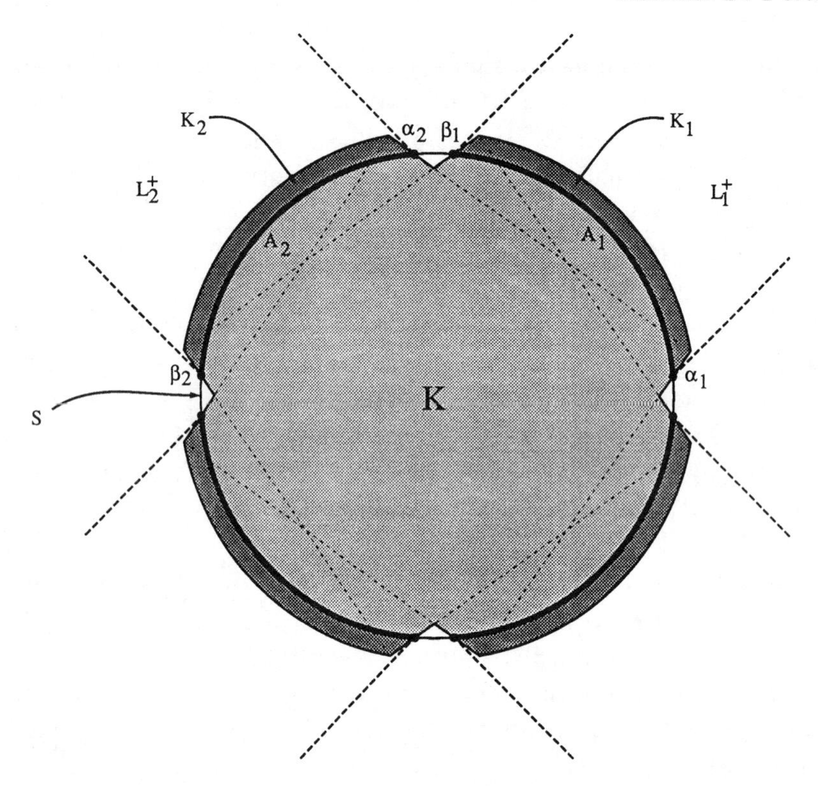

Figure 6.6. The set $K \in \mathcal{M}_G$ (shaded); $G = C_2^n$, $n = m = 4$.

G-invariant polygons[6], so it suffices to establish (5.7) when K itself is a finite union of such polygons.

Again we need consider only the case where $K \cap S = \cup A_j$, a finite disjoint union of closed arcs. Define α_j, θ_j, β_j, φ_j, L_j^+, and K_j as in the proof of Theorem 6.1 but replace Figures 6.4 and 6.5 by Figures 6.6 and 6.7, respectively. If the inclusion (6.9) can be established then (5.7) again follows from Lemma 6.1 and Remark 6.1. All arguments in the previous proof continue to hold with the following two exceptions: (a) since $n \geq 4$, replace π by $\pi/2$ as the upper bound in (6.11) and (6.12); (b) since $K \in \mathcal{M}_G$ need not be convex, the verification of (6.9) in the final paragraph of the proof of Theorem 6.1 must be modified as follows.

[6]Since $G \equiv C_2^n$ is irreducible if $n \geq 3$, Lemma 3.2 of Part I implies that $[C_G(x)]^\circ \neq \emptyset$ if $x \neq 0$, where $^\circ$ denotes "interior". It follows from (6.9) that for each $\lambda > 1$, $\cup\{\lambda[C_G(x)]^\circ | x \in K\}$ is an open covering of the compact set K, hence there exists a finite subcovering $\cup\{\lambda[C_G(x_i)]^\circ | i = 1, \ldots, n\}$. Then $K_\lambda \equiv \cup\{\lambda C_G(x_i) | i = 1, \ldots, n\}$ is a finite union of closed convex G-invariant polygons such that $K \subset K_\lambda \subset \lambda K$. Thus $K_\lambda \to K$ as $\lambda \downarrow 1$. [In case (ii) below, $G \equiv \mathcal{H}_2^n$ is again irreducible if $n \geq 3$ so this argument remains valid. If $n = 2$, then G is not irreducible but again $[C_G(x)]^\circ \neq \emptyset$ unless x lies in the wall of a fundamental region (see Footnote 8) in which case we define $[C_G(x)]^\circ$ to be the relative interior of $C_G(x)$.]

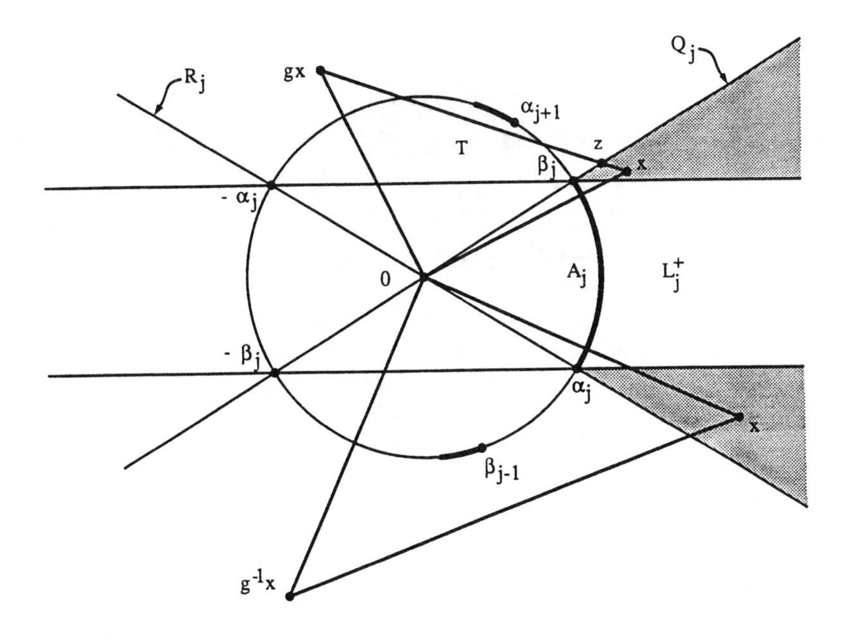

Figure 6.7.

First note that $G \equiv C_2^n = \{I, g, \ldots, g^{n-1}\}$, where g is the rotation through angle $2\pi/n$ about 0 in $\mathbb{R}^2$. Again we may assume that $\alpha_j \neq \beta_j$, i.e., that $\theta_j < \varphi_j$. If (6.9) fails, consider $x \in K_j \backslash L_j^+$. By definition of K_j, x must lie in one of the two shaded wedge-shaped regions indicated in Figure 6.7. If x lies in the upper shaded region, then its image gx must lie in the open region outside B and strictly between the infinite rays Q_j and R_j emanating from 0 and passing through β_j and $-\alpha_j$ respectively[7]. Thus the half-open line segment $[x,\ gx)$ must intersect the ray Q_j at some point z outside B. This implies that the triangle T with vertices 0, x, and gx intersects the *open* arc $(\beta_j, \alpha_{j+1}) \subset S\backslash K$. Since G acts effectively on $\mathbb{R}^2$, however, $0 \in C_G(x)$ (see Lemma 3.1 of Part I), hence $T \subset C_G(x) \subset K$ (recall that $K \in \mathcal{M}_G$), a contradiction. If it is assumed that x lies in the lower shaded region in Figure 6.7, simply replace gx by $g^{-1}x$ and $(\beta_j,\ \alpha_{j+1})$ by $(\alpha_j,\ \beta_{j-1})$ to reach a similar contradiction. Thus (6.9) is again verified.

(ii) $G = \mathcal{H}_2^n$, $n \geq 2$. Since $C_2^n \subset \mathcal{H}_2^n$ this case is covered by (i) when $n \geq 4$, but the following argument is valid for all $n \geq 2$. Note that if $F_1, \ldots, F_{2n}$

---

[7]It is again essential to verify that Figure 6.7 accurately depicts the location of gx, i.e., that gx lies strictly between the rays Q_j and R_j. If we write $x = |x|\exp(i\eta)$ with $\theta_j \leq \eta \leq \varphi_j$ then $gx = |x|\exp i[\eta + (2\pi/n)]$, so it must be verified that $\varphi_j < \eta + (2\pi/n) < \theta_j + \pi$. But this follows from (6.12) with π replaced by $\pi/2$. Similarly, $g^{-1}x$ lies in the open region outside B and strictly between the infinite rays $-R_j$ and $-Q_j$.

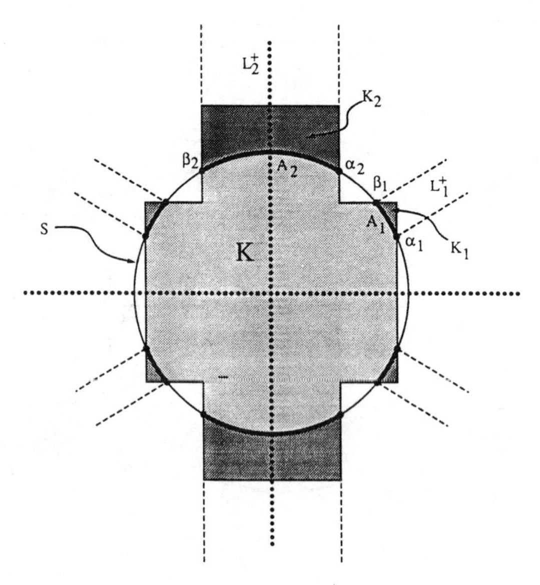

Figure 6.8. The set $K \in \mathcal{M}_G$ (shaded); $G = \mathcal{H}_2^n$, $n = 2$, $m = 6$.

are the fundamental regions[8] for the finite reflection group $G \equiv \mathcal{H}_2^n$, then $\Gamma F_1, \ldots, \Gamma F_{2n}$ are the fundamental regions for the finite reflection group $\tilde{G} \equiv \Gamma G \Gamma'$, where $\Gamma \in \mathcal{O}_2$. Thus, by means of an orthogonal change of basis we may assume that $\tilde{G} = G$.

As in (i), it must be shown that (5.7) holds for every compact $K \in \mathcal{M}_G$ and every contraction D. Again we may assume that K is a finite union of closed convex G-invariant polygons (see Footnote 6).

As before we need consider only the case where $K \cap S = \cup A_j$, a finite disjoint union of closed arcs. Define α_j, θ_j, β_j, φ_j, L_j^+, and K_j as in the proof of Theorem 6.1, but now replace Figures 6.4 and 6.5 by Figures 6.8 and 6.9, respectively[9]. To establish (5.7) it again suffices to verify (6.9). All arguments in the proof of Theorem 6.1 continue to hold (including (6.11) and (6.12) since $\mathcal{H}_2^n \supset \mathcal{C}_2^n$) with the exception of the verification of (6.9), which must be modified as follows.

Again we may assume that $\alpha_j \neq \beta_j$. Neither α_j nor β_j can lie in the wall

[8] The reader may review the elementary geometric structure of the reflection groups $\mathcal{H}_2^n$ in Grove and Benson (1985, pp. 8-9), in particular the representation $\mathbb{R}^2 = \cup\{g\bar{F} | g \in \mathcal{H}_2^n\}$, where $\bar{F}$ is the closure of any fixed *fundamental region* F for $\mathcal{H}_2^n$. Such a region is an open convex cone in $\mathbb{R}^2$ that subtends an angle of π/n at 0 and which is oriented such that the reflections across its two boundary rays, or *walls*, generate the group $\mathcal{H}_2^n$. There are exactly $2n$ disjoint fundamental regions $F_1, \ldots, F_{2n}$, and for each $g \in \mathcal{H}_2^n$, $\{gF_1, \ldots, gF_{2n}\}$ is some permutation of $\{F_1, \ldots, F_{2n}\}$. Additional properties of finite reflections groups utilized in the present paper may be found in Chapter 4 of Grove and Benson (1985) and in Section 3 of Eaton and Perlman (1977).

[9] In Figure 6.8, $n = 2$ and the $2n \equiv 4$ fundamental regions (whose walls are indicated by heavily dotted lines) coincide with the four (open) quadrants of $\mathbb{R}^2$.

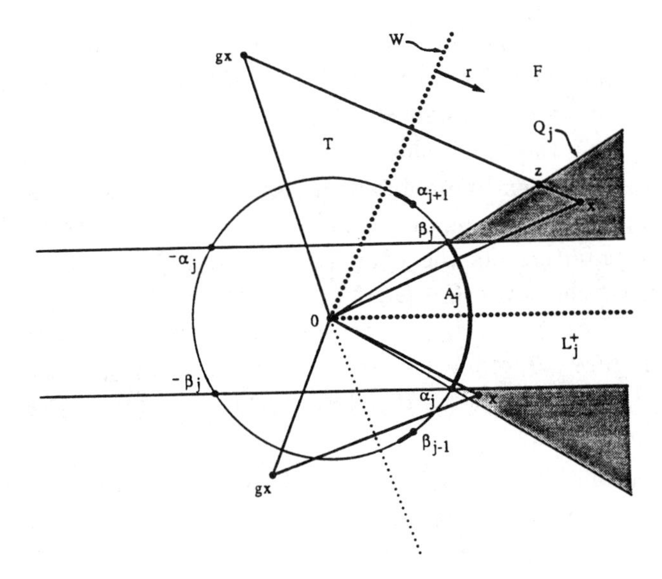

Figure 6.9.

of a fundamental region[10]. Either (a) α_j and β_j lie in the same fundamental region or (b) α_j and β_j lie in adjacent fundamental regions, for otherwise the union of the arc A_j and all its G-images would completely cover S. (Cases (a) and (b) both occur in Figure 6.8.) In case (b), α_j and β_j must be equidistant from the common wall between them (see Figure 6.9).

If (6.9) fails, consider $x \in K_j \backslash L_j^+$. By the definition of K_j, x must lie in one of the two shaded wedge-shaped regions indicated in Figure 6.9. If x lies in the upper shaded region, then

$$(x - \beta_j)'(\beta_j - \alpha_j) > 0 \tag{6.15}$$

and x lies in the same fundamental region (call it F) as β_j. Let W denote the first wall of F encountered when traversing S in a counterclockwise direction starting at β_j and let r denote the unit vector normal to W that points into F. Define $g = I - 2rr'$, i.e., g is the reflection across the wall W, hence $g \in G$. Then

$$\begin{aligned}(gx - \beta_j)'(\beta_j - \alpha_j) = \\ (x - \beta_j)'(\beta_j - \alpha_j) - 2(r'x)(r'\beta_j - r'\alpha_j)) > 0\end{aligned} \tag{6.16}$$

[10]Suppose that α_j lies in the wall of some fundamental region. Since K is G-invariant, the reflection of the closed arc A_j across that wall is contained in K, hence the closed arc consisting of the union of A_j and its reflection is contained in $K \cap S$. But β_j lies in the interior of this closed arc, which contradicts the fact that β_j is not an interior point of $K \cap S$. Similarly, β_j cannot lie in the wall of a fundamental region.

by (6.15) and the two inequalities $r'\beta_j \leq r'\alpha_j$, $r'x > 0$.[11] Thus, by (6.15) and (6.16) both x and its reflected image gx lie strictly on the same side of the strip L_j, and both lie outside B. Because $gx \in gF$ which is disjoint from F, gx cannot lie in the upper shaded region that contains x (see Figure 6.9). Therefore the half-open line segment $[x, gx)$ must intersect the ray Q_j at some point z outside B. As before, this implies that the triangle T with vertices 0, x, and gx intersects the *open* arc $(\beta_j,\ \alpha_{j+1}) \subset S\backslash K$. Since G acts effectively on $\mathbb{R}^2$, however, $0 \in C_G(x)$ (by Lemma 3.1 of Part I), hence $T \subset C_G(x) \subset K$ (since $K \in \mathcal{M}_G$), a contradiction. If x lies in the lower shaded region in Figure 6.9, replace F by the fundamental region containing α_j and replace $(\beta_j,\ \alpha_{j+1})$ by $(\alpha_j,\ \beta_{j-1})$ to reach a similar contradiction. Thus (6.9) is again verified. □

REMARK 6.5 Examples 3.2 and 3.3 in Part I show that the assumption that Σ_1 is G-invariant cannot be discarded in Theorem 6.1. Example 3.4 shows that the conclusion of Theorem 6.2 is false if $G = \mathcal{C}_2^2 \equiv \{\pm I\}$, in which case $\mathcal{M}_G \equiv \mathcal{M}_1$ is the class of centrally symmetric sets that are star-shaped with respect to the origin in $\mathbb{R}^2$. (This counterexample easily may be extended to $G \equiv \{\pm I\}$ acting on $\mathbb{R}^p$ with $p \geq 3$). □

REMARK 6.6 If $G = \mathcal{C}_2^3$ then the crucial inclusion (6.9) fails for some (but not all) sets $K \in \mathcal{M}_G \backslash \mathcal{C}_G$, hence the above proof fails to establish the inequality (5.7) for such sets. It is uncertain, however, whether or not (5.7) (and hence the conclusion of Theorem 6.2) is true for such sets. To see this, consider the three sets K in Figures 6.10-6.12. For the first two sets (6.9) does hold so (5.7) is true, while for the third set (6.9) *fails* but (5.7) is uncertain. We conjecture that (5.7) is true for *every* $K \in \mathcal{M}_G$, hence that Theorem 6.2 is valid also for $G = \mathcal{C}_2^3$. If this is true then Theorem 6.2 *would be valid for every effective subgroup* G *of* $\mathcal{O}_2$ *except* $\{\pm I\}$. With somewhat less confidence we conjecture that when $p \geq 3$, Theorem 6.2 *is valid for every effective subgroup* G *of* $\mathcal{O}_p$ *except those* G *for which there exists a* G-*invariant subspace* $V \subseteq \mathbb{R}^p$ *of dimension* ≥ 2 *such that the restriction of the action of* G *to* V *is* $\{\pm I\}$. As with Theorem 6.1, however, the method of proof used above to establish Theorem 6.2 in the bivariate case does not extend in any obvious way to the multivariate case $p \geq 3$ (recall Remark 6.3). □

[11] The scalar product $r'v$ is the (signed) distance from the vector v to the wall W. Because $\beta_j \in F$ and since the angle subtended by F at 0 is $\leq \pi/2$, β_j is closer to W than α_j (consider the cases (a) and (b) separately), so the first inequality holds. The second is immediate since $x \in F$.

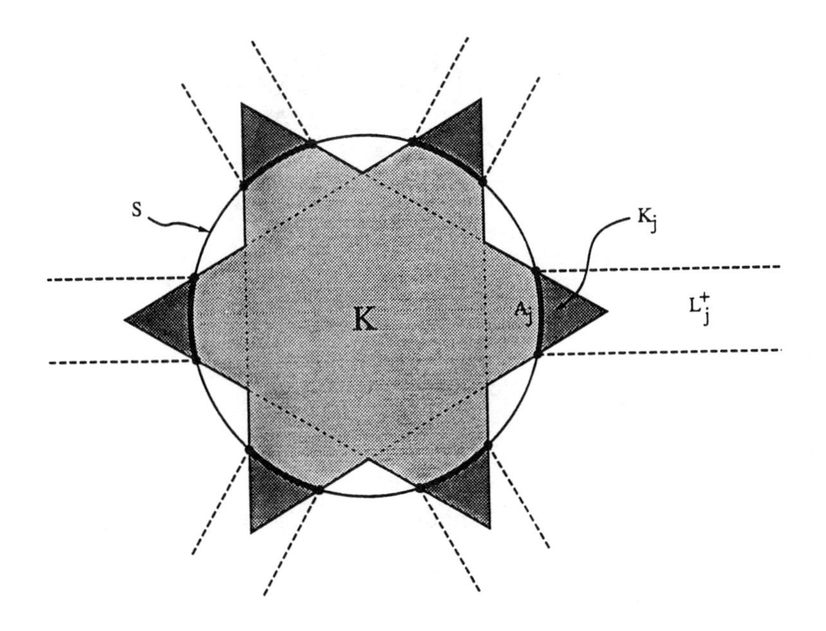

Figure 6.10. $K \in \mathcal{M}_G \backslash \mathcal{C}_G$ (shaded), $G = \mathcal{C}_2^3$; (6.9) holds, (5.7) true.

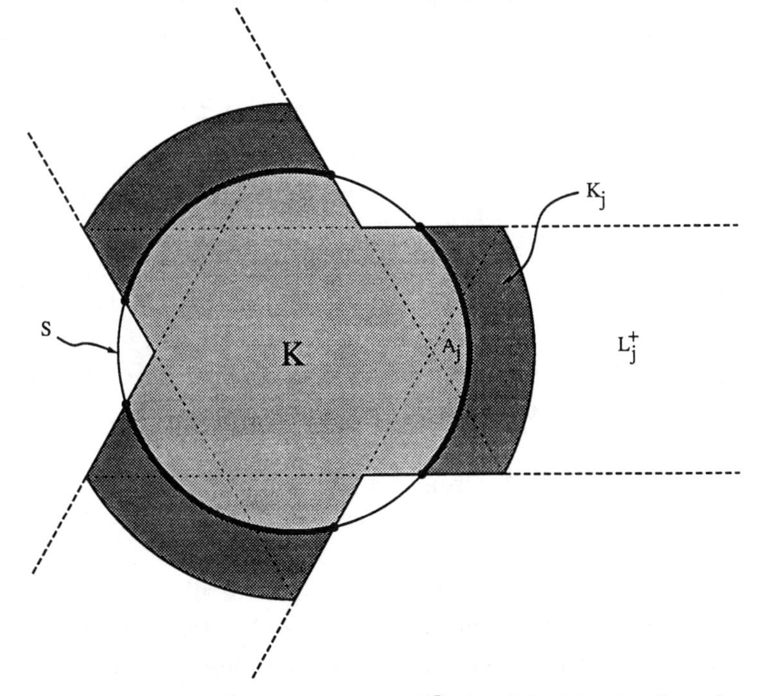

Figure 6.11. $K \in \mathcal{M}_G \backslash \mathcal{C}_G$ (shaded), $G = \mathcal{C}_2^3$; (6.9) holds, (5.7) true.

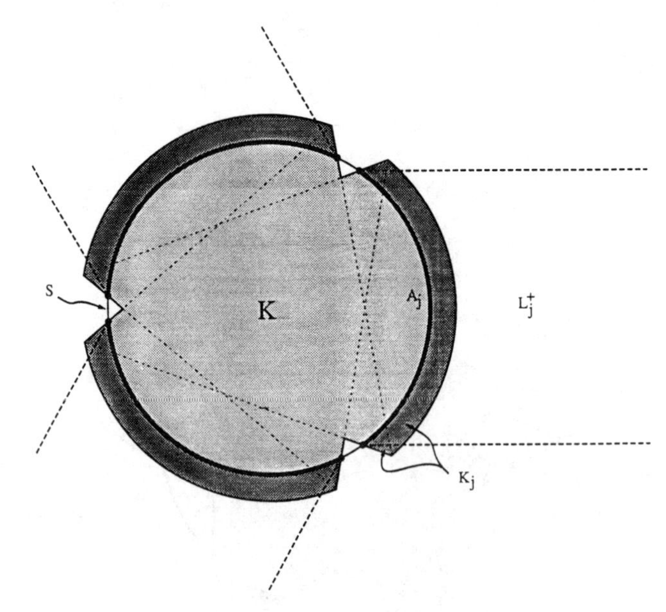

Figure 6.12. $K \in \mathcal{M}_G \backslash \mathcal{C}_G$ (shaded), $G = \mathcal{C}_2^3$; (6.9) fails, (5.7) uncertain.

7. A Sharper Inequality

We return to the general case $p \geq 2$ and let B, S, ν, and D be as defined in Section 5. Theorem 5.1 implies that for every (not necessarily closed) $K \in \mathcal{C}_1$,

$$\nu(\bar{K}) \equiv \nu(K^\circ) + \nu(\partial K) \geq \nu(D\bar{K}), \tag{7.1}$$

where $\bar{K}$, K°, and ∂K denote the closure, interior, and boundary of K, respectively (see (5.7)). Fefferman, Jodeit, and Perlman (1972, Section 3) sharpened this inequality by showing that if $D \neq I$ then (7.1) remains valid with the term $\nu(\partial K)$ deleted, even though $\nu(\partial K)$ may be positive and/or $\nu(K^\circ)$ may be 0. Therefore, when $C \in \mathcal{C}_1$ the contribution of the boundary of C plays no role in the concentration inequality (5.0) for elliptically contoured distributions even though such distributions need not be absolutely continuous with respect to Lebesgue measure on $\mathbb{R}^p$, hence may assign nonzero probability to the boundary of K.

In this section we extend this sharpened result from $\mathcal{C}_1$ to the classes $\mathcal{C}_G$ and $\mathcal{M}_G$ in the bivariate case and show further that if Theorems 6.1 and 6.2 can be extended from $\mathbb{R}^2$ to $\mathbb{R}^p$ for $p \geq 3$ then for many groups G the sharper forms of their concentration inequalities will follow as corollaries. This requires a non-trivial modification of the argument of Fefferman, Jodeit, and Perlman (1972, Theorem 2), again because the transformation $K \to DK$ need not preserve the $\tilde{G}$-invariance of K unless $G = \tilde{G} = \{\pm I\}$ (see the paragraph containing (5.12)).

The following four lemmas contain the technical core of the argument.

Recall that $\overline{AC} = A\bar{C}$, $(AC)^\circ = AC^\circ$, and $\partial(AC) = A(\partial C)$ for any set $C \subseteq \mathbb{R}^p$ and any nonsingular linear transformation A: $\mathbb{R}^p \to \mathbb{R}^p$. If $\{C_t\}$ is a family of subsets of $\mathbb{R}^p$ indexed by a real parameter $t \geq 0$, we write $C_t \uparrow C$ to indicate pointwise monotone convergence of the indicator function of C_t to that of C and $C_t \to C$ a.e. $[\nu]$ to indicate pointwise convergence of the indicator function of $C_t \cap S$ to that of $C \cap S$ a.e. $[\nu]$. If D is a contraction, note that $D_t \equiv D + t(I - D)$ is also a contraction for every $0 < t < 1$ and that $D_t \downarrow D$ as $t \downarrow 0$.

LEMMA 7.1 *Let $\mathcal{C}$ be a family of subsets $C \subseteq \mathbb{R}^p$ with the following four properties:*

(i) $tC \in \mathcal{C}$ $\forall t > 0$.

(ii) If $C^\circ \neq \emptyset$ then $t\bar{C} \uparrow C^\circ$ a.e. $[\nu]$ as $t \uparrow 1$.

(iii) If $C^\circ = \emptyset$ then $\nu(D\bar{C}) = 0$ for every contraction $D \neq I$.

(iv) If $C^\circ \neq \emptyset$ then for each contraction $D \neq I$, $D_tC^\circ \to D\bar{C}$ a.e. $[\nu]$ as $t \downarrow 0$.

Then the following three conditions are equivalent:

(a) $\nu(\bar{C}) \geq \nu(D\bar{C})$ $\forall C \in \mathcal{C}$ and $\forall$ contractions D.

(b) $\nu(C^\circ) \geq \nu(DC^\circ)$ $\forall C \in \mathcal{C}$ and $\forall$ contractions D.

(c) $\nu(C^\circ) \geq \nu(D\bar{C})$ $\forall C \in \mathcal{C}$ and $\forall$ contractions $D \neq I$.

If in addition, (v): $C \in \mathcal{C} \Rightarrow \bar{C} \in \mathcal{C}$, then (a) $\Leftrightarrow$ (a′): $\nu(C) \geq \nu(DC)$ $\forall$ closed $C \in \mathcal{C}$ and $\forall$ contractions D.

PROOF The implications (*c*) $\Rightarrow$ (*a*) and (*c*) $\Rightarrow$ (*b*) are immediate.

(a) $\Rightarrow$ (b): If $C^\circ = \emptyset$ then (*b*) is trivial. If $C^\circ \neq \emptyset$ then $\nu(t\bar{C}) \geq \nu(tD\bar{C})$ by (i) and (a). Now let $t \uparrow 1$ and apply (ii) to obtain (b).

(b) $\Rightarrow$ (c): If $C^\circ = \emptyset$ then (c) is trivial by (iii). If $C^\circ \neq \emptyset$ then $\nu(C^\circ) \geq \nu(D_tC^\circ) \to \nu(D\bar{C})$ as $t \downarrow 0$ by (b) and (iv).

(a) $\Leftrightarrow$ (a′): obviously (a) $\Rightarrow$ (a′); under assumption (v), clearly (a′) $\Rightarrow$ (a). □

LEMMA 7.2 *For any compact subgroup $G \subseteq \mathcal{O}_p$ that acts effectively on $\mathbb{R}^p$, the class $\mathcal{C}_G$ satisfies conditions (i)-(v) of Lemma 7.1.*

PROOF Suppose that $C \in \mathcal{C}_G$. The convexity of C implies the convexity of $\bar{C}$ and tC (cf. Eggleston (1966), p. 9), while the G-invariance of C implies the G-invariance of $\bar{C}$ and tC, so conditions (i) and (v) are satisfied. If $C^\circ = \emptyset$ then C convex $\Rightarrow$ $\bar{C}$ lies in a proper subspace of $\mathbb{R}^p$, hence $D\bar{C}$ is ν-null for every D, so (iii) holds.

To verify (iii), assume that $C^\circ \neq \emptyset$. Since G is effective, $0 \in C$ [12], hence C is star-shaped with respect to 0. Thus tC, and therefore $t\bar{C}$, increases as

[12] For any $x \in C$ define $x_G = \int_G gx \, d\mu(g)$, where μ is the Haar probability measure on G. Clearly $gx_G = x_G$ $\forall g \in G$, so $x_G = 0$ as G is effective. But $x_G \in C_G(x) \subseteq C$ since $C \in \mathcal{C}_G$, so $0 \in C$. (Note that this also provides a proof of Lemma 3.1.)

$t \uparrow 1$. Also, if $x \in C^o$ then $\tau x \in C^o$ for some $\tau > 1$, hence $x \in \tau^{-1}C^o \subseteq \tau^{-1}\bar{C} \subseteq \cup\{t\bar{C}|t \uparrow 1\}$, so $C^o \subseteq \cup\{t\bar{C}|t \uparrow 1\}$. Next we show that $0 \in C^o$. If not then $0 \in \partial C$, so the convex set C must be supported at 0 by some $(p-1)$-dimensional subspace, i.e., $C \subseteq \{y \in \mathbb{R}^p | a'y \geq 0\}$ for some $a \neq 0$. Thus, since C is G-invariant, $a'(gx) \geq 0 \ \forall x \in C$ and $\forall g \in G$, so $a'x_G = \int_G a'(gx)d\mu(g) \geq 0$ (see Footnote 12). But $x_G = 0$ as G is effective, hence $a'(gx) = 0 \ \forall x \in C$ and $\forall g \in G$. In particular $a'x = 0 \ \forall x \in C$, contradicting the assumption that $C^o \neq \emptyset$. Thus it must hold that $0 \in C^o$. Therefore $t\bar{C} \equiv t\bar{C} + (1-t)0 \subset C^o$ if $0 < t < 1$ (cf. Eggleston (1966, Corollary 2, p. 10)), so $\cup\{t\bar{C}|t \uparrow 1\} \subseteq C^o$ and (ii) is verified.

To verify (iv), assume that $C^o \neq \emptyset$ and $D \neq I$. Let χ_t and χ denote the indicator functions of the sets D_tC^o and $D\bar{C}$, respectively. If $x \in DC^o$ then $x \in D_tC^o$ for all t near 0, while if $x \notin D\bar{C}$ then $x \notin D_t\bar{C}$ for all t near 0, so in both cases $\chi_t(x) \to \chi(x)$ as $t \downarrow 0$. If $x \in \partial(DC)(\subseteq D\bar{C})$ then either $x \in D_tC^o$ for all t near 0, in which case again $\chi_t(x) \to \chi(x)$ as $t \downarrow 0$, or else there exists a sequence $t_n \downarrow 0$ such that $x \notin D_{t_n}C^o$ for every n, in which case $\chi_t(x) \not\to \chi(x)$ as $t \downarrow 0$. Therefore, in order to complete the verification of (iv) it must be shown that $\nu(\Delta) = 0$, where

$$\Delta = \{x \in \gamma(DC) | \ \exists t_n \downarrow 0 \text{ such that } x \notin D_{t_n}C^o \text{ for every } n\}. \tag{7.2}$$

Since $\nu(\{x | x_i = 0 \text{ for some } i = 1, \ldots, p\}) = 0$, it suffices to show that $\nu(\Delta \cap \{x | x_i \neq 0, \ i = 1, \ldots, p\}) = 0$. We shall show that $\nu(\Delta \cap \mathbb{R}^+) = 0$ where $\mathbb{R}^+ = \{x \in \mathbb{R}^p | x_i > 0, i = 1, \ldots, p\}$; the other $2^p - 1$ cases follow similarly. Set $K = DC$ and $x_n = D(D_{t_n})^{-1}x$ in (7.2). Since $0 < x_n \uparrow x$ when $x \in \mathbb{R}^+$ (note that $x_n \neq x$ since $D \neq I$) we have that

$$\Delta \cap \mathbb{R}^+ \subseteq \{x \in \partial K \cap \mathbb{R}^+ \mid \exists x_n \uparrow x \text{ such that } x_n \notin K^o \text{ for every } n\} \equiv \Delta^+ \tag{7.3}$$

and shall show that $\nu(\Delta^+) = 0$.

Let $Q \equiv \{x \equiv (x_1, \ldots, x_p) | \ 0 \leq x_i \leq 1, \ i = 1, \ldots, p\}$ denote the closed unit cube in $\mathbb{R}^p$, let θ_i be the unit vector with i-th component 1, and for $\epsilon > 0$ let $L_i(x, \epsilon) \equiv [x - \epsilon\theta_i, x)$ denote the half-open line segment connecting $x - \epsilon\theta_i$ and x. For $\epsilon > 0$ define $Q(x, \epsilon) = (x - \epsilon Q)\backslash\{x\}$ and $L(x, \epsilon) = \cup\{L_i(x, \epsilon) | i = 1, \ldots, p\}$; then

$$L(x, \epsilon) \subset Q(x, \epsilon) \subset \text{ convex hull}[L(x, \epsilon\sqrt{p})]. \tag{7.4}$$

Fix $x \in \Delta^+$. By (7.3), $Q(x, \epsilon) \not\subset K^o \ \forall \epsilon > 0$, hence by (7.4) and the convexity of K^o, $L(x, \epsilon\sqrt{p}) \not\subset K^o \ \forall \epsilon > 0$. Therefore, there exist $i \in \{1, \ldots, p\}$ and a sequence $\{\delta_n\} \downarrow 0$ such that $x - \delta_n\theta_i \notin K^o \ \forall n$, hence there exists $i \in \{1, \ldots, p\}$ such that $L_i(x, x_i) \cap K^o = \emptyset$ (since $x \in \partial K$ and $K^o \notin \emptyset$ – apply Eggleston (1966, Corollary 2, p. 10)). Thus

$$\Delta^+ \subseteq \cup\{\Delta_i | i = 1, \ldots, p\}, \tag{7.5}$$

where $\Delta_i = \{x \in \partial K \cap \mathbb{R}^+ | L_i(x, x_i) \cap K^\circ = \emptyset\}$. (In fact, equality holds in (7.5).) In order to show that $\nu(\Delta^+) = 0$, it therefore suffices to show that $\nu(\Delta_i) = 0$, $i = 1, \ldots, p$.

The remainder of the proof now parallels the treatment of cases (iii) and (iv) in the proof of Theorem 2, Fefferman, Jodeit, and Perlman (1972). First consider Δ_1. Since $\bar{K}$ is convex and $x \in \partial K$, $\Delta_1 = A \cup B$ where

$$A = \{x \in \partial K \cap \mathbb{R}^+ |\ L_1(x, \epsilon) \subset \partial K \text{ for some } \epsilon > 0\}$$

$$B = \{x \in \partial K \cap \mathbb{R}^+ |\ L_1(x, x_1) \cap \bar{K} = \emptyset\}$$

(note that $A \cap B = \emptyset$). Because the projection of A onto $\{x | x_1 = 0\}$ lies in the boundary of the projection of $\bar{K}$ onto $H_0 \equiv \{x | x_1 = 0\}$[13], which boundary has $(p-1)$-dimensional Lebesgue measure 0, it follows that $\nu(A) = 0$. Finally, B is contained in the graph of a positive convex function (the "lower boundary" of $\bar{K} \cap \mathbb{R}^+$) so $\nu(B) = 0$ (apply the Lemma following Theorem 2 in Fefferman, Jodeit, and Perlman (1972)). Similarly, $\nu(\Delta_i) = 0$ for $i = 2, \ldots, p$. □

By (6.13), $C \in \mathcal{M}_G$ iff C is an (arbitrary) union of sets in $\mathcal{C}_G$. Since the boundary of such a set may be irregular, in order to extend Lemma 7.2 to $C \in \mathcal{M}_G$ it is necessary to impose an additional smoothness assumption on C. One such condition, which covers most sets occurring in applications, is the following: define $\hat{\mathcal{M}}_G$ to be the collection of all $C \in \mathcal{M}_G$ such that $\partial C = \partial(\bar{C}) = \cup M_j$, a finite or countable disjoint union of smooth $(p-1)$-dimensional manifolds M_j (hence ∂C is piecewise smooth). Furthermore, it is necessary to impose a stronger assumption on the group G itself.

LEMMA 7.3 *For any compact subgroup $G \subseteq \mathcal{O}_p$ that acts irreducibly on $\mathbb{R}^p$, the class $\hat{\mathcal{M}}_G$ satisfies conditions (i)-(v) of Lemma 7.1.*

PROOF Suppose that $C \in \hat{\mathcal{M}}_G$. By (6.13), C satisfies (i) since $C_G(tx) = tC_G(x)$. To verify (v), consider $x \in \bar{C}$. Then there exists a sequence $\{x_n\} \subset C$ such that $x_n \to x$. Since

$$C_G(x) = \{\alpha_1 g_1 x + \cdots + \alpha_k g_k x | k \geq 1,\ g_i \in G,\ \alpha_i \geq 0,\ \sum \alpha_i = 1\}$$

and $\|gx_n - gx\| = \|x_n - x\|$ for each $g \in G$, it follows that $\delta(C_G(x_n), C_G(x)) \leq \|x_n - x\| \to 0$, where δ denotes the Hausdorff metric (cf. Valentine (1976),

[13] If not, then there would exist $x \in A$ such that the projection $x - x_1\theta_1$ of x onto H_0 lies in the interior of the projection of $\bar{K}$ onto H_0. This would imply that there exist $y \in \bar{K}$ and $\delta > 0$ such that the projection $y - y_1\theta_1$ of y onto H_0 satisfies $y - y_1\theta_1 = (1+\delta)(x - x_1\theta_1)$. Also, since $x \in A$ there would exist $\epsilon > 0$ such that the closed triangle with vertices x, $x - \epsilon\theta_1$, and y is contained in $\bar{K}$. But since $0 \in K^\circ$, this would imply that the open line segment $(x - \epsilon\theta_1, x) \subset K^\circ$, contradicting the fact that $x \in A$.

p. 36). But $C_G(x_n) \subseteq C$ for every n, hence $C_G(x) \subseteq \bar{C}$. Thus $\bar{C} \in \mathcal{M}_G$ and (v) is satisfied.

Since G is irreducible, $C^\circ = \emptyset$ implies that $C = \{0\}$ (apply Lemma 3.2), so (iii) is trivial. To verify (ii), assume that $C^\circ \neq \emptyset$. It follows as in the second paragraph of the proof of Lemma 7.2 that $t\bar{C}$ increases as $t \uparrow 1$ and $C^\circ \subseteq \cup\{t\bar{C}|t \uparrow 1\}$. To show that $\cup\{t\bar{C}|t \uparrow 1\} \subseteq C^\circ$ it suffices to show that $t\bar{C} \subseteq C^\circ$ if $0 < t < 1$. For $x \in \bar{C}\backslash\{0\}$ choose a sequence $\{x_n\} \subset C$ such that $x_n \to x$; then as above, $\delta(C_G(x_n), C_G(x)) \to 0$. Because $C_G(x_n)$ and $C_G(x)$ are bounded convex sets with non-empty interiors (since G is irreducible), it follows that $\delta([C_G(x_n)]^\circ, [C_G(x)]^\circ) = \delta(C_G(x_n), C_G(x)) \to 0$. But $tx \in [C_G(x)]^\circ$ because $0 \in [C_G(x)]^\circ$ (since G is irreducible), so $\exists$ n such that[14] $tx \in [C_G(x_n)]^\circ \subseteq C^\circ$, as claimed.

To verify (iv), as in the proof of Lemma 7.2 it suffices to show that $\nu(\Delta^+) = 0$, where Δ^+ is given by (7.3) and $K = DC$. Since $\partial K = D(\partial C) = \cup(DM_j) \equiv \cup N_j$, it is enough to to show that $\nu(\Delta_j^+) = 0$, where Δ_j^+ is defined as Δ^+ in (7.3) but with ∂K replaced by the relative interior of N_j, a smooth $(p-1)$-dimensional open manifold. Since for every $x \in \Delta_j^+$ it holds that $Q(x,\epsilon) \not\subset K^\circ$ $\forall \epsilon > 0$, it can be shown that

$$\Delta_j^+ \subseteq \{x \in [\text{rel int}(N_j)] \cap \mathbb{R}^+ |\ N(x) \notin \mathbb{R}^+\},$$

where $N(x)$ is the outward normal vector to N_j at x. But $S(x) \in \mathbb{R}^+$ for each $x \in \mathbb{R}^+$, where $S(x)$ denotes the outward normal vector to the sphere S at x. Therefore the sphere S and the manifold Δ_j^+ intersect *transversely*, so their intersection must be a manifold of dimension $\leq p-2$ (cf. Guilleman and Pollack (1974), Theorem, p. 30; Do Carmo (1976), Ex. 17, p. 90), hence $\nu(\Delta_j^+) = 0$ as required. □

LEMMA 7.4 *Suppose that $G = G_1 \times \cdots \times G_t$, a direct product of compact irreducible groups acting on $\mathbb{R}_1 \times \cdots \times \mathbb{R}_t$, where $\sum \dim(\mathbb{R}_i) = p$. Then the class $\hat{\mathcal{M}}_G$ satisfies conditions (i)-(v) of Lemma 7.1.*

PROOF The first and third paragraphs of the proof of Lemma 7.3 carry over to this case without change, while the second paragraph must be modified as follows:

For each $i = 1, \ldots, t$ define $\tilde{\mathbb{R}}_i = \{0\} \times \cdots \times \{0\} \times \mathbb{R}_i \times \{0\} \times \cdots \times \{0\}$ and note that $\nu(\tilde{\mathbb{R}}_i) = 0$. Since each G_i acts irreducibly on $\mathbb{R}_i$, $C^\circ = \emptyset$ implies that $\bar{C} \subseteq \cup \tilde{\mathbb{R}}_i$, so (iii) is immediate. To verify (ii), assume that $C^\circ \neq \emptyset$. As in the second paragraph of the proof of Lemma 7.2, $t\bar{C}$ increases as $t \uparrow 1$ and

[14] This requires the following fact: if A_n, A are nonempty, convex, open sets in $\mathbb{R}^p$ such that $\delta(A_n, A) \to 0$, then $A \subseteq \cup A_n$. (For $y \in A$, choose n such that $\delta(A_n, A) < \|y - \partial A\|$. If $y \notin A_n$ then $\exists$ a hyperplane H that separates A_n and y. This would imply that $\exists z \in A \cap N$, where N is the line normal to H through y, such that $\|z - A_n\| > \|z - y\| \geq \delta(A_n, A)$, contradicting the definition of $\delta(A_n, A)$. Therefore $y \in A_n$.)

$C^\circ \subseteq \cup\{t\bar{C} | t \uparrow 1\}$. To show that $\nu(\cup\{t\bar{C}|t \uparrow 1\} \backslash C^\circ) = 0$ it suffices to show that $t\bar{C} \subseteq C^\circ \cup (\cup \tilde{\mathbb{R}}_i)$ if $0 < t < 1$. For $x \in \bar{C} \backslash (\cup \tilde{\mathbb{R}}_i)$ choose a sequence $\{x_n\} \subset C$ such that $x_n \to x$; then as above, $\delta(C_G(x_n), C_G(x)) \to 0$. Because $C_G(x_n)$ and $C_G(x)$ are bounded convex sets with non-empty interiors (since $x_n, x \notin \cup \tilde{\mathbb{R}}_i$), it follows that $\delta([C_G(x_n)]^\circ, [C_G(x)]^\circ) = \delta(C_G(x_n), C_G(x)) \to 0$. But $tx \in [C_G(x)]^\circ$ because $0 \in [C_G(x)]^\circ$ (since $x \notin \cup \tilde{\mathbb{R}}_i$), so $\exists\, n$ such that (see Footnote 14) $tx \in [C_G(x_n)]^\circ \subseteq C^\circ$, as claimed. □

When $p = 2$, Lemma 7.2 applies to all compact subgroups $G \subseteq \mathcal{O}_2$ except $\mathcal{C}_2^1$ and $\mathcal{H}_2^1$, Lemma 7.3 applies to all compact subgroups of $\mathcal{O}_2$ except $\mathcal{C}_2^1$, $\mathcal{H}_2^1$, $\mathcal{C}_2^2$, and $\mathcal{H}_2^2$, while Lemma 7.2 applies to $\mathcal{H}_2^2$. Thus, from the equivalence of (a′) and (c) in Lemma 7.1 we obtain the following sharpened versions of Theorems 6.1 and 6.2:

THEOREM 7.1 *Suppose that* $X \sim EC_2(\Sigma)$. *If* $C \in \mathcal{C}_G$, *then* $\Sigma_1 \leq \Sigma_2$, $\Sigma_1 \neq \Sigma_2 \Rightarrow P_{\Sigma_1}(C^\circ) \geq P_{\Sigma_2}(\bar{C})$ *provided that* Σ_1 *is* G*-invariant and* G *acts effectively on* $\mathbb{R}^2$ *(i.e.,* $G \neq \mathcal{C}_2^1$ *or* $\mathcal{H}_2^1$*).*

THEOREM 7.2 *Suppose that* $X \sim EC_2(\Sigma)$. *If* $C \in \hat{\mathcal{M}}_G$, *then* $\Sigma_1 \leq \Sigma_2$, $\Sigma_1 \neq \Sigma_2 \Rightarrow P_{\Sigma_1}(C^\circ) \geq P_{\Sigma_2}(\bar{C})$ *provided that* Σ_1 *is* G*-invariant and* G *is irreducible or is the direct product of irreducible compact groups (i.e.,* $G \neq \mathcal{C}_2^1$, $\mathcal{H}_2^1$, *or* $\mathcal{C}_2^2$*), but also* $G \neq \mathcal{C}_2^3$.

Finally, Fefferman, Jodeit, and Perlman (1972, p. 118) presented several sufficient conditions for the *strict* inequality $\nu(C^\circ) > \nu(D\bar{C})$ to hold when $D \neq I$ and $C \in \mathcal{C}_1$. Their discussion remains valid when $C \in \mathcal{C}_G$ and G acts effectively, and when $C \in \hat{\mathcal{M}}_G$ and G is irreducible or is the direct product of irreducible compact groups. (When $C \in \hat{\mathcal{M}}_G$, their argument on p. 118 showing that $d\bar{C} \subseteq C^\circ$ must be replaced by our arguments in the proofs of Lemmas 7.3 and 7.4 showing that $t\bar{C} \subseteq C^\circ$ and $t\bar{C} \subseteq C^\circ \cup (\cup \tilde{\mathbb{R}}_i)$, respectively.)

Acknowledgement

We gratefully thank Victor Klee and James Morrow for helpful suggestions and David Perlman and Alice Kelly for preparing the graphic illustrations.

REFERENCES

DAS GUPTA, S., EATON, M. L., OLKIN, I., PERLMAN, M. D., SAVAGE, L. J. AND SOBEL, M. (1972). Inequalities on the probability content of convex regions for elliptically contoured distributions. In *Proc. Sixth Berkeley Symp. Math. Statist.*

Probab. Vol. II, L. M. LeCam, J. Neyman and E. L. Scott, eds., University of California Press, Berkeley, CA. 241-265.

Do Carmo, M. P. (1976). *Differential Geometry of Curves and Surfaces.* Prentice-Hall, Englewood Cliffs, NJ.

Eaton, M. L. and Perlman, M. D. (1977). Reflection groups, generalized Schur functions, and the geometry of majorization. *Ann. Probab.* **5** 829-860.

Eaton, M. L. and Perlman, M. D. (1991). Concentration inequalities for multivariate distributions: I. Multivariate normal distributions. *Statist. Prob. Lett.* **12** 487-504.

Eggleston, H. G. (1966). *Convexity*. Cambridge University Press, Cambridge.

Fefferman, C., Jodeit, M. and Perlman, M. D. (1972). A spherical surface measure inequality for convex sets. *Proc. Amer. Math Soc.* **33** 114-119.

Grove, L. C. and Benson, C. T. (1985). *Finite Reflection Groups*, 2nd edition, Springer-Verlag, New York.

Guillemin, V. and Pollack, A. (1974). *Differential Topology.* Prentice-Hall, Englewood Cliffs, NJ.

Valentine, F.A. (1976). *Convex Sets.* Robert E. Krieger, Huntington, New York.

Department of Statistics
University of Washington
Seattle, WA 98195

Stochastic Inequalities
IMS Lecture Notes – Monograph Series
Volume 22 (1993)

INEQUALITIES ON EXPECTATIONS BASED ON THE KNOWLEDGE OF MULTIVARIATE MOMENTS[1]

By ANDRÁS PRÉKOPA

Rutgers University

The paper deals with discrete moment problems where the possible values of a random vector form a known finite set. First, some earlier results concerning the one dimensional discrete moment problem are summarized. Then, restricting the discussion to the two-dimensional case, for the sake of simplicity, two different discrete moment problems are formulated: (a) the known moments are those where the exponents of the random variables are chosen between 0 and some upper bounds; (b) the sum of the exponents is less than or equal to a given number. The bounds that can be obtained by our technique include bounds for probabilities and expectation.

1. Introduction

The one-dimensional discrete moment problem can be formulated in the following manner. Given a random variable ξ, the possible values of which are known to be $z_0 < z_1 < \cdots < z_n$ and a function $f(z)$, $z \in \{z_0, z_1, \ldots, z_n\}$. We want to give lower and upper bounds for $E[f(\xi)]$, based on the knowledge of the moments $\mu_k = E[\xi^k]$, $k = 1, 2, \ldots, m$, while the probability distribution of ξ is unknown.

Introducing the notations $p_i = P\{\xi = z_i\}$, $f_i = f(z_i)$, $i = 0, 1, \ldots, n$, $\mu_0 = 1$, we obtain the above mentioned bounds by solving the linear programming problems

$$\min(\max) \sum_{i=0}^{n} f_i p_i \tag{1.1}$$

subject to

$$\sum_{i=0}^{n} z_i^k p_i = \mu_k, \qquad k = 0, 1, \ldots, m$$

$$p_i \geq 0, \qquad i = 1, 2, \ldots, n,$$

[1]Research supported by National Science Foundation Grant No. DMS–9005159.

AMS 1991 *subject classifications.* Primary: 60E99, 26B25, 26D15; Secondary: 41A05, 41A63, 65D05.

Key words and phrases. Discrete moment problem, multivariate moment inequalities, interpolation, linear programming.

where we assume that $m < n$. Problems (1.1) are termed as discrete power moment problems. Replacing z_i^k by $\binom{z_i}{k}$ and μ_k by S_k, where $S_k = E[\binom{\xi}{k}]$, $k = 0, 1, \ldots, m$, we obtain the binomial moment problem which plays an important role in bounding probabilities of logical functions of events (Prékopa (1988, 1990a)). The discrete binomial moment problem can be transformed into the discrete power moment problem; hence we restrict ourselves to problems (1.1). (Note that if instead of the consecutive moments $\mu_1, \mu_2, \ldots, \mu_m$, the moments $\mu_{k_1}, \mu_{k_2}, \ldots, \mu_{k_m}$ would be known where $k_1 < k_2 < \cdots < k_m$ are non-consecutive integers then simple equivalence between the power and the binomial moment problems no longer exists.)

The duals of the problems (1.1) throw new light to this approach of bounding $E[f(\xi)]$. If (1.1) is a minimization problem, then its dual is

$$\max \sum_{k=0}^{m} \mu_k x_k \tag{1.2}$$

subject to

$$\sum_{k=0}^{m} z_i^k x_k \le f_i, \qquad i = 0, 1, \ldots, n,$$

and if (1.1) is a maximization problem, then its dual is

$$\min \sum_{k=0}^{m} \mu_k y_k \tag{1.3}$$

subject to

$$\sum_{k=0}^{m} z_i^k y_k \ge f_i, \qquad i = 0, 1, \ldots, n.$$

The optimum values of these problems are called the sharp lower and upper bounds for $E[f(\xi)]$. Since problems (1.1) have feasible solutions and finite optima, the duality theorem of linear programming ensures that so do problems (1.2), (1.3) and the optimum values of the primal-dual pairs coincide. Thus, we have the inequalities

$$\sum_{k=0}^{m} \mu_k x_k \le E[f(\xi)] \le \sum_{k=0}^{m} \mu_k y_k, \tag{1.4}$$

where $x_0, x_1, \ldots, x_m$ satisfy (1.2) and $y_0, y_1, \ldots, y_m$ satisfy (1.3). The bounds (1.4) are the best in the case of the optimal solutions x, y.

Among the choices of the function f, prominent are the following:

(1) f has positive divided differences of order $m+1$ (for the definition of the divided differences see the next section).

(2) $f_r = 1$ and $f_i = 0$ for $i \neq r$.

(3) $f_0 = \cdots = f_{r-1} = 0$, $f_r = \cdots = f_n = 1$.

In cases (2) and (3) we are bounding $P\{\xi = z_r\}$ and $P\{\xi \geq z_r\}$, respectively.

If we did not know the possible values of ξ but we still would know the moments $\mu_1, \mu_2, \ldots, \mu_m$ then the general moment problem (see, e.g., Krein and Nudelman (1977)):

$$\min(\max) \int_a^b f(z) d\sigma \tag{1.5}$$

subject to

$$\int_a^b z^k d\sigma = \mu_k, \qquad k = 0, 1, \ldots, m$$

would provide us with bounds for $E[f(\xi)]$, where σ is the unknown probability distribution function on $[a, b]$. Assuming $a = z_0$, $b = z_n$, furthermore, designating by L_d, U_d and L_c, U_c the optimum values corresponding to problems (1.1) and (1.5), respectively, we have the relations

$$L_c \leq L_d \leq E[f(\xi)] \leq U_d \leq U_c. \tag{1.6}$$

This means that if the set of possible values of ξ is a known discrete set and we utilize it in the form of solving problems (1.1) then better bounds can be obtained than through solving problems (1.5). This is so even though the optimal solutions of problems (1.5) are discrete distributions. In fact, the supports of these distributions may not be subsets of $\{z_0, z_1, \ldots, z_n\}$.

Recent discovery by Samuels and Studden (1989) and by Prékopa (1988, 1990a, 1990b) of the fact that the sharp Bonferroni inequalities are essentially solutions of discrete moment problems, stresses the importance of the discrete case. A variety of applications of the discrete moment problem, ranging from communication or power system reliability calculations to approximations in queueing systems, can be mentioned.

2. Dual Feasible Bases and Lagrange Polynomials Associated with Problems (1.1)

In this section we further restrict ourselves to that special case of the objective function f where all $m+1$st divided differences are positive. The first order divided differences of f are

$$[z_{i_1}, z_{i_2}]f = \frac{f(z_{i_2}) - f(z_{i_1})}{z_{i_2} - z_{i_1}}, \qquad 0 \leq i_1 < i_2 \leq n. \tag{2.1}$$

The higher order divided differences are defined recursively by

$$[z_{i_1}, z_{i_2}, \ldots, z_{i_k}, z_{i_{k+1}}]f = \frac{[z_{i_2}, \ldots, z_{i_{k+1}}]f - [z_{i_1}, \ldots, z_{i_k}]f}{z_{i_{k+1}} - z_{i_1}}, \tag{2.2}$$

$$0 \le i_1 < i_2 < \cdots < i_k < i_{k+1} \le n$$

for $k \ge 2$. It is known (see, e.g., Jordan (1965)) that if all divided differences of order k, corresponding to consecutive points, are positive then all divided differences of order k are positive, and we have the equality

$$[z_{i_1}, z_{i_2}, \ldots, z_{i_{k+1}}]f = \frac{\begin{vmatrix} 1 & 1 & \cdots & 1 \\ z_{i_1} & z_{i_2} & \cdots & z_{i_{k+1}} \\ \vdots & & \ddots & \\ z_{i_1}^{k-1} & z_{i_2}^{k-1} & \cdots & z_{i_{k+1}}^{k-1} \\ f(z_{i_1}) & f(z_{i_2}) & \cdots & f(z_{i_{k+1}}) \end{vmatrix}}{\begin{vmatrix} 1 & 1 & \cdots & 1 \\ z_{i_1} & z_{i_2} & \cdots & z_{i_{k+1}} \\ \vdots & & \ddots & \\ z_{i_1}^{k-1} & z_{i_2}^{k-1} & \cdots & z_{i_{k+1}}^{k-1} \\ z_{i_1}^{k} & z_{i_2}^{k} & \cdots & z_{i_{k+1}}^{k} \end{vmatrix}}. \tag{2.3}$$

If f is defined for every z in $[z_0, z_n]$ and $f^{(m+1)}(z) > 0$ at every interior point z of this interval, then all $m+1$st divided differences of f are positive on $z_0, z_1, \cdots, z_n$ (see Jordan (1965)). The positivity of the first order divided differences means that f is increasing and the positivity of the second order divided differences means the convexity of the function f, i.e., the polygon connecting the points $(z_i, f(z_i)), i = 0, 1, \ldots, n$ in the plane, is convex. This implies that an equivalent formulation of the positivity of the second order divided differences is:

$$\frac{f(z_{i_3}) - f(z_{i_1})}{z_{i_3} - z_{i_1}} > \frac{f(z_{i_2}) - f(z_{i_1})}{z_{i_2} - z_{i_1}}, \qquad 0 \le i_1 < i_2 < i_3 \le n.$$

We assume that the $m+1$st divided differences of f are positive (while there is no condition on the lower order divided differences). Thus, we handle the type (1) of the objective function f, mentioned in the Introduction. The results corresponding to the others are presented in Prékopa (1990b).

Let $\mathbf{a}_i^{\mathrm{T}} = (1, z_i, \cdots, z_i^m)$, $i = 0, \cdots, n$, and $A = (\mathbf{a}_0, \cdots, \mathbf{a}_n)$. Furthermore let B be an $(m+1) \times (m+1)$ part of the matrix A. Since B is a Vandermonde matrix, it is non-singular and thus, it represents a basis in the linear programming (minimization or maximization) problem (1.1). The columns of B will be called basic vectors.

Let $\mathbf{f}_\mathrm{B}$ designate the vector consisting of those f_i values as components which correspond to basic vectors $\mathbf{a}_i$. By definition, B is dual feasible if

(2.4) $f_j - \mathbf{f}_\mathrm{B}^\mathrm{T} B^{-1}\mathbf{a}_j \geq 0$ for all j, in the minimization problem,

(2.5) $f_j - \mathbf{f}_\mathrm{B}^\mathrm{T} B^{-1}\mathbf{a}_j \leq 0$ for all j, in the maximization problem.

If $\mathbf{a}_j$ is basic then we have equality in the above relations.

Let I be the set of subscripts of the basic vectors and $L_\mathrm{I}(z)$ the Lagrange polynomial corresponding to the points z_i, $i \in I$. Then

$$L_\mathrm{I}(z) = \sum_{i=0}^{m} L_{\mathrm{I}i}(z) f(z_i),$$

where

$$L_{\mathrm{I}i}(z) = \prod_{j \in I - \{i\}} \frac{(z - z_j)}{(z_i - z_j)}$$

is the ith fundamental polynomial. Furthermore let $\mathbf{b}^\mathrm{T}(z) = (1, z, \ldots, z^m)$. Clearly we have $\mathbf{b}(z_i) = \mathbf{a}_i$, $i = 0, 1, \ldots, n$ and

$$f(z) - \mathbf{f}_\mathrm{B}^\mathrm{T} B^{-1}\mathbf{b}(z) = \frac{1}{|B|} \begin{vmatrix} f(z) & \mathbf{f}_\mathrm{B}^\mathrm{T} \\ \mathbf{b}(z) & B \end{vmatrix}. \tag{2.6}$$

From (2.6) we first derive

$$\mathbf{f}_\mathrm{B}^\mathrm{T} B^{-1}\mathbf{b}(z) = L_\mathrm{I}(z). \tag{2.7}$$

In fact, $\mathbf{f}_\mathrm{B}^\mathrm{T} B^{-1}\mathbf{b}(z)$ is an mth degree polynomial that is equal to $f(z_i)$ if $z = z_i$. Another observation is that if $z \notin \{z_i, i \in I\}$ then the second determinant on the right hand side in (2.6) is different from 0. This follows from (2.3) and the assumption that all $m+1$st order divided differences of f are positive. This implies that if $z \notin \{z_i, i \in I\}$ then (2.6) is nonzero, in other words, no basis is dual-degenerate. Hence, in (2.4) and (2.5) we have equalities at basic points, otherwise we have strict inequalities. This means that the basis B is dual feasible in the minimization (maximization) problem (1.1) if and only if the Lagrange polynomial corresponding to the basic points $\{z_i, i \in I\}$ is strictly below (above) the function $f(z)$ at any nonbasic point z.

Using (2.6) and (2.3), we obtain the equation

$$f(z) - L_\mathrm{I}(z) = \prod_{j \in I} (z - z_j)\,[z, z_i, i \in I]\, f, \tag{2.8}$$

which is well-known in interpolation theory. We also mention Newton's form for the interpolating polynomial:

$$L_I(z) = f_0 + \sum_{k=0}^{m} \prod_{j \in I^{(k)}} (z - z_j) \left[z_j, j \in I^{(k)}\right] f,$$

where $I^{(k)}$ is the set of the first $k+1$ elements of I and f_0 is the function value corresponding to the first element in I.

In order to find necessary and sufficient condition that a basis is dual feasible in any of the problems (1.1), equation (2.8) can be used. Since $[z, z_i, i \in I]f > 0$ for every $z \notin \{z_i, i \in I\}$, we see that the basis is dual feasible in the minimization problem, if and only if

$$\prod_{j \in I}(z - z_j) > 0 \qquad \text{for all non-basic } z,$$

and is dual feasible in the maximization problem, if and only if

$$\prod_{j \in I}(z - z_j) < 0 \qquad \text{for all non-basic } z.$$

Thus, the basic vectors have to follow each other according to some patterns that can be best summarized in terms of their subscripts. If the basis B corresponds to the subscript set I, then sometimes we will write $B(I)$ instead of B.

THEOREM 2.1 *A basis $B(I)$ is dual feasible in the minimization (maximization) problem* (1.1) *if and only if the subscript set I, with elements arranged in increasing order, has the following structure:*

	$m+1$ even	*$m+1$ odd*
Minimization problem	$\{j, j+1, \ldots, k, k+1\}$	$\{0, j, j+1, \ldots, k, k+1\}$
Maximization problem	$\{0, j, j+1, \ldots, k, k+1, n\}$	$\{j, j+1, \ldots, k, k+1, n\}$.

3. Bounds for $E[f(\xi)]$

Theorem 2.1 can be used to give bounds for $E[f(\xi)]$. These can be obtained in terms of formulas if m is small ($m \le 4$) or by algorithms if m is large.

Any dual feasible basis B provides us with a bound. If it is dual feasible in the minimization problem, then the corresponding objective function value is smaller than or equal to the optimum value. Hence,

$$\mathbf{f}_{\mathrm{B}}^{\mathrm{T}} B^{-1}\mu = E\left[L_{\mathrm{I}}(\xi)\right] \le E\left[f(\xi)\right], \tag{3.1}$$

where $\mu^{\mathrm{T}} = (1, \mu_1, \ldots, \mu_m)$. If, on the other hand, B is dual feasible in the maximization problem, then

$$\mathbf{f}_{\mathrm{B}}^{\mathrm{T}} B^{-1}\mu = E\left[L_{\mathrm{I}}(\xi)\right] \ge E\left[f(\xi)\right]. \tag{3.2}$$

Note that the inequalities in (3.1) and (3.2) hold also for each possible value of ξ, if we remove the expectations.

A basis B is said to be primal feasible if $B^{-1}\mu \geq 0$. A basis that is both primal and dual feasible, is optimal. Combining these remarks with Theorem 2.1, we have

THEOREM 3.1 *Assume that the function f has all positive divided differences of order $m+1$. The following assertions hold true:*

(a) *If I has one of the two structures*

$$\{j,\, j+1, \ldots, k,\, k+1\}, \qquad \{0,\, j,\, j+1, \ldots, k,\, k+1\},$$

then

$$L_{\mathrm{I}}(z) \leq f(z),$$

with strict inequality for all nonbasic z, and

$$E\left[L_{\mathrm{I}}(\xi)\right] \leq E\left[f(\xi)\right].$$

This bound is sharp if $B(I)$ is a primal feasible basis in problem (1.1).

(b) *If I has one of the structures*

$$\{j,\, j+1, \ldots, k,\, k+1,\, n\}, \qquad \{0,\, j,\, j+1, \ldots, k,\, k+1,\, n\},$$

then

$$L_{\mathrm{I}}(z) \geq f(z),$$

with strict inequality for all nonbasic z, and

$$E\left[L_{\mathrm{I}}(\xi)\right] \geq E\left[f(\xi)\right].$$

This bound is sharp if $B(I)$ is a primal feasible basis in problem (1.1).

In order to obtain the sharp bound we need to check which one is that dual feasible basis $B(I)$ for which we also have primal feasibility, i.e., $[B(I)]^{-1}\mu \geq 0$. If m is small then Theorem 2.1 gives us a key to find this $B(I)$ by a formula (see Boros and Prékopa (1989), Prékopa (1990b)). However, if m is large then the sharp bound can be obtained only by an algorithm.

Instead of a general linear programming algorithm, the following very advantageous dual type algorithm can be used to solve any of the problems (1.1).

Step 0. Pick any dual feasible basis subscript set I, in accordance with Theorem 2.1.

Step 1. Check if $[B(I)]^{-1}\mu \geq 0$. If yes, then stop; optimal basis and optimal solution has been reached. Otherwise pick any j for which $([B(I)]^{-1})_j < 0$, and go to Step 2.

Step 2. Delete the jth vector from $B(I)$ (which is not necessarily the same as $\mathbf{a}_j$) and include that vector which restores the dual feasible basis structure. Go to Step 1.

Since no dual degeneracy occurs, the objective function values are strictly increasing and the algorithm terminates in a finite number of steps.

The above algorithm is of dual type but it is not exactly a special case of the dual algorithm of Lemke (1954). The difference is that here the incoming vector can be found very easily through a logical analysis of the subscript set I, rather than a costly procedure involving reduced prices. For more details of the algorithm see Prékopa (1990a).

NUMERICAL EXAMPLE We present sharp lower and upper bounds for the moment generating function $E\left(e^{t\xi}\right)$ at the point $t = 0.1$. We assume that the possible values of ξ are known to be $z_i = i$, $i = 0, 1, \cdots, 20$ and we know the first three moments of ξ: $\mu_1 = 9.73086229944$, $\mu_2 = 129.5641151$, $\mu_3 = 1903.250122$.

The function $f(z) = e^{0.1z}$ has positive derivative of any order at any z, hence the condition for f (that its fourth order divided differences, on the set of possible values of ξ, are positive) is fulfilled.

Using the algorithm described in this section, both the minimization and maximization problems (1.1) have been solved instantly on a 33MHz/486 PC. The code was written in APL language which is very suitable to these problems. Below we present the subscript sets of the bases encountered in the subsequent iterations, together with the optimal solutions.

Minimization problem

Initial basis	6	7	8	9
	7	8	9	10
	8	9	10	11
	7	8	10	11
	7	8	11	12
	6	7	11	12
	5	6	11	12
	5	6	12	13
	4	5	12	13
	4	5	13	14
	3	4	13	14
	3	4	14	15
Optimal basis	3	4	15	16

The optimal solution is:

$$p_3 = 0.3170498444, \qquad p_4 = 0.1397544435, \qquad p_{15} = 0.4704357076,$$

$$p_{16} = 0.0727600045, \qquad p_i = 0, \quad \text{for any other } i.$$

The optimum value is:

$$e^{(0.1)3}p_3 + e^{(0.1)4}p_4 + e^{(0.1)15}p_{15} + e^{(0.1)16}p_{16} = 3.105190886.$$

Maximization problem

Initial basis	0	1	2	20
	0	2	3	20
	0	3	4	20
	0	4	5	20
	0	5	6	20
	0	6	7	20
	0	7	8	20
	0	8	9	20
	0	9	10	20
Optimal basis	0	10	11	20

The optimal solution is:

$$p_0 = 0.20523691, \qquad p_{10} = 0.2755241346, \qquad p_{11} = 0.3787952662,$$

$$p_{20} = 0.140443686, \qquad p_i = 0, \quad \text{for any other } i.$$

The optimum value is:

$$e^{(0.1)}p_0 + e^{(0.1)10}p_{10} + e^{(0.1)11}p_{11} + e^{(0.1)20}p_{20} = 3.129899305.$$

Thus, we have the sharp bounds:

$$3.105190886 \le E\left(e^{0.1\xi}\right) \le 3.129899305.$$

4. Multivariate Discrete Moment Problems

For the sake of simplicity we restrict ourselves to the discussion of the bivariate case. The results generalize to the multivariate case in a straightforward manner.

Let ξ_1 and ξ_2 be two discrete random variables with known finite supports which are z_{ij}, $j = 0, 1, \ldots, n_i$, $i = 1, 2$, and assume that some of the bivariate moments

$$\mu_{\alpha\beta} = E\left[\xi_1^{\alpha}\xi_2^{\beta}\right] \tag{4.1}$$

are known, where α and β are nonnegative integers, while the probabilities

$$p_{ij} = P\{\, \xi_1 = z_{1i},\, \xi_2 = z_{2j} \,\}$$

are unknown. Note that the support of the random vector (ξ_1, ξ_2) is part of the set $\{z_{10}, \cdots, z_{1n_1}\} \times \{z_{20}, \cdots, z_{2n_2}\}$ but we do not assume any further knowledge about it. Let furthermore $f(z_1, z_2)$ be a function on the set $\{z_{10}, \cdots, z_{1n_1}\} \times \{z_{20}, \cdots, z_{2n_2}\}$. We intend to give lower and upper bounds for $E[f(\xi_1, \xi_2)]$ under some conditions regarding the moments (4.1) and the function f.

As regards the moments $\mu_{\alpha\beta}$, we consider two cases:

(a) there exist positive integers m_1, and m_2 such that $\mu_{\alpha\beta}$ are known for all α and β satisfying $0 \le \alpha \le m_1$, $0 \le \beta \le m_2$;

(b) there exists a positive integer m such that $\mu_{\alpha\beta}$ are known for all $\alpha \ge 0$, $\beta \ge 0$, $\alpha + \beta \le m$.

The corresponding linear programming problems providing us with the sharp lower and upper bounds for $E[f(\xi_1, \xi_2)]$, are the following (let $f_{ij} = f(z_{1i}, z_{2j})$):

$$\min\,(\max) \sum_{i=0}^{n_1} \sum_{j=0}^{n_2} f_{ij} p_{ij} \tag{4.2}$$

subject to

$$\sum_{i=0}^{n_1} \sum_{j=0}^{n_2} z_{1i}^{\alpha} z_{2j}^{\beta} p_{ij} = \mu_{\alpha\beta}$$

$$0 \le \alpha \le m_1, \qquad 0 \le \beta \le m_2$$

$$p_{ij} \ge 0, \qquad 0 \le i \le n_1, \qquad 0 \le j \le n_2$$

and

$$\min\,(\max) \sum_{i=0}^{n_1} \sum_{j=0}^{n_2} f_{ij} p_{ij} \tag{4.3}$$

subject to

$$\sum_{i=0}^{n_1} \sum_{j=0}^{n_2} z_{1i}^{\alpha} z_{2j}^{\beta} p_{ij} = \mu_{\alpha\beta}$$

$$\alpha \ge 0, \qquad \beta \ge 0, \qquad \alpha + \beta \le m$$

$$p_{ij} \ge 0, \qquad 0 \le i \le n_1, \qquad 0 \le j \le n_2.$$

Regarding the function f, the technique developed in Prékopa (1990b) for the univariate discrete moment problem and partly outlined in the previous sections, allows for handling problem (4.2) in three different cases which are analogous with the cases (1), (2) and (3) mentioned in Section 1. In this paper, however, we restrict ourselves to the two-dimensional version of case (1).

Our condition on f is formulated for the case of problem (4.2). Later, we will use the results concerning problem (4.2), to obtain results for problem (4.3). First we introduce some notations. Let $I_1 \subset \{0, 1, \dots, n_1\}$, $I_2 \subset \{0, 1, \dots, n_2\}$, $|I_1| = m_1 + 1$, $|I_2| = m_2 + 1$ be some subscript sets and

$$L_{I_1}^{(1)}(z_1, z_2) = \sum_{i \in I_1} f(z_{1i}, z_2) L_{I_1 i}^{(1)}(z_1)$$

$$L_{I_2}^{(2)}(z_1, z_2) = \sum_{j \in I_2} f(z_1, z_{2j}) L_{I_2 j}^{(2)}(z_2)$$

$$L_{I_1 I_2}(z_1, z_2) = \sum_{i \in I_1} \sum_{j \in I_2} f(z_{1i}, z_{2j}) L_{I_1 i}^{(1)}(z_1) L_{I_2 j}^{(2)}(z_2)$$

where

$$L_{I_1 i}^{(1)}(z_1) = \prod_{j \in I_1 - \{i\}} \frac{z_1 - z_{1j}}{z_{1i} - z_{1j}},$$

$$L_{I_2 i}^{(2)}(z_2) = \prod_{j \in I_2 - \{i\}} \frac{z_2 - z_{2j}}{z_{2i} - z_{2j}}.$$

We use the order (m_1+1, m_2+1) divided differences of f over $\{z_{10}, \dots, z_{1n_1}\} \times \{z_{20}, \dots, z_{2n_2}\}$ which are defined in a natural way through the subsequent applications of the divided difference operations. This property of f is ensured if it is defined on $[z_{10}, z_{1n_1}] \times [z_{20}, z_{2n_2}]$ and

$$\frac{\partial^{m_1+m_2+2} f(z_1, z_2)}{\partial z_1^{m_1+1} \partial z_2^{m_2+1}} > 0$$

for every interior point of the rectangle (see Popoviciu (1945)).

Conditions on f in problems (4.2)

Let I_1, I_2 be a pair of dual feasible subscript sets, both in the minimization (maximization) problem (1.1), using m_1, and m_2, respectively, instead of m. We assume that at least one of the conditions (i), (ii), (iii), (iv), presented below, is satisfied.

$(i\,a)$ For any fixed $z_2 \in \{z_{20}, \dots, z_{2n_2}\}$, the function of the variable z_1 : $f(z_1, z_2)$, has all positive divided differences of order $m_1 + 1$.

$(i\,b)$ For any fixed $z_1 \in \{z_{10}, \dots, z_{1n_1}\}$, the function of the variable z_2 : $L_{I_1}^{(1)}(z_1, z_2)$, has all positive divided differences of order $m_2 + 1$.

$(ii\,a)$ For any fixed $z_1 \in \{z_{10}, \dots, z_{1n_1}\}$, the function of the variable z_2 : $f(z_1, z_2)$, has all positive divided differences of order $m_2 + 1$.

$(ii\,b)$ For any fixed $z_2 \in \{z_{20}, \dots, z_{2n_2}\}$, the function of the variable z_1 : $L_{I_2}^{(2)}(z_1, z_2)$, has all positive divided differences of order $m_1 + 1$.

(iii) The function $f(z_1, z_2)$ has all positive divided differences of order $(m_1 + 1, m_2 + 1)$ and both $(i\,b)$ and $(ii\,b)$ hold.

(iv) Let $I_1^{(i)}$ and $I_2^{(j)}$ designate the sets of the first $i+1$ and $j+1$ elements in I_1 and I_2, respectively and $[z_{1h}, h \in I_1^{(i)}, z_{2k}, k \in I_2^{(j)}]f$ the divided difference of order (i, j) of the function f, corresponding to the points $z_{1h}, h \in I_1^{(i)}; z_{2k}, k \in I_2^{(j)}$. We assume that the inequality (let f_0 be the function value corresponding to the first elements in I_1 and I_2):

$$f(z_1, z_2) > f_0 + \sum_{\substack{i=0 \\ i+j\geq 1}}^{m_1} \sum_{j=0}^{m_2} \left[z_{1h}, h \in I_1^{(i)}; z_{2k}, k \in I_2^{(j)}\right] f \times \prod_{h \in I_1^{(i)}} (z_1 - z_{1h}) \prod_{k \in I_2^{(j)}} (z_2 - z_{2k})$$

holds for every $(z_1, z_2) \notin \{z_{1i}, i \in I_1\} \times \{z_{2j}, j \in I_2\}$, if I_1, I_2 correspond to minimization problems. If I_1, I_2 correspond to maximization problems then the opposite inequality is assumed to hold. Note that the sum has the same value for any ordering of the points in the sets $\{z_{1i}, i \in I\}$ and $\{z_{2j}, j \in I_2\}$, and is obtained so that we write up the Newton's form of the polynomial $L_{I_1 I_2}(z_1, z_2)$ subsequently for z_1 and z_2.

Introducing the notations (let $\mu_{00} = 1$):

$$A_i = \begin{pmatrix} 1 & 1 & \cdots & 1 \\ z_{i0} & z_{i1} & \cdots & z_{in_i} \\ \vdots & & \ddots & \\ z_{i0}^{m_i} & z_{i1}^{m_i} & \cdots & z_{in_i}^{m_i} \end{pmatrix}, \qquad i = 1, 2,$$

$$A = A_1 \otimes A_2 = \begin{pmatrix} A_1 & A_1 & \cdots & A_1 \\ z_{20}A_1 & z_{21}A_1 & \cdots & z_{2n_2}A_1 \\ \vdots & & \ddots & \\ z_{20}^{m_2}A_1 & z_{21}^{m_2}A_1 & \cdots & z_{2n_2}^{m_2}A_1 \end{pmatrix},$$

$$\begin{aligned} \mathbf{b}^T &= E\left[(1, \xi_1, \dots, \xi_1^{m_1}) \otimes (1, \xi_2, \dots, \xi_2^{m_2})\right] \\ &= (\mu_{00}, \mu_{10}, \dots, \mu_{m_1 0}, \mu_{01}, \mu_{11}, \dots), \\ \mathbf{p}^T &= (p_{ij}, 0 \leq i \leq n_1, 0 \leq j \leq n_2), \\ \mathbf{f}^T &= (f_{ij}, 0 \leq i \leq n_1, 0 \leq j \leq n_2), \end{aligned}$$

we can rewrite problems (4.2) in the concise form:

$$\min\,(\max)\,\mathbf{f}^{\mathrm{T}}\mathbf{p} \tag{4.4}$$

subject to

$$\begin{aligned} A\mathbf{p} &= \mathbf{b}, \\ \mathbf{p} &\geq \mathbf{0}. \end{aligned}$$

5. Bounds for $E[f(\xi_1, \xi_2)]$

Let I_1, I_2 be basis subscript sets from $\{0, 1, \ldots, n_1\}$ and $\{0, 1, \ldots, n_2\}$, respectively. The set $I = I_1 \times I_2$ represents a basis subscript set for problem (4.4) and $B(I) = B_1(I_1) \otimes B_2(I_2)$ is a basis from A. Bases of this type will be called rectangular. The vector $\mathbf{f}_{\mathrm{B}} = \mathbf{f}_{\mathrm{B}_1\mathrm{B}_2}$ designates that part of f which corresponds to the basis vectors in B. We prove

THEOREM 5.1 *If f satisfies one of the conditions (i), (ii), (iii), (iv) and I_1, I_2 correspond to minimization (maximization) problem, then we have the relations*

$$f(z_1, z_2) = L_{\mathrm{I}_1\mathrm{I}_2}(z_1, z_2) \tag{5.1}$$

for $(z_1, z_2) \in \{z_{1i}, i \in I_1\} \times \{z_{2j}, j \in I_2\}$, *and*

$$f(z_1, z_2) > (<)\ L_{\mathrm{I}_1\mathrm{I}_2}(z_1, z_2), \tag{5.2}$$

otherwise. Furthermore, $I_1 \times I_2$ is a dual feasible basis subscript set in the minimization (maximization) problem.

PROOF Let $B_1(I_1)$ and $B_2(I_2)$ designate that $(m_1 + 1) \times (m_1 + 1)$ and $(m_2 + 1) \times (m_2 + 1)$ parts of A_1 and A_2 which correspond to the columns with subscript sets I_1 and I_2, respectively. Then, as it is easy to show, we have the relations:

$$\begin{aligned} &\frac{1}{|B_1(I_1) \otimes B_2(I_2)|} \begin{vmatrix} f(z_1, z_2) & \mathbf{f}^{\mathrm{T}}_{\mathrm{B}_1\mathrm{B}_2} \\ \begin{pmatrix} 1 \\ z_1 \\ \vdots \\ z_1^m \end{pmatrix} \otimes \begin{pmatrix} 1 \\ z_2 \\ \vdots \\ z_2^m \end{pmatrix} & B_1(I_1) \otimes B_2(I_2) \end{vmatrix} \\ &= f(z_1, z_2) - L_{\mathrm{I}_1\mathrm{I}_2}(z_1, z_2) \\ &= f(z_1, z_2) - \Bigg(f_0 + \sum_{\substack{i=0 \\ i+j\geq 1}}^{m_1} \sum_{j=0}^{m_2} \left[z_{1h},\, h \in I_1^{(i)},\, z_{2k},\, k \in I_{2(j)} \right] f \\ &\qquad \prod_{h \in I_1^{(i)}} (z_1 - z_{1h}) \prod_{k \in I_2^{(j)}} (z_2 - z_{2k}) \Bigg). \end{aligned} \tag{5.3}$$

Now the expression in the first line of (5.3) is the reduced price (traditionally designated by $c - z$ in linear programming theory) corresponding to the basis with subscript set $I_1 \times I_2$ in problem (4.2) and the point (z_1, z_2). The dual feasibility in the minimization (maximization) problem means that these values are nonnegative (nonpositive) for every nonbasic (z_1, z_2). We will prove positivity (negativity), i.e., also the dual non-degeneracy of the basis. Note that (5.1) holds trivially.

To prove (5.2) under (i) and (ii) is simple. We only have to repeat the reasoning, applied to the one-dimensional case.

The assertion under (iv) is a consequence of the equality between the first and third lines in (5.3).

The assertion under (iii) is a consequence of the equality (see Popoviciu (1945)):

$$\begin{aligned} f(z_1, z_2) - &\left\{ L_{I_1}^{(1)}(z_1, z_2) + L_{I_2}^{(2)}(z_1, z_2) - L_{I_1 I_2}(z_1, z_2) \right\} \\ &= f(z_1, z_2) - L_{I_1 I_2}(z_1, z_2) - \left\{ L_{I_1}^{(1)}(z_1, z_2) \right. \\ &\quad \left. - L_{I_1 I_2}(z_1, z_2) + L_{I_2}^{(2)}(z_1, z_2) - L_{I_1 I_2}(z_1, z_2) \right\} \\ &= \prod_{i \in I_1} (z_1 - z_{1i}) \prod_{j \in I_2} (z_2 - z_{2j}) [z_{1i}, i \in I_1; z_{2j}, j \in I_2] f \end{aligned} \tag{5.4}$$

and the equality between the first and the second lines in (5.3). We only have to apply (2.8) and the rest of the proof is simple. □

THEOREM 5.2 *Suppose that f satisfies one of the conditions (i), (ii), (iii), and (iv). Then the following assertions hold true:*

(a) *If I_1 and I_2 both have one of the structures:*

$$\{j, j+1, \ldots, k, k+1\}, \qquad \{0, j, j+1, \ldots, k, k+1\},$$

then

$$L_{I_1 I_2}(z_1, z_2) \le f(z_1, z_2),$$

with strict inequality for all nonbasic (z_1, z_2), and

$$E\left[L_{I_1 I_2}(\xi_1, \xi_2)\right] \le E\left[f(\xi_1, \xi_2)\right]. \tag{5.5}$$

This bound is sharp if $B(I_1 \times I_2)$ is a primal feasible basis in problem (4.2).

(b) *If I_1 and I_2 both have one of the structures:*

$$\{j, j+1, \ldots, k, k+1, n\}, \qquad \{0, j, j+1, \ldots, k, k+1, n\},$$

then

$$L_{I_1 I_2}(z_1, z_2) \ge f(z_1, z_2),$$

with strict inequality for all nonbasic (z_1, z_2), and

$$E\left[L_{\mathrm{I}_1\mathrm{I}_2}(\xi_1, \xi_2)\right] \geq E\left[f(\xi_1, \xi_2)\right]. \tag{5.6}$$

This bound is sharp if $B(I_1 \times I_2)$ is a primal feasible basis in problem (4.2).

PROOF The theorem is a consequence of Theorem 4.1 and the fact that a both primal and dual feasible basis is optimal. □

REMARK It is not sure that among the bases of the form $B_1(I_1) \otimes B_2(I_2)$ there is one which is primal feasible. This is ensured if ξ_1 and ξ_2 are independent because in this case $\mu_{\alpha\beta} = E[\xi_1^\alpha]E[\xi_2^\beta]$ and the constraints in problem (4.2) split into two separate sets of constraints, where there are primal feasible bases $B_1(I_1)$ and $B_2(I_2)$. If ξ_1 and ξ_2 are dependent random variables then the sharp inequalities may not be among those in (5.5) and (5.6). Still, we can obtain the sharp inequalities if we use $B_1(I_1) \otimes B_2(I_2)$ as an initial dual feasible basis and apply the dual method for the solution of the problem.

For the case of problem (4.3) the bounds obtained for problem (4.2) can be used in the following manner. First we observe that if both I_1 and I_2 are dual feasible basis subscript sets in the minimization problem, $|I_1| + |I_2| = m + 2$, and f satisfies one of the conditions (i), (ii), (iii), and (iv), with $m_1 = |I_1|$, $m_2 = |I_2|$, then

$$L_{\mathrm{I}_1\mathrm{I}_2}(z_1, z_2) \leq f(z_1, z_2), \tag{5.7}$$

for any (z_1, z_2). Similarly, if I_1, I_2 are dual feasible basis subscript sets in the maximization problem, $|J_1| + |J_2| = m + 2$, and f satisfies one of the conditions (i), (ii), (iii), and (iv), with $m_1 = |J_1|$, $m_2 = |J_2|$,then

$$L_{\mathrm{J}_1\mathrm{J}_2}(z_1, z_2) \geq f(z_1, z_2) \tag{5.8}$$

for any (z_1, z_2). Then we replace (z_1, z_2) by (ξ_1, ξ_2), take expectations in (5.7) and (5.8), and let I_1, I_2, J_1, J_2 vary so that the best bounds for $E[f(\xi_1, \xi_2)]$ are obtained. This result is summarized in

THEOREM 5.3 *We have the inequalities*

$$\begin{aligned}
&\max\{E\left[L_{\mathrm{I}_1\mathrm{I}_2}(\xi_1, \xi_2)\right] \,|\, I_1, I_2 \textit{ dual feasible for the min problem}, \\
&\qquad |I_1| + |I_2| = m + 2\} \leq E\left[f(\xi_1, \xi_2)\right] \\
&\quad \leq \max\{E\left[L_{\mathrm{J}_1\mathrm{J}_2}(\xi_1, \xi_2)\right] \,|\, J_1, J_2 \textit{ dual feasible for the max problem}, \\
&\qquad |J_1| + |J_2| = m + 2\},
\end{aligned}$$

where we assume that the function f satisfies the condition mentioned in Section 4, *for all I_1, I_2 and J_1, J_2, respectively, that are allowed in the above inequalities.*

REMARK When constructing the bounds presented in Theorem 5.3 we may restrict ourselves to some of the rectangular bases with $|I_1|+|I_2|=m+2$ ($|J_1|+|J_2|=m+2$). In this case the bounds become weaker but we impose less condition on f.

We illustrate the above bounds in the case of $m=2$. Problem (4.2) is now the following

$$\text{Minimize } \sum_{i=0}^{n_1}\sum_{j=0}^{n_2} f_{ij}p_{ij} \tag{5.9}$$

subject to

$$\sum_{i=0}^{n_1}\sum_{j=0}^{n_2} p_{ij} = \mu_{00} \tag{5.9a}$$

$$\sum_{i=0}^{n_1}\sum_{j=0}^{n_2} z_{1i}p_{ij} = \mu_{10} \tag{5.9b}$$

$$\sum_{i=0}^{n_1}\sum_{j=0}^{n_2} z_{2j}p_{ij} = \mu_{01} \tag{5.9c}$$

$$\sum_{i=0}^{n_1}\sum_{j=0}^{n_2} z_{1i}^2p_{ij} = \mu_{20} \tag{5.9d}$$

$$\sum_{i=0}^{n_1}\sum_{j=0}^{n_2} z_{2j}^2p_{ij} = \mu_{02} \tag{5.9e}$$

$$\sum_{i=0}^{n_1}\sum_{j=0}^{n_2} z_{1i}z_{2j}p_{ij} = \mu_{11} \tag{5.9f}$$

$$p_{ij}\geq 0,\qquad i=0,1,\ldots,n_1,\quad j=0,1,\ldots,n_2.$$

For a given pair I_1, I_2, with $|I_1|+|I_2|=m+2=4$ we pick a subset of the set of constraints of problem (4.3) so that the matrix of the new constraints is a tensor product of two matrices with sizes $|I_1|\times(n_1+1)$ and $|I_2|\times(n_2+1)$, respectively. There are three possibilities to do this concerning problem (5.9).

The first one is to pick the constraints (5.9a), (5.9b), (5.9d). In this case $|I_1|=3$, $|I_2|=1$.

The second one is to pick the constraints (5.9a), (5.9c), (5.9e). In this case $|I_1| = 1$, $|I_2| = 3$.

The third one is to pick the constraints (5.9a), (5.9b), (5.9c), (5.9f). In this case $|I_1| = 2$, $|I_2| = 2$.

In order to simplify the formulation of the next theorem we introduce the notation:

$$Z_1 = \{z_{10}, \cdots, z_{1n_1}\}, \quad Z_2 = \{z_{20}, \cdots, z_{2n_2}\}, \quad Z = Z_1 \times Z_2, \quad z = (z_1, z_2).$$

THEOREM 5.4 *Suppose that f has positive divided differences of orders* $(1,0)$, $(0,1)$, $(2,0)$, $(0,2)$, $(3,0)$, $(0,3)$, $(2,1)$, $(1,2)$. *Then we have the following bounds on $f(z_1, z_2)$:*

$$\sum_{j \in I_2} f(z_{10}, z_{2j}) L^{(2)}_{I_2 j}(z) \leq f(z_1, z_2), \qquad z \in Z \tag{5.10}$$

for any $I_2 = \{0, l, l+1\}$, $1 \leq l \leq n_2 - 1$, with strict inequality for $z \notin \{z_{10}\} \times \{z_{2j}, j \in I_2\}$;

$$\sum_{i \in I_1} f(z_{1i}, z_{20}) L^{(1)}_{I_1 i}(z_1) \leq f(z_1, z_2), \qquad z \in Z, \tag{5.11}$$

for any $I_1 = \{0, k, k+1\}$, $1 \leq k \leq n_1 - 1$, with strict inequality for $z \notin \{z_{1i}, i \in I_1\} \times \{z_{20}\}$;

$$\sum_{i \in K_1, j \in K_2} f(z_{1i}, z_{2j}) L^{(1)}_{k_1 i}(z_1) L^{(2)}_{K_2 j}(z_2) \leq f(z_1, z_2), \qquad z \in K, \tag{5.12}$$

for any $K_1 = \{r, r+1\}$, $0 \leq r \leq n_1 - 1$, $K_2 = \{s, s+1\}$, $0 \leq s \leq n_2 - 1$, with strict inequality for $z \notin \{z_{1i}, i \in I_1\} \times \{z_{2j}, j \in I_2\}$;

$$\sum_{j \in I_2} f(z_{1n_1}, z_{2j}) L^{(2)}_{I_2 j}(z_2) \geq f(z_1, z_2), \qquad z \in Z, \tag{5.13}$$

for any $I_2 = \{l, l+1, n_2\}$, $0 \leq l \leq n_2 - 2$, with strict inequality for $z \notin \{z_{1n_1}\} \times \{z_{2j}, j \in I_2\}$;

$$\sum_{i \in I_1} f(z_{1i}, z_{2n_2}) L^{(1)}_{I_1 i}(z_1) \geq f(z_1, z_2), \qquad z \in Z, \tag{5.14}$$

for any I_1 of the form $I_1 = \{k, k+1, n_1\}$, $0 \leq k \leq n_1 - 1$, with strict inequality for $z \notin \{z_{1i}, i \in I_1\} \times \{z_{2n_2}\}$;

$$\sum_{i \in K_1, j \in K_2} f(z_{1i}, z_{2j}) L^{(1)}_{K_1 i}(z_1) L^{(2)}_{K_2 j}(z_2) \geq f(z_1, z_2), \qquad z \in Z, \tag{5.15}$$

for $K_1 = \{0, n_1\}$, $K_2 = \{0, n_2\}$, with strict inequality for $z \notin K_1 \times K_2$.

Proof Inequalities (5.10), (5.11), (5.13) and (5.14) are consequences of Theorem 3.1 and the assumption that for any z_{1i} the function $f(z_{1i}, z_2)$ is strictly increasing in z_2 and for any z_{2j} the function $f(z_1, z_{2j})$ is strictly increasing in z_1. We have utilized the assumption that all divided differences of orders $(0,1)$, $(1,0)$, $(0,3)$ and $(3,0)$ are positive.

Inequality (5.15) is a simple consequence of the positivity of the $(2,0)$, $(0,2)$ order divided differences.

To prove (5.12) assume first that $z_1 \notin \{z_{1r}, z_{1r+1}\}$, $z_2 \notin \{z_{2s}, z_{2s+1}\}$. Since all (1, 2) and (2, 1) order divided differences are positive, we derive

$$\begin{aligned}
&\frac{f(z_1,z_2)+f(z_{1r},z_{2s})-f(z_{1r},z_2)-f(z_1,z_{2s})}{(z_1-z_{1r})(z_2-z_{2s})} \\
&\quad > \frac{f(z_{1r+1},z_2)+f(z_{1r},z_{2s})-f(z_{1r},z_2)-f(z_{1r+1},z_{2s})}{(z_{1r+1}-z_{1r})(z_2-z_{2s})} \\
&\quad > \frac{f(z_{1r+1},z_{2s+1})+f(z_{1r},z_{2s})-f(z_{1r},z_{2s+1})-f(z_{1r+1},z_{2s})}{(z_{1r+1}-z_{1r})(z_{2s+1}-z_{2s})}.
\end{aligned} \tag{5.16}$$

On the other hand, the positivity of the (0, 2), (2, 0) order divided differences imply that

$$f(z_{1r},z_2) > f(z_{1r},z_{2s}) + \frac{f(z_{1r},z_{2s+1})-f(z_{1r},z_{2s})}{z_{2s+1}-z_{2s}}(z_2-z_{2s}) \tag{5.17}$$

$$f(z_1,z_{2s}) > f(z_{1r},z_{2s}) + \frac{f(z_{1r+1},z_{2s})-f(z_{1r},z_{2s})}{z_{1r+1}-z_{1r}}(z_1-z_{1r}). \tag{5.18}$$

Picking the inequality that exists between the first and third lines in (5.16) and utilizing (5.17) and (5.18), we obtain:

$$\begin{aligned}
f(z_1,z_2) > {} & f(z_{1r},z_{2s}) + \frac{f(z_{1r},z_{2s+1})-f(z_{1r},z_{2s})}{z_{2s+1}-z_{2s}}(z_2-z_{2s}) \\
& + \frac{f(z_{1r+1},z_{2s})-f(z_{1r},z_{2s})}{z_{1r+1}-z_{1r}}(z_1-z_{1r}) \\
& + \frac{f(z_{1r+1},z_{2s+1})+f(z_{1r},z_{2s})-f(z_{1r},z_{2s+1})}{(z_{1r+1}-z_{1r})(z_{2s+1}-z_{2s})} \\
& - \frac{f(z_{1r+1},z_{2s})}{(z_{1r+1}-z_{1r})(z_{2s+1}-z_{2s})}(z_1-z_{1r})(z_2-z_{2s}).
\end{aligned} \tag{5.19}$$

Inequality (5.19) is the same as (5.12).

Considering the case where either $z_1 \in \{z_{1r}, z_{1r+1}\}$ or $z_2 \in \{z_{2s}, z_{2s+1}\}$ holds, we can easily check the validity of (5.19) in all possible cases. Since (5.19) is the same as (5.12), the proof of the theorem is complete. □

Theorem 5.4 provides us with a tool to establish lower and upper bounds for $E[f(\xi_1,\xi_2)]$. We only have to plug ξ_1 and ξ_2 in the place of z_1 and z_2, respectively, in all inequalities and pick the best lower and upper bounds.

Each expectation inequality that we obtain from (5.10), (5.11), (5.13), (5.14) can be optimized by the use of the algorithm presented in Section 3. The expectation inequality that we obtain from (5.15) is already optimal. There is no easy algorithm, however, to find the best pair K_1, K_2 to optimize the expectation inequality that we obtain from (5.12). We may try out each pair K_1, K_2 and pick that one which provides us with the largest lower bound. Instead of doing this, the following may be suggested.

Starting from any rectangular basis $B_1(K_1) \otimes B_2(K_2)$ as initial dual feasible basis of the $4 \times [(n_1+1)(n_2+1)]$ size linear programming (where we minimize the objective function of problem (5.9) subject to (5.9a), (5.9b), (5.9c), (5.9f) and the nonnegativity restrictions), we carry out the dual method and obtain a (not necessarily rectangular) optimal basis. The corresponding optimum value is the best lower bound on $E[f(\xi_1,\xi_2)]$, using μ_{10}, μ_{01}, μ_{11}.

NUMERICAL EXAMPLE Assume that the possible values of any of the random variables ξ_1, ξ_2 are known to be $0, \cdots, 9$. Assume furthermore that

$$\mu_{10} = 4.8, \qquad \mu_{20} = 31.5$$

$$\mu_{01} = 4.1, \qquad \mu_{02} = 27.5, \qquad \mu_{11} = 19.95$$

and let

$$f(z_1, z_2) = e^{0.005(z_1^2+z_2^2+z_1z_2)}.$$

Considering one subproblem of problem (5.9), let (i,j) represent the column vector consisting of the coefficients of $z_{1i}^{\alpha} z_{2j}^{\beta}$, as components, for the allowed α, β values. Thus, any basis of any subproblem is a collection of subscript pairs (i,j).

The results are summarized in the following tables:

	Optimal basis of the reduced problem	Nonzero elements of the optimal probability distribution	Optimum value: Bound
Lower bound based on μ_{10}, μ_{20}	(0,0), (6,0), (7,0)	$p_{00} = 0.2642857143$ $p_{60} = 0.35$ $p_{70} = 0.3857142857$	1.176108584
Upper bound based on μ_{10}, μ_{20}	(2,9), (3,9), (9,9)	$p_{29} = 0.1285714286$ $p_{39} = 0.55$ $p_{99} = 0.3214285714$	2.285735942
Lower bound based on μ_{01}, μ_{02}	(0,0), (0,6), (0,7)	$p_{00} = 0.3857142857$ $p_{06} = 0.2$ $p_{07} = 0.4142857143$	1.154458017
Upper bound based on μ_{01}, μ_{02}	(1,9), (2,9), (9,9)	$p_{19} = 0.05$ $p_{29} = 0.6428571429$ $p_{99} = 0.3071428571$	2.189880833

Table 1. Optimal lower and upper bounds on $E[f(\xi_1,\xi_2)]$ using only univariate first and second order moments.

		Nonzero elements of the corresponding probability distribution	Bound
Lower bound based on μ_{10}, μ_{01}, μ_{11}	Best rectangular basis (4,4), (4,5) (5,4), (5,5)	$p_{00} = 0.45$ $p_{45} = -0.25$ $p_{54} = 0.45$ $p_{55} = 0.35$	1.352634113
Optimal lower bound using μ_{10}, μ_{01}, μ_{11}	Optimal basis (0,0), (4,4) (5,4), (5,5)	$p_{00} = 0.0125$ $p_{44} = 0.1375$ $p_{54} = 0.7$ $p_{55} = 0.15$	1.355182972
Optimal upper bound using μ_{10}, μ_{01}, μ_{11}	Optimal basis (0,0), (0,9) (9,0), (9,9)	$p_{00} = 0.2574074074$ $p_{09} = 0.2092592593$ $p_{90} = 0.287037037$ $p_{99} = 0.2462962963$	1.831596631

Table 2. Lower and upper bounds on $E[f(\xi_1,\xi_2)]$ using first order moments and the expectation of the product of the random variables. Observe that the best rectangular basis is not primal feasible because $p_{45} < 0$.

Using the largest lower bound and smallest upper bound, we obtain the inequalities:

$$1.355182972 \leq E\left[f(\xi_1, \xi_2)\right] \leq 1.831596631.$$

Figures 1, 2, 3 and 4 serve to illustrate the structures of the dual feasible bases that appear in Tables 1 and 2.

9	o	o	□	□	o	o	o	o	o	□
8	o	o	o	o	o	o	o	o	o	o
7	o	o	o	o	o	o	o	o	o	o
6	o	o	o	o	o	o	o	o	o	o
5	o	o	o	o	o	o	o	o	o	o
4	o	o	o	o	o	o	o	o	o	o
3	o	o	o	o	o	o	o	o	o	o
2	o	o	o	o	o	o	o	o	o	o
1	o	o	o	o	o	o	o	o	o	o
0	•	o	o	o	o	o	•	•	o	o
	0	1	2	3	4	5	6	7	8	9

Figure 1. • (□) means: optimal basis producing lower (upper) bound using μ_{10}, μ_{20}

9	o	o	o	o	o	o	o	o	o	□
8	o	o	o	o	o	o	o	o	o	o
7	•	o	o	o	o	o	o	o	o	o
6	•	o	o	o	o	o	o	o	o	o
5	o	o	o	o	o	o	o	o	o	o
4	o	o	o	o	o	o	o	o	o	o
3	o	o	o	o	o	o	o	o	o	o
2	o	o	o	o	o	o	o	o	o	□
1	o	o	o	o	o	o	o	o	o	□
0	•	o	o	o	o	o	o	o	o	o
	0	1	2	3	4	5	6	7	8	9

Figure 2. • (□) means: optimal basis producing lower (upper) bound using μ_{01}, μ_{02}

	0	1	2	3	4	5	6	7	8	9
9	□	o	o	o	o	o	o	o	o	□
8	o	o	o	o	o	o	o	o	o	o
7	o	o	o	o	o	o	o	o	o	o
6	o	o	o	o	o	o	o	o	o	o
5	o	o	o	o	•	•	o	o	o	o
4	o	o	o	o	•	•	o	o	o	o
3	o	o	o	o	o	o	o	o	o	o
2	o	o	o	o	o	o	o	o	o	o
1	o	o	o	o	o	o	o	o	o	o
0	□	o	o	o	o	o	o	o	o	□

Figure 3. • means: best rectangular basis producing lower bound using μ_{10}, μ_{01}, μ_{11}
□ means: optimal basis producing upper bound using μ_{10}, μ_{01}, μ_{11}

	0	1	2	3	4	5	6	7	8	9
9	o	o	o	o	o	o	o	o	o	o
8	o	o	o	o	o	o	o	o	o	o
7	o	o	o	o	o	o	o	o	o	o
6	o	o	o	o	o	o	o	o	o	o
5	o	o	o	o	o	△	o	o	o	o
4	o	o	o	o	△	△	o	o	o	o
3	o	o	o	o	o	o	o	o	o	o
2	o	o	o	o	o	o	o	o	o	o
1	o	o	o	o	o	o	o	o	o	o
0	△	o	o	o	o	o	o	o	o	o

Figure 4. △ means: optimal basis producing lower bound using μ_{10}, μ_{01}, μ_{11}

References

Boros, E. and Prékopa, A. (1989). Closed form two-sided bounds for probabilities that exactly r and at least r out of n events occur. *Math. Oper. Res.* **14**, 317–342.

Jordan, C. (1965). *Calculus of Finite Differences.* Chelsea, New York.

Kall, P. (1987). Stochastic programs with recourse: An upper bound and the related moment problem. *Zeit. Oper. Res.* **8** 74–85.

KLEIN, HANEVELD, W. (1992). Multilinear approximation on rectangles and the related moment problem. *Math. Oper. Res.* To appear.

KREIN, K. AND NUDELMAN, A. (1977). *The Markov moment problem and extremal problems.* Trans. Math. Mono. **50** American Mathematical Society, Providence, RI.

LEMKE, C.E. (1954). The dual method for solving the linear programming problem. *Naval Res. Logist. Quart.* **1** 36–47.

POPOVICIU, T. (1945). *Les Fonctions Convexes. Actualités Scientifiques et Industrielles* **992** Hermann, Paris.

PRÉKOPA, A. (1988). Boole-Bonferroni inequalities and linear programming. *Oper. Res.* **36** 145–162.

PRÉKOPA, A. (1990a). Sharp bounds on probabilities using linear programming. *Oper. Res.* **38** 227–239.

PRÉKOPA, A. (1990b). The discrete moment problem and linear programming. *Discrete Appl. Math.* **27** 235–254.

SAMUELS, S.M. AND STUDDEN, W. J. (1989). Bonferroni-type probability bounds as an application of the theory of Tchebycheff system. *Probability, Statistics and Mathematics, Papers in Honor of Samual Karlin.* T. W. Anderson, K. B. Athreya and D. L. Iglehart, eds. Academic Press, San Diego, CA. 271–289.

RUTGERS CENTER FOR OPERATIONS RESEARCH
P.O. BOX 5062
NEW BRUNSWICK, NJ 08903–5062

Stochastic Inequalities
IMS Lecture Notes – Monograph Series
Volume 22 (1993)

ON FKG–TYPE AND PERMANENTAL INEQUALITIES

By YOSEF RINOTT[1] and MICHAEL SAKS[2]

University of California, San Diego

In this paper we survey results from Rinott and Saks (1990) on a "2m-function" inequality which generalizes the FKG and associated inequalities. We also present related conjectures and partial results on permanents and sums of permutation matrices. We hope that the motivation given in the first part of the paper, and the subsequent discussion will attract the attention of problem solvers to our conjectures.

1. Introduction

The FKG inequality (Fortuin, Kasteleyn and Ginibre (1971)) has been applied in many fields, including statistical mechanics, combinatorics, reliability theory and stochastic inequalities. In order to state it we need the following notation and definition: for $\mathbf{x} = (x_1, x_2, \ldots, x_k)$ and $\mathbf{y} = (y_1, y_2, \ldots, y_k)$ in $\mathbb{R}^k$, $\mathbf{x} \vee \mathbf{y}$ and $\mathbf{x} \wedge \mathbf{y}$ in $\mathbb{R}^k$ are defined to have coordinates $(\mathbf{x} \vee \mathbf{y})_j = \max(x_j, y_j)$ and $(\mathbf{x} \wedge \mathbf{y})_j = \min(x_j, y_j)$, $j = 1, \ldots, k$.

DEFINITION A σ-finite (nonnegative) measure μ on $\mathbb{R}^k$ is said to be an *FKG measure* if μ has a density function ϕ with respect to some product measure $d\sigma$ on $\mathbb{R}^k$, (that is, $d\sigma(\mathbf{x}) = \Pi_{j=1}^k d\sigma_j(x_j)$, and $d\mu(\mathbf{x}) = \phi(\mathbf{x})d\sigma(\mathbf{x})$), satisfying for all $\mathbf{x}$ and $\mathbf{y}$ in $\mathbb{R}^k$,

$$\phi(\mathbf{x})\phi(\mathbf{y}) \leq \phi(\mathbf{x} \vee \mathbf{y})\phi(\mathbf{x} \wedge \mathbf{y}). \tag{1}$$

Condition (1) is referred to as *multivariate total positivity of order 2* (MTP_2) in Karlin and Rinott (1980). It can be shown that if a positive density ϕ is TP_2 in every pair of variables, then (1) holds, i.e., ϕ is MTP_2. We now state the FKG inequality as follows:

[1]Work supported in part by National Science Foundation Grant DMS-9001274.

[2]Work supported in part by National Science Foundation Grant CCR89-11388.

AMS 1991 ***subject classifications.*** Primary 26D15; Secondary 15A15, 05A20, 60E15, 62H05.

Key words and phrases. Correlation inequalities, stochastic ordering, majorization, permutation matrices.

THEOREM 1.1 *Let* $\mathbf{X}$ *be a random vector in* $\mathbb{R}^k$ *whose distribution is an FKG probability measure. Then for any pair of nondecreasing real valued functions* α *and* β *defined on* $\mathbb{R}^k$, *we have*

$$E\{\alpha(\mathbf{X})\beta(\mathbf{X})\} \geq E\alpha(\mathbf{X}) \cdot E\beta(\mathbf{X}). \tag{2}$$

Note that when (2) holds, $\mathbf{X}$ is said to be *associated.*

Sarkar (1969) discovered that if $\mathbf{X}$ is a random vector having a density which is TP_2 in pairs then it is associated, a result which is very close to the FKG inequality.

Holley (1974) proved the following:

THEOREM 1.2 *Let* f_1, f_2, *be probability densities with respect to some product measure* $\mathbb{R}^k$, *satisfying* $f_1(\mathbf{x})f_2(\mathbf{y}) \leq f_1(\mathbf{x} \vee \mathbf{y})f_2(\mathbf{x} \wedge \mathbf{y})$. *Let* $\mathbf{X}$ *and* $\mathbf{Y}$ *be a random vectors in* $\mathbb{R}^k$ *having distributions with the densities* f_1 *and* f_2 *respectively. Then for any nondecreasing function* α *defined on* $\mathbb{R}^k$

$$E\alpha(\mathbf{X}) \geq E\alpha(\mathbf{Y}).$$

Next came the "4-function" Theorem of Ahlswede and Daykin (1978):

THEOREM 1.3 *Let* f_1, f_2 *and* g_1, g_2 *be nonnegative real valued functions defined on* $\mathbb{R}^k$ *that satisfy the following condition:* $f_1(\mathbf{x})f_2(\mathbf{y}) \leq g_1(\mathbf{x}\vee\mathbf{y})g_2(\mathbf{x}\wedge\mathbf{y})$. *Then, for any FKG measure* μ *on* $\mathbb{R}^k$:

$$\int_{\mathbb{R}^k} f_1(\mathbf{x})d\mu(\mathbf{x}) \int_{\mathbb{R}^k} f_2(\mathbf{x})d\mu(\mathbf{x}) \leq \int_{\mathbb{R}^k} g_1(\mathbf{x})d\mu(\mathbf{x}) \int_{\mathbb{R}^k} g_2(\mathbf{x})d\mu(\mathbf{x}). \tag{3}$$

It is an easy exercise to show that Ahlswede and Daykin's result implies Holley's theorem, which in turn implies the FKG inequality. For details, references and some examples and applications, see, e.g., Karlin and Rinott (1980), Graham (1982).

In the presence of theorems involving a single density (FKG), two densities (Holley), four functions (Ahlswede and Daykin), and further studies by Ahlswede and Daykin (1979) and Daykin (1980), it was natural to look for a more general result. This was done in Rinott and Saks (1990). In order to describe the next result we need the following notation. Given vectors $\mathbf{x}^i = (x_1^i, \ldots, x_k^i) \in \mathbb{R}^k$, $i = 1, \ldots, m$, define $\mathbf{x}^{[l]}$ to be the vector in $\mathbb{R}^k$ whose jth coordinate $(1 \leq j \leq k)$ is the lth largest among $\mathbf{x}_j^i$, $i = 1, \ldots, m$. Formally we have $\mathbf{x}^{[l]} = \bigvee_{S:|S|=l} \bigwedge_{i\in S} \mathbf{x}^i$, $l = 1, \ldots, m$. In particular note that $\mathbf{x}^{[1]} = \vee_{i=1}^m \mathbf{x}^i$, $\mathbf{x}^{[m]} = \wedge_{i=1}^m \mathbf{x}^i$.

We can now quote the main result from Rinott and Saks (1990).

THEOREM 1.4 *Let $f_1,\ldots,f_m$ and $g_1,\ldots,g_m$ be nonnegative real valued functions defined on $\mathbb{R}^k$ satisfying the following condition: for every sequence $\mathbf{x}^1,\ldots,\mathbf{x}^m$ of elements from $\mathbb{R}^k$*

$$f_1(\mathbf{x}^1)f_2(\mathbf{x}^2)\cdots f_m(\mathbf{x}^m) \le g_1(\mathbf{x}^{[1]})g_2(\mathbf{x}^{[2]})\cdots g_m(\mathbf{x}^{[m]}). \tag{4}$$

Then, for any FKG measure μ on $\mathbb{R}^k$:

$$\int_{\mathbb{R}^k} f_1(\mathbf{x})d\mu(\mathbf{x}) \cdots \int_{\mathbb{R}^k} f_m(\mathbf{x})d\mu(\mathbf{x}) \le \int_{\mathbb{R}^k} g_1(\mathbf{x})d\mu(\mathbf{x}) \cdots \int_{\mathbb{R}^k} g_m(\mathbf{x})d\mu(\mathbf{x}). \tag{5}$$

We are not concerned with issues of integrability in this paper and so we always assume that integrals are well-defined.

Note that in the case $m = 2$, condition (4) becomes $f_1(\mathbf{x})f_2(\mathbf{y}) \le g_1(\mathbf{x} \vee \mathbf{y})g_2(\mathbf{x} \wedge \mathbf{y})$. In this case Theorem 1.4 reduces to the "4-function" Theorem of Ahlswede and Daykin (1978).

The starting point of this paper is the one-dimensional version of Theorem 1.4 which we now state, with (hopefully) simpler notation.

PROPOSITION 1.1 *Let $f_1, f_2,\ldots,f_m$ and $g_1,g_2,\ldots,g_m$ be nonnegative real valued functions defined on $\mathbb{R}$ that satisfy the following condition: for every sequence $x_1,x_2,\ldots,x_m$ of elements from $\mathbb{R}$,*

$$f_1(x_1)f_2(x_2)\cdots f_m(x_m) \le g_1(x_1^*)g_2(x_2^*)\cdots g_m(x_m^*), \tag{6}$$

where $(x_1^*,x_2^*,\ldots,x_m^*)$ denotes the decreasing rearrangement of $(x_1,x_2,\ldots,x_m)$. Then, assuming integrability,

$$\int_{\mathbb{R}} f_1(x)d\mu(x)\int_{\mathbb{R}} f_2(x)d\mu(x)\cdots\int_{\mathbb{R}} f_m(x)d\mu(x) \le \int_{\mathbb{R}} g_1(x)d\mu(x)\int_{\mathbb{R}} g_2(x)d\mu(x)\cdots\int_{\mathbb{R}} g_m(x)d\mu(x), \tag{7}$$

for any σ-finite measure μ on $\mathbb{R}$.

Theorem 1.4 can be deduced from Proposition 1.1 by induction arguments which may be of interest. However, they are quite standard and will not be reproduced here. They involve showing that the "marginal" functions defined by $p_i(\tilde{\mathbf{x}}) = \int_{\mathbb{R}} f_i(\tilde{\mathbf{x}},x)d\mu_k(x)$ and $q_i(\tilde{\mathbf{x}}) = \int_{\mathbb{R}} g_i(\tilde{\mathbf{x}},x)d\mu_k(x)$ satisfy the hypothesis of Theorem 1.4 as functions of $\tilde{\mathbf{x}} \in \mathbb{R}^{k-1}$. For details see Rinott and Saks (1990). The important part of the proof is the basis case $n = 1$, which is the content of Proposition 1.1.

In order to see the connection to permanents let us consider Proposition 1.1 in the simple case that $m = 2$, and $d\mu(x) = dx$. It says that if for all $x, y \in \mathbb{R}$

$$f_1(x)f_2(y) \le g_1(x \vee y)g_2(x \wedge y),$$

then, assuming integrability,

$$\int_{\mathbb{R}} f_1(x)dx \int_{\mathbb{R}} f_2(x)dx \le \int_{\mathbb{R}} g_1(x)dx \int_{\mathbb{R}} g_2(x)dx\,. \tag{8}$$

Starting on the l.h.s. of (8) we have

$$\begin{aligned}\int_{\mathbb{R}} f_1(x)dx \int_{\mathbb{R}} f_2(x)dx &= \int\int f_1(x)f_2(y)dxdy \\ &= \int\int_{x<y} \{f_1(x)f_2(y) + f_1(y)f_2(x)\}dxdy \\ &= \int\int \mathbf{Per}\begin{pmatrix} f_1(x)\, f_2(x) \\ f_1(y)\, f_2(y)\end{pmatrix} dxdy\,. \end{aligned} \tag{9}$$

In the case that the integrals are taken with respect to a discrete measure $d\mu$, diagonal terms must also be considered, but as we shall see they do not pose any difficulty. Permanental inequalities enter the story because it is now clear that the inequality

$$\mathbf{Per}\begin{pmatrix} f_1(x)\, f_2(x) \\ f_1(y)\, f_2(y)\end{pmatrix} \le \mathbf{Per}\begin{pmatrix} g_1(x)\, g_2(x) \\ g_1(y)\, g_2(y)\end{pmatrix},$$

would imply (8). In the same way, we shall see that in order to prove (7) it would suffice to show that

$$\mathbf{Per}\begin{pmatrix} f_1(x_1) & f_2(x_1) & \dots & f_m(x_1) \\ f_1(x_2) & f_2(x_2) & \dots & f_m(x_2) \\ \vdots & & & \vdots \\ f_1(x_m) & f_2(x_m) & \dots & f_m(x_m) \end{pmatrix} \le$$

$$\mathbf{Per}\begin{pmatrix} g_1(x_1) & g_2(x_1) & \dots & g_m(x_1) \\ g_1(x_2) & g_2(x_2) & \dots & g_m(x_2) \\ \vdots & & & \vdots \\ g_1(x_m) & g_2(x_m) & \dots & g_m(x_m) \end{pmatrix}. \tag{10}$$

The latter inequality and related results and conjectures are the subject of the next section.

2. Permanents: Results and Conjectures

First, observe that the permanent on the l.h.s. of (10) is equal to $\sum_{\pi \in S_m} \prod_{i=1}^m f_i(x_{\pi(i)})$. At the end of the Introduction we said that (10) would imply (7); we now restate this implication formally as

LEMMA 2.1 *Let f_i and g_i be real valued functions defined on $\mathbb{R}$, satisfying*

$$(11) \qquad \sum_{\pi \in S_m} \prod_{i=1}^{m} f_i(x_{\pi(i)}) \leq \sum_{\pi \in S_m} \prod_{i=1}^{m} g_i(x_{\pi(i)})$$

for any x_i in $\mathbb{R}$, $i = 1, \ldots, m$. Then for any σ-finite measure μ on $\mathbb{R}$ we have (assuming integrability)

$$\prod_{i=1}^{m} \int_{\mathbb{R}} f_i(x) d\mu(x) \leq \prod_{i=1}^{m} \int_{\mathbb{R}} g_i(x) d\mu(x).$$

PROOF Simply take an m-fold integral with respect to the measure $\prod_{i=1}^{m} d\mu(x_i)$ on both sides of (11), to obtain $m! \prod_{i=1}^{m} \int_{\mathbb{R}} f_i(x) d\mu(x) \leq m! \prod_{i=1}^{m} \int_{\mathbb{R}} g_i(x) d\mu(x)$.

Lemma 2.1 and the above discussion show that in order to prove Proposition 1.1, it would suffice to prove

CONJECTURE 2.1 *Let $f_1, f_2, \ldots, f_m$ and $g_1, g_2, \ldots, g_m$ be nonnegative real valued functions defined on $\mathbb{R}$ that satisfy the following condition: for every sequence $x_1, x_2, \ldots, x_m$ of elements from $\mathbb{R}$,*

$$(12) \qquad f_1(x_1) f_2(x_2) \cdots f_m(x_m) \leq g_1(x_1^*) g_2(x_2^*) \cdots g_m(x_m^*),$$

where $(x_1^*, x_2^*, \ldots, x_m^*)$ denotes the decreasing rearrangement of $(x_1, x_2, \ldots, x_m)$. Then for any sequence $x_1, x_2, \ldots, x_m$,

$$(13) \qquad \begin{aligned} \mathbf{Per}\|f_j(x_i)\| &= \sum_{\pi \in S_m} \prod_{i=1}^{m} f_i(x_{\pi(i)}) \\ &\leq \sum_{\pi \in S_m} \prod_{i=1}^{m} g_i(x_{\pi(i)}) = \mathbf{Per}\|g_j(x_i)\| \end{aligned}$$

Conjecture 2.1 can be given a more appealing matrix formulation which we now describe. Defining $A_{i,j} = f_j(x_i)$ and $B_{i,j} = g_j(x_i)$, $i, j = 1, 2, \ldots, m$, we have

$$\mathbf{Per}(A) = \sum_{\pi \in S_m} \prod_{i=1}^{m} f_i(x_{\pi(i)})$$

and

$$\mathbf{Per}(B) = \sum_{\pi \in S_m} \prod_{i=1}^{m} g_i(x_{\pi(i)}).$$

Thus (13) is equivalent to $\mathbf{Per}(A) \leq \mathbf{Per}(B)$, where $x_1, x_2, \ldots, x_m$ is an arbitrary sequence of real numbers. By the invariance of permanents under row permutations we may, without loss of generality, assume that $x_1 \geq x_2 \geq$

$\ldots \geq x_m$. Condition (12) applied to the sequence $x_{\pi(1)}, x_{\pi(2)}, \ldots, x_{\pi(m)}$ for any permutation $\pi \in S_m$, becomes

$$\begin{aligned} A_{\pi(1),1}A_{\pi(2),2}\cdots A_{\pi(m),m} &= f_1(x_{\pi(1)})f_2(x_{\pi(2)})\cdots f_m(x_{\pi(m)}) \\ &\leq g_1(x_1)g_2(x_2)\cdots g_m(x_m) = B_{1,1}B_{2,2}\cdots B_{m,m}. \end{aligned} \tag{14}$$

Note that in Conjecture 2.1, (12) is assumed also when $(x_1, x_2, \ldots, x_m)$ is replaced by $(x_{i_1}, x_{i_2}, \ldots, x_{i_m})$ for any nondecreasing sequence $1 \leq i_1 \leq i_2 \leq \ldots \leq i_m \leq m$ of integers. For this choice, which allows equalities among the x_i's, condition (12) becomes

$$\begin{aligned} A_{i_{\pi(1)},1}A_{i_{\pi(2)},2}\cdots A_{i_{\pi(m)},m} &= f_1(x_{i_{\pi(1)}})f_2(x_{i_{\pi(2)}})\cdots f_m(x_{i_{\pi(m)}}) \\ &\leq g_1(x_{i_1})g_2(x_{i_2})\cdots g_m(x_{i_m}) = B_{i_1,1}B_{i_2,2}\cdots B_{i_m,m}. \end{aligned} \tag{15}$$

This leads to

DEFINITION Let A and B be $m \times m$ nonnegative matrices. We say that the relation

$$A \ll B \tag{16}$$

holds if for any nondecreasing sequence $1 \leq i_1 \leq i_2 \leq \ldots \leq i_m \leq m$ of integers and any permuation π of $\{1, 2, \ldots, m\}$:

$$A_{i_{\pi(1)},1}A_{i_{\pi(2)},2}\cdots A_{i_{\pi(m)},m} \leq B_{i_1,1}B_{i_2,2}\cdots B_{i_m,m}\,. \tag{17}$$

Note that in (17) we allow equalities between the i_j's. It is easily seen that the relation $\ll$ is transitive and reflexive, and thus defines a quasi-order on the set of $m \times m$ matrices. Conjecture 2.1 reduces to

CONJECTURE 2.2 *Let A and B be $m \times m$ nonnegative matrices. If $A \ll B$, then*

$$\mathbf{Per}(A) \leq \mathbf{Per}(B)\,.$$

If the support of $d\mu$ in Proposition 1.1 consists of only two points, say $\{0, 1\}$, then the matrix $A_{i,j} = f_j(x_i)$ defined above has only two distinct rows: $(f_1(1), \ldots, f_m(1))$, and $(f_1(0), \ldots, f_m(0))$. In this case we can prove Conjecture 2.2. This is the content of the next lemma:

LEMMA 2.2 *Conjecture 2.2 holds for the case that A is an $m \times m$ nonnegative matrix such that for some r between 1 and m, A consists of r identical rows followed by $m - r$ identical rows, and B has the same structure.*

We briefly discuss Lemma 2.2 before proving it. It turns out that Lemma 2.2 leads to a *complete* proof of Theorem 1.4 and Proposition 1.1. In fact, as we saw, Lemma 2.2 suffices to prove Proposition 1.1 in the case that the measure $d\mu$ has support on $\{0, 1\}$. This is the basis for an induction argument (to which we briefly referred in the Introduction) which proves Theorem 1.4 when the support of $d\mu$ is $\{0,1\}^n$, for any n. In order to prove Proposition 1.1 for the case that $d\mu$ has a finite support of cardinality s, say, we embed this support in $\{0,1\}^{s-1}$ and apply the previous result. Thus, even Proposition 1.1, which is one-dimensional, requires the multivariate result and the induction argument. For $d\mu$ of infinite support a further approximation argument is required. The details of these arguments are given in Rinott and Saks (1990).

Recall that Proposition 1.1 would follow directly from either Conjecture 2.1, or Conjecture 2.2, without any need for induction, embedding, and approximation. The rest of this paper concerns Conjecture 2.2. We hope that the above discussion and the partial results and variations we present next, will generate interest in our conjectures.

For the proof of Lemma 2.2 we need a simple majorization-type lemma.

LEMMA 2.3 *Let α_l and β_l be nonnegative numbers, $l = 1,\dots,n$. Assume that for any $V \subseteq \{1,\dots,n\}$ there exists a set $W \subseteq \{1,\dots,n\}$ with $|V| = |W|$ and $\prod_{l\in V}\alpha_l \le \prod_{l\in W}\beta_l$. Then $\sum_{l=1}^n \alpha_l \le \sum_{l=1}^n \beta_l$.*

PROOF This follows readily from Theorem 3.C.1.b. in Marshall and Olkin (1979) and the fact that the conditions of the Lemma are equivalent to $(\log(\alpha_1),\dots,\log(\alpha_n))$ being weakly majorized by $(\log(\beta_1),\dots,\log(\beta_n))$.

PROOF OF LEMMA 2.2 For $\pi \in S_m$ define $\alpha_\pi = \Pi_{i=1}^m A_{\pi(i),i}$, $\beta_\pi = \Pi_{i=1}^m B_{\pi(i),i}$. Then, $\mathbf{Per}(A) \le \mathbf{Per}(B)$ is equivalent to $\sum_\pi \alpha_\pi \le \sum_\pi \beta_\pi$. By Lemma 2.3 it suffices to show that for any $V \subseteq S_m$ there exists a set $W \subseteq S_m$ with $|V| = |W|$ and

$$\prod_{\pi\in V}\alpha_\pi \le \prod_{\pi\in W}\beta_\pi . \tag{18}$$

Recall that A has (at most) two distinct rows, $(s_1,s_2,\dots,s_m)$ and $(t_1,t_2,\dots,t_m)$, say. Then $\alpha_\pi = \prod_{j=1}^m s_j^{a_j}t_j^{b_j}$, where a_j and b_j are either 0 or 1 and $a_j + b_j = 1$. Therefore

$$\prod_{\pi\in V}\alpha_\pi = \prod_{j=1}^m s_j^{k_j}t_j^{l_j} \tag{19}$$

for some nonnegative integers satisfying $k_j + l_j = |V|, j = 1,\dots,m$. Denote the two distinct rows of B by $\mathbf{u} = (u_1,u_2,\dots,u_m)$ and $\mathbf{v} = (v_1,v_2,\dots,v_m)$, with $\mathbf{u}$ preceding $\mathbf{v}$. Let $\eta \in S_m$ be such that $k_{\eta(1)} \ge k_{\eta(2)} \ge \dots \ge k_{\eta(m)}$. We can rewrite the r.h.s. of (19) as $s_{\eta(1)}^{k_{\eta(1)}}\cdots s_{\eta(m)}^{k_{\eta(m)}}t_{\eta(1)}^{l_{\eta(1)}}\cdots t_{\eta(m)}^{l_{\eta(m)}}$. Now observe that the latter expression can be written as a product of terms of the form $s_{\eta(1)}\cdots s_{\eta(r)}t_{\eta(r+1)}\cdots t_{\eta(m)}$. Condition (17) applied to the present case becomes $s_{\eta(1)}\cdots s_{\eta(r)}t_{\eta(r+1)}\cdots t_{\eta(m)} \le u_1\cdots u_r v_{r+1}\cdots v_m$. We thus obtain

$$\prod_{j=1}^{m} s_j^{k_j} t_j^{l_j} \le \prod_{j=1}^{m} u_j^{k_{\eta(j)}} v_j^{l_{\eta(j)}} . \tag{20}$$

Finally, note that the quantity on the r.h.s. of (20) equals $\prod_{\pi\in W}\beta_\pi$ for the coset $W = V\eta = \{\pi\eta : \pi \in V\}$. Thus (18) is established and the proof is complete.

In view of Lemma 2.3 and the subsequent discussion (see (18)), the following conjecture would imply Conjecture 2.2.

CONJECTURE 2.3 *Let A and B be $m \times m$ nonnegative matrices satisfying $A \ll B$. Then for every subset V of the permutation group S_m there exists a subset W of S_m with $|V| = |W|$ and*

$$\prod_{\pi\in V}\alpha_\pi \le \prod_{\pi\in W}\beta_\pi , \tag{21}$$

where $\alpha_\pi = \Pi_{i=1}^{m} A_{\pi(i),i}$ and $\beta_\pi = \Pi_{i=1}^{m} B_{\pi(i),i}$.

The case $m = 2$ of Conjecture 2.3 is already covered in the proof of Lemma 2.2. However let us verify it directly. As usual, for distinct indices $1 \le i_1, i_2, \ldots, i_m \le m$, let $(i_1, i_2, \ldots, i_m)$ denote the permutation π with $\pi(j) = i_j, j = 1, \ldots, m$. In the case $m = 2$ of Conjecture 2.3 we have to consider three possible sets V: $\{(1,2)\}, \{(2,1)\}$ and $\{(1,2),(2,1)\}$. It is very easy to see that for the first two, Conjecture 2.3 holds with $W = \{(1,2)\}$. For the last take $W = V$ and use the fact that for any i, $\prod_{j=1}^{m} A_{i,j} \le \prod_{j=1}^{m} B_{i,j}$, so $\prod_{\pi\in S_2}\alpha_\pi = A_{1,1}A_{2,2}A_{2,1}A_{1,2} = (A_{1,1}A_{1,2})(A_{2,1}A_{2,2}) \le (B_{1,1}B_{1,2})(B_{2,1}B_{2,2}) = \prod_{\pi\in S_2}\beta_\pi$, where the inequality is obtained by applying (17) twice. We note that for $m > 2$, we have examples showing that the set W in Conjecture 2.3 need not be unique.

In order to discuss Conjecture 2.3 for $m > 2$, we need some notation. A vector $\mathbf{s} = (s_1, \ldots, s_m) \in \mathbb{R}^m$ is an *m-sequence* if all of its entries are integers in the range 1 to m. We use lower case Greek letters to denote m sequences whose entries are distinct, i.e., that represent permutations. Each m-sequence $\mathbf{s}$ is associated to a matrix $\Psi_{\mathbf{s}}$ which has a 1 in position (s_i, i) for each $i = 1, \ldots, m$, and all other entries are 0. In particular, for a permutation π, Ψ_π is its associated permutation matrix. If T is a set of m-sequences, we set $\Psi_T = \sum_{\mathbf{s}\in T}\Psi_{\mathbf{s}}$.

Let $C_{m,k}$ denote the class of $m \times m$ nonnegative integer matrices with all row and column sums equal to k. It can be shown that this class consists of all matrices that can be written as a sum of k $m \times m$ permutation matrices (which need not be *distinct*). This is proved in the same way as the well known theorem of Birkhoff and Von Neumann about doubly stochastic matrices, see, e.g., Marshall and Olkin (1979, p. 36 2.F.1).

We denote by $\mathcal{U}_{m,k}$ the subset of $\mathcal{C}_{m,k}$ consisting of matrices which can be expressed as the sum of k *distinct* permutation matrices. Note that $\mathcal{U}_{m,k} = \{\Psi_V : V \subseteq S_m, |V| = k\}$.

For $P \in \mathcal{C}_{m,k}$, set $A^P = \prod_{i,j} A_{i,j}^{P_{i,j}}$. Note that for a subset V of S_m, we have

$$A^{\Psi_V} = \prod_{\pi \in V} \alpha_\pi . \tag{22}$$

We can now rewrite Conjecture 2.3 as follows:

CONJECTURE 2.4 *Let A and B be $m \times m$ nonnegative matrices satisfying $A \ll B$. Then for every subset V of S_m there exists a subset W of S_m, with $|V| = |W|$, such that*

$$A^{\Psi_V} \leq B^{\Psi_W}, \tag{23}$$

Given an m-sequence $\mathbf{s}$, let $\mathbf{s}_*$ denote its *increasing* (more precisely, non-decreasing) rearrangement. The set of defining conditions for $A \ll B$ is easily seen to be equivalent to the set of conditions: $A^{\Psi_\mathbf{s}} \leq B^{\Psi_{\mathbf{s}_*}}$ for all m-sequences $\mathbf{s}$. Furthermore, it is clear that if $\{(P_i, Q_i)\}$ is any indexed family of matrix pairs for which $A^{P_i} \leq B^{Q_i}$ and α_i are nonnegative constants, then $A^{\Sigma_i \alpha_i P_i} \leq B^{\Sigma_i \alpha_i Q_i}$. This suggests the following

DEFINITION Let $\mathcal{P}$ denote the convex cone (in $\mathbb{R}^{m^2} \times \mathbb{R}^{m^2}$) consisting of all pairs of matrices of the form $(\sum \alpha_\mathbf{s} \Psi_\mathbf{s}, \sum \alpha_\mathbf{s} \Psi_{\mathbf{s}_*})$ where the sum extends over all m-sequences $\mathbf{s}$, and all $\alpha_\mathbf{s}$ are nonnegative.

Our next goal is to show that Conjectures 2.3-2.4 can be recast in terms of the cone $\mathcal{P}$ without involving the relation $\ll$. For this purpose we need the following straightforward Proposition which provides an alternative characterization of the relation $\ll$.

PROPOSITION 2.1 *$A \ll B$ if and only if $A^P \leq B^Q$ for all (P, Q) is in $\mathcal{P}$.*

REMARK If (P, Q) is a pair of matrices such that $A^P \leq B^Q$ holds whenever A and B are matrices satisfying $A \ll B$, then (P, Q) must belong to $\mathcal{P}$. To see this, note that $A \ll B$ if and only if the $2m^2$–dimensional vector $(-\log A, \log B)$ has nonnegative dot-product with every vector in $\mathcal{P}$. Now apply a separating hyperplane argument (e.g., Farkas' Lemma, see Papadimitriou and Steiglitz (1982, p. 74)): If C is a cone and D is the set of vectors whose dot-product with each vector in C is non-negative, then any vector $\mathbf{v}$ whose dot-product with every vector in D is non-negative belongs to C.

In view of Proposition 2.1 and the remark following it, verifying Conjecture 2.4 is equivalent to proving that for any $P = \Psi_V$ in $\mathcal{U}_{m,k}$ there exists a matrix $Q = \Psi_W$ in $\mathcal{U}_{m,k}$ such that $(P, Q) \in \mathcal{P}$. We thus avoid the condition $\ll$ and reduce Conjectures 2.3-2.4 to a conjecture on sums of permutation matrices:

CONJECTURE 2.5 *For every matrix P in $\mathcal{U}_{m,k}$ there is a matrix Q in $\mathcal{U}_{m,k}$ such that $(P,Q) \in \mathcal{P}$.*

Clearly, this last conjecture would imply all the previous ones.

In order to prove Conjecture 2.5 it would be sufficient to show:

(*) any matrix P in $\mathcal{U}_{m,k}$ has a representation $P = \sum_{l=1}^{k} \Psi_{\mathbf{s}^{(l)}}$, where $\mathbf{s}^{(l)}$ are m-sequences, such that the matrix $Q = \sum_{l=1}^{v} \Psi_{\mathbf{s}_*^{(l)}}$ is also in $\mathcal{U}_{m,k}$.

It is easy to see that any P in $\mathcal{C}_{m,k}$ can be represented as $P = \sum_{l=1}^{k} \Psi_{\mathbf{s}^{(l)}}$. Generally the representation is not unique. However we do not have a way of constructing a representation that guarantees that if P is in $\mathcal{U}_{m,k}$, then so is $Q = \sum_{l=1}^{v} \Psi_{\mathbf{s}_*^{(l)}}$. Thus, we cannot prove Conjecture 2.5 in general. For $m = 2$ and $m = 3$ we can verify (*) by exhausting all cases of $P \in \mathcal{U}_{m,k}$, and constructing for each P a suitable matrix Q. Since $\cup_k \mathcal{U}_{m,k} = \{\Psi_V : V \subseteq S_m\}$, an exhaustive search of $\mathcal{U}_{m,k}$ for all k requires checking $2^{m!}$ subsets V of S_m. Such a search is not hard for $m = 3$, but becomes very time consuming for larger values of m.

REMARK It is not hard to show that any matrix P in $\mathcal{C}_{m,k}$ has a (unique) representation of the form $P = \sum_{l=1}^{k} \Psi_{\mathbf{s}^{(l)}}$ where $\mathbf{s}^{(l)}$ are m-sequences satisfying $\mathbf{s}^{(1)} \leq \mathbf{s}^{(2)} \leq \ldots \leq \mathbf{s}^{(k)}$. This representation appeared to us to be a natural candidate for satisfying (*). However, we have examples showing that the resulting Q matrix need not be in $\mathcal{U}_{m,k}$,

REMARK Results similar to ours were obtained independently by R. Aharoni and U. Keich about a year after we obtained our results. We benefited from discussions of the subject with them.

REFERENCES

AHARONI, R. AND KEICH, U. (1991). A generalization of the Ahlswede–Daykin inequality. Preprint.

AHLSWEDE, I. R. AND DAYKIN, D. E. (1978). An inequality for the weights of two families of sets, their unions and intersections. *Z. W. Gebiete* **43** 183-185.

AHLSWEDE, I. R. AND DAYKIN, D. E. (1979). Inequalities for a Pair of Maps $S \times S \to S$ with S a finite set. *Math. Z.* **165** 267-289.

DAYKIN, D. E. (1980). A hierarchy of inequalities. *Stud. Appl. Math.* **63** 263-264.

FORTUIN, C. M., KASTELEYN, P. W. AND GINIBRE, J. (1971). Correlation inequalities on some partially ordered sets. *Comm. Math. Phys.* **22** 89-103.

GRAHAM, R. L. (1982). Linear extensions of partial orders and the FKG inequality. In *Ordered Sets*, I. Rival ed., D. Reidel Pub. Co. Dordrecht, Holland 213-236.

Holley, R. (1974). Remarks on the FKG inequalities. *Comm. Math. Phys.* **36** 227-231.

Karlin, S. and Rinott, Y. (1980). Classes of orderings of measures and related correlation inequalities. I. Multivariate totally positive distributions. *J. Multivariate Anal.* **10** 467-498.

Marshall, A. W. and Olkin, I. (1979). *Inequalities: Theory of Majorization and its Applications*, Academic Press, New York.

Minc, H. (1978). *Permanents.* Addison-Wesley, Reading, MA.

Papadimitriou, C.H. and Steiglitz, K. (1982). *Combinatorial Optimization.* Prentice Hall, Englewood Cliffs, NJ.

Sarkar, T. P. (1969). Some lower bounds of reliability. Tech. Report No. **124**, Department of Operations Research, Stanford University.

Rinott, Y. and Saks, M. (1990). Correlation inequalities and a conjecture for permanents. To appear in *Combinatorica.*

Rinott, Y. and Saks, M. (1991). Some conjectures on permanents and sums of permutation matrices. Preprint.

Department of Mathematics
University of California, San Diego
La Jolla, CA 92093

Department of Computer Science and Engineering
University of California, San Diego
La Jolla, CA 92093

Stochastic Inequalities
IMS Lecture Notes - Monograph Series
Volume 22 (1993)

OPTIMAL STOPPING VALUES AND PROPHET INEQUALITIES FOR SOME DEPENDENT RANDOM VARIABLES

By YOSEF RINOTT[1] and ESTER SAMUEL-CAHN

University of California, San Diego and Hebrew University

This paper concerns results on comparisons of stopping values, and prophet inequalities for dependent random variables. We describe general results for negatively dependent random variables, and some examples for the case of positive dependence.

1. Introduction

Let $\mathbf{Z} = (Z_1, \dots, Z_n)$ be a finite sequence of random variables, having a known distribution, and such that $E|Z_i| < \infty$. As usual, a random variable t taking values in $\{1, 2, \dots\}$ is said to be a *stopping rule* for $\mathbf{Z}$ if the event $\{t = i\}$ is determined by $Z_1, \dots, Z_i$, $i = 1, 2, \dots$, and $P(t \leq n) = 1$. (Infinite sequences and unbounded stopping rules have been studied by the methods described below, with minor technical modifications. For simplicity we consider only finite sequences in this paper.) The *optimal stopping value* corresponding to $\mathbf{Z}$ is defined by $V(\mathbf{Z}) = \sup_t EZ_t$, where the supremum is taken over all stopping rules for $\mathbf{Z}$. $V(\mathbf{Z})$ can be regarded as the best expected value attainable by a statistician who is restricted to stopping on the basis of observations which have already been taken. On the other hand, if one could decide when to stop on the basis of complete information about the whole sequence, including future observations, the relevant value would be EZ^*, where $Z^* = \max(Z_1, \dots, Z_n)$. The quantity EZ^* is thus the value for a *prophet* who can foresee future observations. Clearly $V(\mathbf{Z}) \leq EZ^*$. Inequalities of the type

$$EZ^* \leq cV(\mathbf{Z}), \tag{1}$$

for $\mathbf{Z}$ in some collection of finite sequences, with constant c depending only on this subclass, are called *ratio prophet inequalities.* For a recent survey on such inequalities, with history and bibliography, see Hill and Kertz (1992).

We shall be interested mainly in two problems:

---

[1]Work supported in part by National Science Foundation Grant DMS-9001274.

AMS 1991 *subject classifications.* Primary 60G40; secondary 60E15, 62H05.

Key words and phrases. Negative dependence, positive dependence, random replacement schemes.

1. Determine sequences $\mathbf{X}$ and $\mathbf{Y}$ of dependent random variables for which the optimal stopping values comparison

$$V(\mathbf{X}) \leq V(\mathbf{Y}) \tag{2}$$

   is valid.

2. Obtain prophet inequalities for collections of dependent sequences.

Qualitatively, if the Y's tend to be larger than the X's then one may expect (2) to hold. However, this is not obvious in the presence of dependence, where the possibility of prediction of future values aids in obtaining a high optimal stopping value for the statistician. Thus, $\mathbf{X} \leq_{st} \mathbf{Y}$ (meaning $Eh(\mathbf{X}) \leq Eh(\mathbf{Y})$ for any nondecreasing function h defined on $\mathbf{R}^n$) does not necessarily imply $V(\mathbf{X}) \leq V(\mathbf{Y})$. For example consider

$$(X_1, X_2) = \begin{cases} (2,10) & \text{w.p. } 1/2 \\ (0,-10) & \text{w.p. } 1/2 \end{cases} \tag{3}$$

and (Y_1, Y_2) independent with $P(Y_1 = 2) = 1$, $P(Y_2 = 10) = P(Y_2 = -10) = 1/2$. Then $\mathbf{X} \leq_{st} \mathbf{Y}$, but $V(\mathbf{X}) = (1/2)\cdot 10 + (1/2)\cdot 0 = 5$, whereas $V(\mathbf{Y}) = 2$. While dependence may work to increase the value through prediction, it also affects the value (both for the statistician and the prophet) directly. For example, for the prophet value, it is well known that EZ^* for independent Z's would be smaller than EZ^* for the same marginal Z's satisfying suitable negative dependence conditions. As we shall show, this also applies to the optimal stopping value. Thus it is natural to expect (2) to hold, for example, when the Y's are in some sense more negatively dependent than the X's.

A good portion of this paper contains a survey and reorganization of previous work of the authors on value comparisons and prophet inequalities for dependent random variables. In the next section we shall bring results from Rinott and Samuel-Cahn (1987) on value comparisons for negatively dependent random variables. In Section 3 we discuss examples of such comparisons under positive dependence. It should be clear from the above discussion that in this case one should anticipate difficulties, because, while the dependence tends to increase the value through prediction, the positive nature of the dependence works to decrease the prophet's value or the statistician's optimal stopping value. Section 5 concerns random replacement schemes. We discuss some results and a conjecture which appeared in Rinott and Samuel-Cahn (1991) and some further partial results on the conjecture. Finally, in Section 5, we reorganize and unify results from our aforementioned two papers, on prophet inequalities for certain classes of dependent random variables. Prophet inequalities for other classes are given in Hill and Kertz (1992).

2. Value Comparisons Under Negative Dependence

DEFINITION The random variables $Z_1, \ldots, Z_n$ are said to be *Negatively lower orthant dependent in sequence* (NLODS) if

$$(4) \qquad P(Z_i < a_i | Z_1 < a_1, \ldots, Z_{i-1} < a_{i-1}) \leq P(Z_i < a_i)$$

for $i = 2, 3, \ldots, n$, and all constants $a_1, \ldots, a_n$ for which the conditional probability in (4) is defined.

It is easy to see that condition (4) is weaker than most of the well known conditions of negative dependence. Thus if $Z_1, \ldots, Z_n$ are *Negatively associated* (NA), i.e., $\mathrm{cov}\{f_1(Z_i, i \in A_1), f_2(Z_j, j \in A_2)\} \leq 0$, for any pair of *disjoint* subsets A_1, A_2 of $\{1, \ldots, n\}$ and any nondecreasing functions f_1, f_2, then they are also NLODS. Likewise, if $Z_1, \ldots, Z_n$ are *Negatively dependent in sequence* (NDS), meaning that $Z_1, \ldots, Z_{i-1} | Z_i = a_i$ is decreasing stochastically in a_i, or if $Z_1, \ldots, Z_n$ are *Conditionally decreasing in sequence* (CDS), i.e., $Z_i | Z_1 = a_1, \ldots, Z_{i-1} = a_{i-1}$ is decreasing stochastically in $a_1, \ldots, a_{i-1}$, for $i = 2, 3, \ldots, n$, then they are NLODS. On the other hand it is easy to see that (4) implies *Negative lower orthant dependence*, that is, $P(Z_1 < a_1, \ldots, Z_n < a_n) \leq \prod_{i=1}^n P(Z_i < a_i)$.

Examples of distributions satisfying (4) include the multinomial, multivariate normal with negative correlations, and permutation distributions, including sampling without replacement, all of which are NA. See Joag-Dev and Proschan (1983) for further details.

THEOREM 2.1 (Rinott and Samuel-Cahn (1987)) *Let $Y_1, \ldots, Y_n$ be NLODS random variables, and let $X_1, \ldots, X_n$ be independent random variables such that for each i, X_i and Y_i have the same marginal distribution, $i = 1, \ldots, n$. Then $V(\mathbf{X}) \leq V(\mathbf{Y})$.*

PROOF Given a sequence of random variables $\mathbf{Z} = (Z_1, \ldots, Z_n)$, and a vector $\mathbf{c} = (c_1, \ldots, c_n)$, with $c_n = -\infty$, and possibly $c_i = -\infty$ for some $i < n$, define the stopping rule $t(\mathbf{c}) = \min\{i \leq n : Z_i \geq c_i\}$. Since $c_n = -\infty$, we have $t(\mathbf{c}) \leq n$. Then for $i > 1$,

$$(5)\ Z_{t(\mathbf{c})} = c_1 + [Z_1 - c_1]^+ + \sum_{i=2}^{n} \{c_i - c_{i-1} + [Z_i - c_i]^+\} \cdot I(t(\mathbf{c}) > i - 1),$$

where $\{c_i - c_{i-1} + [Z_i - c_i]^+\} = Z_i - c_{i-1}$ if $c_i = -\infty$.

Recall (or see, e.g., Chow, Robbins and Siegmund (1971, Theorem 3.2)) that for independent $X_1, \ldots, X_n$, the optimal stopping rule is of the form $t(\mathbf{c}^*)$, with

$$(6) \quad c^*_{i-1} = E(X_i \vee c^*_i) = c^*_i + E[X_i - c^*_i]^+,\ i = 2, \ldots, n, \quad c^*_n = -\infty.$$

One can see directly, or from (5), that

$$V(\mathbf{X}) = EX_{t(\mathbf{c}^*)} = c_1^* + E[X_1 - c_1^*]^+. \tag{7}$$

The constants of (6), which are optimal for the sequence $X_1, \ldots, X_n$, need not be optimal for the sequence $Y_1, \ldots, Y_n$, and therefore

$$V(\mathbf{Y}) \geq EY_{t(\mathbf{c}^*)}. \tag{8}$$

Next note that for $Y_1, \ldots, Y_n$ which are NLODS, see (4), we have

$$E\{h(Y_i) \cdot I(Y_1 < a_1, \ldots, Y_{i-1} < a_{i-1})\} \geq \\ E\{h(Y_i)\} \cdot E\{I(Y_1 < a_1, \ldots, Y_{i-1} < a_{i-1})\}$$

for any nondecreasing function h. In particular, since $I(t(\mathbf{c}) > i-1) = I(Y_1 < c_1, \ldots, Y_{i-1} < c_{i-1})$, we have

$$E\{(c_i - c_{i-1} + [Y_i - c_i]^+) \cdot I(t(\mathbf{c}) > i-1)\} \geq \\ (c_i - c_{i-1} + E[Y_i - c_i]^+) \cdot EI(t(\mathbf{c}) > i-1). \tag{9}$$

To prove the theorem combine (8) with (5) applied to $Y_1, \ldots, Y_n$, and (9), to obtain

$$V(\mathbf{Y}) \geq c_1^* + E[Y_1 - c_1^*]^+ \\ + \sum_{i=2}^{n}\{c_i^* - c_{i-1}^* + E[Y_i - c_i^*]^+\} \cdot EI(t(\mathbf{c}^*) > i-1). \tag{10}$$

Because X_i and Y_i have the same marginal distributions, we can replace $E[Y_i - c_i^*]^+$ by $E[X_i - c_i^*]^+$. Then, by (6), the r.h.s. of (10) reduces to $c_1^* + E[X_1 - c_1^*]^+ = V(\mathbf{X})$, the last equality following from (7), and the proof is complete. □

Theorem 2.1 generalizes the next result due to O'Brien (1983). Our attempts to generalize this result in a different direction are described in Section 4 on random replacement schemes.

COROLLARY 2.1 *Let* $(I_1, \ldots, I_n)$ *and* $(J_1, \ldots, J_n)$ *denote random sampling with and without replacement, respectively, from* $\{1, \ldots, N\}$, $n \leq N$. *Let* $X_k = r_k(I_k)$ *and* $Y_k = r_k(J_k)$, *where for all* $k = 1, \ldots, n$, $r_k(i) \leq r_k(j)$ *if* $1 \leq i < j \leq N$. *Then* $V(\mathbf{X}) \leq V(\mathbf{Y})$.

PROOF This follows from the fact that $X_1, \ldots, X_n$ are independent, while $Y_1, \ldots, Y_n$ are NA, hence NLODS, and Theorem 2.1 applies. □

In Rinott and Samuel-Cahn (1991), we consider (among other things) the following problem. Given independent $Z_1, \ldots, Z_n$, let

$$V(s) = V(\mathbf{Z} | \sum_{i=1}^{n} Z_i = s)$$

denote the optimal stopping value with respect to observations having the conditional distribution of $\mathbf{Z}$ given $\sum_{i=1}^{n} Z_i = s$. In trying to understand the interaction between dependence and stochastic ordering, and how they affect optimal stopping values, it is natural to seek conditions under which $V(s)$ increases as a function of s. We found that if each of the Z_i's has a log-concave density, or probability function, then indeed $V(s)$ is increasing. In this case the observations are also NA, see Joag-Dev and Proschan (1983).

3. Value Comparisons Under Positive Dependence

In view of Theorem 2.1 and previous discussions, it appears natural to look for structures of positively dependent random variables $(X_1, \ldots, X_n)$, such that for independent $(Y_1, \ldots, Y_n)$ with X_i and Y_i having the same marginal distribution for each $i = 1, \ldots, n$, we have,

$$V(\mathbf{X}) \le V(\mathbf{Y}). \tag{11}$$

Association in the sense of Esary, Proschan and Walkup (1967) is an example of a well-known strong condition of positive dependence. The variables $X_1, \ldots, X_n$ are said to be associated if $\operatorname{cov}(f_1(X_1, \ldots, X_n), f_2(X_1, \ldots, X_n)) \ge 0$ for any pair of nondecreasing functions f_1 and f_2. While in Theorem 2.1, a suitable (and rather weak) notion of negative dependence was sufficient for the value comparison, this is not the case for comparisons under positive dependence. For example, the variables (X_1, X_2) of (3) are easily shown to be associated. However, if we set Y_i to be independent having the same marginal distribution as X_i, $i = 1, 2$, then $V(\mathbf{Y}) = 1 < V(\mathbf{X}) = 5$.

With the lack, so far, of general results of comparisons of the type (11) under positive dependence, we shall settle for a few examples. In the first three examples it is easy to see that the X's are associated.

EXAMPLE 3.1 *Let Z_i be independent random variables, and let $0 \le \alpha_i \le 1$ be constants. Set $X_1 = Z_1$ and $X_i = \alpha_i X_{i-1} + (1 - \alpha_i) Z_i$, $i = 2, \ldots, n$. Let $(Y_1, \ldots, Y_n)$ be independent random variables with Y_i having the same marginal distribution as X_i, $i = 1, \ldots, n$. Then*

$$V(\mathbf{X}) \le V(\mathbf{Y}).$$

In the special case that $\alpha_i = (i-1)/i$, we obtain the averages $X_i = \frac{1}{i}\sum_{j=1}^{i} Z_j$.

PROOF The proof is by induction on n. For $n = 2$, set $a = EZ_2$. We have

$$V(X_1, X_2) = E\{X_1 \vee [\alpha_2 X_1 + (1 - \alpha_2) E Z_2]\}$$

$$
\begin{aligned}
&= E\{X_1 I(X_1 \geq a)\} + \alpha_2 E\{X_1 I(X_1 < a)\} \\
&\quad +(1-\alpha_2) a P(X_1 < a) \\
&\leq E\{X_1 I(X_1 \geq a)\} + \alpha_2 E X_1 \cdot P(X_1 < a) \\
&\quad +(1-\alpha_2) a P(X_1 < a) \\
&= E\{X_1 I(X_1 \geq a)\} + E X_2 \cdot P(X_1 < a) \\
&\leq \sup_b \{ E\{X_1 I(X_1 \geq b)\} + E X_2 \cdot P(X_1 < b) \} \\
&= E\{X_1 \vee E X_2\} = E\{Y_1 \vee E Y_2\} = V(Y_1, Y_2).
\end{aligned}
$$

Now set $V^{(n-1)}(X_1) = \sup_{2 \leq t \leq n} E(X_t | X_1)$, and $\tilde{V}^{(n-1)}(Y_1) = \sup_{2 \leq t \leq n} E(Y_t | Y_1)$, which actually does not depend on Y_1. Note that the Markov structure of the X's implies $EV^{(n-1)}(X_1) = V(X_2, \ldots, X_n)$, and clearly $E\tilde{V}^{(n-1)}(Y_1) = V(Y_2, \ldots, Y_n)$. The induction hypothesis can be expressed in the form

$$
EV^{(n-1)}(X_1) = V(X_2, \ldots, X_n) \leq V(Y_2, \ldots, Y_n) = E\tilde{V}^{(n-1)}(Y_1).
$$

Note that $V^{(n-1)}(X_1)$ is a nondecreasing function of X_1. From the structure of the sequence $(X_1, \ldots, X_n)$ with $0 \leq \alpha_i \leq 1$, it is not hard to see that there exists a value $-\infty \leq c \leq \infty$ such that $X_1 \geq V^{(n-1)}(X_1)$ if and only if $X_1 \geq c$. We obtain

$$
\begin{aligned}
V(X_1, \ldots, X_n) &= E\{X_1 \vee V^{(n-1)}(X_1)\} \\
&= E\{X_1 I(X_1 \geq c)\} + E\{V^{(n-1)}(X_1) I(X_1 < c)\} \\
&\leq E\{X_1 I(X_1 \geq c)\} + EV^{(n-1)}(X_1) \cdot P(X_1 < c) \\
&\leq E\{X_1 I(X_1 \geq c)\} + E\tilde{V}^{(n-1)}(Y_1) \cdot P(X_1 < c) \\
&\leq \sup_b \{ E\{X_1 I(X_1 \geq b)\} + E\tilde{V}^{(n-1)}(Y_1) \cdot P(X_1 < b) \} \\
&= E\{X_1 \vee E\tilde{V}^{(n-1)}(Y_1)\} = E\{Y_1 \vee E\tilde{V}^{(n-1)}(Y_1)\} \\
&= V(Y_1, \ldots, Y_n),
\end{aligned}
$$

where the first inequality follows from the monotonicity of $V^{(n-1)}(X_1)$, the second inequality follows from the induction hypothesis, and the last equality follows by the independence of the Y_i's. □

The next example generalizes Example 3.1. Note that a real valued Markov chain can always be represented in the form $X_1 = Z_1$, $X_i = f_i(X_{i-1}, Z_i)$ with independent Z's. Note also that if the functions $f_i(x, z)$ are increasing in x, then the sequence $X_1, \ldots, X_n$ is *Conditionally Increasing in Sequence*, i.e., $P(X_{i+1} > x \,|\, X_1 = x_1, \ldots, X_i = x_i)$ is nondecreasing in $x_1, \ldots, x_i$, for all x and $i = 1, \ldots, n-1$. This implies that $X_1, \ldots, X_n$ are associated. For details see Barlow and Proschan (1975, Theorem 4.7).

EXAMPLE 3.2 *Let Z_i be independent random variables, and let $X_1, \ldots, X_n$ have the Markov structure $X_1 = Z_1$, $X_i = f_i(X_{i-1}, Z_i)$, where Z_i are independent, and $f_i(x,z)$ are functions satisfying $0 \le \frac{\partial f_i}{\partial x} \le 1$, $i = 2, \ldots, n$ (or $0 \le f_i(x',z) - f_i(x,z) \le x' - x$ for any $x' > x$ in the nondifferentiable case). Let $(Y_1, \ldots, Y_n)$ be independent random variables with Y_i having the same marginal distribution as X_i, $i = 1, \ldots, n$. Then*

$$V(\mathbf{X}) \le V(\mathbf{Y}).$$

PROOF As in the proof of Example 3.1 it suffices to show that

1. $V^{(n-1)}(X_1) = \sup_{2 \le t \le n} E(X_t | X_1)$, is nondecreasing in X_1,
2. there exists a value $-\infty \le c \le \infty$ such that $X_1 \ge V^{(n-1)}(X_1)$ if and only if $X_1 \ge c$.

1. is readily shown by induction using the monotonicity of f_i in x. We prove 2. by showing that for $X_1' > X_1$, $V^{(n-1)}(X_1') - V^{(n-1)}(X_1) \le X_1' - X_1$. For $n = 2$, this follows readily from $V^{(1)}(X_1) = E[f_2(X_1, Z_2) | X_1] = \int f_2(X_1, z) dF(z)$, and $\frac{\partial f_2}{\partial x} \le 1$, where F denotes the distribution of Z_2.

The proof now requires induction; we prefer to demonstrate the case $n = 3$, and leave the details of the induction to the reader. For $n = 3$ we have $V^{(2)}(X_1) = E\{[f_2(X_1, Z_2) \vee E(f_3(X_2, Z_3)|X_2)] | X_1\}$, and $f_3(X_2, Z_3) = f_3(f_2(X_1, Z_2), Z_3)$. We have $E(f_3(X_2, Z_3)|X_2) = h(X_2)$, say, where (by arguments as above), $X_2' > X_2$ implies $h(X_2') - h(X_2) \le X_2' - X_2$. Define $g(X_1, Z_2) = h(X_2) = h(f_2(X_1, Z_2))$. Then, replacing f_2 by f for brevity, we have $V^{(2)}(X_1) = \int [f(X_1, z) \vee g(X_1, z)] dF(z)$ where F denotes the cdf of Z_2, $0 \le \frac{\partial f}{\partial x} \le 1$, and if g is differentiable $\frac{\partial g}{\partial x} \le 1$ (by the chain rule), and in any case $X_1' > X_1$ implies $g(X_1', z) - g(X_1, z) \le X_1' - X_1$. Note that $f(X_1', z) \vee g(X_1', z) - f(X_1, z) \vee g(X_1, z) \le [f(X_1', z) - f(X_1, z)] \vee [g(X_1', z) - g(X_1, z)]$, so that for $X_1' > X_1$

$$f(X_1', z) \vee g(X_1', z) - f(X_1, z) \vee g(X_1, z) \le X_1' - X_1,$$

and substituting Z_2 for z, and taking expectations, we obtain for $n = 3$, $V^{(n-1)}(X_1') - V^{(n-1)}(X_1) \le X_1' - X_1$. □

EXAMPLE 3.3 *Let $Z_0, Z_1, \ldots, Z_n$ be independent random variables and let $X_i = Z_0 + Z_i$, $i = 1, \ldots, n$. Let $Y_1, \ldots, Y_n$ be independent random variables with Y_i having the same marginal distribution as X_i, $i = 1, \ldots, n$. Then*

$$V(\mathbf{X}) \le V(\mathbf{Y}).$$

PROOF It is not hard to verify the relations

$$V(\mathbf{X}) \le EZ_0 + V(Z_1, \ldots, Z_n) \le V(\mathbf{Y}).$$

The first inequality above is left to the reader. In order to prove the second, set $Y_i = Z_{0i}+Z_i$, where all the Z's are independent, and Z_{0i} is distributed like Z_0, $i = 1,\ldots,n$. We then have $V(\mathbf{Y}) = Ef(Z_{01},\ldots,Z_{0n-1},Z_1,\ldots,Z_{n-1})$ where f is a convex function (which depends on the constant $E(Z_0 + Z_n)$). For example, for $n = 3$ we have $V(\mathbf{Y}) = E\{(Z_{01} + Z_1) \vee E[(Z_{02} + Z_2) \vee E(Z_{03} + Z_3)]\}$. By Jensen's inequality and the independence of the Z's, we obtain a lower bound to the latter expression by replacing the variables Z_{0i} by their expectation EZ_0. The lower bound thus obtained is readily seen to equal $EZ_0 + V(Z_1,\ldots,Z_n)$. □

EXAMPLE 3.4 *Let $X_1,\ldots,X_n$ be a martingale, and let $Y_1,\ldots,Y_n$ be independent random variables with Y_i having the same marginal distribution as X_i, $i = 1,\ldots,n$. Then*

$$V(\mathbf{X}) \leq V(\mathbf{Y}).$$

PROOF Simply note that $V(\mathbf{X}) = EX_1 = EY_1 \leq V(\mathbf{Y})$. □

Note that being a martingale, $X_1,\ldots,X_n$ are nonnegatively correlated, but need not be associated.

4. Value Comparisons for Random Replacement Schemes

Random replacement schemes were introduced by Karlin (1974). Consider sampling from a finite population, say $\mathcal{N} = \{1,\ldots,N\}$; when the ith observation is taken, it is returned to the population with some probability, say π_i, independently of observation values, and removed with probability $1 - \pi_i$. The observations are taken at each step at random, that is, with equal probability for every number present in the population. Clearly sampling with and without replacement are special cases, and one might look for a hierarchy of comparisons, or ordering, generalizing the comparison in Corollary 2.1.

We now define random replacement schemes more formally. Set $\pi = (\pi_1,\ldots,\pi_{n-1})$ with $0 \leq \pi_i \leq 1$, and let U_i be independent Bernoulli variables, $P(U_i = 1) = \pi_i$, $i = 1,\ldots,n-1$. Consider an urn (or population) containing the values $\{1,\ldots,N\}$. Select a value J_1 at random from the urn; return it if $U_1 = 1$, and remove it from the urn if $U_1 = 0$. Now select J_2 at random from the resulting urn, and return it if and only if $U_2 = 1$. Continue in this manner until a sample $(J_1,\ldots,J_n)$ is obtained. Now define $X_k = r_k(J_k)$, where the real valued functions $r_k(i)$, $i \in \{1,\ldots,N\}$, $k = 1,\ldots,n$ are monotone nondecreasing in i for each k. This monotonicity will always be assumed in the sequel. Other conditions on $r_k(i)$ will appear later. The functions $r_k(i)$ may be seen as the reward for drawing the value i at step k, and at each step the rewards increase with the value drawn. Define the optimal stopping value to be

$$V_\pi^{(n)} = \sup_t EX_t,$$

where the supremum is taken over stopping rules with respect to the fields $\mathcal{F}_k(J_1, \ldots, J_k, U_1, \ldots, U_k)$. This means that the content of the urn at the time of the (possible) next draw is always known.

In these terms, Corollary 2.1 can be recast in the form:

$$V_{\mathbf{1}}^{(n)} \le V_{\mathbf{0}}^{(n)}, \tag{12}$$

where $\mathbf{1}$ ($\mathbf{0}$) denotes the $n-1$-vector of 1's (0's).

The following generalization of (12) holds.

THEOREM 4.1 (Rinott and Samuel-Cahn (1991)). *For any π and all $n \le N$,*

$$V_\pi^{(n)} \le V_{\mathbf{0}}^{(n)}. \tag{13}$$

We shall review the proof of this theorem at the end of this section.

One may conjecture that (12) can also be generalized to:

$$V_{\mathbf{1}}^{(n)} \le V_\pi^{(n)}. \tag{14}$$

However this is not true in general. For $N = n = 3$, $r_1(\cdot) \equiv 0$, $r_2(1) = 0$, $r_2(2) = r_2(3) = 3$, $r_3(1) = r_3(2) = 0$, $r_3(3) = 4$, we have, $V_{11}^{(3)} = 22/9 > V_{01}^{(3)} = 21/9$. It is possible that (14) holds if the functions $r_k(i)$ do not depend on k, or perhaps also when they are decreasing in k, for each i, i.e., values are discounted in time of observation.

Note that in general the sequence $\mathbf{X}$ obtained in random replacement schemes is not NLODS, even when $r_k(i)$ does not depend on k. For example, if $n = N = 3$, and $r_k(i) = i$, $i, k = 1, 2, 3$, and $\pi_1 = 0, \pi_2 = 1$, then it is easily seen that $P(X_3 < 3 | X_2 < 3) = 3/4 > P(X_3 < 3) = 2/3$. Thus, (14) cannot be derived from Theorem 2.1. For $n = 2$, (14) is easy:

LEMMA 4.1 *For $n = 2$, (14) holds.*

NOTATION Define $V_{\pi_1,\ldots,\pi_{k-1}}^{(k)}(\mathcal{M})$ to be the optimal stopping value when initially the urn contains the elements of an ordered set $\mathcal{M}$ where $|\mathcal{M}| \ge k$, at most k draws are allowed, and the replacement probabilities are $\pi_1, \ldots, \pi_{k-1}$. The functions r_k are suppressed in this notation. We may use $V_{\pi_1,\ldots,\pi_{k-1}}^{(k)}(\mathcal{M})$ with $r_1, \ldots, r_k$, and also with $r_2, \ldots, r_{k+1}$. We shall comment on this point when the latter case occurs, although the notation should be clear from the context.

PROOF OF LEMMA 4.1 For $n = 2$, we have

$$\begin{aligned} V^{(2)}_{\pi_1} = V^{(2)}_{\pi_1}(\mathcal{N}) &= \pi_1 E\{X_1 \vee V^{(1)}(\mathcal{N})\} \\ &\quad +(1-\pi_1)E\{X_1 \vee V^{(1)}(\mathcal{N} - \{J_1\})\} \\ &= \pi_1 V^{(2)}_1(\mathcal{N}) + (1-\pi_1)V^{(2)}_0(\mathcal{N}) \geq V^{(2)}_1(\mathcal{N}), \end{aligned} \tag{15}$$

where the final inequality follows from $V^{(2)}_1(\mathcal{N}) \leq V^{(2)}_0(\mathcal{N})$, which is a simple case of (12). Here, $V^{(1)}$ was used with respect to the function r_2. □

For $n > 3$, we are unable to prove (14) even in the seemingly simple case of $r_k(i)$ not depending on k. However, if it is true, then the following lemma could be a step in the right direction. It simplifies (14), which involves a random replacement scheme on the r.h.s., to a comparison between two deterministic schemes: complete replacement, and removal of the first draw followed by complete replacement. The case of $n = 3$ of (14), with some restrictions on r_k, will be derived from this lemma later.

LEMMA 4.2 *Fix* $m \geq 3$. $V^{(n)}_{\mathbf{1}} \leq V^{(n)}_{\pi}$ *for all* $3 \leq n \leq m$ *and* N *satisfying* $n \leq N$, *if and only if*

$$V^{(n)}_{1,\ldots,1} \leq V^{(n)}_{0,1,\ldots,1}, \tag{16}$$

holds for all $3 \leq n \leq m$ *and* $n \leq N$.

PROOF Clearly, (16) is necessary. To prove sufficiency, we shall make use of the following straightforward generalization of (15):

$$\begin{aligned} V^{(n)}_{\pi_1,\ldots,\pi_{n-1}}(\mathcal{N}) &= \pi_1 E\{X_1 \vee V^{(n-1)}_{\pi_2,\ldots,\pi_{n-1}}(\mathcal{N})\} \\ &\quad +(1-\pi_1)E\{X_1 \vee V^{(n-1)}_{\pi_2,\ldots,\pi_{n-1}}(\mathcal{N} - \{J_1\})\}. \end{aligned} \tag{17}$$

Assuming (16) holds, we now prove $V^{(n)}_{\mathbf{1}} \leq V^{(n)}_{\pi}$ by induction on n. In the present notation the latter inequality is expressed as

$$V^{(n)}_{1,\ldots,1}(\mathcal{N}) \leq V^{(n)}_{\pi_1,\ldots,\pi_{n-1}}(\mathcal{N}), \tag{18}$$

which we consider for all n, N such that $n \leq N = |\mathcal{N}|$. The induction hypothesis, see (18), for $n-1$ is

$$V^{(n-1)}_{1,\ldots,1}(\mathcal{N}) \leq V^{(n-1)}_{\pi_2,\ldots,\pi_{n-1}}(\mathcal{N}),$$

It holds for $n = 3$ $(n-1 = 2)$ by Lemma 4.1. Applying the induction hypothesis to the r.h.s. of (17) twice, the second time with the population being $\mathcal{N} - \{J_1\}$ instead of $\mathcal{N}$ we obtain the first inequality below:

$$\begin{aligned} V^{(n)}_{\pi_1,\ldots,\pi_{n-1}}(\mathcal{N}) &\geq \pi_1 E\{X_1 \vee V^{(n-1)}_{1,\ldots,1}(\mathcal{N})\} \\ &\quad +(1-\pi_1)E\{X_1 \vee V^{(n-1)}_{1,\ldots,1}(\mathcal{N} - \{J_1\})\} \\ &= \pi_1 V^{(n)}_{1,\ldots,1}(\mathcal{N}) + (1-\pi_1)V^{(n)}_{0,1,\ldots,1}(\mathcal{N}) \geq V^{(n)}_{1,\ldots,1}(\mathcal{N}), \end{aligned} \tag{19}$$

where the equality follows from $E\{X_1 \vee V^{(n-1)}_{1,\ldots,1}(\mathcal{N})\} = V^{(n)}_{1,\ldots,1}(\mathcal{N})$, and $E\{X_1 \vee V^{(n-1)}_{1,\ldots,1}(\mathcal{N}-\{J_1\})\} = V^{(n)}_{0,1,\ldots,1}(\mathcal{N})$, and the last inequality follows from (16). In this proof, $V^{(n-1)}$ was always used with respect to $r_2, \ldots, r_n$. □

We cannot prove (16) even for $r_k(i)$ not depending on k, but it appears like a more tractable conjecture than (14). A computer search with a variety of functions $r_k(i)$ (not depending on k), and $N \leq 15$, did not produce a counterexample to (16). In the case $n = 2$, (14) is already established in Lemma 4.1. For $n = 3$ we have

PROPOSITION 4.1 *Let $n = 3$, $N \geq 3$ and $r_1(i) \geq r_2(i)$, for $i = 1, \ldots, N$. Then*

$$V^{(3)}_{1,1} \leq V^{(3)}_{0,1}.$$

Clearly, our assumption holds if for all i, $r_k(i)$ is decreasing in k. This is a natural assumption which says that the earlier you observe a certain element i, the higher its value. In other words, there is a cost for time, or for taking more observations. By Lemma 4.2 we conclude that under the conditions of Proposition 4.1, (14) holds for $n = 3$, i.e.,

$$V^{(3)}_{1,1} \leq V^{(3)}_{\pi_1,\pi_2}.$$

PROOF OF PROPOSITION 4.1 Define $\bar{r}_k = \frac{1}{N}\sum_{j=1}^N r_k(j)$ and

$$\bar{r}_k[i] = \frac{1}{N-1} \sum_{j:i\neq j=1}^{N} r_k(j).$$

Let $A(i) = \frac{1}{N}\sum_{j=1}^N \{r_2(j) \vee \bar{r}_3[i]\}$, and $B(i) = \frac{1}{N-1}\sum_{j:i\neq j=1}^N \{r_2(j) \vee \bar{r}_3[i]\}$. Note that if for some i, $A(i) > B(i)$, then it is readily seen that $r_2(i) > \bar{r}_3[i]$ and $r_2(i) > A(i)\,(> B(i)\,)$. Since $r_1(i) \geq r_2(i)$, we conclude that $A(i) > B(i)$ implies $r_1(i) > A(i) > B(i)$. It is now easy to see that

$$\frac{1}{N}\sum_{i=1}^{N}\{r_1(i) \vee B(i)\} \geq \frac{1}{N}\sum_{i=1}^{N}\{r_1(i) \vee A(i)\}. \tag{20}$$

Note that the l.h.s. of (20) equals $V^{(3)}_{0,1}$. In order to proceed we now need a simple lemma whose proof is given in Rinott and Samuel-Cahn (1991, Lemma 3.3).

LEMMA 4.3 *Let $h(x), g(x), x \in \mathbb{R}$, be an increasing and a decreasing function, respectively. If X is a random variable such that the expectations below exist, then*

$$E\{h(X) \vee g(X)\} \geq E\{h(X) \vee Eg(X)\}. \tag{21}$$

By Lemma 4.3, the r.h.s. of (20) is $\geq$ than

$$\frac{1}{N}\sum_{j=1}^{N}\{r_1(j) \vee \frac{1}{N}\sum_{i=1}^{N} A(i)\}. \tag{22}$$

Finally, it suffices to show that the r.h.s. of (22) is $\geq$ than $V_{1,1}^{(3)}$. We have,

$$\frac{1}{N}\sum_{i=1}^{N} A(i) = \frac{1}{N}\sum_{j=1}^{N}\frac{1}{N}\sum_{i=1}^{N}\{r_2(j) \vee \bar{r}_3[i]\}. \tag{23}$$

Applying Lemma 4.3, or simple convexity, to the inner sum in the r.h.s. of (23) and noting the relation $\frac{1}{N}\sum_{i=1}^{N}\bar{r}_3[i] = \bar{r}_3$ we conclude that the r.h.s. of (23) is $\geq \frac{1}{N}\sum_{j=1}^{N}\{r_2(j) \vee \bar{r}_3\}$. Denote the latter quantity by C. Thus the r.h.s. of (22) is $\geq \frac{1}{N}\sum_{j=1}^{N}\{r_1(j) \vee C\}$, which is exactly $V_{1,1}^{(3)}$ and the proof of Proposition 4.1 is complete. □

For the proof of Theorem 4.1 we shall need a simple lemma whose proof can be found in Rinott and Samuel-Cahn (1991).

LEMMA 4.4 *Let J be a random element of $\mathcal{N}$. Then for any $m \leq N-1$,*

$$V_{\mathbf{0}}^{(m)}(\mathcal{N}) \leq EV_{\mathbf{0}}^{(m)}(\mathcal{N} - \{J\}).$$

In words, removing a random (known) element from the population before sampling, increases the average stopping value for sampling without replacement.

Perhaps this lemma is best explained by an example. For $N = 3$, $m = 2$, and $r_k(i) = i$, we have $V_{\mathbf{0}}^{(2)}(\{1,2,3\}) = (1/3)\cdot(2+3)/2+(1/3)\cdot 2+(1/3)\cdot 3 = 5/2$, corresponding to the first sampled item being 1,2, or 3, respectively. If prior to sampling, a random element J is removed from $\mathcal{N} = \{1,2,3\}$, we have $EV_{\mathbf{0}}^{(2)}(\{1,2,3\}-\{J\}) = (1/3)\cdot 3+(1/3)\cdot 3+(1/3)\cdot 2 = 8/3$, corresponding to the removed element J being 1,2, or 3, respectively.

PROOF OF THEOREM 4.1 It is easy to see that arguments similar to those given for Lemma 4.2 imply also that in order to prove $V_{\pi}^{(n)} \leq V_{\mathbf{0}}^{(n)}$ it suffices to prove $V_{1,0,\ldots,0}^{(n)} \leq V_{0,\ldots,0}^{(n)}$. In order to prove the latter inequality, note that $V_{\mathbf{0}}^{(n-1)}(\mathcal{N} - \{J_1\})$ is decreasing in J_1, while $X_1 = r_1(J_1)$ is increasing in J_1. Applying Lemma 4.3 and then Lemma 4.4 to obtain the inequalities below, we have

$$\begin{aligned} V_{0,\ldots,0}^{(n)} &= E\{X_1 \vee V_{\mathbf{0}}^{(n-1)}(\mathcal{N} - \{J_1\})\} \\ &\geq E\{X_1 \vee EV_{\mathbf{0}}^{(n-1)}(\mathcal{N} - \{J_1\})\} \\ &\geq E\{X_1 \vee V_{\mathbf{0}}^{(n-1)}(\mathcal{N})\} = V_{1,0,\ldots,0}^{(n)}, \end{aligned}$$

and the proof is complete. In this proof, $V^{(n-1)}$ was used with respect to $r_2, \ldots, r_n$. □

5. Prophet Inequalities

In this section we review certain prophet inequalities for some of the models discussed above. We start with simple technical lemmas.

LEMMA 5.1 *Let* $(X_1, \ldots, X_n)$ *be either NLODS random variables (including independent random variables), or observations arising under any random replacement scheme of the type described in Section 4, with* $r_k(i) \geq r_{k+1}(i)$ *for all* i, k. *Then for any constant* c,

$$E\{[X_k - c]^+ \,|\, X_1 \vee \cdots \vee X_{k-1} < c\} \geq E[X_k - c]^+,\; k = 2, \ldots, n.$$

PROOF For independent random variables the result is obvious (with equality), and the inequality follows easily from the definition of NLODS variables. For random replacement schemes the result follows from the fact that conditionally on any values of $X_1 < c, \ldots X_{k-1} < c$, and any replacement indicators $U_1, \ldots, U_n$, X_k is distributed as $r_k(J)$, where J is drawn from an urn from which some elements I with $r_k(I) < c$ have been removed. Here the monotonicity of $r_k(i)$ in k was used. □

Henceforth we shall consider only nonnegative random variables, and exclude (without further mention) the trivial case that they are all identically zero. For such a sequence $(X_1, \ldots, X_n)$, and $b \geq 0$, let $t(b)$ denote the stopping time: $t(b) = \inf\{k : X_k \geq b\} \wedge n$.

LEMMA 5.2 *Let* $(X_1, \ldots, X_n)$ *be nonnegative random variables. Let* $b > 0$ *be the unique constant satisfying* $\sum_{k=1}^n E[X_k - b]^+ = b$. *Suppose*

$$E\{[X_k - b]^+ \,|\, X_1 \vee \cdots \vee X_{k-1} < b\} \geq E[X_k - b]^+,\; k = 2, \ldots, n. \tag{24}$$

Then

$$b < EX_{t(b)}. \tag{25}$$

PROOF

$$\begin{aligned} EX_{t(b)} &\geq E\{X_{t(b)} I(X_1 \vee \cdots \vee X_n \geq b)\} \\ &= E\{\, bI(X_1 \vee \cdots \vee X_n \geq b) \\ &\qquad + \sum_{k=1}^n [X_k - b]^+ I(X_1 \vee \cdots \vee X_{k-1} < b)\} \\ &\geq bP(X_1 \vee \cdots \vee X_n \geq b) \end{aligned}$$

$$
\begin{aligned}
&+\sum_{k=1}^{n} E[X_k - b]^+ P(X_1 \vee \cdots \vee X_{k-1} < b) \\
> \ & bP(X_1 \vee \cdots \vee X_n \geq b) \\
&+P(X_1 \vee \cdots \vee X_n < b) \sum_{k=1}^{n} E[X_k - b]^+ = b\,;
\end{aligned}
$$

here the first inequality holds because $X_i \geq 0$, the second inequality follows from (24), and the last inequality from $P(X_1 \vee \cdots \vee X_{k-1} < b) \geq P(X_1 \vee \cdots \vee X_n < b)$, with strict inequality for the first k such that $P(X_k > b) > 0$. □

LEMMA 5.3 *Let* $(Y_1, \ldots, Y_n)$ *be any nonnegative random variables, and let* $b \geq 0$ *be the unique constant satisfying* $\sum_{k=1}^{n} E[Y_k - b]^+ = b$. *Then* $E\{Y_1 \vee \cdots \vee Y_n\} \leq 2b$.

PROOF Simply take expectations on both sides of the simple relation: $Y_1 \vee \cdots \vee Y_n \leq b + \sum_{k=1}^{n} [Y_k - b]^+$. □

THEOREM 5.1 *Let* $(X_1, \ldots, X_n)$ *be nonnegative random variables which are either NLODS (including independent random variables), or observations arising under any random replacement scheme of the type described in Section 4, with* $r_k(i) \geq r_{k+1}(i)$ *for all* i, k, *or any other random variables which satisfy for every constant* c,

$$E\{[X_k - c]^+ \mid X_1 \vee \cdots \vee X_{k-1} < c\} \geq E[X_k - c]^+, \; k = 2, \ldots, n.$$

Then the prophet inequality

$$E\{X_1 \vee \cdots \vee X_n\} < 2V(\mathbf{X}) \tag{26}$$

holds. Moreover, if $(Y_1, \ldots, Y_n)$ *are* any *nonnegative random variables such that for each* i, X_i *and* Y_i *have the same (marginal) distribution,* $i = 1, \ldots, n$, *then*

$$E\{Y_1 \vee \cdots \vee Y_n\} < 2V(\mathbf{X}).$$

PROOF It clearly suffices to prove the second part of the theorem. Noting that the quantity b defined in Lemmas 5.2 - 5.3 depends on marginal distributions only, and applying Lemmas 5.1 - 5.3 we have,

$$\frac{1}{2} E\{Y_1 \vee \cdots \vee Y_n\} \leq b < EX_{t(b)} \leq V(\mathbf{X}). \tag{27}$$ □

For *independent* random variables the inequality (26) was obtained by Krengel and Sucheston (1978). This latter article provided the inspiration to a large body of results on prophet inequalities. For independent $0 \leq X_k \leq 1$, Hill (1983) sharpened the result to

$$E\{X_1 \vee \cdots \vee X_n\} < 2V(\mathbf{X}) - V(\mathbf{X})^2\,.$$

The negative dependence condition of Theorem 5.2 below, which generalizes Hill's result, is stronger than the NLODS condition; however it is weaker than CDS (see Section 2).

THEOREM 5.2 (Samuel-Cahn (1991)) *Let* $0 \leq X_k \leq 1$, $k = 1, \ldots, n$, *and suppose that* $X_1, \ldots, X_n$ *are negatively dependent in the sense that* $P(X_k < a_k | X_1 < a_1, \ldots, X_{k-1} < a_{k-1})$ *is nondecreasing in* $a_1, \ldots, a_{k-1}$, *for all* $k = 2, \ldots, n$. *Then*

$$E\{X_1 \vee \cdots \vee X_n\} < 2V(\mathbf{X}) - V(\mathbf{X})^2.$$

For positively dependent random variables we quote a result for averages (recall Example 3.1).

THEOREM 5.3 (Hill (1986)) *Let* Z_i *be independent nonnegative random variables, and consider the averages* $X_i = \frac{1}{i}\sum_{j=1}^{i} Z_j$, $i = 1, \ldots, n$. *Then*

$$E\{X_1 \vee \cdots \vee X_n\} < 2V(\mathbf{X}).$$

REFERENCES

BARLOW, R. E. AND PROSCHAN, F. (1975). *Statistical Theory of Reliability and Life Testing Probability Models.* Holt, Rinehart and Winston, New York.

CHOW, Y. S., ROBBINS, H. AND SIEGMUND, D. (1971). *Great Expectations: The Theory of Optimal Stopping.* Houghton Mifflin, Boston, MA.

ESARY, J. D., PROSCHAN, F. AND WALKUP, D. W. (1967). Association of random variables, with applications. *Ann. Math. Statist.* **44** 1466-1474.

HILL, T. P. (1983). Prophet inequalities and order selection in optimal stopping problems. *Proc. Amer. Math. Soc.* **88** 131-137.

HILL, T. P. (1986). Prophet inequalities for averages of independent non-negative random variables. *Math. Zeit.* **192** 427-436.

HILL, T. P. AND KERTZ, R. P. (1992). A survey of prophet inequalities in optimal stopping theory. In *Strategies for Sequential Search and Selection in Real-Time.* F.T. Bruss, T.S. Ferguson, and S.M. Samuels, eds. American Mathematical Society, Providence, RI. 191-207.

JOAG-DEV, K. AND PROSCHAN, F. (1983). Negative association of random variables, with applications. *Ann. Statist.* **11** 286-295.

KARLIN, S. (1974). Inequalities for symmetric sampling plans I. *Ann. Statist.* **2** 1065-1094.

KRENGEL, U. AND SUCHESTON, L. (1978). On semiamarts, amarts, and processes with finite value. In *Probability on Banach Spaces,* J. Kuelbs, ed. Marcel Dekker, New York. 197-266.

O'BRIEN, G. L. (1983). Optimal stopping when sampling with and without replacement. *Z. Wahrsch. verw. Gebiete.* **64** 125-128.

RINOTT, Y. AND SAMUEL-CAHN, E. (1987). Comparisons of optimal stopping values and prophet inequalities for negatively dependent random variables. *Ann. Statist.* **15** 1482-1490.

RINOTT, Y. AND SAMUEL-CAHN, E. (1991). Orderings of optimal stopping values and prophet inequalities for certain multivariate distributions. *J. Multivariate Anal.* **37** 104-114.

SAMUEL-CAHN, E. (1991). Prophet inequalities for bounded negatively dependent random variables. *Statist. Prob. Lett.* **12** 213-216.

DEPARTMENT OF MATHEMATICS
UNIVERSITY OF CALIFORNIA, SAN DIEGO
LA JOLLA, CA 92093

DEPARTMENT OF STATISTICS
HEBREW UNIVERSITY
JERUSALEM 91905, ISRAEL

Stochastic Inequalities
IMS Lecture Notes – Monograph Series
Volume 22 (1993)

SOME APPLICATIONS OF MONOTONE TRANSFORMATIONS IN STATISTICS[1]

By ALLAN R. SAMPSON

University of Pittsburgh

A number of results concerning monotone transformations of random variables are reviewed. Particular attention is paid to the effects of choice of monotone scaling in two settings: (a) describing and quantifying dependence between two random variables, and (b) comparing two populations with ordinal categorical responses.

Properties of the concordant and discordant monotone correlation coefficients (Kimeldorf, May and Sampson (1982)) between random variables X and Y are discussed, and computational approaches are considered.

The two sample problem is explored where responses are ordinal categories and typical statistical procedures involve the arbitrary choice of monotone scales. The effects of the choice of scaling upon the resultant analyses are examined in detail.

1. Introduction

In a variety of situations, it is of interest to consider how the results of the analyses change when we transform the relevant random variables by monotone functions. The usual purpose of this is to study the effects of monotonically rescaling the measured random quantities. Depending on our needs, we might want a statistical procedure that is invariant to monotone scale changes, or we might want to choose an appropriate scaling in situations where the natural choice of scales is not clear. The purpose of this paper is to review some results in this area focusing on a somewhat less than standard usage of monotone transformations.

Traditional concerns about monotone invariance can lead to various notions. In some settings it leads to considering statistical procedures which depend only on ranks of the data. For jointly distributed random variables, it can lead to a discussion of procedures which depend solely on the copula

[1]Research supported by National Security Agency Grant No. MDA–904–90–H–4036. Reproduction in whole or part is permitted for any purpose of the United States Government.

AMS 1991 *subject classifications.* Primary 62F03, 62H02; Secondary 62A05.

Key words and phrases. Scaling, monotone dependence, two–sample, ordinal variables, contingency table, concordant monotone correlation, correspondence analysis.

(Sklar (1959)) or uniform representation (Kimeldorf and Sampson (1975)), or in the case of data, upon the multivariate empirical rank distribution (e.g., Block, Chhetry, Fang and Sampson (1990)). For ordinal contingency tables, invariance leads to other related notions.

The focus of this paper is somewhat different than these preceding traditional concerns about invariance. We are interested in describing how certain probabilistic concepts and statistical notions depend on the choice of monotone scales, and utilizing this knowledge for assessing appropriateness of scales. This idea of sensitivity to scales and rescaling is particularly important in statistical usages where there is no natural choice of scales. Ordinal contingency tables offer such an example where, for instance, one variable might be an evaluative response such as excellent, very good, good, etc., and the other is degree of involvement such as none, some, etc. In Section 2, we discuss the effects of monotone scaling on certain measures of dependence and in Section 3 we consider how various two-sample tests are affected by choice of monotone scores.

2. Measures of Dependence and Scaling for Bivariate Random Variables

Lancaster (1969) interweaves several lines of research to present a set of techniques concerning bivariate random variables X and Y which describe their structure and measure their degree of relationship. Many of these results rely on the canonical decomposition of a bivariate p.d.f. See also Kendall and Stuart (1979, Chapter 33). We review some relevant definitions below.

DEFINITION 1

(i) Random variables X and Y are *mutually completely dependent (MCD)* if there exists a one–to–one function γ so that $Y = \gamma(X)$ w.p. 1.

(ii) The *sup–correlation* between random variables X and Y, denoted by $\rho'(X,Y)$, is defined as $\sup \rho(f(X), g(Y))$, taken over all suitable functions f and g.

(iii) The support $S \times T$ of (X,Y) is said to consist of (at least) k *disjunct pieces* if there exists partitions $S_1, \ldots, S_k$ of S and $T_1, \ldots, T_k$ of T such that

$$P((X,Y) \in S_i \times T_i) > 0, \quad i = 1, \ldots, k$$

and

$$P((X,Y) \in S_i \times T_j) = 0 \qquad \text{for all } i \neq j.$$

Intuitively MCD was thought to be an antithesis of independence and the sup–correlation was a method of measuring how dependent random variables

X and Y are. Clearly, X and Y are independent if and only if $\rho'(X,Y) = 0$; and if X and Y are MCD, then $\rho'(X,Y) = 1$. However, the converse to the latter is not true. If $\rho'(X,Y) = 1$, then under suitable regularity conditions, it can be shown that the support of X,Y consists of at least two or more disjunct pieces.

Lancaster (1969) discusses in great detail the above notions as well as many other related ones. The sup–correlation has been discussed in further detail as a measure of dependence, particularly for contingency tables when it is readily computable. Additionally, these notions form, in part, the basis for correspondence analysis. More recently, the notion of the canonical expansion of a joint p.m.f. has been explored statistically by Gilula (1984).

Kimeldorf and Sampson (1978) and others, including Vitale (1990), have shown that MCD is not an appropriate antithesis to independence. In fact, there exists $\{X_n, Y_n\}$ all with the same respective univariate marginals such that X_n, Y_n are MCD for each n and yet, X_n, Y_n converge in distribution to independent random variables X, Y. Obviously, for large enough n, if one were to take a random sample of such X_n, Y_n, this sample for all intents and purposes would look like a sample from independent random variables.

To counter this difficulty Kimeldorf and Sampson (1978) introduce the notion of X and Y being *monotone dependent* if in Definition 1(i), γ is a monotone function. (If γ is increasing we say X and Y are *increasing dependent*, and similarly for γ decreasing). Moreover, they show that if X_n, Y_n are monotone dependent for each n, and X_n, Y_n converge in distribution to X, Y, then X, Y are monotone dependent. This, and other reasons, suggest that monotone dependence serves as a suitable antithesis to independence.

To measure the degree of monotone dependence between a pair of random variables X and Y, Kimeldorf and Sampson (1978) introduce the notion of monotone correlation $\rho^*(X,Y)$, where f, g are required in Definition 1(ii) to be monotone functions. Additionally, Kimeldorf and Sampson (1978) observe the following straightforward properties: (a) $\rho^*(X,Y) = 0$, if and only if X and Y are independent; (b) $|\rho(X,Y)| \le \rho^*(X,Y) \le \rho'(X,Y)$; and (c) for the bivariate normal, the inequalites in (b) become equalities.

The monotone correlation is further refined by Kimeldorf, May and Sampson (1982). If in Definition 1(ii), f and g are both required to be increasing (or both decreasing) the resulting measure of dependence is called the *concordant monotone correlation (CMC)*. Motivated to study the monotone correlation notion when f is increasing and g is decreasing (or equivalently, vice versa) Kimeldorf, May and Sampson (1982) observe that

$$\sup_{f\uparrow, g\downarrow} \rho(f(X), \quad g(Y)) = \sup_{f\uparrow, g\uparrow} \rho(f(X), \quad -g(Y) = -\inf_{f\uparrow, g\uparrow} \rho(f(X), g(Y)).$$

This observation leads them to define the *discordant monotone correlation (DMC)* by $\inf \rho(f(X), g(Y))$, taken over all f, g both increasing or both decreasing. Clearly there is the following relationship: $\rho^*(X,Y) = \max(CMC(X,Y), -DMC(X,Y))$.

Kimeldorf, May and Sampson (1982) also introduce the *iso-CMC (ICMC)* and *iso-DMC (IDMC)* when f is restricted to equal to g in the preceding definitions of the CMC and DMC, respectively. (Interestingly they show that if X and Y have an exchangeable distribution, it is *not* necessarily true that $ICMC(X,Y) = CMC(X,Y)$.)

While both the CMC and DMC are measures of association, respectively, for increasing monotone dependence and decreasing monotone dependence, they serve another very useful purpose: for any increasing functions f, g

$$DMC(X,Y) \leq \rho(f(X), g(Y)) \leq CMC(X,Y). \tag{1}$$

The implications of (1) will be discussed shortly.

When X and Y are jointly discrete random variables with finite support, a number of additional useful results can be obtained. If in Definition 1(iii), $S \leq S_2 \leq \cdots \leq S_k$ and $T_1 \leq (\geq) T_2 \leq (\geq) \cdots \leq (\geq) T_k$, where $U \leq V$ means for all $u \in U$, $v \in V$, $u \leq v$, we say the support consists of *increasing (decreasing) disjunct pieces.* (Obviously, this definition extends beyond the discrete case.) Chhetry, DeLeeuw and Sampson (1990) show that the $CMC = 1$, if and only if the support of X and Y consists of two or more increasing disjunct pieces (and equivalently $DMC = -1$, if and only if the support consists of two or more decreasing disjunct pieces). To compare $\rho'(X,Y)$ and $CMC(X,Y)$, we recall Lehmann's (1966) notion of Y being *positively regressive dependent (PRD)* on X if $P(Y > y \mid X = x)$ is nondecreasing in x for all y. Schriever (1983) showed that if Y is PRD on X and X is PRD on Y, then $\rho'(X,Y) = CMC(X,Y)$. However, Chhetry and Sampson (1987) give an example where X and Y are not mutually PRD, yet $\rho'(X,Y) = CMC(X,Y)$. Lastly, under the assumption of finite discrete support (with the number of support points of X being three or more, or of $Y \geq 3$) we have $CMC(X,Y) = DMC(X,Y)$, if and only if X and Y are independent random variables.

To explore the uses of the inequalites in (1) for arbitrary X and Y, we begin by noting that various measures of dependence are defined as correlations between certain increasing functions. Spearman's rho, ρ_s, can be defined by $\rho(F(X), G(Y))$, where F and G are, respectively, the marginal c.d.f.'s of X and of Y. Another possible measure is given by $\rho(\Phi^{-1}F(X), \Phi^{-1}G(Y))$, where Φ is the standard normal c.d.f. Let $m(X,Y)$ generically denote a measure of monotone dependence between X and Y of the form $\rho(\phi(X), \psi(Y))$, where ϕ, ψ are particular increasing functions. Such measures are discussed by Agresti (1984) and Williams (1952). Then from (1), any such $m(X,Y)$ must lie in the interval $[DMC(X,Y), CMC(X,Y)]$. If in addition we require $\phi = \psi$, denoting the resulting measure by $m_I(X,Y)$, then $IDMC \leq m_I \leq$

$ICMC$. Now consider a situation where we are not certain which measure of monotone dependence is most meaningful for a problem being analyzed. If the CMC is relatively close to the DMC, the problem is rendered moot, in that the inequality of (1) yields that *all* measures of monotone dependence must be close to each other.

This observation is more pertinent when measuring dependence for a bivariate ordinal contingency table (or equivalently dealing with bivariate discrete random variables with finite support). Here there is an extensive literature for measures of positive dependence (e.g., Agresti (1984) or Schriever (1985)), with a number in the form of a measure of monotone dependence. In this situation, oftentimes appropriate scalings for the ordinal X and Y variables are not available. Thus, having the knowledge that the CMC and DMC are close indicates little sensitivity of a measure of monotone dependence to the choice of scalings.

To illustrate these notions, we consider in Table 1 the "father–son British social mobility data" (Glass and Hall (1954)). In this case since the variable being measured for both father and for son is the same occupational status variable which is ordinal in nature, it is appropriate to use *iso–scaling*, (that is, requiring $\phi = \psi$). In this table Status S1 is professional, and high administrative; Status S2 is managerial, executive and higher grade supervisory; Status S3 is lower grade supervisory; Status S4 is skilled manual; and Status S5 is semi–skilled and unskilled manual. A son in Status j whose father is in Status i is said to be upwardly mobile if $j < i$ (e.g., Bishop, Fienberg, and Holland (1975, p. 321)).

Father's Occupational Status	Son's Occupational Status				
	S1	S2	S3	S4	S5
S1	50	45	8	18	8
S2	28	174	84	154	55
S3	11	78	110	223	96
S4	14	150	185	714	447
S5	3	42	72	320	411

Table 1. British mobility data (3,500 father–son data values). (Glass and Hall (1954))

In this case the $ICMC = .496$ and $IDMC = .242$, (see Kimeldorf, May and Sampson (1982)) indicating that regardless of the monotone scaling for these five ordinal categories, the resulting correlation is between .242 and .496. (The monotone scales corresponding to the $ICMC$ assign S5, S4, S3, S2 and S1, respectively, the values 0, .077, .158, .373, and 1. The monotone scales corresponding to the $IDMC$ assign to S5, S4, S3, S2, and S1, respectively, 0, 1, 1, 1, and 1.)

Another focus of this development is examining and interpreting the resulting monotone scales for the CMC. In dual scaling (Nishisato (1980)) or in correspondence analysis (Benzecri (1973)), the scalings derived for contingency tables are those which maximize $\rho(a(X), b(Y))$ over all a, b, whether or not they are monotone. Obviously these $a(\ )$ and $b(\ )$ are the functions which yield the sup–correlation between the row and column classifications. If the classifications are categorical, these resulting scalings have standard interpretations. However, when the classifications are ordinal, a natural requirement is that the scales be monotone. There is no guarantee for this in the usual dual scalings, whereas, our scales obviously guarantee monotonicity.

Using a random generation procedure described in their paper, Kimeldorf, May and Sampson (1982) generated the following random 10×10 probability matrix with a slight amount of positive dependence.

		Y									
	b_1	b_2	b_3	b_4	b_5	b_6	b_7	b_8	b_9	b_{10}	
	a_1	0.0331	0.0111	0.0092	0.0049	0.0016	0.0028	0.0009	0.0108	0.0096	0.0007
	a_2	0.0101	0.0361	0.0057	0.0081	0.0133	0.0062	0.0121	0.0066	0.0003	0.0020
	a_3	0.0102	0.0059	0.0347	0.0027	0.0055	0.0020	0.0104	0.0046	0.0069	0.0056
	a_4	0.0144	0.0018	0.0065	0.0342	0.0006	0.0071	0.0055	0.0066	0.0084	0.0113
X	a_5	0.0006	0.0016	0.0087	0.0132	0.0435	0.0061	0.0100	0.0046	0.0044	0.0053
	a_6	0.0022	0.0035	0.0151	0.0015	0.0056	0.0427	0.0062	0.0035	0.0089	0.0125
	a_7	0.0002	0.0084	0.0026	0.0020	0.0005	0.0086	0.0387	0.0007	0.0034	0.0111
	a_8	0.0084	0.0100	0.0079	0.0036	0.0100	0.0128	0.0044	0.0303	0.0121	0.0065
	a_9	0.0028	0.0079	0.0141	0.0008	0.0133	0.0077	0.0064	0.0139	0.0402	0.0068
	a_{10}	0.0009	0.0149	0.0042	0.0108	0.0022	0.0144	0.0130	0.0151	0.0146	0.0438

Table 2. Random 10×10 probability matrix.
(Kimeldorf, May and Sampson (1982))

For this joint distribution the $CMC = .443$ and the monotone scales are given in Table 3.

	1	2	3	4	5	6	7	8	9	10
X	0.	.461	.461	.461	.872	.872	.872	.872	.873	1.
Y	0.	.537	.541	.541	.842	.842	.842	.842	.842	1.

Table 3. Monotone Scales for Table 2.

It is interesting to observe that for the X variable, there are only 5 distinct scores for the 10 ordinal categories and for the Y variable, the same is true. This suggests that for cross–prediction purposes the appropriately collapsed 5×5 distribution with the noted monotone scales allows for the best linear predictability. Although this approach to obtaining monotone scales requires further development, it appears to provide a notion of dual scalings for ordinal contingency tables or may form a basis for "ordinal correspondence analysis."

For bivariate discrete distributions on an $m \times n$ lattice $\{u_1, \ldots, u_m\} \times \{v_1, \ldots, v_n\}$, the dual scalings can be computed from a spectral decomposition of the matrix $Q^* = D_r^{-1/2} Q D_c^{-1/2}$, where $Q = \{q_{ij}\} \equiv \{\text{Prob}(X = u_i, Y = v_j)\}$, $D_r = \text{Diag}(q_{1+}, \ldots, q_{m+})$ and $D_c = \text{Diag}(q_{+1}, \ldots, q_{+n})$. For instance, the sup–correlation is the square root of the second largest eigenvalue of $Q^* Q^{*'}$. On the other hand, to compute the CMC and the corresponding monotone scores is a more difficult computational problem. An analogy of this increased difficulty, is the comparative difficulty of the following two optimization problems.

$$\sup_{x'x=1} x'Sx \tag{2}$$

and

$$\sup_{\substack{x'x=1 \\ x_1 \leq \cdots \leq x_p}} x'Sx, \tag{3}$$

where S is a symmetric $p \times p$ matrix.

Kimeldorf, May and Sampson (1982) express the monotone correlation problem as a nonlinear programming problem with linear constraints and then employ a numerical optimization algorithm of May (1979). The resulting software package, called MONCOR, is described by Kimeldorf, May and Sampson (1981) and is available from the author (written in FORTRAN and requiring IMSL routines.)

When dealing with continuous bivariate random variables, the problems are obviously compounded. To compute the sup–correlation and corresponding canonical variables requires solving a continuous eigenfunction problem, although for some bivariate distributions, certain classical bivariate expansions yield these. We are unaware of any technique like these to bring to bear for the CMC and the corresponding monotone scalings.

Based upon a random sample from a bivariate continuous distribution, the ACE algorithm of Breiman and Friedman (1985) can be applied to estimate the sup–correlation and corresponding canonical variables. There appear to be monotonicity constraints within the ACE algorithm which would allow the estimation of the monotone correlation and corresponding monotone scalings.

No explicit sampling results are applicable to the CMC and DMC, even in the case of multinomial sampling. Perhaps bootstrapping may be effective in some of these cases. An asymptotic distribution for the sample sup–correlation has been obtained by Sethuraman (1990).

No extensions of monotone dependence notions to three or more dimensions are available.

3. Two–Sample Ordinal Data

Suppose we have random samples from two populations or treatments where each observation falls into one of k levels of an ordinal categorization. There are a variety of standard statistical procedures for comparing the two populations or treatments based upon these data.

An example of this type of problem is in clinical trials, where there is an experimental procedure (E) and a control (C), and the evaluations of each patient are the physician's global rating. These ratings are typically 5–point or 7–point scales, e.g., very improved, moderately improved, no change, moderately deteriorated, and very deteriorated. The standard analysis procedure typically involves: scoring the responses, on an equal–spaced scale, or using rank scores, or even dichotomizing the response. The interpretative difficulty is that the scorings are in some sense arbitrary, as would be any dichotomization.

More specifically, let $L_1 < \cdots < L_k$ denote the ordinal categorical levels' labels where "$<$" denotes the underlying experimental order. The arbitrariness in many of these two sample procedures comes from the choice of increasing scores $x_1 \leq \cdots \leq x_k$ ($x_1 \neq x_k$) that one can assign to the respective levels $L_1, \ldots, L_k$. Among the standardly used scoring systems are: (i) $1, \ldots, k$, (ii) $R_1, \ldots, R_k$, where R_i is the marginal mid–rank score for level i; (iii) ridit scores (and modified ridits), and (iv) $0, \ldots, 0, 1, \ldots, 1$ which proves a dichotomization of the levels.

As Kimeldorf, Sampson and Whitaker (1992), hereafter denoted by KSW, note, the commonly used procedures: (a) Wilcoxon–Mann–Whitney, (b) χ^2 on the dichotomization, (c) scored two sample t–test, (d) Cochran–Armitage test, and (e) appropriate log–linear models all share a common feature. For all sample sizes (or asymptotically in case (e)), the resultant test statistics are monotonically increasing functions of a certain correlation $r(x_1, \ldots, x_k)$. Denote the resulting data in the form:

	Levels				
	L_1	L_2	$\cdots$	L_k	Total
C	m_1	m_2	$\cdots$	m_k	m
E	n_1	n_2	$\cdots$	n_k	n
Total	$m_1 + n_1$	$m_2 + n_2$	$\cdots$	$m_k + n_k$	N

Then $r(x_1, \ldots, x_k)$ is the Pearson correlation coefficient based on the scores $x_1, \ldots, x_k$ and the values 0 assigned to C, and 1 to E.

The issue as discussed by KSW (1992) is how does the choice of scores effect $r(x_1, \ldots, x_k)$ and, consequently, the noted standardly employed procedures.

The approach taken by KSW (1992) is to find

$$r_{\text{MAX}} = \max_{\substack{x_1 \leq \cdots \leq x_k \\ x_1 \neq x_k}} r(x_1, \ldots, x_k)$$

and

$$r_{\text{MIN}} = \min_{\substack{x_1 \leq \cdots \leq x_k \\ x_1 \neq x_k}} r(x_1, \ldots, x_k)$$

and, thus, by monotonicity obtain the max and min of any of the related test statistics.

For example, suppose that our goal is to test $H_0 : E = C$ versus the alternative that E produces "larger" values than C, and we plan to use a scored t–test and the t distribution as the approximate null hypothesis distribution. If the resulting minimum t–statistic, t_{MIN}, is greater than t^α, the appropriate one–sided α–level critical value, then *all* scoring systems produce a statistically siginificant result. Similarly if $t_{\text{MAX}} < t^\alpha$, then *no* scoring system can produce a significant result. If we face the situation $t_{\text{MIN}} < t^\alpha < t_{\text{MAX}}$, which we call the "straddling" case, then the results depend on the choice of scoring system. In this case, much care must be taken in the choice of scales and in justifying them.

The computational approach of MONCOR could be used to compute r_{MIN} and r_{MAX}. However, an analytical solution is possible in this setting. We now briefly describe our approach to computing r_{MIN} and r_{MAX}, noting that since r is location and scale invariant, we can, if convenient, assume that $x_1 = 0$ and $x_k = 1$.

RESULT 1 *The empirical distribution for treatment E is stochastically greater (smaller) than that of C, if and only if* $r(X_1, \ldots, X_k) \geq 0$ (≤ 0) *for all possible* $x_1 \leq \cdots \leq x_k$ $(x_1 \neq x_k)$.

The applications of Result 1 are immediate. If treatments E and C are not stochastically comparable, then there exist scores, so that the one–sided or two–sided t–test, etc. will *not* reject $H_0 : E = C$. In fact, in this situation there exist scorings which yield both *positive* and *negative* t–values.

RESULT 2A *If E is stochastically incomparable to C, then* r_{MAX} *occurs at the scores* $y_1^* \leq \cdots \leq y_k^*$ *that minimize*

$$\sum_{i=1}^{k} (m_i + n_i)\{n_i/(m_i + n_i) - y_i\}^2 \tag{4}$$

among $y_1 \leq \cdots \leq y_k$; *and* r_{MIN} *occurs at the scores* $z_1^* \leq \cdots \leq z_k^*$ *that minimize*

$$\sum_{i=1}^{k} (m_i + n_i)\{m_i/(m_i + n_i) - y_i\}^2 \tag{5}$$

among $y_1 \leq \cdots \leq y_k$.

RESULT 2B *If E is stochastically greater than C,* r_{MAX} *occurs at the* $y_1^* \leq \cdots \leq y_k^*$ *given in (2A) and* r_{MIN} *occurs at one of the* $k-1$ *monotone extreme points, namely,* $(0,1,\ldots,1),(0,0,1,\ldots,1),\ldots,(0,\ldots,0,1)$.

RESULT 2C *If E is stochastically smaller than C, then* r_{MAX} *occurs at a monotone extreme point and* r_{MIN} *occurs at the* $z_1^* \leq \cdots \leq z_k^*$ *given in (2A).*

The solutions to (4) and (5) can be obtained, respectively, from the isotonic regression of $n_i/(m_i+n_i)$ and of $m_i/(m_i+n_i)$, both with weights (m_i+n_i). Robertson, Wright and Dykstra (1988) give a variety of algorithms to compute these isotonic regressions, including the simple Pool–Adjacent–Violators–Algorithm (PAVA). KSW (1992) illustrate the application of the PAVA technique to solve a number of examples. For moderate k, it is straightforward to directly compute r_{MIN} and r_{MAX}.

KSW (1992) provide proofs for Result 1 and Result 2. They also discuss the further interpretation of these results in data analysis but no distributional results have been obtained for the statistics: r_{MIN} and r_{MAX}. Further research on this class of problems is considered by Gautam (1991) in his dissertation.

4. Discussion

For both the problems, correlation between ordinal variables and testing two ordinal populations, we have considered the effects due to arbitrary monotone scorings. In each case we obtain min and max bounds on the appropriate statistics. The correlation problem requires extensive computation and the two–sample problem is solved simply.

Further discussion of the problematic "straddling" case for the two sample problem is given by KSW (1992). The extension of the two sample case to the one–way analysis of variance and the multivariate setting are being established by Guatam, Kimeldorf and Sampson.

Acknowledgement

The author wishes to thank both referees for their helpful comments.

REFERENCES

AGRESTI, A. (1984). *Analysis of Ordinal Categorical Data.* Wiley, New York.

BENZECRI, J. P. (1973). *L'analyse des donnees II: l'analyse des correspondences.* Dunod, Paris.

BISHOP, Y., FIENBERG, S. E. AND HOLLAND, P. (1975). *Discrete Multivariate Analysis: Theory and Practice.* MIT Press, Cambridge, MA.

BLOCK, H. W., CHHETRY, D., FANG, Z., AND SAMPSON, A. R. (1990). Partial orderings on permutations and dependence orderings on bivariate empirical distributions. *Ann. Statist.* **18** 1840–1850.

BREIMAN, L. AND FRIEDMAN, J. (1985). Estimating optimal transformations for multiple regression and correlation. *J. Amer. Stat. Assoc.* **80** 580–598.

CHHETRY, D. AND SAMPSON, A. R. (1987). A projection decomposition for bivariate discrete probability distributions. *SIAM J. Alg. Discrete Math.* **8** 501–509.

CHHETRY, D., DE LEUUW, J. AND SAMPSON, A. R. (1990). Monotone correlation and monotone disjunct pieces. *SIAM J. Matrix Anal. Appl.* **11** 361–368.

GAUTAM, S. P. (1991). Application of the t, T^2 and F statistics to ordinal categorical data. Ph.D. Dissertation, Department of Mathematical Sciences, University of Texas at Dallas.

GILULA, Z. (1984). On some similarities between canonical correlation models and latent class models for two–way contingency tables. *Biometrika* **71** 523–529.

GLASS, D. V. AND HALL, J. R. (1954). Social Mobility in Britain: A study of inter–generation changes in status. In *Social Mobility in Britain,* D. V. Glass, ed. Routledge and Kogan Paul, London.

KENDALL, M. G. AND STUART, A. (1979). *The Advanced Theory of Statistics* Vol. 2, 4th edition, Griffen, London.

KIMELDORF, G. AND SAMPSON, A. R. (1975). Uniform representations of bivariate distributions. *Commun. Statist.* **4** 617–627.

KIMELDORF, G. AND SAMPSON, A. R. (1978). Monotone dependence. *Ann. Statist.* **6** 895–903.

KIMELDORF, G., MAY, J. H., AND SAMPSON, A. R. (1981). A program to compute concordant and other monotone correlations. In *Proceedings of Computer Science and Statistics: 13th Symposium on the Interface,* W. F. Eddy, ed. Springer–Verlag, 348–351.

KIMELDORF, G., MAY, J. H. AND SAMPSON, A. R. (1982). Concordant and discordant monotone correlations and their evaluation using nonlinear optimization. In *Optimization in Statistics,* S. H. Zanakis and J. S. Rustagi, eds., North–Holland, Amsterdam. 117–130.

KIMELDORF, G., SAMPSON, A. R. AND WHITAKER, L. (1992). Min and max scorings for two–sample ordinal data. *J. Amer. Statist. Assoc.* **87** 241–247.

LANCASTER, H. O. (1969). *The Chi–Squared Distribution.* Wiley, New York.

LEHMANN, E. (1966). Some concepts of dependence. *Ann. Math. Statist.* **37** 1137–53.

MAY, J. H. (1979). Solving nonlinear programs without using analytic derivatives. *Oper. Res.* **27** 457–484.

NISHISATO, S. (1980). *Analysis of Categorical Data: Dual Scaling and Its Applications.* University of Toronto Press, Toronto.

ROBERTSON, T., WRIGHT, F. T. AND DYKSTRA, R. L. (1988). *Order Restricted Statistical Inference.* Wiley, New York.

SCHRIEVER, B. F. (1983). Scaling of order dependent categorical variables with correspondence analysis. *Intern. Statist. Rev.* **51** 225–238.

SCHRIEVER, B. F. (1985). Order Dependence. Ph.D. Dissertation, Free University of Amsterdam.

SETHURAMAN, J. (1990). The asymptotic distribution of the Renyi maximal correlation. *Commun. Statist. - Theo. Meth.* **19** 4291–4298.

SKLAR, A. (1959). Fonctions de répartition a n dimensions et leurs marges. *Publ. Inst. Statist. Univ. Paris 8,* 229–231.

VITALE, R. A. (1990). On stochastic dependence and class of degenerate distributions. In *Topics of Statistical Dependence*, H. W. Block, T. Savits, and A. R. Sampson, eds. Institute of Mathematical Statistics, Hayward, CA. 459–469.

WILLIAMS, E. J. (1952). Use of scores for the analysis of association in contingency tables. *Biometrika* **39** 274–289.

DEPARTMENT OF MATHEMATICS AND STATISTICS
UNIVERSITY OF PITTSBURGH
PITTSBURGH, PA 15260

Stochastic Inequalities
IMS Lecture Notes – Monograph Series
Volume 22 (1993)

SECRETARY PROBLEMS AS A SOURCE OF BENCHMARK BOUNDS

By STEPHEN M. SAMUELS

Purdue University

Secretary problems are those sequential selection problems in which the payoff (or cost) depends on the observations only through their ranks. A subclass of such problems allows only selection rules based on relative ranks. The performance of such rules provides readily accessible lower bounds for procedures based on more information. Included here are familiar bounds, like $1/e$; well-known bounds, like 3.8695; and brand-new bounds, like 2.6003.

1. Googol

Although there is a *pre-history* associated with Secretary Problems—which some have traced back into the 19th century—the generally agreed-upon "big bang" took place with Martin Gardner's presentation of the following problem in his Mathematical Games column in the February, 1960 *Scientific American.*

> Ask someone to take as many slips of paper as he pleases, and on each slip write a different positive number. The numbers may range from small fractions of one to a number the size of a *googol* (1 followed by a hundred zeros) or even larger. These slips are turned face-down and shuffled over the top of a table. One at a time you turn the slips face up. The aim is to stop turning when you come to the number that you guess to be the largest of the series. You cannot go back and pick a previously turned slip. If you turn over all the slips, then of course you must pick the last one turned.

The "solution" in the March, 1960 column, treated googol as though it were the classical best-choice problem. That is to say, it was taken for granted that only stopping rules based on the relative ranks of the numbers need be considered. (That was, after all, the point of calling it "googol," wasn't it?)

AMS 1991 ***subject classifications.*** Primary 60G40, Secondary 62L15.

Key words and phrases. Optimal stopping, best choice, backward induction, dynamic programming, googol, relative ranks.

The optimal rule based on relative ranks is well-known to be among rules of the form: let $k-1$ go by and then select the first relatively best one (if any). For such stopping rules, the probability of best choice is

$$\phi_n(k) = \sum_{i=k}^{n} \frac{1}{n} \cdot \frac{k-1}{i-1}. \tag{1}$$

where n is the total number of slips of paper. A well-known elementary argument (which need not be repeated here) shows that, for k equal to half of n, this probability is at least $1/4$, no matter how large n is. If $k(n)/n \to x$ as $n \to \infty$, then

$$\phi_n(k(n)) \to -x \log x, \tag{2}$$

which is about .35 if $x = .5$, but is maximized at $x = 1/e \approx .3679$; and the celebrated maximum is also $1/e$.

Here we have the most famous secretary problem benchmark bound. How sharp is it?

1.1. *Full Information Problem*

Suppose the numbers are known to be a random sample (i.e., i.i.d.) from some specified continuous distribution. Whatever the distribution, the optimal probability of selecting the largest number, call it w_n, decreases with n to

$$\begin{aligned} \lim_{n\to\infty} w_n &= e^{-c} - (e^c - c - 1)\int_1^\infty x^{-1}e^{-cx}\,dx. \\ &\approx .5802 \end{aligned} \tag{3}$$

where $c \approx .804$ is the solution to

$$\sum_{j=1}^{\infty} c^j/j!j = 1. \tag{4}$$

See Gilbert and Mosteller (1966, Section 3) and Samuels (1982).

1.2. *Partial Information Problems from the Minimax Point of View*

Now suppose that only some parametric family (e.g. normal) is specified, perhaps together with a prior distribution on the parameters.

For location and scale parameter families, Petruccelli (1978) found sufficient conditions for the existence of a sequence of invariant stopping rules for which the probability of best choice converges to the full-information limiting value as $n \to \infty$. The normal family satisfies his conditions, but the uniforms do not.

For the family of uniforms on $(\theta-\frac{1}{2},\theta+\frac{1}{2})$, Petruccelli (1980) showed that the limiting probability of best choice for the best invariant rules (which are minimax by a version of the Hunt-Stein theorem) is .4352, a value intermediate between the full-information value of .5802 and the "no-information" value of $1/e = .3679$.

For the family of all uniforms, Samuels (1981) showed that the best-choice problem solution is minimax, so, at least from the minimax point of view, knowing that the distribution is uniform is "no information." A simplification of the original proof was given by Ferguson (1989), and can also be found in Samuels (1991).

1.3. *Partial Information Problems from a Bayesian Point of View*

This minimax result still begs the question of whether there is *any* exchangeable distribution for which it is optimal to consider only the relative ranks. (It is implicit in the statement of Googol that the numbers are exchangeable.) Samuels (1989) posed this question and called it Ferguson's Secretary Problem because Ferguson (1989) showed that for any $\varepsilon > 0$ and for any n, there is a two-parameter Pareto prior distribution on θ such that, when sequentially observing n uniform r.v.'s on $[0,\theta]$, the best rule based only on relative ranks comes within ϵ of being optimal. Recently, Hill and Krengel (1991b) have extended Ferguson's result to the case where the number of items, n, is unknown with a known upper bound. In this context, "optimal" is replaced by "minimax-optimal." The same authors had already found the minimax rules based on relative ranks in Hill and Krengel (1991a).

For the case $n = 2$, the answer to the above question is NO; there is *no* exchangeable distribution for which it is optimal to consider only the relative ranks. This is easily seen by the following simple and well-known argument:

> Let X_1, X_2 be the first and second numbers examined, respectively. Now, pick any number, x, between the inf and the sup of the support of the X's, and choose X_1 if $X_1 > x$; otherwise choose X_2. If both X_1 and X_2 turn out to be bigger than x, or if both are smaller than x, then (by exchangeability) this rule selects the larger of the two with probability 1/2, while, if one random variable is larger than x while the other is smaller, the larger one is sure to be chosen. Thus, setting the unknown $P\{\min(X_1,X_2) < x < \max(X_1,X_2)\}$ equal to c, say, we have
>
> $$P\{X_\tau = \max(X_1,X_2)\} = c + (1-c)/2 = (1/2)(1+c),$$
>
> which is strictly greater than 1/2. This beats rules based only on relative ranks, which, for $n = 2$, are necessarily constants, so have probability 1/2 of success.

Recently Silverman and Nádas (1991) have shown, to the surprise of many, that for $n = 3$, there *are* such distributions. They began by giving the following necessary and sufficient conditions for achieving optimality with a rule based on relative ranks:

$$\begin{aligned} &P(X_1 = \max(X_1, X_2, X_3)|X_1) \\ &\quad +(1/4)P(X_1 = \min(X_1, X_2, X_3)|X_1) \leq 1/2 \quad a.s. \end{aligned} \tag{5}$$

and

$$P(\max(X_1, X_2) = \max(X_1, X_2, X_3)|X_1, X_2) \geq 1/2 \quad a.s. \tag{6}$$

(This corrected an error in Samuels (1989) which had (6) all right, but omitted the second term of (5).) Then they let X_1, X_2, X_3, given θ, be conditionally i.i.d., uniform on $[0, \theta]$, with prior density on θ of the form:

$$g(\theta) = tI_{\{0<\theta\leq 1\}} + (1-t)\frac{\alpha}{\theta^{1+\alpha}}I_{\{\theta>1\}} \qquad \alpha > 0,\ 0 \leq t \leq 1. \tag{7}$$

This family includes—for $t = 0$—the Pareto priors used in Ferguson (1989) to "come within ε;" see above. By enlarging the class of available priors, Silverman and Nádas (1991) were able to find a subclass for which (5) and (6) are both satisfied; namely those with $3t/(2-3t) \leq \alpha \leq 2t/(1-2t)$.

If this result can be extended to all n, then it can truly be said that the famous $1/e$ benchmark bound is sharp.

2. A General Class of Problems

Problems involving arbitrary loss functions, a random number of arrivals, or a sampling cost—generally considered separately in the literature—can be combined into a single model, as follows:

Let N denote the number of rankable items which appear in random order, $X_1, X_2, \ldots, X_N$ be their ranks, and $Y_1, Y_2, \ldots, Y_N$ be the corresponding relative ranks. Conditional on $\{N = n\}$, $X_1, X_2, \ldots, X_n$ are a random permutation of $\{1, 2, \ldots, n\}$, hence the Y_i's are independent, uniform on $\{1, 2, \ldots, i\}$. All stopping rules, τ, are based on the Y's but suitably modified to contend with the possible randomness of N. And there are risks, $A(i, j)$, for stopping at time i with an item of relative rank j, of the form

$$A(i,j) = H(i) + K(i,j). \tag{8}$$

(The use of two terms, where one would do, is for clarity in what follows.) An optimal rule is one which minimizes $EA(\tau, Y_\tau)$.

For example, here are three problems, all of which have

$$H^{(n^*)}(i) = (i-1)/n^*,$$

$$K^{(n^*)}(i,1) = \sum_{m=i+1}^{n^*} (1 - i/m)(1/n^*),$$

$$K^{(n^*)}(i,j) \equiv 1 \qquad j > 1. \tag{9}$$

• Best-choice problem with N uniform on 1 to n^*: subject to the "boundary condition" that, when the N-th item is best, we only get it by actually selecting it (Presman and Sonin (1972)).

• Payoff equals proportion of time holding the relatively best: $N \equiv n^*$; and if we "stop" with the τ-th item, a relatively best one, and the next relatively best item is the σ-th item, our reward is $(\sigma - \tau)/n^*$; or, if there is no subsequent relatively best item, the reward is $(n^* - \tau)/n^*$ (Ferguson, Hardwick and Tamaki (1991)).

• Linear sampling cost plus oddball risk term: $N \equiv n^*$; $H^{(n^*)}(i)$ is the linear sampling cost, and $K^{(n^*)}(i,1)$ is the hard-to-interpret risk when selecting a relatively best item at stage i.

2.1. *General Loss Function*

Let us now specialize. Let $N \equiv n$ be fixed, and prescribe a non-decreasing loss function, $q(\cdot)$, where $q(i)$ is the loss for selecting the item which turns out to be i-th best overall. Then we can take $H(i) \equiv 0$ and $K(i,j) = R^{(n)}(i,j)$, where

$$R^{(n)}(i,j) = E_n[q(X_i) \mid Y_i = j]$$

$$= \sum_{k=j}^{n} \frac{\binom{k-1}{j-1}\binom{n-k}{i-j}}{\binom{n}{i}} q(k). \tag{10}$$

Using backward induction plus the independence of the sequence of relative ranks, we can conclude that the quantities

$$c_i^{(n)} \equiv \inf\nolimits_{\tau > i} E_n(R^{(n)}(\tau, Y_\tau) \mid Y_1, \ldots, Y_i) \tag{11}$$

are constants, and that the formula

$$c_{i-1}^{(n)} = \frac{1}{i}\sum_{j=1}^{i} \min[R^{(n)}(i,j),\, c_i^{(n)}] \qquad i = n-1,\, n-2, \ldots, 1, \tag{12}$$

holds, with boundary condition

$$c_{n-1}^{(n)} = \frac{1}{n}\sum_{j=1}^{n}[R^{(n)}(n,j)] \tag{13}$$

and minimal risk, over all stopping rules,

$$v_n = c_0^{(n)} \tag{14}$$

Equation (12) can be rewritten in difference equation form as

$$\frac{c_i^{(n)} - c_{i-1}^{(n)}}{1/n} = \frac{1}{i/n}\sum_{j=1}^{i}\left[c_i^{(n)} - R^{(n)}(i,j)\right]^+ . \tag{15}$$

For fixed j, the risk (10) is decreasing in i to $q(j)$, so, from (12), if we let

$$i_k^{(n)} = \begin{cases} \min\{i : R^{(n)}(i,k) \le c_i^{(n)}\} \\ n \qquad \text{if no such } i \end{cases}, \tag{16}$$

then an optimal rule stops at the first i for which, for some k, $i_k \le i < i_{k+1}$ and $Y_i \le k$. Its risk is

$$v_n = c_0^{(n)} = c_1^{(n)} = \ldots = c_{i_1-1}^{(n)}. \tag{17}$$

In addition, the risks have a limit as $n \to \infty$, namely

$$i/n \to t \Rightarrow R^{(n)}(i,j) \to R_j(t) = \sum_{k=1}^{\infty} q(k)\binom{k-1}{j-1} t^j (1-t)^{k-j}, \tag{18}$$

and, as Mucci (1973a and 1973b) showed for a large class of $q(\cdot)$'s, $c_i^{(n)} \approx f(i/n)$, where

$$f'(t) = \frac{1}{t}\sum_{j=1}^{\infty}[f(t) - R_j(t)]^+ \quad 0 \le t < 1 \tag{19}$$

with boundary condition, $f(1) = \sup q(\cdot)$, and a non-decreasing sequence of *thresholds,*

$$t_k : \ f(t_k) = R_k(t_k) \qquad k = 1, 2, \ldots. \tag{20}$$

$f(\cdot)$ is constant on $[0, t_1]$, so the limiting optimal risk is

$$v = \lim v_n = f(0) = f(t_1), \tag{21}$$

and (19) can be rewritten as the piecewise differential equation,

$$\left(\frac{f(t)}{t^k}\right)' = -\frac{1}{t^{k+1}}\sum_{j=1}^{k} R_j(t) \qquad t_k \le t \le t_{k+1}. \tag{22}$$

2.1.1. Some special cases

BEST-CHOICE PROBLEM: $q(1) = 0$ and $q(i) \equiv 1$ for all $i > 1$; $R^{(n)}(i,1) = 1 - i/n$, $R_1(t) = 1 - t$,

$$f(t) = \begin{cases} 1 - e^{-1} & 0 \le t \le e^{-1} \\ 1 - |t \ln t| & e^{-1} \le t \le 1 \end{cases} . \tag{23}$$

POLYNOMIAL-IN-RANKS PROBLEM: If $q(i) \equiv i(i+1)\cdots(i+a-1)$ for some integer $a \ge 1$, then (10) becomes

$$R^{(n)}(i,j) = j(j+1)\cdots(j+a-1)\cdot\left(\frac{(n+1)\cdots(n+a)}{(i+1)\cdots(i+a)}\right), \tag{24}$$

so

$$R_j(t) = j(j+1)\cdots(j+a-1)t^{-a} \tag{25}$$

and $v = a!/t_1^a = a!B_a^a$, where

$$B_a = \prod_{j=1}^{\infty}\left(\frac{j+a+1}{j}\right)^{1/(j+a)} \to \exp(\pi^2/6) \;\text{ as } a \to \infty. \tag{26}$$

Chow *et al* (1964) derived $B_1 \simeq 3.8695$ directly from the difference equation, without using the piecewise differential equation, and, recently, Robbins (1989) reported the above generalization. This remarkable result of a finite risk (less than four, in fact, for $a = 1$) despite an unbounded loss function can also be obtained in a more transparent way using memory-length one rules (see Section 3.1).

2.1.2. Monotonicity of the optimal risk, v_n

For any non-decreasing loss function, $q(\cdot)$, the optimal risk, v_n, is non-decreasing in n, the number of items to be observed. And v_n is strictly increasing whenever $q(\cdot)$ is. One way to prove this is to consider an $n+1$-arrival problem in which we are told when the current item is *worst* of all $n+1$. Since an optimal rule never selects a relatively worst item unless it is the last one, we can use the optimal $n+1$-arrival rule to select one of the other n arrivals. This is easily seen to be simply a *randomized* n-arrival rule, hence suboptimal for n arrivals, but possibly super-optimal for $n+1$ since it never selects the worst one. See, e.g., Chow *et al* (1964).

2.1.3. A risk 'paradox'

Since, for *fixed* sample size, the optimal risk, v_n, is increasing, one might naively expect the optimal risk for a bounded (by n^*) arrival distribution to be less than v_{n^*}. That this is quite false was demonstrated in Gianini-Pettitt (1979) for the arrival distributions defined by

$$P(N = i \mid N \geq i) = (n^* - i + 1)^{-\alpha} \qquad i = 1, 2, \ldots, n^* \tag{27}$$

(which includes, for $\alpha = 1$, the uniform distributions). For loss equal to the rank, the limiting optimal risk is infinite if $\alpha < 2$ and equal to the fixed sample size limit (3.8695) if $\alpha > 2$. The "paradox" disappears when we realize (as shown at the beginning of this section) that uncertainty about the number of arrivals is like imposing a sampling cost.

2.1.4. Minimax quantile problem

Take a random sample of size n from a member of some family, $\{F_\theta\}$, of distributions and try to minimize $\max_\theta EF_\theta(X_\tau)$. This is not a secretary problem but, in the special case where the F_θ's are uniform, the solution to the rank problem is relevant, because the expectations of the successive order statistics are proportional to the ranks. Thus, from the solution to the Rank Problem (26), the minimax risk is at most $\approx 3.8695/n$—for a rule based only on the relative ranks!—which is not far from the asymptotically optimal value of $(3 + 2\sqrt{2})^{1/\sqrt{2}}/n \simeq 3.4780/n$ (Samuels (1981)).

2.1.5. Full information rank problem

A rank problem which *is* a secretary problem is the one where we sample from a known continuous distribution and wish to maximize the expected rank of an item selected by a stopping rule. This problem is currently under study by Assaf and Samuel-Cahn (1991) and by Bruss and Ferguson (1991). The optimal expected rank, as $n \to \infty$ has, so far, been shown to be somewhere between 1.85 and 2.33.

3. An Infinite Model

Let the best, second best, etc., of an *infinite* sequence of rankable items arrive at times $U_1, U_2, \ldots$, which are i.i.d., uniformly distributed on the unit interval, $[0, 1]$. For each t in this interval, let $V_i(t)$ be the arrival time of the item which is i-th best among all those which arrive before time t. Let $\mathcal{F}_t$ be the sigma field generated by $V_1(t), V_2(t), \ldots$, and consider the class of all stopping rules, τ, adapted to the $\mathcal{F}_t$'s and taking values in the set $\{U_i\} \cup \{0, 1\}$; the values 0 and 1 are included to allow for the possibility of not starting or not stopping.

As in Section 2, let $A(t, j)$ be the prescribed loss for stopping at time t with an item of relative rank j; the goal is to minimize $EA(\tau, Y_\tau)$, among all stopping rules, τ, where Y_τ is the relative rank, at its arrival time, of the item which arrives at time τ. For example, as in Section 2.1, a non-decreasing loss function, $q(\cdot)$, may be given, and

$$A(t,j) = R_j(t) \equiv E[q(X_t) \mid Y_t = j] \tag{28}$$

where, in a slight abuse of notation, we are letting X_t and Y_t denote the absolute and relative ranks at time t, of an arrival at time t. This can easily be made rigorous and the result is that $R_j(t)$ is precisely the limit of $R^{(n)}(i,j)$ given in (18).

This model was first proposed in an abstract by Rubin (1966) and worked out in detail in Gianini and Samuels (1976), Gianini (1977), and Lorenzen (1979). It is appealing for a number of reasons, among them

- it is consistent with the "finite model" of the previous section, because if the ranks of the n successive arrivals are a random permutation of 1 to n, then the arrival times of the best, second best, etc., are also a random permutation of 1 to n and *vice versa;*
- it yields upper bounds for finite problem risks in an elementary way;
- it is, in several ways, the limit of the finite problem—in particular, backward induction yields the differential equation (19) directly;
- it is the natural setting for a number of important applications.

3.1. *Easy Upper Bounds with Memory-Length One Rules*

If the stopping risk is given by (28)—i.e. there is no sampling cost—then the risk using any stopping rule in the infinite problem is an upper bound for the optimal risks for all n in the corresponding (i.e. same $q(\cdot)$ function) finite problem of Section 2.1. This is because, if we augment the sigma-fields, $\mathcal{F}_t$ to include information about which arrivals by time t are among the n best overall, and modify an arbitrary infinite problem stopping rule by having it select the last arrival among the n best whenever the original rule does not select one of the n best, then the modified rule has reduced risk (because $q(\cdot)$ is non-decreasing); but it is also a *randomized* rule (hence suboptimal) for the n-arrival problem, because, regardless of their arrival times, the successive ranks of the n best arrivals are a random permutation of 1 to n.

The infinite model includes stopping rules which are much more tractable than any in the finite problem. For example, suppose we choose an infinite sequence of numbers

$$0 = R_0 < A_1 < R_1 < A_2 < \cdots < A_k < R_k < \cdots < 1 \tag{29}$$

(that's R as in *Remember* and A as in *Accept*), and stop at τ equal to the first time we have an arrival in an interval of the form $[A_i, R_i)$ which is better than the best arrival in the previous interval, $[R_{i-1}, A_i)$. Let the A's and R's be chosen so that for all $i = 0, 1, \ldots,$

$$(R_{i+1} - R_i) = R_1(1 - R_1)^i \tag{30}$$

and

$$(A_{i+1} - R_i)/(R_{i+1} - R_i) \equiv p. \tag{31}$$

(The idea here is to make the problem recursive.) Then $P(\tau > R_i) = p^i$, so $\tau < 1$ a.s. Moreover, since the k-th best arrival in $(R_1, 1)$ has expected rank $k/(1 - R_1)$, we easily conclude that

$$EX_\tau = EX_\tau I_{\{\tau < R_1\}} + P(\tau > R_1)[EX_\tau/(1 - R_1)], \tag{32}$$

which is finite if and only if

$$P(\tau > R_1) = p < 1 - R_1. \tag{33}$$

Thus, for the Rank Problem of Section 2.1, the infinite model provides an easy demonstration of the finiteness of the limiting risk. Rubin and Samuels (1977) studied these memory-length one rules and showed that, for the optimal rule of this type, the expected rank is about 7.4. This same argument can be used for any polynomial loss. In particular, for losses of the form $q(i) = i(i+1)\cdots(i+m)$, the above rules can be shown to have finite risk if and only if $p < (1 - R_1)^m$.

3.2. *Infinite Problem as Limit of Finite Problem*

When the stopping loss satisfies (28), [and under mild conditions otherwise], if there is any stopping rule with finite risk, then, using backward induction, we can conclude, as in Section 2 that the quantities

$$f(t) \equiv \inf_{\tau > t} E(A(\tau, Y_\tau) \mid \mathcal{F}_t) \tag{34}$$

are constants satisfying the differential equation (19), with boundary condition $f(1) = \sup q(\cdot)$, minimal risk, v, given by (21), and optimal rule which is the (appropriately scaled) limit of the finite problem optimal rules, namely, if we haven't stopped before time t_k, given by (20), then we stop with the first arrival in $[t_k, t_{k+1}]$, if any, which has relative rank $\leq k$.

In addition, (21) *always* holds, i.e., the infinite problem minimal risk, finite or infinite, is always the limit of the finite problem minimal risks. We have already seen, in Section 3.1, that v is finite whenever $q(\cdot)$ grows no faster than a polynomial. On the other hand, if $\sum[\log q(k)]/k^2 = \infty$, then v is infinite (Gianini (1977)).

3.3. *Easy Upper Bounds for Unknown Number of Arrivals*

Instead of prescribing a prior distribution of the number of arrivals, as in Section 2, and having them appear in discrete time, we may have the arrivals occur in continuous time as some kind of stochastic process. Suppose for example that the best, second best, etc., of N (a random variable) items arrive at times which are I.I.D. with some continuous distribution, F, on the time interval, $(0, \infty)$. A much-studied special case is the best-choice problem with a Poisson arrival process, which is equivalent to $N \sim$ Poisson and $F \sim$ uniform on $(0, T)$ for some T (Cowan and Zabczyk (1978) and Bruss (1987)).

Since, without loss of generality, the arrival times can be taken to be uniform on $(0, 1)$, the infinite model provides a unifying framework for such problems (Bruss (1984), Bruss and Samuels (1987 and 1990), and Sakaguchi (1989)). Suppose, in the infinite model, instead of observing the entire infinite collection of arrivals, we can only observe the N best, where N has distribution G. What effect does this "censoring" have on, say, an optimal rule, τ^*, for the general loss problem, (28)? Clearly censoring delays stopping, and—as long as the loss for not stopping, $Q(N)$, is no bigger than $q(N+1)$—censoring is guaranteed to reduce the risk. Thus, the particular G, be it Poisson or whatever, is nearly irrelevant: The optimal infinite problem stopping rule for the given loss function $q(\cdot)$ provides an upper bound for the optimal risk for *all* distributions of N. Moreover, it can easily be shown to be nearly optimal itself for all stochastically large N. Specifically, letting $v^{(N)}$ denote the minimal risk, we have

$$v^{(N)} \leq E^{(N)} q(R_{\tau^*}) \leq v \tag{35}$$

and

$$N \uparrow \infty \text{ in distribution} \Rightarrow v^{(N)} \uparrow v. \tag{36}$$

(The result applies whenever v is finite.)

For loss functions that are eventually constant, as in choosing one of the r best, a more logical loss for not stopping is $Q(N) \equiv c$ (a constant). If $c \leq \max q(\cdot)$, the above results must be modified slightly; they apply only to N's for which

$$P\{q(N+1) = \max q(\cdot) | N > 0\} = 1. \tag{37}$$

This, by the way, is guaranteed in the *best choice problem*: $r = 1$.

3.4. *Best-Choice Problem with Recall*

Suppose we relax the stopping-rule requirement of no recall to allow "backward solicitation" of previously observed items. Specifically, let $\alpha \epsilon [0, 1)$ be a *recall time;* an item which arrives at time t can be held until time $t + \alpha$, when it must be either selected or discarded (Rocha (1988)). As one would expect, the optimal rule is of the form τ_t: ignore all arrivals up to time t, and thereafter select the first "candidate" which is still relatively best at the end of the recall period (or at time 1, whichever comes first). This rule selects the overall best item with probability

$$\psi_\alpha(t) \equiv P[\tau_t = \min(U_1 + \alpha, 1)] \equiv P[\tau_t \text{ "picks best"}]. \tag{38}$$

The goal is to find—for each $\alpha \in [0,1]$—the optimal t, say t_α, and the corresponding probability, $\psi_\alpha(t_\alpha) \equiv v(\alpha)$.

For $t \geq 1 - \alpha$ it is trivial that

$$\psi_\alpha(t) = P[U_1 > t] = 1 - t \qquad 1 - \alpha \leq t \leq 1, \tag{39}$$

and the classical no-recall case, $\alpha = 0$, is also immediate:

$$\psi_0(t) = P[\{U_1 > t\} \cap \{V_1(U_1) < t\}] = \int_t^1 \frac{t}{z} dz = -t \ln t. \tag{40}$$

(Both U and V were defined at the beginning of this section.) But, the recall case, with $t < 1 - \alpha$ is more complicated. For $0 < \delta \leq \alpha$ and $t + \delta \leq 1 - \alpha$,

$$\{\tau_t \text{ "picks best" but } \tau_{t+\delta} \text{ "doesn't"}\} = \{t < U_1 \leq t + \delta\} \tag{41}$$

while

$$\begin{aligned} &\{t < V_1(t+\alpha) < t + \delta\} \cap \{\tau_{t+\alpha} \text{ "picks best"}\} \\ &\supset \{\tau_{t+\delta} \text{ "picks best" but } \tau_t \text{ "doesn't"}\} \supset \\ &\{t < V_1(t+\alpha+\delta) < t+\delta\} \cap \{\tau_{t+\alpha+\delta} \text{ "picks best"}\}. \end{aligned} \tag{42}$$

This leads to the useful inequalities,

$$\delta\left[1 - \frac{\psi_\alpha(t+\alpha)}{t+\alpha}\right] < \psi_\alpha(t) - \psi_\alpha(t+\delta) < \delta\left[1 - \frac{\psi_\alpha(t+\alpha+\delta)}{t+\alpha+\delta}\right]. \tag{43}$$

which tell us that $\psi_\alpha(t)/t$ is decreasing; and that $\psi_\alpha(\cdot)$ is unimodal, with its maximum at

$$t_\alpha \equiv t^*_\alpha - \alpha \tag{44}$$

where

$$t^*_\alpha = \begin{cases} \alpha & \text{if } \alpha \geq \frac{1}{2} \\ \psi_\alpha(t^*_\alpha) & \text{if } \alpha \leq \frac{1}{2} \end{cases} \tag{45}$$

and satisfies the differential equation

$$\psi'_\alpha(t) = \frac{\psi_\alpha(t+\alpha)}{t+\alpha} - 1 \qquad 0 < t < 1 - \alpha. \tag{46}$$

It's easy to solve the differential equation on $[(1-2\alpha)^+, 1-\alpha]$ by substituting the known value $\psi_\alpha(t+\alpha) = 1-t-\alpha$, from (39), into the right side of (46). The solution is

$$\psi_\alpha(t) = 2 - 2t - \alpha + \ln(t+\alpha) \qquad (1-2\alpha)^+ \le t \le 1-\alpha. \tag{47}$$

Thus the complete solution for $\alpha \ge \frac{1}{2}$ is

$$\left.\begin{array}{rcl} t_\alpha & = & 0 \\ v(\alpha) & = & 2 - \alpha + \ln\alpha \end{array}\right\} \quad \alpha \ge \tfrac{1}{2}. \tag{48}$$

And, for $\psi_\alpha(1-2\alpha) \ge (1-2\alpha)$—which holds for $\alpha \ge .260303$—we can solve for t^*_α from (47). But then the trouble starts. For $t < 1-2\alpha$ we face integrals of the form

$$\int \frac{\ln(t+2\alpha)}{t+\alpha}dt \tag{49}$$

which cannot be expressed in closed form. An attractive alternative to numerical integration (which works quite well) is to obtain, probabilistically, improved upper and lower bounds for $\psi(\cdot)$—better than those in (43)—and use them to get upper and lower bounds for t^*_α and for $\psi_\alpha(t^*_\alpha)$.

Equations (48) are the limiting solutions for the corresponding finite problem studied by Smith and Deely (1975).

4. Multiple Criteria

4.1. *Best Choice Problem with Independent Criteria*

Suppose each item's relative ranks with respect to m independent criteria are to be observed and we want to maximize the probability of choosing an item which is best in an at least one criterion. The finite problem was studied by Gnedin (1982). Asymptotically, an m-dimensional infinite model applies, leading to the differential equation

$$f'(t) = \frac{m}{t}[f(t) - (1-t)]^+ \qquad 0 \le t < 1 \tag{50}$$

where $f(t)$ is, as usual, the minimal risk for rules which do not stop before t and the boundary condition is $f(1) = 1$. The solution for $m > 1$ is

$$\begin{array}{rcll} f(t) & = & 1 - \dfrac{m}{m-1}t(1-t^{m-1}) & \quad t > t^* \\ 1 - f(t^*) & = & t^* = (\dfrac{1}{m})^{1/(m-1)}. & \end{array} \tag{51}$$

In particular, for $m = 2$, an asymptotically optimal rule lets half of the items go by before being willing to stop, and has probability 1/2 of success.

4.2. *Best Rank Problem with Independent Criteria*

Govindarajulu (1991) studied a version of the secretary problem in which there are two independent streams of candidates. At stage i, the relative ranks of the i-th item—relative to its predecessors in its own stream—are observed. Only one item, from one stream or the other, will be selected, and the goal is to minimize its rank in its own stream.

If selection is made after seeing both items, this problem is equivalent to a two independent criteria problem; if not, they are still asymptotically equivalent.

For m criteria, asymptotically, the same m-dimensional model as in Section 4.1 applies, leading to the piecewise differential equation

$$f'(t) = \frac{mk}{t} f(t) - \frac{m}{2} \frac{k(k+1)}{t^2} \qquad t_k \le t \le t_{k+1} \tag{52}$$

with boundary condition $f(1) = \infty$, where the thresholds, t_k, are increasing and satisfy $f(t_k) = k/t_k$. Dividing both sides of (52) by t^{mk} leads to the solution

$$f(t_1) = 1/t_1 = \prod_{k=1}^{\infty} \left(1 + \frac{2(mk+1)}{k(mk+2-m)}\right)^{1/(mk+1)}. \tag{53}$$

	Best Choice	Best Rank	
m	t^*	t_1	$f(t_1)$
1	.3679	.2584	3.8695
2	.5000	.3846	2.6003
3	.5774	.4670	2.1413
4	.6300	.5267	1.8987
5	.6687	.5724	1.7469
10	.7743	.7040	1.4205
20	.8541	.8088	1.2363
50	.9233	.9011	1.1097
100	.9545	.9425	1.0610

Table 1. Asymptotic Results for m-Criteria Problems.

4.3. *Sum of the Ranks Problem with Independent Criteria*

Suppose, as before, that we observe the relative ranks with respect to m independent criteria, but now want to minimize the sum of the ranks. This is a multivariate extension of the rank problem in Section 2.1; however, for $m \ge 2$, we cannot expect the risk to remain bounded as the number of items becomes infinite. Samuels and Chotlos (1986) give both the optimal stopping and extreme value results for m independent permutations,

$$\langle X_1^{(1)}, \ldots, X_n^{(1)} \rangle, \; \cdots \; , \langle X_1^{(m)}, \ldots, X_n^{(m)} \rangle, \tag{54}$$

of the integers 1 to n. As $n \to \infty$,

$$\frac{1}{n^{1-1/m}} E\left[\min_{1\leq i\leq n} \sum_{j=1}^{m} X_i^{(j)}\right] \to (m!)^{1/m}\Gamma(1+1/m) \tag{55}$$

and

$$\frac{1}{n^{1-1/m}} E\left[\min_{\tau\leq n} \sum_{j=1}^{m} X_\tau^{(j)}\right] \to \left[\frac{(m+1)!}{m}\right]^{1/m}. \tag{56}$$

The ratio of these two limits—the right side of (56) divided by the right side of (55)—decreases to one as $m \uparrow \infty$.

4.3.1. Sums of i.i.d uniforms

It is worth noting that (55) and (56) also hold if all of the $X_i^{(j)}$'s are i.i.d., uniform on $(0,1)$. In fact the solution to the above secretary problem was obtained in Samuels and Chotlos (1986) by showing that it can be approximated by this non-secretary problem. The approximation works for $m > 1$ but not for $m = 1$. Lindley (1961), in effect, tried it for $m = 1$ and got an asymptotic risk of 2, which he noted was incorrect. The correct limit is 3.8695, as in (26).

REFERENCES

ASSAF, D. AND SAMUEL-CAHN, E. (1991). The secretary problem: minimizing the expected rank with i.i.d. random variables. Preprint.

BRUSS, F. T. (1984). A unified approach to a class of best choice problems with an unknown number of options. *Ann. Probab.* **12** 882-889.

BRUSS, F. T. (1987). On an optimal selection problem by Cowan and Zabczyk. *J. Appl. Prob.* **24** 918-928.

BRUSS, F. T. AND FERGUSON, T. S. (1991). Minimizing the expected rank with full-information. Preprint.

BRUSS, F. T. AND SAMUELS, S. M. (1987). A unified approach to a class of optimal selection problems with an unknown number of options. *Ann. Probab.* **15** 824-830.

BRUSS, F. T. AND SAMUELS, S. M. (1990). Conditions for quasi-stationarity of the Bayes rule in selection problems with an unknown number of rankable options. *Ann. Probab.* **18** 877-886.

CHOW, Y. S., ROBBINS, H., MORIGUTI, S. AND SAMUELS, S. M. (1964). Optimal selection based on relative rank (the "secretary problem"). *Israel J. Math.* **2** 81-90.

COWAN, R. AND ZABCZYK, J. (1978). An optimal selection problem associated with the Poisson process. *Theor. Probab. Appl.* **23** 584-592.

FERGUSON, T. S. (1989). Who solved the secretary problem? (with discussion) *Statist. Sci.* **4** 282-289.

FERGUSON, T. S., HARDWICK, J. P. AND TAMAKI, M. (1991). Maximizing the duration of owning a relatively best object. In *Strategies for Sequential Search and Selection in Real-Time.* F. T. Bruss, T. S. Ferguson and S. M. Samuels, eds. American Mathematical Society, Providence, RI. 37–57.

GARDNER, M. (1960). Mathematical Games. *Sci. Amer.* **202** 135, 178.

GIANINI, J. (1977). The infinite secretary problem as the limit of the finite problem. *Ann. Probab.* **5** 636-644.

GIANINI, J. AND SAMUELS, S. M. (1976). The infinite secretary problem. *Ann. Probab.* **4** 418-432.

GIANINI-PETTITT, J. (1979). Optimal selection based on relative ranks with a random number of individuals. *Adv. Appl. Prob.* **11** 720-736.

GILBERT, J. AND MOSTELLER, F. (1966). Recognizing the maximum of a sequence. *J. Amer. Statist. Assoc.* **61** 35-73.

GNEDIN, A. V. (1982). Multicriterial problem of optimum stopping of the selection process. (translated from Russian). *Automation Rem. Cont.* **42** 981-986.

GOVINDARAJULU, Z. (1991). The secretary problem: optimal selection from two streams of candidates. In *Strategies for Sequential Search and Selection in Real-Time.* F. T. Bruss, T. S. Ferguson and S. M. Samuels, eds. American Mathematical Society, Providence, RI. 65–75.

HILL, T. P. AND KRENGEL, U. (1991a). Minimax-optimal stop rules and distributions in secretary problems. *Ann. Probab.* **19** 342-353.

HILL, T. P. AND KRENGEL, U. (1991b). On the game of googol. Preprint.

LINDLEY, D. V. (1961). Dynamic programming and decision theory. *Appl. Statist.* **10** 39-52.

LORENZEN, T. J. (1979). Generalizing the secretary problem. *Adv. Appl. Probab.* **11** 384-396.

MUCCI, A. G. (1973a). Differential equations and optimal choice problems. *Ann. Statist.* **1** 104-113.

MUCCI, A. G. (1973b). On a class of secretary problems. *Ann. Probab.* **1** 417-427.

PETRUCCELLI, J. D. (1978). Some best choice problems with partial information. Unpublished thesis, Department of Statistics, Purdue University.

PETRUCCELLI, J. D. (1980). On a best choice problem with partial information. *Ann. Statist.* **8** 1171-1174.

PRESMAN, E. L. AND SONIN, I. M. (1972). The best choice problem for a random number of objects. *Theor. Prob. Appl.* **17** 657-668.

ROBBINS, H. (1989). [Comment on Ferguson (1989)] *Statist. Sci.* **4** 291.

ROCHA, A. (1988). The secretary problem with recall. Preprint.

RUBIN, H. (1966). The "secretary" problem. *Ann. Math. Statist.* **37** 544.

RUBIN, H. AND SAMUELS, S. M. (1977). The finite-memory secretary problem. *Ann. Probab.* **5** 627-635.

SAKAGUCHI, M. (1989). Some infinite problems in classical secretary problems. *Math. Japonica* **34** 307-318.

SAMUELS, S. M. (1981). Minimax stopping rules when the underlying distribution is uniform. *J. Amer. Stat. Assoc.* **76** 188-197.

SAMUELS, S. M. (1982). Exact solutions for the full information best choice problem. *Purdue Univ. Stat. Dept. Mimeo Series* **82-17**.

SAMUELS, S. M. (1989). Who will solve the secretary problem? [Comment on Ferguson (1989)] *Statist. Sci.* **4** 289-291.

SAMUELS, S. M. (1991). Secretary Problems. In *Handbook of Sequential Analysis.* B. K. Ghosh and P. K. Sen, eds. Marcel Dekker, New York. 381-405 (Chapter 16).

SAMUELS, S. M. AND CHOTLOS, B. (1986). A multiple criteria optimal selection problem. In *Adaptive Statistical Procedures and Related Topics,* J. Van Ryzin, ed., Institute of Mathematical Statistics, Hayward, CA. 62-78.

SILVERMAN, S. AND NÁDAS, A. (1991). On the game of googol as *the* secretary problem. In *Strategies for Sequential Search and Selection in Real-Time,* F. T. Bruss, T. S. Ferguson and S. M. Samuels, eds. American Mathematical Society, Providence, RI. 77-83.

SMITH, M. H. AND DEELY, J. J. (1975). A secretary problem with finite memory. *J. Amer. Statist. Assoc.* **70** 357-361.

DEPARTMENT OF STATISTICS
PURDUE UNIVERSITY
WEST LAFAYETTE, IN 47907–1399

Stochastic Inequalities
IMS Lecture Notes – Monograph Series
Volume 22 (1993)

COMPARISON OF EXPERIMENTS OF SOME MULTIVARIATE DISTRIBUTIONS WITH A COMMON MARGINAL

By MOSHE SHAKED[1] and Y. L. TONG[2]

University of Arizona and Georgia Institute of Technology

In this paper we review some current work on comparison of experiments of some multivariate distributions. First we describe some results regarding comparison of experiments of univariate distributions that belong to two-parameters exponential families and that satisfy the semi-group property. Then we discuss comparison of experiments of vectors that arise from an additive model based on univariate two-parameter exponential families of random variables. These models give rise to vectors of random variables which are positively dependent and these are compared to vectors of independent random variables with the same marginals. It is shown that positively dependent random variables contain less information than independent random variables. Finally we describe some results regarding the comparison of experiments of exchangeable and nonexchangeable normal random vectors. In particular, we show how the majorization ordering can be used to identify various information orderings of multivariate normal random vectors which have a common marginal density.

1. Introduction

Let $\mathbf{X} = (X_1, X_2, \ldots, X_n)$ and $\mathbf{Y} = (Y_1, Y_2, \ldots, Y_n)$ be two random vectors such that $X_1 =_d X_2 =_d \cdots =_d X_n =_d Y_1 =_d Y_2 =_d \cdots =_d Y_n$, where '$=_d$' denotes equality in law. That is, $\mathbf{X}$ and $\mathbf{Y}$ have the same univariate marginal distributions and all these marginals are equal to each other. Let $\theta \in \Theta$ be an unknown parameter and assume that the distributions of $\mathbf{X}$ and $\mathbf{Y}$ depend on θ. Denote the distributions of $\mathbf{X}$ and $\mathbf{Y}$ by F_θ and G_θ, respectively. In this paper we will be concerned with the amount of information

[1]Research supported in part by U.S. Air Force Office of Scientific Research Grant AFOSR-84-0205. Reproduction in whole or in part is permitted for any purpose by the United States Government.

[2]Research supported in part by a grant from U.S. Air Force Office of Scientific Research through National Science Foundation Grant DMS-9149151.

AMS 1991 *subject classifications.* 62B15, 62H20.

Key words and phrases. Comparison of experiments, information inequalities, majorization, positive dependence, multivariate normal distribution.

about θ that is contained in F_θ and in G_θ. We will review some current work regarding comparison of experiments which are based on vectors which may have dependent random variables. The basic intuitive conjecture, that we can prove in some instances, is that if $X_1, X_2, \ldots, X_n$ are more 'positively dependent' than $Y_1, Y_2, \ldots, Y_n$ then they should contain less information about θ than the Y_j's. This is clearly the case in the extreme case where the Y_j's are independent and identically distributed and the X_j's are all equal to each other with probability one. But we will show that in some cases, even if the X_j's are not necessarily totally dependent, but only positively dependent, then they contain less information about θ than the Y_j's. We will show that in other cases the information content is a monotone function of the 'strength of dependence' of the underlying random variables. In still other cases the information content is shown to be monotone in the amount of 'homogeneity' of the underlying random variables.

2. Background and Preliminaries

The notion of comparison of experiments, as introduced by Blackwell (1951, 1953) and others, concerns a partial ordering of the information contained in the experiments (or in the distributions of the underlying random variables). For a review of the basic ideas and related results see Goel and DeGroot (1979), Lehmann (1988) and Torgersen (1991).

DEFINITION 1 The experiment associated with $\mathbf{Y}$ is said to be *at least as informative* for θ as that associated with $\mathbf{X}$, in symbols $\mathbf{X} \leq_{(i)} \mathbf{Y}$ or $F_\theta \leq_{(i)} G_\theta$, if for every decision problem involving θ and every prior distribution on Θ the expected Bayes risk from F_θ is not less than that from G_θ.

Recently many useful results have been obtained by researchers on the comparison of various types of experiments. For example, Hansen and Torgersen (1974) considered the comparison of normal experiments, Torgersen (1984) and Stepniak, Wang and Wu (1984) studied the comparison of linear experiments, Hollander, Proschan and Sconing (1985, 1987) and Goel (1988) gave results comparing experiments with censored data, and Lehmann (1988) discussed the comparison of location parameter experiments. Recently Eaton (1991) discussed a group action on covariances with applications to the comparison of linear normal experiments.

Lehmann (1959, p. 75) noted the following sufficient condition:

PROPOSITION 2 *The information inequality* $\mathbf{X} \leq_{(i)} \mathbf{Y}$ *holds if there exists a function* $\psi : \mathbb{R}^{n+r} \rightarrow \mathbb{R}^n$ *and an* r*-dimensional random vector* $\mathbf{Z}$ $(r \geq 1)$, *which is independent of* $\mathbf{Y}$ *and having a distribution which does not depend on* θ, *such that* $\mathbf{X} =_d \psi(\mathbf{Y}, \mathbf{Z})$.

Proposition 2 is the basic technical tool that we will use throughout this paper. LeCam (1964) noticed that the condition of Proposition 2 is necessary as well when the family $\{F_\theta, \theta \in \Theta\}$ is dominated.

In certain cases an ordering, via comparison of experiments, can be obtained for a given family of distributions. For example:

(a) Let $X_1, X_2, \ldots, X_n$ be independent and identically distributed Poisson random variables with mean $t_1\theta$ and let $Y_1, Y_2, \ldots, Y_n$ be independent and identically distributed Poisson random variables with mean $t_2\theta$, where $t_1 \leq t_2$ are known real numbers and θ is an unknown parameter. Then $\mathbf{X} \leq_{(i)} \mathbf{Y}$. This easily follows from Lehmann (1959, p. 77).

(b) Let $X_1, X_2, \ldots, X_n$ $[Y_1, Y_2, \ldots, Y_n]$ be independent and identically distributed normal random variables with mean θ and standard deviation σ_2 $[\sigma_1]$, where θ is unknown and σ_1 and σ_2 are fixed such that $0 < \sigma_1 \leq \sigma_2$. Then $\mathbf{X} \leq_{(i)} \mathbf{Y}$ (Goel and DeGroot (1979)).

(c) Let $X_1, X_2, \ldots, X_n$ $[Y_1, Y_2, \ldots, Y_n]$ be independent and identically distributed gamma random variables with shape parameter a and scale parameter $b\theta^{k_1}$, $[b\theta^{k_2}]$ where $a > 0, b > 0, k_1 > 0$, and $k_2 > 0$ are fixed and $\theta > 0$ is an unknown parameter. If $k_1 \leq k_2$ then $\mathbf{X} \leq_{(i)} \mathbf{Y}$ (Goel and DeGroot (1979)).

Note that in each one of these examples the distributions of the X_i's and of the Y_i's belong to a particular univariate family of distributions. More explicitly, in each of the examples we have a family of univariate densities $f_{t,\theta}$, with respect to the Lebesgue or the counting measure, depending on a parameter $(t, \theta) \in \Theta_1 \times \Theta_2 \subset \mathbb{R}^2$ of the form

$$f_{t,\theta}(x) = c(t, \theta)\alpha(x, t)e^{\phi(\theta)x} \tag{1}$$

where c, α, and ϕ are some real-valued Borel-measurable functions defined on $\Theta_1 \times \Theta_2$, $\mathbb{R} \times \Theta_1$, and Θ_2 respectively. In addition to the Poisson, normal and gamma densities also the binomial and the negative binomial densities, among others, are of the form (1). See Shaked and Tong (1990) for the particular explicit expressions of the functions c, α, and ϕ for these densities.

The families of density functions mentioned above also have the semi-group property. Formally a family of density functions $f_{t,\theta}$, with respect to the measure μ, is said to have the semi-group property in the parameter $t \in \Theta_1$, where $\Theta_1 = (0, \infty)$ or $\Theta_1 = \{1, 2, \ldots\}$, if

$$f_{t_1,\theta} * f_{t_2,\theta} = f_{t_1+t_2,\theta}, \qquad t_1 \in \Theta_1, \quad t_2 \in \Theta_1$$

Here $*$ denotes the convolution operation:

$$f_{t_1,\theta} * f_{t_2,\theta}(x) = \int f_{t_1,\theta}(y) f_{t_2,\theta}(x - y) d\mu(y).$$

Roughly speaking, if $f_{t,\theta}$ has the representation (1) and also satisfies the semi-group property then the parameter t can be thought of as a 'sample size'. The following result then is not surprising (for a proof see Shaked and Tong (1990)):

PROPOSITION 3 *Let $f_{t,\theta}$ be a density of the form* (1) *with $\Theta_1 = (0, \infty)$ or $\Theta_1 = \{1, 2, \ldots\}$, which has the semi-group property in t. Let $F_{t,\theta}$ denote the distribution function associated with the density $f_{t,\theta}$. Then*

$$F_{t_2,\theta} \geq_{(i)} F_{t_1,\theta} \text{ whenever } t_2 \geq t_1.$$

By Blackwell (1953), $(X_1, X_2, \ldots, X_n) \leq_{(i)} (Y_1, Y_2, \ldots, Y_n)$ when $X_j \leq_{(i)} Y_j$, $j = 1, 2, \ldots, n$, and the coordinates of these two vectors are independent. Thus from Proposition 3 we obtain:

COROLLARY 4 *Let X_j and Y_j have the densities $f_{s,\theta}$ and $f_{\tau,\theta}$ of the form* (1) *($j = 1, 2, \ldots, n$) for some functions c, α, and ϕ with $\Theta_1 = (0, \infty)$ or $\Theta_1 = \{1, 2, \ldots\}$. Suppose that the X_j's are independent and that the Y_j's are independent. If $s_j \leq \tau_j$, $j = 1, 2, \ldots, n$, and $f_{t,\theta}$ satisfies the semi-group property in t, then*

$$(X_1, X_2, \ldots, X_n) \leq_{(i)} (Y_1, Y_2, \ldots, Y_n).$$

It follows, using Proposition 2, that $X_1 + X_2 + \cdots + X_n \leq_{(i)} Y_1 + Y_2 + \cdots + Y_n$.

3. Comparison of Vectors of Independent and Dependent Random Variables

In examples (a) - (c), or more generally in Corollary 4, the vectors $\mathbf{X}$ and $\mathbf{Y}$ consist of independent random variables. The main thrust of this section is to obtain results in which the assumption of mutual independence of the X_j's is relaxed. In the next section we will also relax the assumption of independence of the Y_j's.

It is well known that in certain statistical applications the assumption of independence is not realistic. For example, in many reliability problems the lifetimes of components in a coherent system are positively dependent. This happens when the system involves several common units or when the components are subjected to the same set of stresses. In statistical decision theory, if the random variables $X_1, X_2, \ldots, X_n$ are conditionally independent and identically distributed given some random quantity Γ, and if the distribution of Γ is nonsingular, then, after unconditioning, the joint distribution of $X_1, X_2, \ldots, X_n$ is positively dependent by mixture (Shaked (1977)). Thus the random variables are not independent.

In this section we let $\mathbf{X} = (X_1, X_2, \ldots, X_n)$ and $\mathbf{Y} = (Y_1, Y_2, \ldots, Y_n)$ be two random vectors, the distribution of each depending on some parameter θ. We consider random vectors such that marginally $X_j =_d Y_j$ whatever the value of θ is. We suppose that for each θ the Y_j's are independent but the X_j's may be positively dependent in a certain fashion. Then one would expect $\mathbf{Y}$ to be at least as informative as $\mathbf{X}$. This can be easily seen in the extreme case where $P\{X_1 = X_2 = \cdots = X_n\} = 1$. Then the information contained in $\mathbf{X}$ is the same as the information contained in one observation, Y_1, say, whereas the information contained in $\mathbf{Y}$ is larger, since it is based on n observations.

Shaked and Tong (1990) proved that this is indeed the case for a special model of dependence. They considered the following model. Let

$$\mathbf{X} = \begin{pmatrix} U_1' + V_1' \\ U_2' + V_1' \\ \vdots \\ U_n' + V_1' \end{pmatrix} \quad \text{and } \mathbf{Y} = \begin{pmatrix} U_1 + V_1 \\ U_2 + V_2 \\ \vdots \\ U_n + V_n \end{pmatrix},$$

where $U_1, U_2, \ldots, U_n$ are mutually independent with distributions depending on θ; $V_1, V_2, \ldots, V_n$ are independent and identically distributed, independent of the U_j's, and with a common distribution depending on θ; $U_1', U_2', \ldots, U_n', V_1'$ are all independent; and for each θ,

$$U_j' =_d U_j \text{ and } V_1' =_d V_1, \quad j = 1, 2, \ldots, n.$$

It is then seen that the Y_j's are independent whereas the X_j's are positively associated [e.g., in the sense of Esary, Proschan and Walkup (1967)].

In the following theorem it is assumed that all the univariate random variables mentioned above have densities of the form (1) with respect to a σ-finite measure μ and for some fixed functions c, α, and ϕ. More specifically we suppose that the density of U_j and U_j' is $f_{t_1,\theta}$ and that the density of V_j is $f_{t_2,\theta}$, $j = 1, 2, \ldots, n$ (thus, in particular, the density of V_1 is $f_{t_2,\theta}$). As before, μ will be either the Lebesgue measure or the counting measure. The proof of the theorem can be found in Shaked and Tong (1990).

THEOREM 5 *Let $\mathbf{X}$ and $\mathbf{Y}$ be two random vectors as described in the preceding paragraph. Then*

$$\mathbf{X} \leq_{(i)} \mathbf{Y}. \tag{2}$$

Theorem 5 complements Corollary 4. In the latter two vectors of independent random variables are compared. In the former only one of the vectors consists of independent random variables, but the two random vectors have equal marginals.

Theorem 5 shows how some multivariate Poisson random vectors [as described, e.g., in Johnson and Kotz (1969, Chapter 11, Section 4) and references therein] can be compared in the comparison-of-experiments ordering. Similarly, multivariate gamma distributions [as, e.g., in Johnson and Kotz (1972, Chapter 40, Section 2)] and multivariate negative-binomial distributions can be compared. Theorem 5 also shows (after some calculations) that a normal random vector with independent and identically distributed components, all with mean θ, is more informative than a similar normal random vector with the same marginals but with positively correlated permutation symmetric components, when the common variance is known. In the next section we discuss even a more general result for normal random vectors.

4. Comparison of Normal Vectors with a Common Marginal Distribution

Shaked and Tong (1990) proved the following monotonicity result. Note that in the following result not only it is seen that independence is more informative than positive dependence, but more than that, it is seen there that negative dependence is even more informative than independence.

PROPOSITION 6 *Let* $\mathbf{X}$ *and* $\mathbf{Y}$ *be two vectors of* n *normal random variables with means* θ, *a common known variance* $\sigma^2 > 0$, *and common correlation coefficients* ρ_2 *and* ρ_1, *respectively. Then*

$$\mathbf{X} \leq_{(i)} \mathbf{Y} \text{ for all } -\frac{1}{n-1} \leq \rho_1 \leq \rho_2 \leq 1.$$

Let $\mathbf{X}_\rho$ denote a multivariate normal random vector with means θ, a common variance σ^2, and a common correlation coefficient $\rho \geq -1/(n-1)$. Here the unknown parameter is θ and the parameter ρ indexes the distribution of $\mathbf{X}_\rho$. From the proof of Proposition 6 in Shaked and Tong (1990) it is seen that, for a fixed ρ, the permutation symmetric multivariate normal distribution is monotone in the sense $\leq_{(i)}$ as a function of the sample size n (the larger n is, the more informative is the vector $\mathbf{X}_\rho$ provided $\rho < 1$).

The Fisher's information corresponding to the distribution of $\mathbf{X}_\rho$ is

$$E\left(\frac{\partial}{\partial\theta}\log f_\rho(\mathbf{X})\right)^2 = \frac{1}{\mathrm{Var}(\bar{X}_\rho)}, \quad \rho \in (-\frac{1}{n-1}, 1),$$

where $\bar{X}_\rho$ is the average of the components of $\mathbf{X}_\rho$ and

$$f_\rho(\mathbf{x}) = \frac{1}{(2\pi\sigma)^{n/2}|\mathbf{R}(\rho)|^{1/2}}\exp\{-\frac{1}{2}(\mathbf{x}-\theta\mathbf{1})'\mathbf{R}^{-1}(\rho)(\mathbf{x}-\theta\mathbf{1})\}.$$

Here $\mathbf{R}(\rho)$ is the covariance matrix of $\mathbf{X}_\rho$. Thus, in this case, the two experiments based on $\mathbf{X}_{\rho_1}$ and $\mathbf{X}_{\rho_2}$ can be ordered in the ordering $\leq_{(i)}$ if, and only if, they can be ordered according to the Fisher's information. In general the equivalence of these two orderings is not always true (see, e.g., Hansen and Torgersen (1974)).

In Shaked and Tong (1985) some notions of partial ordering of exchangeable random variables by positive dependence are introduced, and for exchangeable normal variables the orderings reduce to the ordering of the correlation coefficients. Consequently, from Proposition 6 it follows that if exchangeable normal variables are more positively dependent then the experiment is less informative. A question of interest is then what can be said for normal variables which are not exchangeable.

Shaked and Tong (1992) provide an answer to this question by showing how a more general partial ordering of positive dependence yields a monotonicity result for nonexchangeable normal variables with a common marginal distribution.

In order to introduce the partial ordering of positive dependence for multivariate normal vectors we first need some preliminaries. Consider an n-dimensional vector of nonnegative integers given by

$$\text{(3)} \qquad \mathbf{k} = (k_1, \ldots, k_r, 0, \ldots, 0), \; k_1 \geq k_2 \geq \cdots \geq k_r \geq 1, \; \sum_{s=1}^{r} k_s = n,$$

for some $r \leq n$. (The assumption of monotonicity of k_s in s is not an essential restriction. If it does not hold then the random variables can always be relabeled, if necessary, yielding the assumed monotonicity.) For arbitrary but fixed $0 \leq \rho_1 \leq \rho_2 \leq 1$ let us define a correlation matrix $\mathbf{R}(\mathbf{k})$ given by

$$\text{(4)} \qquad \rho_{ij}(\mathbf{k}) = \begin{cases} 1 & \text{for } i = j, \\ \rho_2 & \text{for } i \neq j \text{ and } \sum_{s=0}^{m} k_s + 1 \leq i, j \leq \sum_{s=0}^{m+1} k_s, \\ & \qquad m \in \{1, 2, \ldots, r-1\}, \\ \rho_1 & \text{otherwise,} \end{cases}$$

where $k_0 \equiv 0$.

If $\mathbf{X}$ has a correlation matrix $\mathbf{R}(\mathbf{k})$ then its components belong to r groups, with group sizes $k_1, k_2, \ldots, k_r$, respectively, such that the correlations within groups are ρ_2 and the correlations between groups are ρ_1. This type of correlation matrices arise in many applied problems in linear models and multivariate analysis. For example, if in a family of four children the first two [the last two] are brothers/sisters, but any pair between the two groups are half brothers/sisters, then under the additive genetic model the vector of measurements (X_1, X_2, X_3, X_4), of a certain biological variable, will have means θ, variances σ^2, and a correlation matrix $\mathbf{R}(\mathbf{k})$ with

$k = (2,2,0,0)$. For references on the applications of such a correlation matrix in an agricultural genetic selection problem see, e.g., Tong (1990, pp. 129-130).

Let $\mathbf{k}^*$ be another vector of nonnegative integers such that

$$(5) \quad \mathbf{k}^* = (k_1^*, \ldots, k_{r^*}^*, 0, \ldots, 0), \; k_1^* \geq k_2^* \geq \cdots \geq k_{r^*}^* \geq 1, \; \sum_{s=1}^{r} k_s^* = n,$$

for some $r^* \leq n$, and let $\mathbf{R}(\mathbf{k}^*)$ be defined similarly. Let $\mathbf{X}$ and $\mathbf{Y}$ have, respectively, the multinormal distributions

$$(6) \qquad \mathcal{N}_n(\theta\mathbf{1}, \sigma^2\mathbf{R}(\mathbf{k})) \text{ and } \mathcal{N}_n(\theta\mathbf{1}, \sigma^2\mathbf{R}(\mathbf{k}^*))$$

for some $\mathbf{k}, \mathbf{k}^*$ satisfying (3) and (5), respectively, where $\theta \in \mathbb{R}$ is the common mean, $\sigma^2 > 0$ is the common known variance, and $\mathbf{1} = (1, \ldots, 1)$. Clearly the X_j's and the Y_j's defined in (6) have a common univariate $\mathcal{N}(\theta, \sigma^2)$ distribution. In the special case $\mathbf{k} = (n, 0, \ldots, 0)$ and $\mathbf{k}^* = (1, 1, \ldots, 1)$ both $\mathbf{X}$ and $\mathbf{Y}$ are exchangeable normal vectors with correlation coefficients ρ_2, ρ_1, respectively. However they are not exchangeable otherwise. A result of Tong (1989) states that if $\mathbf{k} \succ \mathbf{k}^*$, where '$\succ$' denotes the majorization ordering, then the X_j's tend to "hang together" more than the Y_j's, hence are more positively dependent in the sense that

$$(7) \qquad E\prod_{i=1}^{n} \phi(X_i) \geq E\prod_{i=1}^{n} \phi(Y_i) \text{ for all } \phi : \mathbb{R} \to [0, \infty),$$

provided the expectations exist. [Note that (7) implies that $\text{Corr}(\phi(X_i), \phi(X_j)) \geq \text{Corr}(\phi(Y_i), \phi(Y_j))$ for all ϕ.] The question of interest is whether this partial ordering of positive dependence also provides a partial ordering for information on θ in the sense of Definition 1. Shaked and Tong (1992) answered this question in the following theorem:

THEOREM 7 *Assume that* $\mathbf{X}$ *and* $\mathbf{Y}$ *satisfy* (6) *where* $\theta \in \mathbb{R}$ *is the unknown parameter,* $\sigma^2 > 0$ *is the common known variance, and* $0 \leq \rho_1 \leq \rho_2 \leq 1$ *are arbitrary but fixed. If* $\mathbf{k} \succ \mathbf{k}^*$*, then* $\mathbf{X} \leq_{(i)} \mathbf{Y}$.

The proof of Theorem 7 given by Shaked and Tong (1992) depends on an application of Torgersen (1984). Eaton (1991, Remark 2.3) noted that it is also possible to give an alternative proof of the theorem based on a direct verification.

A related result which follows from the remarks after Proposition 2.2 of Eaton (1991) is the following. Let $\mathbf{R}(\mathbf{k})$ be as defined in (4), but now, in order to point out the dependence of $\mathbf{R}(\mathbf{k})$ on ρ_1 and on ρ_2 we write it as $\mathbf{R}_{\rho_1,\rho_2}(\mathbf{k})$ where $0 \leq \rho_1 \leq \rho_2 \leq 1$.

THEOREM 8 *Assume that* $\mathbf{X}$ *has the multinormal distribution* $\mathcal{N}_n(\theta\mathbf{1}, \sigma^2\mathbf{R}_{\rho_1,\rho_2}(\mathbf{k}))$ *and that* $\mathbf{Y}$ *has the multinormal distribution* $\mathcal{N}_n(\theta\mathbf{1}, \sigma^2\mathbf{R}_{\rho_1',\rho_2'}(\mathbf{k}))$ *where* $\theta \in \mathbb{R}$ *is the unknown parameter,* $\sigma^2 > 0$ *is the common known variance and* $\mathbf{k}$ *is fixed partition vector as described in* (3).
(a) If $\rho_2 = \rho_2' \geq \rho_1 \geq \rho_1'$ *then* $\mathbf{X} \leq_{(i)} \mathbf{Y}$.
(b) If $\rho_2 \geq \rho_2' \geq \rho_1 = \rho_1'$ *then* $\mathbf{X} \leq_{(i)} \mathbf{Y}$.

EXAMPLE 9 In order to illustrate the result in Theorem 7 let us consider the special case $n = 4$. Let $\mathbf{R}_4$ $[\mathbf{R}_1]$ be the 4×4 correlation matrix with all the correlation coefficients being ρ_2 $[\rho_1]$, and let

$$\mathbf{R}_3 = \begin{pmatrix} 1 & \rho_2 & \rho_2 & \rho_1 \\ \rho_2 & 1 & \rho_2 & \rho_1 \\ \rho_2 & \rho_2 & 1 & \rho_1 \\ \rho_1 & \rho_1 & \rho_1 & 1 \end{pmatrix} \quad \text{and} \quad \mathbf{R}_2 = \begin{pmatrix} 1 & \rho_2 & \rho_1 & \rho_1 \\ \rho_2 & 1 & \rho_1 & \rho_1 \\ \rho_1 & \rho_1 & 1 & \rho_2 \\ \rho_1 & \rho_1 & \rho_2 & 1 \end{pmatrix}.$$

If $0 \leq \rho_1 \leq \rho_2 \leq 1$ and $\mathbf{X}$ and $\mathbf{Y}$ have the distributions $\mathcal{N}_4(\theta\mathbf{1}, \sigma^2\mathbf{R}_{j+1})$ and $\mathcal{N}_4(\theta\mathbf{1}, \sigma^2\mathbf{R}_j)$ respectively, where $\sigma^2 > 0$ is known, then $\mathbf{X} \leq_{(i)} \mathbf{Y}$ holds for $j = 1, 2, 3$. This follows from Theorem 7 and the fact that $(4,0,0,0) \succ (3,1,0,0) \succ (2,2,0,0) \succ (1,1,1,1)$.

When we combine Theorem 7 with existing results, other useful results can be obtained. For example, if $\mathbf{X}$ and $\mathbf{Z}$ have the $\mathcal{N}_n(\theta\mathbf{1}, \sigma^2\mathbf{R}(\mathbf{k}))$ and $\mathcal{N}_n(\theta\mathbf{1}, \mathbf{\Sigma})$ distributions, respectively, and if there exists a correlation matrix $\mathbf{R}(\mathbf{k}^*)$ such that $\mathbf{k} \succ \mathbf{k}^*$ and $\sigma^2\mathbf{R}(\mathbf{k}^*) - \mathbf{\Sigma}$ is either positive definite or positive semidefinite, then $\mathbf{X} \leq_{(i)} \mathbf{Z}$ holds.

It is worthwhile to note that if $\mathbf{X}$ and $\mathbf{Y}$ satisfy (6) then, by a simple calculation, the amounts of Fisher's information on θ in the density functions of $\mathbf{X}$ and $\mathbf{Y}$ are, respectively, $\sigma^{-2}(\mathbf{1}(\mathbf{R}(\mathbf{k}))^{-1}\mathbf{1}')$ and $\sigma^{-2}(\mathbf{1}(\mathbf{R}(\mathbf{k}^*))^{-1}\mathbf{1}')$. Furthermore, the variances of the least-squares estimators of θ based on $\mathbf{X}$ and $\mathbf{Y}$ are, respectively, $\sigma^2(\mathbf{1}(\mathbf{R}(\mathbf{k}))^{-1}\mathbf{1}')^{-1}$ and $\sigma^2(\mathbf{1}(\mathbf{R}(\mathbf{k}^*))^{-1}\mathbf{1}')^{-1}$. Thus Theorem 7 yields a partial ordering for the Fisher's information and for the variances of the least-squares estimators via a majorization ordering of $\mathbf{k}$ and $\mathbf{k}^*$ in the correlation matrices.

Finally we point out that smaller correlations are not necessarily indicators of larger amounts of information. More specifically, if $\mathbf{X}$ and $\mathbf{Y}$ are multivariate normal random vectors with the same marginal distributions and with correlation matrices $\{\rho_{i,j}\}_{i,j=1}^n$ and $\{\eta_{i,j}\}_{i,j=1}^n$, respectively, such that $\rho_{i,j} \geq \eta_{i,j}$ for all i and j, then, in general, when σ^2 is known, it is not necessarily true that $\mathbf{X} \leq_{(i)} \mathbf{Y}$; some conditions must be imposed on the structures of the correlation matrices in order to assure that $\mathbf{X} \leq_{(i)} \mathbf{Y}$.

For example, such conditions can be found in Proposition 6 and in Theorem 8. Eaton (1991, Example 2.1) has shown that there exist multivariate normal random vectors $\mathbf{X}$ and $\mathbf{Y}$ such that $\mathbf{X}$ and $\mathbf{Y}$ have the same univariate marginals, the coordinates of $\mathbf{Y}$ are independent and identically distributed, the coordinates of $\mathbf{X}$ are positively correlated, but $\mathbf{Y} \leq_{(i)} \mathbf{X}$.

REFERENCES

BLACKWELL, D. (1951). Comparison of experiments. In *Proc. Second Berkeley Symp. Math. Statist. Probab.*, J. Neyman, ed. University of California Press, Berkeley, CA. 93-102.

BLACKWELL, D. (1953). Equivalent comparison of experiments. *Ann. Math. Statist.* **24** 265-272.

EATON, M. L. (1991). A group action on covariances with applications to the comparison of linear normal experiments. Tech. Report, School of Statistics, University of Minnesota.

ESARY, J. D., PROSCHAN, F. AND WALKUP, D. W. (1967). Association of random variables with applications. *Ann. Math. Statist.* **38** 1966-1974.

GOEL, P. K. (1988). Comparison of experiments and information in censored data. *Statistical Decision Theory and Related Topics VI*, Vol. **2**, S. S. Gupta and J. O. Berger, eds. Springer-Verlag, New York. 335-349.

GOEL, P. K. AND DEGROOT, M. H. (1979). Comparison of experiments and information measures. *Ann. Statist.* **7** 1066-1077.

HANSEN, O. H. AND TORGERSEN, E. N. (1974). Comparison of linear experiments. *Ann. Statist.* **2** 367-373.

HOLLANDER, M., PROSCHAN, F. AND SCONING, J. (1985). Information in censored models. Tech. Report No. M701, Department of Statistics, Florida State University.

HOLLANDER, M., PROSCHAN, F. AND SCONING, J. (1987). Measuring information in right-censored models. *Naval Res. Log. Quart.* **34** 669-681.

JOHNSON, N. L. AND KOTZ, S. (1969). *Distributions in Statistics: Discrete Distributions.* Wiley, New York.

JOHNSON, N. L. AND KOTZ, S. (1972). *Distributions in Statistics: Continuous Multivariate Distributions.* Wiley, New York.

LECAM, L. (1964). Sufficiency and approximate sufficiency. *Ann. Math. Statist.* **35** 1419-1455.

LEHMANN, E. L. (1959). *Testing Statistical Hypotheses.* Wiley, New York.

LEHMANN, E. L. (1988). Comparing location experiments. *Ann. Statist.* **16** 521-533.

SHAKED, M. (1977). A concept of positive dependence for exchangeable random variables. *Ann. Statist.* **5** 505-515.

SHAKED, M. AND TONG, Y. L. (1985). Some partial orderings of exchangeable random variables by positive dependence. *J. Multivariate Anal.* **17** 333-349.

SHAKED, M. AND TONG, Y. L. (1990). Comparison of experiments for a class of positively dependent random variables. *Canad. J. Statist.* **18** 79-86.

SHAKED, M. AND TONG, Y. L. (1992). Comparison of experiments via dependence of normal variables with a common marginal distribution. *Ann. Statist.* **20** 614–618.

STEPNIAK, C., WANG, S.-G. AND WU, C. F. J. (1984). Comparison of linear experiments with known covariances. *Ann. Statist.* **12** 358-365.

TONG, Y. L. (1989). Inequalities for a class of positively dependent random variables with a common marginal. *Ann. Statist.* **17** 429-435.

TONG, Y. L. (1990). *The Multivariate Normal Distribution.* Springer-Verlag, New York.

TORGERSEN, E. N. (1984). Orderings of linear models. *J. Statist. Plann. Inf.* **9** 1-17.

TORGERSEN, E. N. (1991). *Comparison of Statistical Experiments.* Cambridge University Press, Cambridge, England.

DEPARTMENT OF MATHEMATICS
UNIVERSITY OF ARIZONA
TUCSON, AZ 85721

SCHOOL OF MATHEMATICS
GEORGIA INSTITUTE OF TECHNOLOGY
ATLANTA, GA 30332–0160

Stochastic Inequalities
IMS Lecture Notes – Monograph Series
Volume 22 (1993)

ON THE BIAS OF THE JACKKNIFE ESTIMATE OF VARIANCE[1]

By RICHARD A. VITALE

University of Connecticut

Using machinery developed earlier for the covariances of symmetric statistics, we consider various aspects of the bias of the jackknife estimate of variance.

1. Introduction

The jackknife estimate of variance (Quenouille (1949, 1956), Tukey (1958)) can be described as follows. Given a symmetric function h of iid arguments $X_1, X_2, \ldots, X_m$, it is desired to estimate $\sigma^2 = \text{Var } h$. With an augmented supply $X_1, X_2, \ldots, X_n$ where $n = m+1$, or more generally $n \geq m+1$, one forms $Q = \binom{n-1}{m}^{-1} \sum_{|I|=m} [h(X_I) - \overline{h}]^2$ where $\overline{h} = \binom{n}{m}^{-1} \sum_{|I|=m} h(X_I)$ and $X_I \equiv (X_{i_1}, X_{i_2}, \ldots, X_{i_m})$ with $I = \{i_1, i_2, \ldots, i_m\}$. Several papers (Efron and Stein (1981), Karlin and Rinott (1982), Bhargava (1983), Vitale (1984), Steele (1986)) have considered the bias relation

$$\sigma^2 \leq EQ, \tag{1.1}$$

which has come to be known as the Efron–Stein inequality. Our purpose here is to investigate aspects of (1.1) including (i) an alternate proof with variant forms of the condition for equality, (ii) a sharpening, (iii) a complementary upper bound, and (iv) a consideration of Q as an estimator which is "contaminated" by estimators of other parameters.

2. Preliminaries

If $X_1, X_2, \ldots, X_n$ are iid random variables and h is a symmetric function of m of them with $Eh^2 < \infty$, then we set

$$r_k = \text{Cov}[h(X_I), h(X_J)]$$

[1]Work supported in part under grants Office of Naval Research N00014–90–J–1641 and National Science Foundation DMS-9002665.

AMS 1991 *subject classifications.* Primary 60B15; Secondary 62M10, 62M15.

Key words and phrases. Efron–Stein inequality, jackknife estimate of variance, symmetric statistic, U–statistic.

where $k = |I \cap J|$, $I = \{i_1, \ldots, i_m\}$, $J = \{j_1, \ldots, j_m\}$. Setting

$$h_k(X_1, \ldots, X_k) = E[h(X_1, \ldots, X_m) \mid X_1, \ldots, X_k] \qquad k = 1, \ldots, m$$

and

$$g_{|J|}(X_J) = \sum_{I \subseteq J} \{(-1)^{|J|-|I|}\} h_{|I|}(X_I)$$

leads to Hoeffding's (1948) ANOVA–type expansion:

$$h(X_1, \ldots, X_m) = \sum_{J \subseteq \{1, \ldots, m\}} g_{|J|}(X_J).$$

Here different terms are uncorrelated and

$$v_k \equiv \mathrm{Var}\, g_k(X_1, \ldots, X_k) = \sum_{\ell=1}^{k} (-1)^{k-\ell} \binom{k}{\ell} r_\ell$$

with the inverse relation

$$r_k = \sum_{\ell=1}^{k} \binom{k}{\ell} v_\ell.$$

The *index* and *dual index* of h are defined to be $\min\{k \mid v_k > 0\}$ and $\max\{k \mid v_k > 0\}$ respectively (Vitale (1992)). These parameters bracket the orders of interaction among $X_1, \ldots, X_m$ which appear in the expansion for h. This is seen clearly, for example, when h is the kth $(1 \leq k \leq m)$ symmetric polynomial in $\varphi(X_1), \ldots, \varphi(X_m)$, where $E\varphi(X_1) = 0$.

3. The Efron–Stein Inequality

As noted in the introduction, several proofs have been given for the inequality. Here we give one based on an explicit representation for the bias (cf. Karlin and Rinott (1982, Eqn. 5.4) for an alternate form).

THEOREM 1 *If h has index c and dual index c', then the bias of the jackknife estimate of variance is nonnegative and given by*

$$EQ - \sigma^2 = \frac{1}{\binom{n-1}{m}} \sum_{k=1}^{m-1} \binom{m}{k} \binom{n-m}{m-k} \left(\frac{k}{m} r_m - r_k \right), \tag{3.1}$$

equivalently,

$$= \sum_{\ell=c}^{c'} v_\ell \binom{m}{\ell} \left[\frac{m}{n-m} - \frac{\binom{n-\ell}{m-\ell}}{\binom{n-1}{m}} \right]. \tag{3.2}$$

The bias is zero if and only if $c' = 1$, which is equivalent to each of the following three conditions:

$$\frac{r_k}{k} = \frac{r_m}{m} \qquad k = 1, \ldots, m-1 \tag{3.3}$$

$$v_\ell = 0 \qquad \ell = 2, \ldots, m \tag{3.4}$$

$$h(X_1, \ldots, X_m) = h^*(X_1) + h^*(X_2) + \cdots + h^*(X_m) \text{ for some } h^*. \tag{3.5}$$

PROOF The identity $\sum_{|I|=m}(h(X_1) - \bar{h})^2 = \left[2\binom{n}{m}\right]^{-1} \sum_{|I|=|J|=m}[h(X_I) - h(X_J)]^2$ allows Q to be written as $\left[2\binom{n-1}{m}\binom{n}{m}\right]^{-1} \sum_{|I|=|J|=m}[h(X_I) - h(X_J)]^2$ and thus

$$EQ = \left[\binom{n-1}{m}\binom{n}{m}\right]^{-1} \sum_{|I|=|J|=m} [r_m - r_{|I\cap J|}].$$

Making use of the fact that the number of pairs (I, J) with $|I \cap J| = k$ is $\binom{n}{m}\binom{m}{k}\binom{n-m}{m-k}$, we have

$$EQ = \binom{n-1}{m}^{-1} \left[\binom{n}{m} r_m - \sum_{k=1}^{m} \binom{m}{k}\binom{n-m}{m-k} r_k\right]. \tag{3.6}$$

Adding and subtracting the value

$$\frac{m\binom{n}{m}}{n} r_m = \left(\sum_{k=1}^{m} \binom{m}{k}\binom{n-m}{m-k}\frac{k}{m}\right) r_m$$

and re-arranging terms yields (3.1). To get (3.2), we insert the expression $r_k = \sum_{\ell=1}^{k} \binom{k}{\ell} v_\ell$ into (3.1) and reverse the resulting double summation to obtain

$$\binom{n-1}{m}^{-1} \sum_{\ell=1}^{m} v_\ell \sum_{k=1}^{m} \binom{m}{k}\binom{n-m}{m-k}\left[\binom{m}{\ell}\frac{k}{m} - \binom{k}{\ell}\right].$$

The inner summation, which vanishes for $\ell = 1$, is the difference of $\binom{m}{\ell}\sum_{k=1}^{m}\binom{m}{k}\binom{n-m}{m-k}\frac{k}{m} = \binom{m}{\ell}\frac{m}{n}\binom{n}{m}$ and $\sum_{k=1}^{m}\binom{m}{k}\binom{n-m}{m-k}\binom{k}{\ell} = \binom{m}{\ell}\binom{n-\ell}{m-\ell}$, and (3.2) follows. The bracketed expression in (3.2) is easily checked to be nonnegative, and thus the bias is nonnegative. Conditions for equality follow from Vitale (1992, Theorem 5.2).

4. A Sharpening, and Complementary Inequality

Here we show that by refining the proof of Theorem 1 it is possible to get more precise results.

THEOREM 2 *Suppose that h has index greater than or equal to c and dual index less than or equal to c'. Then* (1.1) *can be improved to*

$$\left[\frac{n}{n-m} - \frac{\binom{n-c}{m-c}}{\binom{n-1}{m}}\right]\sigma^2 \le EQ \tag{4.1}$$

with the complementary inequality

$$EQ \le \left[\frac{n}{n-m} - \frac{\binom{n-c'}{m-c'}}{\binom{n-1}{m}}\right]\sigma^2. \tag{4.2}$$

PROOF Without loss of generality, assume that h has index and dual index precisely c and c' respectively. As before,

$$EQ - \sigma^2 = \sum_{\ell=c}^{c'} v_\ell \binom{m}{\ell}\left[\frac{m}{n-m} - \frac{\binom{n-\ell}{m-\ell}}{\binom{n-1}{m}}\right].$$

We observe now that the bracketed factor is not simply nonnegative but *nondecreasing in ℓ* (vanishing at $\ell = 1$). Accordingly, we have the bound

$$EQ - \sigma^2 \ge \left[\sum_{\ell=c}^{c'} v_\ell \binom{m}{\ell}\right] \cdot \left[\frac{m}{n-m} - \frac{\binom{n-c}{m-c}}{\binom{n-1}{m}}\right] \ge \sigma^2 \left[\frac{m}{n-m} - \frac{\binom{n-c}{m-c}}{\binom{n-1}{m}}\right],$$

which becomes (4.1) upon re–arrangement. The upper bound (4.2) is found in a similar manner.

It can be easily verified that (4.1) coincides with (1.1) when $c = 1$ and gives a strict improvement when $c > 1$. As another case, consider $m = n - 1$ (the situation originally treated by Efron and Stein). Then (4.1) and (4.2) become

$$c\sigma^2 \le EQ \le c'\sigma^2.$$

5. "Explaining" the Bias

We conclude by showing that the bias of Q can be understood as the result of an unbiased estimator of σ^2 being weighted against estimators of lower order covariances. Once again write Q in the form

$$Q = \left[2\binom{n-1}{m}\binom{n}{m}\right]^{-1} \sum_{|I|=|J|=m} [h(X_I) - h(X_J)]^2. \tag{5.1}$$

Let $\sum^{(k)}$ stand for summation over pairs (I, J) such that $|I \cap J| = k$, and recall that $N_k = \sum^{(k)} 1 = \binom{n}{m}\binom{m}{k}\binom{n-m}{m-k}$. The relations $E[h(X_I) - h(X_J)]^2 = 2[r_m - r_{|I\cap J|}]$ provide $\hat{r}_k$ as an unbiased estimator of r_k, where

$$\hat{r}_m = (2N_0)^{-1} \sum^{(0)} [h(X_I) - h(X_J)]^2$$

and

$$r_k = \hat{r}_m - (2N_k)^{-1} \sum^{(k)} [h(X_I) - h(X_J)]^2 \qquad k = 1, \ldots, m-1.$$

Substituting these expressions into (5.1) gives

$$\begin{aligned} Q &= \hat{r}_m + \sum_{k=1}^{m-1} \binom{m}{k}\binom{n-m}{m-k}\left(\frac{k}{m}\hat{r}_m - \hat{r}_k\right) \\ &= \frac{n}{n-m}\hat{r}_m - \binom{n-1}{m}^{-1} \sum_{k=1}^{m-1} \binom{m}{k}\binom{n-m}{m-k}\hat{r}_k, \end{aligned}$$

which displays the effects of the lower-order estimators.

References

Bhargava, R. P. (1983). A property of the jackknife estimation of the variance when more than one observation is omitted. *Sankhyā Ser. A* **45** 112–119.

Efron, B. and Stein, C. (1981). The jackknife estimate of variance. *Ann. Statist.* **9** 586–596.

Hoeffding, W. (1984). A class of statistics with asymptotically normal distribution. *Ann. Math. Statist.* **19** 293–325.

Karlin, S. and Rinott, Y. (1982). Applications of ANOVA type decompositions for comparisons of conditional variance statistics including jackknife estimates. *Ann. Statist.* **10** 485–501.

Quenouille, M. H. (1949). Approximate tests of correlation in time-series. *J. Royal Statist. Soc. B* **11** 68–84.

Quenouille, M. H. (1956). Notes on bias in estimation. *Biometrika* **43** 353–360.

Steele, J. M. (1986). An Efron–Stein inequality for nonsymmetric statistics. *Ann. Statist.* **14** 753–758.

Tukey, J. W. (1958). Bias and confidence in not–quite large samples (abstract). *Ann. Math. Statist.* **29** 614.

Vitale, R. A. (1984). An expansion for symmetric statistics and the Efron–Stein inequality. In *Inequalities in Statistics and Probability*. Y. L. Tong, ed. Institute of Mathematical Statistics, Hayward, CA. 112–114.

Vitale, R. A. (1992). Covariances of symmetric statistics. *J. Multivariate Anal.* **41** 14–26.

Department of Statistics
University of Connecticut
Storrs, CT 06269

Author Citation Index[1]

Aharoni, R., 332
Ahlswede, I. R., 332
Agresti, A., 359
Akemann, C. A., 116
Akgiray, V., 219
Aldous, D. J., 1, 100
Alam, K., 38, 219
Alon, N., 116
Anderson, J., 116
Anderson, T. W., 133
Armstrong, T., 116
Arnold, B. C., 17, 66, 100, 145, 159
Arratia, R., 100
Artstein, Z., 116
Assaf, D., 371
Aven, T., 100

Baccelli, F., 66, 253
Bailey, N. T. J., 235
Balanda, K. P., 17
Balakrishnan, N., 100
Ball, F., 235
Baringhaus, L., 274
Barlow, R. E., iv, 66, 91, 219, 253, 343
Barton, D. E., 100
Bauer, P., 100
Beck, A., 116
Benzecri, J. P., 359
Berg, C., 182
Berger, J. O., 145
Begrmann, R., 235
Benson, C. T., 284
Bentkus, V., 274
Berman, M., 100
Bhandari, S. K., 145
Bhargava, R. P., 399
Birnbaum, Z. W., 91, 133
Bishop, Y., 359
Bjerve, S., 219
Blackwell, D., 76, 116, 145, 196, 388
Block, H. W., iv, 100, 359
Bochner, S., 219
Boland, P. J., 25, 91, 145, 159
Bolker, E. D., 116
Bonferroni, C. E., 100
Boole, G., 100
Booth, G. G., 219
Boros, E., 309
Bradley, R. A., 38
Breimam, L., 359
Brindley, J. E. C., 219
Brown, M., 1, 25
Bruss, F. T., 371
Buja, A., 274
Buzacott, J. A., 253

Cambanis, S., 219
Cassier, G., 182
Chan, L. K., 66
Chan, W. T., 133, 145
Chen, L. H. Y., 100
Chhetry, D., 359
Chong, K.-M., 145
Chotlos, B., 371
Chow, Y. S., 343, 371
Clark, P. K., 219
Clarkson, D. B., 145
Cogburn, R., 1

[1]The cited page is the first page of the article in which the author's name is cited.

Cohen, A., 33
Cohen, J. W., 219, 253
Costigan, T., 100
Cowan, R., 371
Cox, D. R., 145
Crawford, V. P., 116
Cressie, N., 100

Dall'Aglio, G., 211
Darling, D. A., 100
Darroch, J. N., 100
Das Gupta, S., 145, 284
David, F. N., 17, 100
David, H. A., 38, 100, 274
Davis, P. J., 100
Daykin, D. E., 332
Deely, J. J., 371
De Groot, M. H., 76, 145, 388
De Haan, L., 66
Demko, S., 116
Dharmadhikari, S., iv, 38
Diaconis, P., 1, 145
Diananda, P. H., 182
Do Carmo, M. P., 284
Doksum, K., 219
Donnelly, P., 219
Downey, P. J., 66
Dubins, L. E., 116, 196
Du Mouchel, W. H., 219
Dvoretzki, A., 116
Dykstra, R. L., 359

Eagleson, G. K., 100
Eaton, M. L., iv, 76, 133, 145, 284, 388
Efron, B., 399
Eggleston, H. G., 284
Ehrensfeld, S., 76
El-Neweihi, E., 25, 91
Elton, J., 116
Erikkson, E. A., 145
Esary, J. D., 100, 219, 235, 343, 388
Even, S., 116

Fama, E., 219
Fan, Y.-A., 145
Fang, Z., 359
Fefferman, C., 284
Feldman, R., 219
Feller, W., 116, 219
Ferguson, T. S., 371
Fernique, X., 219
Ferron, R., 211
Fienberg, S. E., 145, 359
Fink, A. M., 116
Fisher, R. A., 116
Flannery, B. R., 1
Fortuin, C., 219, 332
Fréchet, M., 211
Freedman, D., 196
Friedman, J., 359
Fussell, J. B., 91

Galton, F., 219
Gamow, G., 116
Gardner, M., 116, 371
Gass, S. I., 159
Gautam, S. P., 359
Gelenbe, E., 66
Gianini, J. 371
Gilat, D., 196
Gilbert, J., 371
Gilula, Z., 359
Ginibre, J., 219, 332
Giovagnoli, A., 145
Glass, D. V., 359
Glaz, J., 100
Glenn, W. A., 38
Gnedin, A. V., 371
Gnedenko, B. V., 66
Goel, P. K., 76, 388
Goldstein, L., 100
Gordon, L., 100
Gouweleeuw, J., 116
Govindarajulu, Z., 371

Graham, R. L., 332
Gravey, A., 100
Graybill, F. A., 211
Greenwood, P., 219
Groeneveld, R., 100
Grove, L. C., 284
Guillemin, V., 284
Gumbel, E. J., 211
Gun, L., 253
Gupta, S., 50

Hackl, P., 100
Haldane, J. B. S., 145
Hall, J. R., 359
Hall, Jr., M., 182
Halmos, P. R., 116, 133
Hardwick, J. P., 371
Hardy, G. H., iv, 38, 145, 196
Hartley, H. O., 274
Hansen, O. H., 76, 388
Harrus, G., 66, 153
Henze, N., 274
Hickey, R. J., 145
Hill, T. P., 116, 343, 371
Hinkley, D. V., 145
Hodges, J. L., 50
Hoeffding, W., 211, 399
Holland, P., 359
Hollander, M., 50, 91, 159, 253, 388
Holley, R., 332
Holst, L., 100
Hoover, D. R., 100
Hoppe, F., 100
Hougaard, P., 219
Hsu, J. C., 50
Huang, J. S., 66
Hunter, D., 100
Huntington, R. J., 100

Iscoe, I., 1

Jackson, J. R., 253
Jacobson, D. H., 182
Janson, S., 100
Jaynes, E. T., 145
Jean-Marie, A., 253
Jeffreys, H., 145
Jensen, D. R., 133
Joag-Dev, K., iv, 38, 219, 343
Jodeit, M., 284
Joe, H., 145, 159
Johnson, B. McK., 100
Johnson, N. L., 17, 133, 388
Johnson, R. W., 145
Jones, M., 116
Jordan, C., 309

Kall, P., 309
Kamoun, F., 253
Karlin, S., iv, 116, 219, 332, 343, 399
Kastelyn, P., 219, 332
Keich, U., 332
Keilson, J., 1
Kemperman, J. H. B., 33
Kendall, M. G., 38, 359
Kennedy, D. P., 116
Kenyon, J. R., 100
Kertz, R. P., 116, 196, 343
Kimball, A. W., 100
Kimeldorf, G., 359
Kirby, R., 116
Kirman, A., 116
Klefsjö, B., 235
Klein, Haneveld, W., 309
Kleinrock, L., 253
Knaster, B., 116
Korolev, V. Yu., 219
Korolyuk, D. V., 1
Kotz, S., 133, 388
Krein, K., 309
Krengel, U., 116, 343, 371
Kuhn, H. W., 116
Kuo, L., 100

Lai, T. L., 66, 100
Lancaster, H. O., 359
LaSalle, J. P., 116
Leadbetter, M. R., 100
LeCam, L., 388
Lee, M.-L. T., 219
Lefèvre, Cl., 235
Legut, J., 116
Lehmann, E. L., 50, 76, 219, 359, 388
Lemke, C. E., 309
de Leuuw, J., 359
Liggett, T. M., 1
Lindenstrauss, J., 116
Lindgren, G., 100
Lindley, D. V., 371
Littlewood, J. E., iv, 38, 145, 196
Liu, J., 38
Loève, M., 196
Loewner, C., 133
Logan, B. F., 274
Lorenzen, T. J., 371
Lyapunov, A., 116

McDonald, D., 1
MacGillivray, H. L., 17
McClure, D. E., 100
Maier, R. S., 66
Makowski, A. M., 253
Mandelbrot, B. B., 219
Mann, B. L., 50
Mardia, K. V., 211
Margolies, D., 116
Marsaglia, G., 50
Marshall, A. W., iv, 25, 38, 76, 91, 133, 145, 159, 253, 332
Martin, D. H., 182
Maserick, P. H., 182
Mattner, L., 274
May, J. H., 359
Mead, C. A., 145
Mehta, C. R., 145
Melamed, J. A., 219
Micchelli, A. M., 182
Michaletzky, G., 235
Minc, H., 332
Mitronović, D. S., iv
Mittnik, S., 219
Moriguti, S., 274, 371
Mosler, K., iv
Mosteller, F., 371
Mucci, A. G., 371
Muntz, R., 253

Nádas, A., 371
Nagel, K., 50
Nataf, A., 211
Natvig, B., 25, 91
Naus, J., 100
Neff, N. D., 100
Newell, G. F., 100
Neyman, J., 116
Nishisato, ., 359
Nudelman, A., 309

O'Brien, G. L., 343
Oja, H., 17
Olkin, I., iv, 25, 33, 38, 76, 91, 133, 145, 159, 253, 284, 332

Panchapakesan, S., 50
Papadimitriou, C. H., 332
Park, D. H., 133
Parzen, E., 100
Patel, N. R., 145
Paz, A., 116
Pearson, E., 116
Perlman, M., 33, 77, 284
Pestein, V., 116
Petruccelli, J. D., 371
Phelps, R. R., 145
Picard, Ph., 235
Pickands, III., J., 66
Pinkus, A., 182
Pirie, W. R., 50
Pitt, L. D., 33, 219

Plackett, R. L., 274
Plateau, B., 66, 253
Pledger, G., 91
Pollack, A., 284
Polonsky, I., 100
Pólya, G., iv, 38, 145
Popoviciu, T., 309
Powell, M. J. D., 182
Prékopa, A., 100, 309
Presman, E. L., 371
Press, S. J., 76
Prikry, K., 116
Proschan, F., iv, 25, 66, 91, 100, 133, 145, 159, 219, 235, 253, 343, 388
Pyke, R., 100

Quenouille, M. H., 399

Rachev, S. T., 219
Raiffa, H., 145
Ravinshaker, N., 100
Rebman, K., 116
Reeds, J. R., 274
Reiss, R. D., 100
Resnick, S. L., 66, 219
Rinott, Y., 332, 343, 399
Robbins, H., 66, 343, 371
Roberts, A. W., 196
Robertson, T., 359
Rocha, A., 371
Rockafellar, R. T., 196
Rootzen, H., 100
Rösler, U., 196
Ross, S. M., 25, 66, 91, 145, 253
Rubin, H., 371
Ruch, E., 145
Rüselendorf, L., 219
Rvačeva, E. L., 219
Ryff, J. V., 145

Sackrowitz, H. B., 33
Sakaguchi, M., 371
Saks, M., 332
Samorodnitsky, G., 219
Sampson, A. R., iv, 100, 359
Samuel-Cahn, E., 100, 343, 371
Samuels, S. M., 309, 371
Samuelson, W., 116
Sarkar, T. P., 332
Savage, L. J., 284
Savits, T. H., iv
Saxena, K. M. L., 219
Scarsini, M., iv, 145
Schervish, M. J., 100
Schlaifer, R., 145
Schranner, R., 145
Schriever, B. F., 359
Sconing, J., 388
Seligman, T. H., 145
Seneta, E., 100
Sengupta, A., 219
Sethuraman, J., 91, 145, 159, 253, 359
Shahshahani, M., 91
Shaked, M., 25, 76, 219, 253, 388
Shanthikumar, J. G., 25, 145, 253
Sherman, S., 133
Shen, K., 25
Shenton, L. R., 235
Shepp, L. A., 274
Shore, J. E., 145
Shwartz, A., 253
Sidak, Z., 100
Siegmund, D., 343
Silverman, S., 371
Sil'vestrov, D. S., 1
Simonis, A., 145
Sklar, A., 211, 359
Smith, B. B., 38
Smith, M. H., 371
Sobel, M., 284
Solomon, H., 25
Sonin, I. M., 371
Spanier, E. H., 116

Spurrier, J. D., 50
Stam, A. J., 219
Stecke, K., 253
Steele, J. M., 399
Steiglitz, K., 332
Stein, C., 100, 399
Steinhaus, H., 116
Stepniak, C., 76, 388
Stern, M., 116
Stone, A. H., 116
Stormquist, W., 116
Stoyan, D., 145, 235
Stuart, A., 359
Studden, W., 116, 309
Sucheston, L., 116
Svensson, L. G., 116, 343
Szasz, D., 219
Szynal, D., 219

Tamaki, M., 371
Taqqu, M., 219
Taylor, H. M., 219
Terrell, G. R., 274
Terry, M. E., 38
Thompson, J. W. A., 219
Thurstone, L. L., 38
Tikhomirov, V. M., 274
Tomescu, I., 100
Tong, Y. L., iv, 25, 33, 76, 100, 116, 133, 159, 219, 388
Torgersen, E. N., 76, 388
Tsang, W., 50
Tukey, J. W., 116, 399
Tweedie, R. L., 100

Urbanik, K., 116

Valentine, F. A., 284
Van Zwet, W. R., 17
Varberg, D. E., 196
Vesley, W. E., 91
Vitale, R. A., 359, 399
Vulikh, B. Z., 133

Wald, A., 116
Walkup, D. W., 100, 133, 219, 235, 343, 388
Wallenstein, S., 100
Wang, S.-G., 76, 388
Weiss, G., 91
Weller, W., 116
West, D. B., 116
Wets, R. J. B., 133
Whitaker, L., 359
Whitemore, G. A., 219
Whitt, W., 235
Widder, D. V., 66
Wilczynski, M., 116
Williams, E. J., 359
Wolff, R. W., 253
Wolfowitz, J., 116
Wooddall, D. R., 116
Worsley, K. J., 100
Wright, F. T., 359
Wu, C. F. J., 76, 388
Wu, L. S. Y., 100
Wu, W., 219
Wynn, H. P., 145

Xie, M., 25

Yao, D. D., 253
Yiannoutsos, C., 100
Ylvisaker, D., 76
Young, H. P., 116

Zabczyk, J., 371
Zabell, S. A., 145
Zacks, S., 133
Zermelo, E., 38

Key Words and Phrases Index[1]

active and standby redundancy, 25
α–stable, 219
applications, 133
association, 33, 219
asymmetry, 17

backward induction, 371
best choice, 371
bisection, 116
Bonferroni–type inequalities, 100

cake–cutting theorems, 116
characterizations of probability distributions, 274
classification problem, 116
collective epidemic model, 235
comparison of experiments, 77, 388
compartmental urn model, 235
completely monotone, 1
concordant monotone correlation, 359
contingency table, 359
convex order, 17, 196
convex set, 284
convexity, 159
copositive matrices, 182
correlation inequalities, 332
correspondence analysis, 359
covariance spaces, 182
cyclic group, 284

death process, 235
dependence, 211, 219
dihedral group, 284
discrete moment problem, 309
dispersion, 159
dual cones, 182
dynamic programming, 371

Efron–Stein inequality, 399
elliptically contoured distributions, 284
entropy, 145
expectation inequalities, 274
exponential approximation, 1
extremal problems, 274
extreme order statistics, 66

fair–division problems, 116
family of polynomials, 235
finite multivariate moment problem, 182
fixed marginals, 211
flexible manufacturing systems, 253
Fréchet bounds, 211
generalized convex functions, 196
googol, 371
group induced orderings, 76
group invariance, 284

ham–sandwich theorems, 116
Hardy–Littlewood maximal function, 196
heavy traffic limit, 66
hitting time, 1
hyperbolic function, 196
hypothesis testing, 116

importance of components, 91
information inequalities, 388
interpolation, 309

jackknife estimate of variance, 399
$\lambda(F)$ inequalities, 50

[1]This index utilizes authors' key words and phrases. The cited page is the first page of the article in which that word or phrase appears.

Lagrange multiplier rule, 274
likelihood ratio ordering, 25
linear normal experiments, 76
linear preference model, 38
linear programming, 159, 309
Lorenz order, 17
Lyapounov convexity, 116

majorization, 38, 145, 332, 388
Markov chain, 1
martingales, 196
matrix inequalities, 159
matrix orderings, 133
minimal repair, 253
mixed factorial moments, 235
moment of inertia, 211
monotone dependence, 359
monotone functions, 133
moving sums, 100
$\mu(F)$ inequalities, 50
multivariate concentration inequalities, 284
multivariate dependence, 211
multivariate moment inequalities, 309
multivariate normal distribution, 211, 388

negative dependence, 343
nonnegative polynomials, 182

optimal stopping, 116, 359
order statistics, 38, 100
ordered alternatives, 33
ordered alternatives test, 50
ordering, 145
ordinal variables, 359
orthogonal group, 284

partitioning inequalities, 116
peakedness, 17
permutation matrices, 332
positive correlation, 76
positive definite and rectangular matrices, 133
positive dependence, 33, 343, 388
probability distribution, 145
product–type approximations, 100

quasi–stationary, 100

random replacement schemes, 343
relative ranks, 371
reliability/performability, 253
resequencing delay, 66
resequencing queue, 253
role of module, 91

scaling, 359
Schur functions, 91
second order moment space, 182
selecting the best treatment, 50
series and parallel systems, 25
skewness functionals, 17
star order, 17
stochastic allocation, 253
stochastic bounds, 133
stochastic convexity, 253
stochastic dependence, 211
stochastic inequalities, 38, 66
stochastic order, 196
stochastic order relations, 235
stochastic ordering, 25, 33, 66, 332
stochastic Schur convexity, 253
stochastic transposition increasingness, 253
symmetric statistic, 399

threshold model, 38
tied comparisons, 38
two–sample, 359

U–statistic, 399
unbiased tests, 33
uniform spacings, 100
unimodality, 38

waiting time, 1